Lecture Notes in Computer Science 12086

More information about this series at http://www.springer.com/series/7407

Nataša Jonoska · Dmytro Savchuk (Eds.)

Developments in Language Theory

24th International Conference, DLT 2020
Tampa, FL, USA, May 11–15, 2020
Proceedings

 Springer

Editors
Nataša Jonoska (iD)
University of South Florida
Tampa, FL, USA

Dmytro Savchuk
University of South Florida
Tampa, FL, USA

ISSN 0302-9743 ISSN 1611-3349 (electronic)
Lecture Notes in Computer Science
ISBN 978-3-030-48515-3 ISBN 978-3-030-48516-0 (eBook)
https://doi.org/10.1007/978-3-030-48516-0

LNCS Sublibrary: SL1 – Theoretical Computer Science and General Issues

This Springer imprint is published by the registered company Springer Nature Switzerland AG
The registered company address is: Gewerbestrasse 11, 6330 Cham, Switzerland

Preface

The year 2020 will be remembered in history by the pandemic of COVID-19 caused by a novel corona virus that triggered the cancellation of many events around the world. The International Conference on Developments in Language Theory (DLT 2020) was one of those events; the conference was supposed to be held during May 11–15, 2020, at the University of South Florida, USA, but due to the pandemic outbreak it was canceled. The schedule of the conference allowed for the paper submission and refereeing process to be finished before the outbreak, therefore this volume contains the accepted papers of DLT 2020. The authors of these papers will be invited to present the work covered in this volume at the next installation of the conference (DLT 2021), which, if the world pandemic events allow, will be held in Porto, Portugal, in September 2021.

There were 38 papers submitted with authors from 18 countries. The Program Committee accepted 24 of these submissions. The papers on average received 3.7 reviews. The production of this volume was made possible through the diligent work of the Program Committee and the expert referees. We are very grateful to all of the Program Committee and the superb sub-reviewers for their conscientious work.

As this volume indicates, should the world events have allowed the conference to convene, it would have been a very successful and scientifically gratifying event.

We would like to acknowledge the flexibility of National Science Foundation and National Security Agency, as well as the University of South Florida, in allowing partial support of the authors of this volume's papers to participate at DLT 2021. The review process and preparation of this volume was greatly facilitated by EasyChair.

April 2020
Nataša Jonoska
Dmytro Savchuk

Organization

Program Committee

Valerie Berthe	CNRS, IRIF, France
Miroslav Ciric	University of Niš, Serbia
Volker Diekert	University of Stuttgart, Germany
Jérôme Durand-Lose	Université d'Orléans, France
Murray Elder	University of Technology Sydney, Australia
Paweł Gawrychowski	University of Wroclaw, Poland
Yo-Sub Han	Yonsei University, South Korea
Tomohiro I.	Kyushu Institute of Technology, Japan
Oscar Ibarra	University of California, Santa Barbara, USA
Nataša Jonoska	University of South Florida, USA
Juhani Karhumaki	University of Turku, Finland
Jarkko Kari	University of Turku, Finland
Lila Kari	University of Waterloo, Canada
Jetty Kleijn	Leiden University, The Netherlands
Markus Lohrey	University of Siegen, Germany
Andrei Paun	University of Bucharest, Romania
Giovanni Pighizzini	University of Milan, Italy
Igor Potapov	The University of Liverpool, UK
Svetlana Puzynina	Saint Petersburg State University, Russia
Agustín Riscos-Núñez	University of Seville, Spain
Kai Salomaa	Queen's University, Canada
Dmytro Savchuk	University of South Florida, USA
Shinnosuke Seki	The University of Electro-Communications, Japan
Michał Skrzypczak	University of Warsaw, Poland
Pascal Weil	CNRS, University of Bordeaux, France
Reem Yassawi	The Open University, UK
Hsu-Chun Yen	National Taiwan University, Taiwan
Zoran Šunić	Hofstra University, USA

Steering Committee

Marie-Pierre Béal	Université Paris-Est Marne-la-Vallée, France
Mikolaj Bojanczyk	University of Warsaw, Poland
Cristian S. Calude	The University of Auckland, New Zealand
Volker Diekert	University of Stuttgart, Germany
Yo-Sub Han	Yonsei University, South Korea
Juraj Hromkovic	ETH Zürich, Switzerland
Oscar H. Ibarra	University of California, Santa Barbara, USA
Nataša Jonoska	University of South Florida, USA

Juhani Karhumäki (Chair 2010)	University of Turku, Finland
Martin Kutrib	University of Giessen, Germany
Giovanni Pighizzini	University of Milan, Italy
Michel Rigo	University of Liège, Belgium
Antonio Restivo	University of Palermo, Italy
Wojciech Rytter	University of Warsaw, Poland
Kai Salomaa	Queen's University, Canada
Shinnosuke Seki	The University of Electro-Communications, Japan
Mikhail Volkov	Ural Federal University, Russia
Takashi Yokomori	Waseda University, Japan

Honorary Members

| Grzegorz Rozenberg | Leiden University, The Netherlands |
| Arto Salomaa | University of Turku, Finland |

Additional Reviewers

Basold, Henning	Hertrampf, Ulrich	Niskanen, Reino
Bishop, Alex	Imai, Katsunobu	Ochem, Pascal
Bodini, Olivier	Ivanov, Sergiu	Orellana-Martín, David
Campbell, Graham	Janczewski, Wojciech	Peltomäki, Jarkko
Carton, Olivier	Keeler, Chris	Place, Thomas
Ceccherini-Silberstein, Tullio	Kim, Hwee	Poupet, Victor
Charlier, Emilie	Ko, Sang-Ki	Prigioniero, Luca
Cho, Da-Jung	Konstantinidis, Stavros	Rampersad, Narad
Choffrut, Christian	Kuske, Dietrich	Rao, Michael
Coons, Michael	Kutrib, Martin	Reeves, Lawrence
Czerwiński, Wojciech	Leroy, Julien	Rosenfeld, Matthieu
Dolce, Francesco	Leupold, Peter	Sebastien, Labbe
Dudek, Bartlomiej	Lombardy, Sylvain	Seelbach Benkner, Louisa
Fabiański, Grzegorz	Malcher, Andreas	Senizergues, Geraud
Fernau, Henning	Maneth, Sebastian	Shur, Arseny
Ferrari, Margherita Maria	McQuillan, Ian	Smith, Taylor
Fici, Gabriele	Michielini, Vincent	Stipulanti, Manon
Fleischer, Lukas	Minamide, Yasuhiko	Szykuła, Marek
Ganardi, Moses	Mishna, Marni	van Vliet, Rudy
Gauwin, Olivier	Mohammed, Abdulmelik	Ventura, Enric
Giammarresi, Dora	Moreira, Nelma	Verlan, Sergey
Hader, Daniel	Nakashima, Yuto	Zeitoun, Marc
	Ng, Timothy	

Contents

Equational Theories of Scattered and Countable Series-Parallel Posets

Amrane Amazigh[(✉)] and Nicolas Bedon

LITIS (EA 4108), Université de Rouen, Rouen, France
Amazigh.Amrane@etu.univ-rouen.fr, Nicolas.Bedon@univ-rouen.fr

Abstract. In this paper we consider two classes of posets labeled over an alphabet A. The class $SP^\diamond(A)$ is built from the letters and closed under the operations of series finite, ω and $\overline{\omega}$ products, and finite parallel product. In the class $^\omega SP(A)$, ω and $\overline{\omega}$ products are replaced by ω and $\overline{\omega}$ powers. We prove that $SP^\diamond(A)$ and $^\omega SP(A)$ are freely generated in their respective natural varieties of algebras $\mathcal{V}$ and $\mathcal{V}'$, and that the equational theory of $\mathcal{V}'$ is decidable.

Keywords: Transfinite N-free posets · Series-parallel posets · Variety · Free algebra · Series product · Parallel product · ω-power · $\overline{\omega}$-power · Decidability

1 Introduction

In his generalization of the algebraic approach of recognizable languages from finite words to ω-words, Wilke [22] introduced *right binoids*, that are two-sorted algebras equipped with a binary product and an ω-power. The operations are linked together by equalities reflecting their properties. These equalities define a variety of algebras. This algebraic study of ω-words have since been extended to more general structures, such as for example partial words (or equivalently, labeled posets) or transfinite strings (long words). In [8], *shuffle binoids* are right binoids equipped with a shuffle operation that enables to take into consideration N-free posets with finite antichains and ω-chains instead of ω-words. In [3,5], the structure of right binoids in two parts is modified in order to enable products to extend over ω, ie. small ordinals ($\leq \omega^n$, $n \in \mathbb{N}$) and countable ordinals. The latter algebras are enriched in [10,11] with operations such as for example reverse ω-power in order to take into account countable linear orderings (scattered in some cases). Some of the previous algebraic enrichments were also applied to shuffle binoids [4,12]. The motivations in [3–5,10,11,17,22] are mainly the study of the links between automata, rational expressions, algebraic recognition and monadic second-order logic. In [7–9,12,22] the authors focus essentially on varieties of algebras; for example, free algebras are characterized in the corresponding varieties, and decisions algorithms for equivalence of terms are provided.

Let us denote by $\overline{\omega}$ the reverse ordering of ω. In this paper we focus on algebras equipped with a parallel product, series product, and either ω and $\overline{\omega}$

© Springer Nature Switzerland AG 2020
N. Jonoska and D. Savchuk (Eds.): DLT 2020, LNCS 12086, pp. 1–13, 2020.
https://doi.org/10.1007/978-3-030-48516-0_1

products or ω and $\overline{\omega}$ powers. For example, the class $SP^\diamond(A)$ of N-free posets in which antichains are finite and chains are scattered and countable orderings lies in this framework. In [2,6] this class has been studied from the point of view of automata, rational expression and logic. We prove here that $SP^\diamond(A)$ is the free algebra in a variety $\mathcal{V}$ of algebras equipped with a parallel product, series product, and ω and $\overline{\omega}$ products. By removing the parallel product, it follows that $A^\diamond$, the class of scattered and countable words over A, is also a free algebra in the corresponding variety. We also consider the class $^\omega SP(A)$ where the ω and $\overline{\omega}$ products are replaced by ω and $\overline{\omega}$ powers, and show that it is freely generated in the corresponding variety $\mathcal{V}'$. Relying of decision results of [2] we prove that the equality of terms of $\mathcal{V}'$ is decidable.

2   Linear Orderings and Posets

We let $|E|$ denote the cardinality of a set E, and $[n]$ the set $\{1, \ldots, n\}$, for any non-negative integer $n \in \mathbb{N}$.

Let J be a set equipped with a strict order $<$. The ordering J is *linear* if either $j < k$ or $k < j$ for any distinct $j, k \in J$. We denote by $\overline{J}$ the backward linear ordering obtained from the set J with the reverse ordering. A linear ordering J is *dense* if for any $j, k \in J$ such that $j < k$, there exists an element i of J such that $j < i < k$. It is *scattered* if it contains no infinite dense sub-ordering. The ordering ω of natural integers is scattered as well as the ordering ζ of all integers (negative, 0 and positive). Ordinals are also scattered orderings. We let $\mathcal{N}$, $\mathcal{O}$ and $\mathcal{S}$ denote respectively the class of finite linear orderings, the class of countable ordinals and the class of countable scattered linear orderings. We also let 0 and 1 denote respectively the empty and the singleton linear ordering. We refer to [20] for more details on linear orderings and ordinals.

A *poset* $(P, <)$ is a set P partially ordered by $<$. For short we often denote the poset $(P, <)$ by P. The *width* of P is $\mathrm{wd}(P) = \sup\{|E| : E \text{ is an antichain of } P\}$ where sup denotes the least upper bound of the set. In this paper, we restrict to posets with finite antichains and countable and scattered chains.

Let $(P, <_P)$ and $(Q, <_Q)$ be two disjoint posets. The *union* (or *parallel composition*) $P \cup Q$ of $(P, <_P)$ and $(Q, <_Q)$ is the poset $(P \cup Q, <_P \cup <_Q)$. The *sum* (or *sequential composition*) $P + Q$ is the poset $(P \cup Q, <_P \cup <_Q \cup P \times Q)$. The sum of two posets can be generalized to a J-sum of any linearly ordered sequence $((P_j, <_j))_{j \in J}$ of pairwise disjoint posets by $\sum_{j \in J} P_j = (\bigcup_{j \in J} P_j, (\bigcup_{j \in J} <_j) \cup (\bigcup_{j, j' \in J, \ j < j'} P_j \times P_{j'}))$. The sequence $((P_j, <_j))_{j \in J}$ is called a J-*factorization*, or (sequential) *factorization* for short, of the poset $\sum_{j \in J} P_j$. A poset P is *sequential* if it admits a J-factorization where J contains at least two elements $j \neq j'$ with $P_j, P_{j'} \neq 0$, or P is a singleton. It is *parallel* when $P = P_1 \cup P_2$ for some $P_1, P_2 \neq 0$. A poset is *sequentially irreducible* (resp. *parallelly irreducible*) when P is either a singleton or a parallel poset (resp. a singleton or a sequential poset). A sequential factorization $((P_j, <_j))_{j \in J}$ of $P = \sum_{j \in J} P_j$ is *irreducible* when all the P_j are sequentially irreducible. It is *non-trivial* if all the P_j are non-empty.

The notions of irreducible and non-trivial parallel factorization are defined similarly. A poset is *scattered* if all its chains are scattered. The class $SP^\diamond$ of *series-parallel scattered and countable posets* is the smallest class of posets containing 0, the singleton, and closed under finite parallel composition and sum indexed by countable scattered linear orderings. By extension of a well-known result on finite posets [18,21], it has a nice characterization in terms of graph properties: $SP^\diamond$ coincides with the class of scattered countable N-free posets without infinite antichains [6]. Recall that $(P, <)$ is *N-free* if there is no $X = \{x_1, x_2, x_3, x_4\} \subseteq P$ such that $< \cap X^2 = \{(x_1, x_2), (x_3, x_2), (x_3, x_4)\}$.

When $P \in SP^\diamond$ and $P = R + P' + S$ or $P = P' \cup R$ for some $R, S, P' \in SP^\diamond$ then P' is a *factor* of P; the factors of P', R and S are also factors of P.

F. Hausdorff proposed in [16] an inductive definition of scattered linear orderings. In fact, each countable and scattered linear ordering is obtained using sums indexed by finite linear orderings, ω and $\bar{\omega}$. This has been adapted in [6] to $SP^\diamond$.

We let $\mathcal{C}_{\cup,+}(E)$ denote the closure of a set E of posets under finite disjoint union and finite disjoint sum.

Definition 1. *The classes of countable and scattered posets (equivalent up to isomorphism) V_α and W_α are defined inductively as follows:*

$$V_0 = \{0, 1\}$$
$$W_\alpha = \mathcal{C}_{\cup,+}(V_\alpha)$$
$$V_\alpha = \left\{ \sum_{i \in J} P_i : J \in \{\omega, \bar{\omega}\} \text{ and } \forall i \in J, P_i \in \bigcup_{\beta < \alpha} W_\beta \right\} \cup \bigcup_{\beta < \alpha} W_\beta \text{ when } \alpha > 0$$

and the class S_{sp} of countable and scattered posets by $S_{sp} = \bigcup_{\alpha \in \mathcal{O}} W_\alpha$.

The following theorem extends a result of Hausdorff on linear orderings [16].

Theorem 1 ([6]). *$S_{sp} = SP^\diamond$.*

For every $\alpha \in \mathcal{O}$, W_α can be decomposed as the closure of V_α by finite disjoint union and finite disjoint sum:

Theorem 2 ([6]). *For all $\alpha \in \mathcal{O}$, $i \in \mathbb{N}$, let*

$$X_{\alpha,0} = V_\alpha$$
$$Y_{\alpha,i} = \left\{ P : \exists n \in \mathbb{N} \ P = \sum_{j \leq n} P_j \text{ such that } P_j \in X_{\alpha,i} \text{ for all } j \leq n \right\}$$
$$X_{\alpha,i+1} = \left\{ P : \exists n \in \mathbb{N} \ P = \bigcup_{j \leq n} P_j \text{ such that } P_j \in Y_{\alpha,i} \text{ for all } j \leq n \right\}$$

Then $W_\alpha = \bigcup_{i \in \mathbb{N}} X_{\alpha,i}$.

Example 1. W_0 is the set of all finite N-free posets. Its subset $Y_{0,0}$ is the set of all finite linear orderings. The linear orderings ω and $\overline{\omega}$ are contained in V_1. Each poset of $V_1 \setminus W_0$ has some chain isomorphic to either ω or $\overline{\omega}$, but can not have a chain isomorphic to ω and another isomorphic to $\overline{\omega}$. The ordering ζ of all integers is in $Y_{1,0}$. For all $\alpha \in \mathcal{O}$, $\omega^\alpha \in V_\alpha$.

Define a well-ordering on $\mathcal{O} \times \mathbb{N}$ by $(\beta, j) < (\alpha, i)$ if and only if $\beta < \alpha$ or $\beta = \alpha$ and $j < i$. As a consequence of Theorems 1 and 2, for any $P \in SP^\diamond$ there exists a unique pair $(\alpha, i) \in \mathcal{O} \times \mathbb{N}$ as small as possible such that $P \in X_{\alpha,i}$.

Definition 2. *The* rank $r(P)$ *of* $P \in SP^\diamond$ *is the smallest pair* $(\alpha, i) \in \mathcal{O} \times \mathbb{N}$ *such that* $P \in X_{\alpha,i}$.

Example 2. The linear ordering ζ has rank $r(\zeta) = (1,1)$. Each linear ordering I of $\mathcal{S}$ has rank $r(I) \in \{(\alpha, 0), (\alpha, 1)\}$ for some $\alpha \in \mathcal{O}$. For all $\alpha \in \mathcal{O}$, $r(\omega^\alpha) = \alpha$.

Remark 1. Let $P \in SP^\diamond$ with $r(P) = (\alpha, 0)$, $\alpha > 0$. Assume that $P = \sum_{j \in J} P_j$ is a non-trivial J-factorization of P for some $J \in \{\omega, \overline{\omega}\}$. If $J = \omega$ (resp. $J = \overline{\omega}$), then, for all $j \in J$, $r(P_j) < r(P)$. In addition, for all $(\beta, i) \in \mathcal{O} \times \mathbb{N}$ such that $(\beta, i) < (\alpha, 0)$, for all $j \in J$ there exists $k \in J$ such that $k > j$ (resp. $k < j$) and

$$(\beta, i) \leq r(P_k) < r(P)$$

This implies that, for all $j \in J$, $\sum_{j' \geq j} P_j$ (resp. $\sum_{j' \leq j} P_j$) is of rank $(\alpha, 0)$.

Lemma 1. *Let* $P \in SP^\diamond$ *be a sequential poset such that* $r(P) = (\alpha, 0)$, $\alpha > 0$. *Let* $\sum_{j \in J} P_j$ *and* $\sum_{j \in J'} P'_j$ *be some non-trivial* J- *and* J'-*factorizations of* P *where* $J, J' \in \{\omega, \overline{\omega}\}$. *Then* $J = J'$.

Proof. Assume by contradiction that $J \neq J'$. Assume wlog that $J = \omega$ and $J' = \overline{\omega}$. Let $L = \sum_{j \leq k} P_j$ and $R = \sum_{k < j < \omega} P_j$ for some $k \in \omega$. Then $P = L + R$. As a consequence of Remark 1, $r(R) = (\alpha, 0)$. Observe that there exists $k' \in \overline{\omega}$ such that R is a sequential factor of $R' = \sum_{j' \geq k'} P'_{j'}$. Let $L' = \sum_{\overline{\omega} < j' < k'} P'_{j'}$. As a consequence of Remark 1, $r(L') = r(R) = (\alpha, 0)$. Furthermore, $r(R) \leq r(R') \leq r(P)$. Thus $r(R') = (\alpha, 0)$ too. We have $P = L' + R'$, and by Theorem 2, $r(P) = (\alpha, 1)$, which is a contradiction.

In [6] an equivalence relation $\sim$ over the elements of a poset of $SP^\diamond$ is given, such that $P/\sim$ is isomorphic to a countable and scattered linear ordering (Lemma 9), and such that each equivalence class is a sequentially irreducible factor of P (Lemma 10). This leads to the following proposition.

Proposition 1 ([6]). *Each poset of* $SP^\diamond$ *admits a unique irreducible sequential factorization.*

Definition 1 and Theorem 1 provide a well-founded definition of $SP^\diamond$ which we consider from now as a set, although originally defined as a class.

3   Labeled Posets

An *alphabet* A is a non-empty set (not necessarily finite) whose elements are called *letters* or *labels*. In the literature a *word* over A is a totally ordered sequence of elements of A. The sequence may have properties depending on the context, for example it can be finite, an ordinal, or a countable scattered linear ordering. The notion of a finite word has early been extended to partial orderings (finite *partial words* or *pomsets* [14,15,23]). In this paper we consider a mixture between the notions of finite partial words and words indexed by scattered and countable linear orderings.

A poset P is *labeled* by A when it is equipped with a *labeling* total map $l \colon P \to A$. Also, the finite labeled posets of width at most 1 correspond to the usual notion of words. We let ϵ denote the empty labeled poset. For short, the singleton poset labeled by $\{a\}$ is denoted by a, and we often make no distinction between a poset and a labeled poset, except for operations.

The *sequential product* (or *concatenation*, denoted by $P \cdot P'$ or PP' for short) and the *parallel product* $P \parallel P'$ of two labeled posets are respectively obtained by the sequential and parallel compositions of the corresponding (unlabeled) posets. By extension, the sequential product $\prod_{j \in J} P_j$ of a linearly ordered sequence of labeled posets is the poset $\sum_{j \in J} P_j$ in which the label of the elements is kept. In particular, the ω-*product* (resp. $\overline{\omega}$-*product*) of an ω-sequence (resp. $\overline{\omega}$-sequence) of labeled posets $(P_i)_{i \in \omega}$ (resp. $(P_i)_{i \in \overline{\omega}}$) is denoted by $\prod_{i \in \omega} P_i$ (resp. $\prod_{i \in \overline{\omega}} P_i$). The ω-*power* (resp. $\overline{\omega}$-*power*) P^ω (resp. $P^{\overline{\omega}}$) of the poset P is the ω-product (resp. $\overline{\omega}$-product) of an ω-sequence (resp. $\overline{\omega}$-sequence) of posets that are all isomorphic to P. As usual, in this paper we consider two labeled posets to be identical if they are isomorphic. By extension, the rank $r(P)$ of a labeled poset P is the rank of its underlying unlabeled poset.

Let A and B be two alphabets and let P be a poset labeled by A. For all $a \in A$, let G_a be some poset labeled by B, and let $G = (G_a)_{a \in A}$. The poset labeled by B consisting of P in which each element labeled by the letter a is replaced by G_a, for all $a \in A$, is denoted by $G \circ_A P$. If the underlying posets of P and of all the G_a are in $SP^\diamond$, then so is $G \circ_A P$.

Definition 3. *Let A be an alphabet. We define:*

- *$SP^\diamond(A)$, the smallest set of posets labeled by A containing ϵ, a for all $a \in A$, and closed under operations of sequential, parallel, ω and $\overline{\omega}$-products. According to Theorem 1, the underlying posets are precisely those of $SP^\diamond$;*
- *$^\omega A$, the smallest subset of $SP^\diamond(A)$ containing ϵ, a for all $a \in A$, and closed under operations of sequential product, ω-power and $\overline{\omega}$-power;*
- *$A^\diamond$, the smallest subset of $SP^\diamond(A)$ containing ϵ, a for all $a \in A$, and closed under operations of sequential product, ω-product and $\overline{\omega}$-product;*
- *$^\omega SP(A)$, the smallest subset of $SP^\diamond(A)$ containing ϵ, a for all $a \in A$, and closed under operations of sequential and parallel product, ω-power and $\overline{\omega}$-power;*

– $\omega SP(A)$, *the smallest subset of* $SP^\diamond(A)$ *containing* ϵ, *a for all* $a \in A$, *and closed under operations of sequential product, parallel product and* ω-*product (note that there is no* $\overline{\omega}$-*product here).*

Note that $^\omega A = \{P \in SP^\diamond(A) : r(P) \in \mathbb{N} \times \{0,1\}\}$ and $^\omega SP(A) = \{P \in SP^\diamond(A) : r(P) \in \mathbb{N} \times \mathbb{N}\}\}$.

4 Varieties

In this section we define the different varieties studied throughout this paper by listing the axioms they satisfy. The usual notions and results of universal algebra apply to our case, even if we use here for example operations of infinite arity. For more details about universal algebra, we refer the reader to [1]. In the following 1 is considered as a neutral element (the interpretation of a constant).

$$x \cdot (y \cdot z) = (x \cdot y) \cdot z \tag{A_1}$$
$$x \parallel (y \parallel z) = (x \parallel y) \parallel z \tag{A_2}$$
$$x \parallel y = y \parallel x \tag{A_3}$$
$$(x \cdot y)^\omega = x \cdot (y \cdot x)^\omega \tag{A_4}$$
$$(x^n)^\omega = x^\omega, \qquad\qquad n \geq 1 \tag{A_5}$$
$$(x \cdot y)^{\overline{\omega}} = (y \cdot x)^{\overline{\omega}} \cdot y \tag{A_6}$$
$$(x^n)^{\overline{\omega}} = x^{\overline{\omega}}, \qquad\qquad n \geq 1 \tag{A_7}$$
$$x \cdot 1 = x \tag{A_8}$$
$$1 \cdot x = x \tag{A_9}$$
$$x \parallel 1 = x \tag{A_{10}}$$
$$1^\omega = 1 \tag{A_{11}}$$
$$1^{\overline{\omega}} = 1 \tag{A_{12}}$$

for all ω-sequences $x_0, x_1, \ldots, x_i, \ldots$ and all decompositions
$(x_0, \ldots, x_{n_0-1}), (x_{n_0}, \ldots, x_{n_1-1}), \ldots, (x_{n_i}, \ldots, x_{n_i-1}), \ldots$

$$\omega((x_0, \ldots, x_{n_0-1}), (x_{n_0}, \ldots, x_{n_1-1}), \ldots) = \omega(x_0, x_1, \ldots) \tag{A_{13}}$$
$$x_0 \cdot \omega(x_1, x_2, \ldots) = \omega(x_0, x_1, x_2, \ldots) \tag{A_{14}}$$
$$\omega(1, 1, \ldots) = 1 \tag{A_{15}}$$

for all $\overline{\omega}$-sequences $\ldots, x_i, \ldots, x_1, x_0$ and all decompositions
$\ldots, (x_{n_i-1}, \ldots, x_{n_i}), \ldots, (x_{n_1-1}, \ldots, x_{n_0}), (x_{n_0-1}, \ldots, x_0)$

$$\overline{\omega}(\ldots, (x_{n_1-1}, \ldots, x_{n_0}), (x_{n_0-1}, \ldots, x_0)) = \overline{\omega}(\ldots, x_1, x_0) \tag{A_{16}}$$
$$\overline{\omega}(\ldots, x_2, x_1) \cdot x_0 = \overline{\omega}(\ldots, x_2, x_1, x_0) \tag{A_{17}}$$
$$\overline{\omega}(\ldots, 1, 1) = 1 \tag{A_{18}}$$

Definition 4. *We define*

- *$\mathcal{V}$, the collection of algebras $(S, \cdot, \|, \omega, \overline{\omega}, 1)$ satisfying the axioms (A_1)–(A_3), (A_8)–(A_{10}) and (A_{13})–(A_{18});*
- *$\mathcal{V}_0$, the collection of algebras $(S, \cdot, \omega, \overline{\omega}, 1)$ satisfying the axioms (A_1), (A_8), (A_9) and (A_{13})–(A_{18});*
- *$\mathcal{V}_1$, the collection of algebras $(S, \cdot, \|, \omega, 1)$ satisfying the axioms (A_1)–(A_3), (A_8)–(A_{10}) and (A_{13})–(A_{15});*
- *$\mathcal{V}'$, the collection of algebras $(S, \cdot, \|, {}^{\omega}, {}^{\overline{\omega}}, 1)$ satisfying the axioms (A_1)–(A_{12});*
- *$\mathcal{V}'_0$, the collection of algebras $(S, \cdot, {}^{\omega}, {}^{\overline{\omega}}, 1)$ satisfying the axioms (A_1), (A_4)–(A_9) and (A_{11}),(A_{12}).*

In order to simplify the notation, an algebra whose set of elements is S is sometimes denoted by S when there is no ambiguity.

5   Freeness

Throughout this section, A denotes an alphabet. We start by proving the freeness of $SP^{\diamond}(A)$.

Theorem 3. *$SP^{\diamond}(A)$ is freely generated by A in $\mathcal{V}$.*

Proof. For all $(\alpha, i) \in \mathcal{O} \times \mathbb{N}$, let $X_{\alpha,i}$ denote the set of posets of $SP^{\diamond}(A)$ of rank (α, i) or less. Let $\mathcal{M} = (M, \cdot, \|, \omega, \overline{\omega}, 1)$ be any algebra of $\mathcal{V}$ and let $h\colon A \to M$ be any function. We show that h can be extended into a homomorphism of $\mathcal{V}$-algebras $h^{\sharp}\colon SP^{\diamond}(A) \to M$ in a unique way. Define $h^{\sharp}$ as $h^{\sharp} = \bigcup\limits_{(\alpha,i)\in\mathcal{O}\times\mathbb{N}} h_{\alpha,i}$ where each $h_{\alpha,i}\colon X_{\alpha,i} \to M$ is defined by induction over (α, i) as follows. Let us denote by $h_{<(\alpha,i)} = \cup_{(\beta,j)<(\alpha,i)} h_{\beta,j}$. Let $P \in X_{\alpha,i}$. If $r(P) < (\alpha, i)$ then $h_{\alpha,i}(P) = h_{<(\alpha,i)}(P)$. Otherwise

- if $\alpha = 0$ and $i = 0$ then $h_{\alpha,i} = h \cup (\epsilon \to 1)$;
- if $\alpha > 0$ and $i = 0$ then P admits a non-trivial J-factorization

$$P = \prod_{j \in J} P_j \tag{19}$$

where $J \in \{\omega, \overline{\omega}\}$ (see Remark 1) and $r(P_j) < r(P)$ for all $j \in J$. Define $h_{\alpha,i}(P)$ by

$$h_{\alpha,i}(P) = \prod_{j \in J} h_{<(\alpha,i)}(P_j)$$

- if $i > 0$:
  - if P is a sequential poset then it has a factorization

$$P = \prod_{j \in [n]} P_j \tag{20}$$

where each P_j is a non-empty poset of rank lower than (α, i) and $n \in \mathcal{N} \setminus \{0, 1\}$. Define $h_{\alpha,i}(P)$ by

$$h_{\alpha,i}(P) = \prod_{j \in [n]} h_{\alpha, i-1}(P_j)$$

- otherwise, P is a parallel poset. Write

$$P = \|_{s \in [n]} P_s$$

where each P_s is a sequential poset and $n \geq 2$. Then, define $h_{\alpha,i}(P)$ by

$$h_{\alpha,i}(P) = \|_{s \in [n]} h_{\alpha,i}(P_s)$$

By Theorem 2, the factorizations used in the definition of $h_{\alpha,i}$ exist. However, observe that the sequential ones ((19) and (20)) are not unique. This would question the fact that $h_{\alpha,i}$ is a well-defined function. For all $P \in SP^\circ(A)$ of rank (α, i), we show that:

1. $h_{\alpha,i}(P)$ does not depend on the factorization of P and thus is well-defined;
2. $h_{\alpha,i}$ commutes with all the operations of $\mathcal{V}$:
   (a) $h_{\alpha,i}(\prod_{j \in J} P_j) = \prod_{j \in J} h_{\alpha,i}(P_j)$, for some $J \in \mathcal{N} \cup \{\omega, \overline{\omega}\}$;
   (b) $h_{\alpha,i}(\|_{s \in [n]} P_s) = \|_{s \in [n]} h_{\alpha,i}(P_s)$, for some $n \in \mathbb{N}$.

We proceed by induction on (α, i). Let us start by proving that $h_{\alpha,i}$ maps $P \in X_{\alpha,i}$ to the same element of M regardless of the factorization of P. If $(\alpha, i) = (0, 0)$ the theorem follows immediately. Otherwise, assume first that $i = 0$. By Lemma 1, all the possible factorizations of P as in (19) are either all ω-factorizations or all $\overline{\omega}$-factorizations. Assume wlog that P admits only ω-factorizations as in (19). Let $P = \prod_{j \in \omega} P_j$ and $P = \prod_{j \in \omega} Q_j$ be two different such ω-factorizations. By definition of $h_{\alpha,i}$

$$h_{\alpha,i}(\prod_{j \in \omega} P_j) = \prod_{j \in \omega} h_{<(\alpha,i)}(P_j) \text{ and } h_{\alpha,i}(\prod_{j \in \omega} Q_j) = \prod_{j \in \omega} h_{<(\alpha,i)}(Q_j)$$

There exists a sequence $(R_j)_{j \in \omega}$ of non-empty posets such that $P = \prod_{k \in \omega} R_k$ and for all $j \in \omega$ there exist $k_{P_j}, k'_{P_j}, k_{Q_j}, k'_{Q_j} \in \omega$ such that

$$P_j = \prod_{k_{P_j} \leq l \leq k'_{P_j}} R_l \text{ and } Q_j = \prod_{k_{Q_j} \leq l \leq k'_{Q_j}} R_l$$

By induction hypothesis $h_{<(\alpha,i)}$ commutes with all the operations of $\mathcal{V}$. Then, we have for all $j \in \omega$:

$$h_{<(\alpha,i)}(P_j) = \prod_{k_{P_j} \leq l \leq k'_{P_j}} h_{<(\alpha,i)}(R_l) \text{ and } h_{<(\alpha,i)}(Q_j) = \prod_{k_{Q_j} \leq l \leq k'_{Q_j}} h_{<(\alpha,i)}(R_l)$$

Thus $\prod_{j \in \omega} h_{<(\alpha,i)}(P_j)$ can be written as

$$\omega(h_{<(\alpha,i)}(R_{k_{P_0}}) \cdot \ldots \cdot h_{<(\alpha,i)}(R_{k'_{P_0}}), h_{<(\alpha,i)}(R_{k_{P_1}}) \cdot \ldots \cdot h_{<(\alpha,i)}(R_{k'_{P_1}}), \cdots)$$

$$\overset{(A_{13})}{=} \omega(h_{<(\alpha,i)}(R_{k_{P_0}}), \ldots, h_{<(\alpha,i)}(R_{k'_{P_0}}), h_{<(\alpha,i)}(R_{k_{P_1}}), \ldots, h_{<(\alpha,i)}(R_{k'_{P_1}}), \cdots)$$

$$\overset{(A_{13})}{=} \omega(h_{<(\alpha,i)}(R_{k_{Q_0}}) \cdot \ldots \cdot h_{<(\alpha,i)}(R_{k'_{Q_0}}), h_{<(\alpha,i)}(R_{k_{Q_1}}) \cdot \ldots \cdot h_{<(\alpha,i)}(R_{k'_{Q_1}}), \cdots)$$

We have $\prod_{j \in \omega} h_{<(\alpha,i)}(P_j) = \prod_{j \in \omega} h_{<(\alpha,i)}(Q_j)$. The case where P admits only $\overline{\omega}$-factorizations as in (19) is proved symmetrically using (A_{16}) instead of (A_{13}). In addition, using (A_1) instead of (A_{13}) and arguments similar to those of the previous case, we prove that when P is sequential and $i > 0$, $h_{\alpha,i}(P)$ does not depend on the factorization of P.

Thus, we have proved that $h_{\alpha,i}$ is well-defined for sequential posets of rank $(\alpha, i) \in \mathcal{O} \times \mathbb{N}$. In addition, the irreducible parallel factorization is unique modulo the commutativity of $\parallel$. Thus $h_{\alpha,i}$ is well-defined for all posets of rank (α, i), for all $(\alpha, i) \in \mathcal{O} \times \mathbb{N}$. Furthermore, proving that $h_{\alpha,i}$ commutes with all the operations in $X_{\alpha,i}$ can be done by induction on $r(P)$ too. The arguments are very similar to those used to prove that $h_{\alpha,i}$ is well-defined. It follows that $h^\sharp$ is a homomorphism of $\mathcal{V}$-algebras. In addition, since $h^\sharp$ relies on h then $h^\sharp$ is unique.

The proofs of the following theorems rely on the same arguments. It suffices to restrict $h^\sharp$ to the operations of the corresponding variety. In particular, this provides a new proof of Theorem 5.

Theorem 4. A° *is freely generated by A in $\mathcal{V}_0$.*

Theorem 5 ([12]). *$\omega SP(A)$ is freely generated by A in $\mathcal{V}_1$.*

In the remainder of this section, we prove the freeness of $^\omega SP(A)$ in $\mathcal{V}'$. The arguments are similar to those of the proof of Theorem 6.1 in [12] in which the variety considered is $\mathcal{V}'$ without $\overline{\omega}$-power. We need the following result.

Theorem 6 ([9]). *$^\omega A$ is freely generated by A in $\mathcal{V}'_0$.*

Lemma 2. *Let A and B be two alphabets. Let $S \subseteq {}^\omega SP(B)$ such that S is closed under sequential product, ω-power and $\overline{\omega}$-power. Let $f \colon A \to G$ be some function defined by $f(a) = G_a \in G$ for some $G \subseteq S$. Then, the function $f^\sharp \colon {}^\omega A \to S$ extending f defined by $f^\sharp(u) = (G_a)_{a \in A} \circ_A u$, for all $u \in {}^\omega A$, is a homomorphism from $({}^\omega A, \cdot, {}^\omega, {}^{\overline{\omega}}, \epsilon)$ to $(S, \cdot, {}^\omega, {}^{\overline{\omega}}, 1)$.*

Furthermore, if f is bijective, S is generated by G, and G contains only sequentially irreducible posets then $f^\sharp$ is bijective.

Proof. Let $u \in {}^\omega A$ whose irreducible sequential factorization is $\prod_{j \in J} u_j$ for some $J \in \mathcal{S}$, where each $u_j \in A$. Note that

$$f^\sharp(u) = (G_a)_{a \in A} \circ_A u = \prod_{j \in J} (G_a)_{a \in A} \circ_A u_i = \prod_{j \in J} f(u_i)$$

Let $v \cdot w, x^{\omega}$ and $y^{\overline{\omega}}$ be some sequential factorizations of u. Then, one can prove easily that

$$f^{\sharp}(u) = f^{\sharp}(v) \cdot f^{\sharp}(w) = f^{\sharp}(x)^{\omega} = f^{\sharp}(y)^{\overline{\omega}}$$

relying on the uniqueness of the irreducible sequential factorization of u (Proposition 1).

Let us prove now that when f is bijective and S is generated by a set of sequentially irreducible posets then $f^{\sharp}$ is bijective. Let $u, v \in {}^{\omega}A$ and assume that $f^{\sharp}(u) = P$ and $f^{\sharp}(v) = Q$. Let $\prod_{i \in I} u_i$ and $\prod_{j \in J} v_j$ be the irreducible sequential factorizations of respectively u and v, for some $I, J \in S$, where each u_i and v_j are in A. By definition of $f^{\sharp}$, $P = \prod_{i \in I} P_i$ and $Q = \prod_{j \in J} Q_j$ where each $P_i = (G_a)_{a \in A} \circ_A u_i$ and $Q_j = (G_a)_{a \in A} \circ_A v_j$. Then, for all $i \in I$ and for all $j \in J$, P_i and Q_j are sequentially irreducible posets of G. Assume that $P = Q$. Then $I = J$ and, for all $i \in I$, $P_i = Q_i$. We have, for all $i \in I$, $u_i = v_i$ since f is injective by hypothesis. In addition, as G generates S, each element P of S can be written as $\prod_{j \in J} P_j$ where each $P_j \in G$, for some $J \in S$. Since f is surjective by hypothesis, for all $j \in J$ there exists $u_j \in A$ such that $f(u_j) = P_j$. Then $f^{\sharp}(\prod_{j \in J} u_j) = P$.

As a consequence of HSP Birkhoff's Theorem (see eg. [1, Theorem 1.3.8]) and Lemma 2:

Corollary 1. *For all $S \subseteq {}^{\omega}SP(A)$ closed under sequential product, ω-power and $\overline{\omega}$-power and generated by a set of sequentially irreducible posets of ${}^{\omega}SP(A)$, $(S, \cdot, {}^{\omega}, \overline{\omega}, 1)$ is a $\mathcal{V}_0'$-algebra.*

In addition, as a consequence of Theorem 6 and Lemma 2:

Corollary 2. *For all $S \subseteq {}^{\omega}SP(A)$ closed under sequential product, ω-power and $\overline{\omega}$-power and generated by a set G of sequentially irreducible posets of ${}^{\omega}SP(A)$, $(S, \cdot, {}^{\omega}, \overline{\omega}, 1)$ is freely generated by G in $\mathcal{V}_0'$.*

We are now ready to prove the following theorem.

Theorem 7. *${}^{\omega}SP(A)$ is freely generated by A in $\mathcal{V}'$.*

Proof. For all $i \in \mathbb{N}$, let ${}^{\omega}SP(A)_i$ be the subset of ${}^{\omega}SP(A)$ consisting all its posets of width lower or equal to i. Then ${}^{\omega}SP(A) = \bigcup_{i \in \mathbb{N}} {}^{\omega}SP(A)_i$. Note that ${}^{\omega}SP(A)_0 = \{\epsilon\}$ and ${}^{\omega}SP(A)_1 = {}^{\omega}A$. Observe that for all $i \in \mathbb{N}$, ${}^{\omega}SP(A)_i$ is closed under sequential product, ω-power and $\overline{\omega}$-power. In addition, for all $i \in \mathbb{N}$, ${}^{\omega}SP(A)_i$ is generated by its sequentially irreducible posets. By Corollary 1, for all $i \in \mathbb{N}$, ${}^{\omega}SP(A)_i$ can be considered as a $\mathcal{V}_0'$-algebra. In addition, by Corollary 2, for all $i \in \mathbb{N}$, ${}^{\omega}SP(A)_i$ is freely generated by its sequentially irreducible posets in $\mathcal{V}_0'$. Then, for all $i \in \mathbb{N}$ and $S \in \mathcal{V}_0'$, a function $h' \colon A \to S$ can be extended in a unique homomorphism of $\mathcal{V}_0'$-algebras $h_i' \colon {}^{\omega}SP(A)_i \to S$.

Let S be some $\mathcal{V}'$-algebra and let $h \colon A \to S$ be some function. We show that h can be extended into a homomorphism of $\mathcal{V}'$-algebras $h^{\sharp} \colon {}^{\omega}SP(A) \to S$ in a unique way. Indeed, we define $h^{\sharp}$ as $h^{\sharp} = \bigcup_{i \in \mathbb{N}} h_i$ where each $h_i \colon {}^{\omega}SP(A)_i \to S$ is defined, by induction on i, as follows:

- when $i = 0$, h_0 is defined by $\epsilon \to 1$;
- when $i = 1$, h_1 is the unique homomorphism of $\mathcal{V}_0'$-algebras $^\omega A \to S$ extending h (Theorem 6);
- when $i \geq 2$, h_i is defined as follows:
  - on posets P of width lower than i, $h_i(P)$ is $h_{i-1}(P)$;
  - on sequential posets P of width i, $h_i(P)$ is $h_i'(P)$;
  - on parallel posets P of width i, $h_i(P)$ is defined relying on the irreducible parallel factorization $\|_{j \in [n]} P_j$ of P, for some $n \in \mathbb{N}$, by:

$$h_i(P) = \|_{j \in [n]} h_{i-1}(P_j)$$

Proving that $h^\sharp$ is a homomorphism of $\mathcal{V}'$-algebras is routine. Furthermore, the uniqueness of $h^\sharp$ comes from the facts that $h^\sharp$ extends h and that A is a generating set of $^\omega SP(A)$.

6   Decidability

Throughout this section, A denotes an alphabet. The set of terms of some signature over A is the smallest set of finite words built from A using the operations of the corresponding signature. In this section we prove the decidability of the equational theory of $\mathcal{V}'$.

Let τ be the signature of $\mathcal{V}'$-algebras. We start by defining the set of terms in which we are interested.

Definition 5. *The set of terms T_A over A is the smallest set satisfying the following conditions:*

- *$A \cup \{1\} \subseteq T_A$;*
- *if $t_1, t_2 \in T_A$ then $t_1 \cdot t_2, t_1 \| t_2 \in T_A$;*
- *if $t \in T_A$ then $t^\omega, t^{\overline{\omega}} \in T_A$.*

By equipping T_A with the operations of τ, we define a structure called the term algebra $\mathcal{T}(A) = (T_A, \cdot, \|, {}^\omega, {}^{\overline{\omega}}, 1)$ over A. Note that T_A can be considered also as the set of trees whose leaves are labeled by $A \cup \{1\}$ and whose internal nodes are labeled by the operations of τ where the out-degree of each internal node coincides with the arity of the corresponding operation.

Two terms $t, t' \in T_A$ are equivalent if t' can be derived from t using the axioms which $\mathcal{V}'$ satisfy (denoted $t \equiv t'$). This equivalence relation is actually a congruence. It is well-known that $\mathcal{T}(A)$ is absolutely free i.e. it is freely generated by A in the class containing all the algebras of signature τ. In addition, as a consequence of Theorem 7, $\mathcal{T}(A)/\equiv$ is isomorphic to $^\omega SP(A)$ (see eg. [1, Theorem 1.3.2]). This isomorphism can be defined by $[\![1]\!] = \epsilon$ and $[\![a]\!] = a$ for all $a \in A$.

Then we have:

Proposition 2. *Let $t, t' \in T_A$. Then $[\![t]\!] = [\![t']\!]$ if and only if $t \equiv t'$ holds in $\mathcal{V}'$.*

As a consequence, proving the decidability of the equational theory of $\mathcal{V}'$ can be reduced to decide whether $[\![t]\!] = [\![t']\!]$.

Theorem 8. *Let* $t, t' \in T_A$. *It is decidable whether* $[\![t]\!] = [\![t']\!]$.

We now give a quick outline of the proof. The terms t and t' can be interpreted as particular forms of rational expressions over languages of $SP^\diamond(A)$, see [6]. By extension of a well-known result of Büchi on ordinals, it is known from [2] that a language of $SP^\diamond(A)$ is rational if and only if it is definable in an extension, named *P-MSO*, of the so-called monadic second-order logic. Two P-MSO formulæ ψ_t and $\psi_{t'}$ such that $L(\psi_t) = [\![t]\!]$ and $L(\psi_{t'}) = [\![t']\!]$ can effectively be built from t and t'. We have $L(\psi_t \wedge \psi_{t'}) = \emptyset$ if and only if $[\![t]\!] \neq [\![t']\!]$. Theorem 8 follows from the decidability of the P-MSO theory of $SP^\diamond(A)$ [2, Theorem 6].

This decision procedure has a non-elementary complexity. Another proof with an exponential complexity (in the size of t, t') can be derived from the proof of [12, Theorem 7.6], in which the $\overline{\omega}$-power is not considered, by replacing the use of [12, Theorem 7.3] by [9, Corollary 3.19].

Acknowledgements. We would like to thank the anonymous referees for their comments on this work. One of them pointed out that Theorem 3 can be deduced from Theorem 1 using the theory of categories, and in particular works by Fiore and Hur [13], Robinson [19], Adámek, Rosicky, Velbil et al.

References

1. Almeida, J.: Finite Semigroups and Universal Algebra. Series in Algebra, vol. 3. World Scientific, Singapore (1994)
2. Amrane, A., Bedon, N.: Logic and rational languages of scattered and countable series-parallel posets. Theor. Comput. Sci. **809**, 538–562 (2020). https://doi.org/10.1016/j.tcs.2020.01.015. http://www.sciencedirect.com/science/article/pii/S0304397520300426
3. Bedon, N.: Automata, semigroups and recognizability of words on ordinals. Int. J. Algebra Comput. **8**(1), 1–21 (1998)
4. Bedon, N.: Complementation of branching automata for scattered and countable N-free posets. Int. J. Found. Comput. Sci. **19**(25), 769–799 (2018). https://doi.org/10.1142/S0129054118420042
5. Bedon, N., Carton, O.: An Eilenberg theorem for words on countable ordinals. In: Lucchesi, C.L., Moura, A.V. (eds.) LATIN 1998. LNCS, vol. 1380, pp. 53–64. Springer, Heidelberg (1998). https://doi.org/10.1007/BFb0054310
6. Bedon, N., Rispal, C.: Series-parallel languages on scattered and countable posets. Theor. Comput. Sci. **412**(22), 2356–2369 (2011)
7. Bloom, S., Choffrut, C.: Long words: the theory of concatenation and ω-power. Theor. Comput. Sci. **259**(1–2), 533–548 (2001)
8. Bloom, S., Ésik, Z.: Shuffle binoids. RAIRO-Theor. Inform. Appl. **32**(4–6), 175–198 (1998)
9. Bloom, S., Ésik, Z.: Axiomatizing omega and omega-op powers of words. RAIRO-Theor. Inform. Appl. **38**(1), 3–17 (2004)
10. Carton, O., Colcombet, T., Puppis, G.: Regular languages of words over countable linear orderings. CoRR abs/1702.05342 (2017). http://arxiv.org/abs/1702.05342

11. Carton, O., Rispal, C.: Complementation of rational sets on countable scattered linear orderings. Int. J. Found. Comput. Sci. **16**(4), 767–786 (2005)
12. Choffrut, C., Ésik, Z.: Two equational theories of partial words. Theor. Comput. Sci. **737**, 19–39 (2018)
13. Fiore, M., Hur, C.K.: On the construction of free algebras for equational systems. Theor. Comput. Sci. **410**(18), 1704–1729 (2009). https://doi.org/10.1016/j.tcs.2008.12.052. http://www.sciencedirect.com/science/article/pii/S0304397508009353. Automata, Languages and Programming (ICALP 2007)
14. Gischer, J.: The equational theory of pomsets. Theor. Comput. Sci. **61**(2–3), 199–224 (1988)
15. Grabowski, J.: On partial languages. Fundam. Inform. **4**(1), 427–498 (1981)
16. Hausdorff, F.: Grundzüge einer theorie der geordneten mengen. Mathematische Annalen **65**(4), 435–505 (1908)
17. Kuske, D.: Towards a language theory for infinite N-free pomsets. Theor. Comput. Sci. **299**, 347–386 (2003)
18. Rival, I.: Optimal linear extension by interchanging chains. Proc. AMS **89**(3), 387–394 (1983)
19. Robinson, E.: Variations on algebra: monadicity and generalisations of equational theories. Formal Aspects Comput. **13**(3–5), 308–326 (2002). https://doi.org/10.1007/s001650200014
20. Rosenstein, J.G.: Linear Orderings. Academic Press, Cambridge (1982)
21. Valdes, J., Tarjan, R.E., Lawler, E.L.: The recognition of series parallel digraphs. SIAM J. Comput. **11**, 298–313 (1982)
22. Wilke, T.: An algebraic theory for regular languages of finite and infinite words. Int. J. Algebra Comput. **3**(4), 447–489 (1993)
23. Winkowski, J.: An algebraic approach to concurrence. In: Bečvář, J. (ed.) MFCS 1979. LNCS, vol. 74, pp. 523–532. Springer, Heidelberg (1979). https://doi.org/10.1007/3-540-09526-8_53

Scattered Factor-Universality of Words

Laura Barker[1], Pamela Fleischmann[1(⊠)], Katharina Harwardt[1],
Florin Manea[2], and Dirk Nowotka[1]

[1] Kiel University, Kiel, Germany
{stu97347,stu120568}@mail.uni-kiel.de, {fpa,dn}@informatik.uni-kiel.de
[2] University of Göttingen, Göttingen, Germany
florin.manea@informatik.uni-goettingen.de

Abstract. A word $u = u_1 \ldots u_n$ is a scattered factor of a word w if u can be obtained from w by deleting some of its letters: there exist the (potentially empty) words $v_0, v_1, \ldots, v_n$ such that $w = v_0 u_1 v_1 \ldots u_n v_n$. The set of all scattered factors up to length k of a word is called its full k-spectrum. Firstly, we show an algorithm deciding whether the k-spectra for given k of two words are equal or not, running in optimal time. Secondly, we consider a notion of scattered-factors universality: the word w, with $\mathrm{alph}(w) = \Sigma$, is called k-universal if its k-spectrum includes all words of length k over the alphabet Σ; we extend this notion to k-circular universality. After a series of preliminary combinatorial results, we present an algorithm computing, for a given k'-universal word w the minimal i such that w^i is k-universal for some $k > k'$. Several other connected problems are also considered.

1 Introduction

A scattered factor (also called subsequence or subword) of a given word w is a word u such that there exist (possibly empty) words $v_0, \ldots, v_n, u_1, \ldots, u_n$ with $u = u_1 \ldots u_n$ and $w = v_0 u_1 v_1 u_2 \ldots u_n v_n$. Thus, scattered factors of a word w are imperfect representations of w, obtained by removing some of its parts. As such, there is considerable interest in the relationship between a word and its scattered factors, both from a theoretical and practical point of view (cf. e.g., the chapter *Subwords* by J. Sakarovitch and I. Simon in [27, Chapter 6] for an introduction to the combinatorial properties). Indeed, in situations where one has to deal with input strings in which errors may occur, e.g., sequencing DNA or transmitting a digital signal, scattered factors form a natural model for the processed data as parts of the input may be missing. This versatility of scattered factors is also highlighted by the many contexts in which this concept appears. For instance, in [16,24,37], various logic-theories were developed around the notion of scattered factors which are analysed mostly with automata theory tools and discussed in connection to applications in formal verification. On an even

F. Manea—Supported by the DFG grant MA 5725/2-1. F.M. thanks Paweł Gawrychowski for his comments and suggestions.

more fundamental perspective, there have been efforts to bridge the gap between the field of combinatorics on words, with its usual non-commutative tools, and traditional linear algebra, via, e.g., subword histories or Parikh matrices (cf. e.g., [30,33,34]) which are algebraic structures in which the number of specific scattered factors occurring in a word are stored. In an algorithmic framework, scattered factors are central in many classical problems, e.g., the longest common subsequence or the shortest common supersequence problems [1,28], the string-to-string correction problem [36], as well as in bioinformatics-related works [10].

In this paper we focus, for a given word, on the sets of scattered factors of a given length: the (full) k-spectrum of w is the set containing all scattered factors of w of length exactly k (up to k resp.). The total set of scattered factors (also called downward closure) of $w = \mathsf{aba}$ is $\{\varepsilon, \mathsf{a}, \mathsf{aa}, \mathsf{ab}, \mathsf{aba}, \mathsf{b}, \mathsf{ba}\}$ and the 2-spectrum is $\{\mathsf{aa}, \mathsf{ab}, \mathsf{ba}\}$. The study of scattered factors of a fixed length of a word has its roots in [35], where the relation $\sim_k$ (called Simon's congruence) defines the congruence of words that have the same full k-spectra. Our main interest here lies in a special congruence class w.r.t. $\sim_k$: the class of words which have the largest possible k-spectrum. A word w is called k-*universal* if its k-spectrum contains all the words of length k over a given alphabet. That is, k-universal words are those words that are as rich as possible in terms of scattered factors of length k (and, consequently, also scattered factors of length at most k): the restriction of their downward closure to words of length k contains all possible words of the respective length, i.e., is a *universal* language. Thus $w = \mathsf{aba}$ is not 2-universal since bb is not a scattered factor of w, while $w' = \mathsf{abab}$ is 2-universal. Calling a words *universal* if its k-spectrum contains all possible words of length k, is rooted in formal language theory. The classical universality problem (cf. e.g., [18]) is whether a given language L (over an alphabet Σ) is equal to Σ^*, where L can be given, e.g., as the language accepted by an automaton. A variant of this problem, called length universality, asks, for a natural number ℓ and a language L (over Σ), whether L contains all strings of length ℓ over Σ. See [14] for a series of results on this problem and a discussion on its motivation, and [14,23,31] and the references therein for more results on the universality problem for various types of automata. The universality problem was also considered for words [6,29] and, more recently, for partial words [2,15] w.r.t. their factors. In this context, the question is to find, for a given ℓ, a word w over an alphabet Σ, such that each word of length ℓ over Σ occurs exactly once as a contiguous factor of w. De Bruijn sequences [6] fulfil this property, and have been shown to have many applications in various areas of computer science or combinatorics, see [2,15] and the references therein. As such, our study of scattered factor-universality is related to, and motivated by, this well developed and classical line of research.

While $\sim_k$ is a well studied congruence relation from language theoretic, combinatorial, or algorithmic points of view (see [11,27,35] and the references therein), the study of universality w.r.t. scattered factors seems to have been mainly carried out from a language theoretic point of view. In [20] as well as in [21,22] the authors approach, in the context of studying the height of piecewise testable languages, the notion of ℓ-rich words, which coincides with the

ℓ-universal words we define here; we will discuss the relation between these notions, as well as our preference to talk about universality rather than richness, later in the paper. A combinatorial study of scattered factors universality was started in [5], where a simple characterisation of k-universal binary words was given. In the combinatorics on words literature, more attention was given to the so called binomial complexity of words, i.e., a measure of the multiset of scattered factors that occur in a word, where each occurrence of such a factor is considered as an element of the respective multiset (see, e.g., [12,25,26,32]). As such, it seemed interesting to us to continue the work on scattered factor universality: try to understand better (in general, not only in the case of binary alphabets) their combinatorial properties, but, mainly, try to develop an algorithmic toolbox around the concept of (k-)universal words.

Our Results. In the preliminaries we give the basic definitions and recall the arch factorisation introduced by Hebrard [17]. Moreover we explain in detail the connection to richness introduced in [20].

In Sect. 3 we show one of our main results: testing whether two words have the same full k-spectrum, for given $k \in \mathbb{N}$, can be done in optimal linear time for words over ordered alphabets and improve and extend the results of [11]. They also lead to an optimal solution over general alphabets.

In Sect. 4 we prove that the arch factorisation can be computed in time linear w.r.t. the word-length and, thus, we can also determine whether a given word is k-universal. Afterwards, we provide several combinatorial results on k-universal words (over arbitrary alphabets); while some of them follow in a rather straightforward way from the seminal work of Simon [35], other require a more involved analysis. One such result is a characterisation of k-universal words by comparing the spectra of w and w^2. We also investigate the similarities and differences of the universality if a word w is repeated or w^R and $\pi(w)$ resp. are appended to w, for a morphic permutation of the alphabet π. As consequences, we get a linear run-time algorithm for computing a minimal length scattered factor of ww that is not a scattered factor of w. This approach works for arbitrary alphabets, while, e.g., the approach of [17] only works for binary ones. We conclude the section by analysing the new notion of k-circular universality, connected to the universality of repetitions.

In Sect. 5 we consider the problem of modifying the universality of a word by repeated concatenations or deletions. Motivated by the fact that, in general, starting from an input word w, we could reach larger sets of scattered factors of fixed length by iterative concatenations of w, we show that, for a word w a positive integer k, we can compute efficiently the minimal ℓ such that w^ℓ is k-universal. This result is extensible to sets of words. Finally, the shortest prefix or suffix we need to delete to lower the universality index of a word to a given number can be computed in linear time. Interestingly, in all of the algorithms where we are concerned with reaching k-universality we never effectively construct a k-universal word (which would take exponential time, when k is given as input via its binary encoding, and would have been needed when solving these problems using, e.g., [10,11]). Our algorithms run in polynomial time w.r.t. $|w|$, the length of the input word, and $\log_2 k$, the size of the representation of k.

2 Preliminaries

Let $\mathbb{N}$ be the set of natural numbers and $\mathbb{N}_0 = \mathbb{N} \cup \{0\}$. Define $[n]$ as the set $\{1, \ldots, n\}$, $[n]_0 = [n] \cup \{0\}$ for an $n \in \mathbb{N}$, and $\mathbb{N}_{\geq n} = \mathbb{N} \setminus [n-1]$. An alphabet Σ is a nonempty finite set of symbols called *letters*. A *word* is a finite sequence of letters from Σ, thus an element of the free monoid Σ^*. Let $\Sigma^+ = \Sigma^* \setminus \{\varepsilon\}$, where ϵ is the empty word. The *length* of a word $w \in \Sigma^*$ is denoted by $|w|$. For $k \in \mathbb{N}$ define $\Sigma^k = \{w \in \Sigma^* \mid |w| = k\}$ and $\Sigma^{\leq k}, \Sigma^{\geq k}$ analogously. A word $u \in \Sigma^*$ is a *factor* of $w \in \Sigma^*$ if $w = xuy$ for some $x, y \in \Sigma^*$. If $x = \varepsilon$ (resp. $y = \epsilon$), u is called a *prefix* (resp. *suffix* of w). Let $\mathrm{Pref}_k(w)$ be the prefix of w of length $k \in \mathbb{N}_0$. The ith letter of $w \in \Sigma^*$ is denoted by $w[i]$ for $i \in [|w|]$ and set $w[i..j] = w[i]w[i+1] \ldots w[j]$ for $1 \leq i \leq j \leq |w|$. Define the *reversal* of $w \in \Sigma^n$ by $w^R = w[n] \ldots w[1]$. Set $|w|_\mathtt{a} = |\{i \in [|w|] \mid w[i] = \mathtt{a}\}|$ and $\mathrm{alph}(w) = \{\mathtt{a} \in \Sigma \mid |w|_\mathtt{a} > 0\}$ for $w \in \Sigma^*$. For a word $u \in \Sigma^*$ we define $u^0 = \varepsilon$, $u^{i+1} = u^i u$, for $i \in \mathbb{N}$. A word $w \in \Sigma^*$ is called *power* (repetition) of a word $u \in \Sigma^*$, if $w = u^t$ for some $t \in \mathbb{N}_{\geq 2}$. A word $u \in \Sigma^*$ is a *conjugate* of $w \in \Sigma^*$ if there exist $x, y \in \Sigma^*$ with $w = xy$ and $u = yx$. A function $\pi : \Sigma^* \to \Sigma^*$ is called *morphic permutation* if π is bijective and $\pi(uv) = \pi(u)\pi(v)$ for all $u, v \in \Sigma^*$.

Definition 1. *A word $v = v_1 \ldots v_k \in \Sigma^*$ is a* scattered factor *of $w \in \Sigma^*$ if there exist $x_1, \ldots, x_{k+1} \in \Sigma^*$ such that $w = x_1 v_1 \ldots x_k v_k x_{k+1}$. Let $\mathrm{ScatFact}(w)$ be the set of all scattered factors of w and define $\mathrm{ScatFact}_k(w)$ (resp., $\mathrm{ScatFact}_{\leq k}(w)$) as the set of all scattered factors of w of length (resp., up to) $k \in \mathbb{N}$. A word $u \in \Sigma^*$ is a* common scattered factor *of $w, v \in \Sigma^*$, if $u \in \mathrm{ScatFact}(w) \cap \mathrm{ScatFact}(v)$; the word u is an* uncommon scattered factor *of w and v (and* distinguishes *them) if u is a scattered factor of exactly one of them.*

For $k \in \mathbb{N}_0$, the sets $\mathrm{ScatFact}_k(w)$ and $\mathrm{ScatFact}_{\leq k}(w)$ are also known as the k-spectrum and the full-k-spectrum of w resp.. Simon [35] defined the congruence $\sim_k$ in which $u, v \in \Sigma^*$ are congruent if they have the same full k-spectrum and thus the same k-spectrum. The *shortlex normal form* of a word $w \in \Sigma^*$ w.r.t. $\sim_k$, where Σ is an ordered alphabet, is the shortest word u with $u \sim_k w$ which is also lexicographically smallest (w.r.t. the given order on Σ) amongst all words $v \sim_k w$ with $|v| = |u|$. The maximal cardinality of a word's k-spectrum is $|\Sigma|^k$ and as shown in [5] this is equivalent in the binary case to $w \in \{\mathtt{ab}, \mathtt{ba}\}^k$. The following definition captures this property of a word in a generalised setting.

Definition 2. *A word $w \in \Sigma^*$ is called k-universal (w.r.t. Σ), for $k \in \mathbb{N}_0$, if $\mathrm{ScatFact}_k(w) = \Sigma^k$. We abbreviate 1-universal by universal. The universality-index $\iota(w)$ of $w \in \Sigma^*$ is the largest k such that w is k-universal.*

Remark 3. Notice that k-universality is always w.r.t. a given alphabet Σ: the word $\mathtt{abcba}$ is 1-universal for $\Sigma = \{\mathtt{a}, \mathtt{b}, \mathtt{c}\}$ but it is not universal for $\Sigma \cup \{\mathtt{d}\}$. If it is clear from the context, we do not explicitly mention Σ. The universality of the factors of a word w is considered w.r.t. $\mathrm{alph}(w)$.

Karandikar and Schnoebelen introduced in [21, 22] the notion of richness of words: $w \in \Sigma^*$ is *rich* (w.r.t. Σ) if $alph(w) = \Sigma$ (and *poor* otherwise) and w is ℓ-rich if w is the concatenation of $\ell \in \mathbb{N}$ rich words. Immediately we get that a word is universal iff it is rich and moreover that a word is ℓ-rich iff it is ℓ-universal and a rich-factorisation, i.e., the factorisation of an ℓ-rich word into ℓ rich words, can be efficiently obtained. However, we will use the name ℓ-*universality* rather than ℓ-*richness*, as richness defines as well, e.g. the property of a word $w \in \Sigma^n$ to have $n + 1$ distinct palindromic factors, see, e.g., [7, 9]. As w is ℓ-universal iff w is the concatenation of $\ell \in \mathbb{N}$ universal words it follows immediately that, if w is over the ordered alphabet $\Sigma = \{1 < 2 < \ldots < \sigma\}$ and it is ℓ-universal then its shortlex normal form w.r.t. $\sim_\ell$ is $(1 \cdot 2 \cdots \sigma)^\ell$ (as this is the shortest and lexicographically smallest ℓ-universal word).

The following observation leads to the next definition: the word $w = \mathsf{abc} \in \{\mathsf{a}, \mathsf{b}, \mathsf{c}\}^*$ is 1-universal and w^s is s-universal for all $s \in \mathbb{N}$. But, $v^2 = (\mathsf{ababcc})^2 \in \{\mathsf{a}, \mathsf{b}, \mathsf{c}\}^*$ is 3-universal even though v is only 1-universal. Notice that the conjugate abccab of v is 2-universal.

Definition 4. *A word $w \in \Sigma^*$ is called k-circular universal if a conjugate of w is k-universal (abbreviate 1-circular universal by circular universal). The circular universality index $\zeta(w)$ of w is the largest k such that w is k-circular universal.*

Remark 5. It is worth noting that, unlike the case of factor universality of words and partial words [2, 6, 15, 29], in the case of scattered factors it does not make sense to try to identify a k-universal word $w \in \Sigma^*$, for $k \in \mathbb{N}_0$, such that each word from Σ^k occurs *exactly* once as scattered factor of w. Indeed for $|\Sigma| = \sigma$, if $|w| \geq k + \sigma$ then there exists a word from Σ^k which occurs at least twice as a scattered factor of w. Moreover, the shortest word which is k-universal has length $k\sigma$ (we need $\mathsf{a}^k \in \mathrm{ScatFact}_k(w)$ for all $\mathsf{a} \in \Sigma$). As $k\sigma \geq k + \sigma$ for $k, \sigma \in \mathbb{N}_{\geq 2}$, all k-universal words have scattered factors occurring more than once: there exists $i, j \in [\sigma + 1]$ such that $w[i] = w[j]$ and $i \neq j$. Then $w[i]w[\sigma + 2..\sigma + k], w[j]w[\sigma + 2..\sigma + k] \in \mathrm{ScatFact}_k(w)$ and $w[i]w[\sigma + 2.\sigma + k] = w[j]w[\sigma + 2..\sigma + k]$.

We now recall the arch factorisation, introduced by Hebrard in [17].

Definition 6 (*[17]*). *For $w \in \Sigma^*$ the* arch factorisation *of w is given by $w = \mathrm{ar}_w(1) \ldots \mathrm{ar}_w(k)r(w)$ for a $k \in \mathbb{N}_0$ with $\mathrm{ar}_w(i)$ is universal and $\mathrm{ar}_w(i)[|\,\mathrm{ar}_w(i)|] \notin alph(\mathrm{ar}_w(i)[1 \ldots |\,\mathrm{ar}_w(i)| - 1])$ for all $i \in [n]$, and $alph(r(w)) \subset \Sigma$. The words $\mathrm{ar}_w(i)$ are called* archs *of w, $r(w)$ is called the* rest. *Set $m(w) = \mathrm{ar}_w(1)[|\,\mathrm{ar}_w(1)|] \ldots \mathrm{ar}_w(k)[|\,\mathrm{ar}_w(k)|]$ as the word containing the unique last letters of each arch.*

Remark 7. If the arch factorisation contains $k \in \mathbb{N}_0$ archs, the word is k-universal, thus the equivalence of k-richness and k-universality becomes clear. Moreover if a factor v of $w \in \Sigma^*$ is k-universal then w is also k-universal: if v has an arch factorisation with k archs then w's arch factorisation has at least k archs (in which the archs of v and w are not necessarily related).

Finally, our main results are of algorithmic nature. The computational model we use is the standard unit-cost RAM with logarithmic word size: for an input of size n, each memory word can hold $\log n$ bits. Arithmetic and bitwise operations with numbers in $[n]$ are, thus, assumed to take $O(1)$ time. Arithmetic operations on numbers larger than n, with ℓ bits, take $O(\ell/\log n)$ time. For simplicity, when evaluating the complexity of an algorithm we first count the number of steps we perform (e.g., each arithmetic operation is counted as 1, no matter the size of the operands), and then give the actual time needed to implement these steps in our model. In our algorithmic problems, we assume that the processed words are sequences of integers (called letters or symbols, each fitting in $O(1)$ memory words). In other words, we assume that the alphabet of our input words is *an integer alphabet*. In general, after a linear time preprocessing, we can assume that the letters of an input word of length n over an integer alphabet Σ are in $\{1, \ldots, |\Sigma|\}$ where, clearly, $|\Sigma| \leq n$. For a more detailed discussion see, e.g., [4].

3   Testing Simon's Congruence

Our first result extends and improves the results of Fleischer and Kufleitner [11].

Theorem 8. *(1) Given a word w over an integer alphabet Σ, with $|w| = n$, and a number $k \leq n$, we can compute the shortlex normal form of w w.r.t. $\sim_k$ in time $O(n)$. (2) Given two words w', w'' over an integer alphabet Σ, with $|w'| \leq |w''| = n$, and a number $k \leq n$, we can test if $w' \sim_k w''$ in time $O(n)$.*

Proof. The main idea of the algorithm is that checking $w' \sim_k w''$ is equivalent to checking whether the shortlex normal forms w.r.t. $\sim_k$ of w' and w'' are equal. To compute the shortlex normal form of a word $w \in \Sigma^n$ w.r.t. $\sim_k$ the following approach was used in [11]: firstly, for each position of w the x- and y-coordinates were defined. The x-coordinate of i, denoted x_i, is the length of the shortest sequence of indices $1 \leq i_1 < i_2 < \ldots < i_t = i$ such that i_1 is the position where the letter $w[i_1]$ occurs w for the first time and, for $1 < j \leq t$, i_j is the first position where $w[i_j]$ occurs in $w[i_{j-1}+1..i]$. Obviously, if a occurs for the first time on position i in w, then $x_i = 1$ (see [11] for more details). A crucial property of the x-coordinates is that if $w[\ell] = w[i] = $ a for some $i > \ell$ such that $w[j] \neq $ a for all $\ell + 1 \leq j \leq i - 1$, then $x_i = \min\{x_\ell, x_{\ell+1}, \ldots, x_{i-1}\} + 1$. The y-coordinate of a position i, denoted y_i, is defined symmetrically: y_i is the length of the shortest sequence of indices $n \geq i_1 > i_2 > \ldots > i_t = i$ such that i_1 is the position where the letter $w[i_1]$ occurs last time in w and, for $1 < j \leq t$, i_j is the last position where $w[i_j]$ occurs in $w[i..i_{j-1} - 1]$. Clearly, if $w[\ell] = w[i] = $ a for some $i < \ell$ such that $w[j] \neq $ a for all $\ell - 1 \geq j \geq i + 1$, then $y_i = \min\{y_{i+1}, \ldots, y_{\ell-1}, y_\ell\} + 1$.

   Computing the coordinates is done in two phases: the x-coordinates are computed and stored (in an array x with elements $x_1, \ldots, x_n$) from left to right in phase 1a, and the y-coordinates are stored in an array y with elements $y_1, \ldots, y_n$ and computed from right to left in phase 1b (while dynamically deleting a position whenever the sum of its coordinates is greater then $k+1$ (cf. [11, Prop. 2])). Then, to compute the shortlex normal form, in a third phase, labelled phase 2, if letters b $>$ a occur consecutively in this order, they are interchanged whenever they have the same x- and y-coordinates and the sum of these coordinates is $k + 1$ (until this situation does not occur anymore).

We now show how these steps can be implemented in $O(n)$ time for input words over integer alphabets. For simplicity, let $x[i..j]$ denote the sequence of coordinates $x_i, x_{i+1}, \ldots, x_j$; $\min(x[i..j])$ denotes $\min\{x_i, \ldots, x_j\}$. It is clear that in $O(n)$ time we can compute all values $last[i] = \max(\{0\} \cup \{j < i | w[j] = w[i]\})$.

Firstly, phase 1a. For simplicity, assume that $x_0 = 0$. While going with i from 1 to n, we maintain a list L of positions $0 = i_0 < i_1 < i_2 < \ldots < i_t = i$ such that the following property is invariant: $x_{i_{\ell-1}} < x_{i_\ell}$ for $1 \le \ell \le t$ and $x_p \ge x_{i_\ell}$ for all $i_{\ell-1} < p \le i_\ell$. After each i is read, if $last[i] = 0$ then set $x_i = 1$; otherwise, determine $x_i = \min(x[last[i]..i-1]) + 1$ by L, then append i to L and update L accordingly so that its invariant property holds. This is done as follows: we go through the list L from right to left (i.e., inspect the elements $i_t, i_{t-1}, \ldots$) until we reach a position $i_{j-1} < last[i]$ or completely traverse the list (i.e., $i_{j-1} = 0$). Let us note now that all elements x_ℓ with $i - 1 \ge \ell \ge last[i]$ fulfill $x_\ell \ge x_{i_j}$ and $i_j \ge last[i]$. Consequently, $x_i = x_{i_j} + 1$. Moreover, $x_{i_{j+1}} \ge x_{i_j} + 1$. As such, we update the list L so that it becomes $i_1, \ldots, i_j, i$ (and x_i is stored in the array x).

Note that each position of w is inserted once in L and once deleted (but never reinserted). Also, the time needed for the update of L caused by the insertion of i is proportional to the number of elements removed from the list in that step. Accordingly, the total time needed to process L, for all i, is $O(n)$. Clearly, this procedure computes the x-coordinates of all the positions of w correctly.

Secondly, phase 1b. We cannot proceed exactly like in the previous case, because we need to dynamically delete a position whenever the sum of its coordinates is greater than $k+1$ (i.e., as soon as we finished computing its y-coordinate and see that it is $> k + 1$; this position does not influence the rest of the computation). If we would proceed just as above (right to left this time), it might be the case that after computing some y_i we need to delete position i, instead of storing it in our list and removing some of the elements of the list. As such, our argument showing that the time spent for inspecting and updating the list in the steps where the y-coordinates are computed amortises to $O(n)$ would not work.

So, we will use an enhanced approach. For simplicity, assume that $y_{n+1} = 0$ and that every time we should eliminate position i we actually set y_i to $+\infty$. Also, let $y[i..j]$ denote the sequence of coordinates $y_i, y_{i+1}, \ldots, y_j$; note that some of these coordinates can be $+\infty$. Let $min(y[i..j])$ denote the minimum in the sequence $y[i..j]$. Similarly to what we did in phase 1a, while going with i from n to 1, we maintain a list L' of positions $n + 1 = i_0 > i_1 > i_2 > \ldots > i_t \ge i$ such that the following property is invariant: $y_{i_{\ell-1}} < y_{i_\ell}$ for $1 \le \ell \le t$ and $y_p \ge y_{i_\ell}$ for all $i_{\ell-1} > p \ge i_\ell$. In the current case, we also have that $y_p = +\infty$ for all $i_t > p \ge i$. The numbers $i_0, i_1, i_2, \ldots, i_t \ge i$ contained in the list L' at some moment in our computation define a partition of the universe $[1, n]$ in intervals: $\{1\}, \{2\}, \ldots,$ $\{i-1\}, [i, i_{t-1} - 1], [i_{t-1}, i_{t-2} - 1], \ldots, [i_1, i_0 - 1]$ for which we define an *interval union-find* data structure [13,19]; here the singleton $\{a\}$ is seen as the interval $[a, a]$. According to [19], in our model of computation, such a structure can be initialized in $O(n)$ time such that we can perform a sequence of $O(n)$ **union** and find operations on it in $O(n)$ time, with the crucial restriction that one can only unite neighbouring intervals. We assume that find(j) returns the bounds of the interval stored in our data structure to which j belongs. From the definition of the list L', it is clear that, before processing position i (and after finishing processing position $i+1$), $y_{i_\ell} = \min(y[i+1..i_{\ell-1}-1])$ holds. We maintain a new array next$[\cdot]$ with $|\Sigma|$ elements: before processing position i, next$[w[i]]$ is the smallest position $j > i$ where $w[i]$ occurs after position i, which was not eliminated (i.e., smallest $j > i$ with $y_j \neq +\infty$), or 0 if there is no such position. Position i is now processed as follows: let $[a,b]$ be the interval returned by find(next$[i]$). If $a = i+1$ then let min $= y_{i_t}$; if $a > i+1$ then there exists j such that $[a,b] = [i_j, i_{j-1} - 1]$ and $t > j > 0$, so let min $= y_j$. Let now $y = \min +1$, and note that we should set $y_i = y$, but only if $x_i + i \leq k+1$. So, we check whether $x_i + i \leq k+1$ and, if yes, let $y_i = y$ and set next$[w[i]] = i$; otherwise, set $y_i = +\infty$ (note that position i becomes, as such, irrelevant when the y-coordinate is computed for other positions). If $y_i = +\infty$ then make the union of the intervals $\{i\}$ and $[i+1, i_{t-1} - 1]$ and start processing $i-1$; L' remains unchanged. If $y_i \neq +\infty$ then make the union of the intervals $\{i\}, [i+1, i_{t-1} - 1], \ldots, [i_{j+1}, i_j - 1]$ and start processing $i-1$; L' becomes $i, i_{j-1}, \ldots, i_0$.

As each position of w is inserted at most once in L', and then deleted once (never reinserted), the number of list operations is $O(n)$. The time needed for the update of L', caused by the insertion of i in L', is proportional to the number of elements removed from L' in that step, so the total time needed (exclusively) to process L is $O(n)$. On top of that, for each position i, we run one find operation and a number of union operations proportional to the number of elements removed from L' in that step. Overall we do $O(n)$ union and find operations on the *union-find* data structure. This takes in total, for all i, $O(n)$ time (including the initialisation). Thus, the time complexity of phase 1b is linear.

Thirdly, phase 2. Assume that w_0 is the input word *of this phase*. Clearly, $|w_0| = m \leq n$, and we have computed the coordinates for all its positions (and maybe eliminated some positions of the initial input word w). We partition in linear time $O(n)$ the interval $[1, m]$ into $2t+1$ (possibly empty) lists of positions $L_1, \ldots, L_{2t+1}$ such that the following conditions hold. Firstly, all elements of L_i are smaller than those of L_{i+1} for $1 \leq i \leq 2t$. Secondly, for i odd, the elements j in L_i have $x_j + y_j < k+1$; for each i even, there exist a_i, b_i such that $a_i + b_i = k+1$ and for all j in L_i we have $x_j = a_i, y_j = b_i$. Thirdly, we want t to be minimal with these properties. We now produce, also in linear time, a new list U: for each $i \leq t$ and $j \in L_{2i}$ we add the triplet $(i, w[j], j)$ in U. We sort the list of triples U (cf. [11, Prop. 10]) with radix sort in linear time [3]. After sorting it, U can be decomposed in t consecutive blocks $U_1, U_2, \ldots, U_t$, where U_i contains the positions of L_{2i} sorted w.r.t. the order on Σ (i.e., determined by the second component of the pair). As such, U_i induces a new order on the positions of w_0 stored in L_{2i}. We can now construct a word w_1 by just writing in order the letters of w_0 corresponding to the positions stored in L_i, for i from 1 to $2t+1$, such that the letters of L_i are written in the original order, for i odd, and in the order induced by U_i, for i even. Clearly, this is a correct implementation of phase 2 which runs in linear time. The word w_1 is the shortlex normal form of w.

Summing up, we have shown how to compute the shortlex normal form of a word in linear time (for integer alphabets). Both our claims follow.   □

This improves the complexity of the algorithm reported in [11], where the problem was solved in $O(n|\Sigma|)$ time. As such, over integer alphabets, testing Simon's congruence for a given k can be done in optimal time, that does not depend on the input alphabet or on k. When no restriction is made on the input alphabet, we can first sort it, replace the letters by their ranks, and, as such, reduce the problem to the case of integer alphabets. In that case, testing Simon's congruence takes $O(|\Sigma| \log |\Sigma| + n)$ time which is again optimal: for $k = 1$, testing if $w_1 \sim_1 w_2$ is equivalent (after a linear time processing) to testing whether two subsets of Σ are equal, and this requires $\Theta(|\Sigma| \log |\Sigma|)$ time [8].

4   Scattered Factor Universality

In this section we present several algorithmic and combinatorial results.

Remark 9. Theorem 8 allows us to decide in linear time $O(n)$ whether a word w over $\Sigma = \{1 < 2 < \ldots < \sigma\}$ is k-universal, for a given $k \le n, \sigma \in \mathbb{N}$. We compute the shortlex normal form of w w.r.t. $\sim_k$ and check whether it is $(1 \cdot 2 \cdots \sigma)^k$.

We can actually compute $\iota(w)$ efficiently by computing its arch factorisation in linear time in $|w|$. Moreover this allows us to check whether w is k-universal for some given k by just checking if $\iota(w) \ge k$ or not.

Proposition 10. *Given a word $w \in \Sigma^n$, we can compute $\iota(w)$ in time $O(n)$.*

Proof. We actually compute the number ℓ of archs in the arch factorisation. For a lighter notation, we use $u_i = \mathrm{ar}_w(i)$ for $i \in [\ell]_0$. The factors u_i can be computed in linear time as follows. We maintain an array C of $|\Sigma|$ elements, whose all elements are initially 0, and a counter h, which is initially $|\Sigma|$. For simplicity, let $m_0 = 0$. We go through the letters $w[j]$ of $w[m_{i-1} + 1..n]$, from left to right, and if $C[w[j]]$ equals 0, we decrement h by 1 and set $C[w[j]] = 1$. Intuitively, we keep track of which letters of Σ we meet while traversing $w[m_{i-1} + 1..n]$ using the array C, and we store in h how many letters we still need to see. As soon as $h = 0$ or $j = n$, we stop: set $m_i = j$ (the position of the last letter of w we read), $u_i = w[m_{i-1} + 1..m_i]$ (the ith arch), and $h = |\Sigma|$ again. If $j < n$ then reinitialise all elements of C to 0 and restart the procedure for $i + 1$. Note that if $j = n$ then u_i is $r(w)$ as introduced in the definition of the arch factorization. The time complexity of computing u_j is $O(|u_j|)$, because we process each symbol of $u_i = w[m_{i-1} + 1..m_i]$ in $O(1)$ time, and, at the end of the procedure, we reinitialise C in $O(|\Sigma|)$ time iff u_i contained all letters of Σ, so $|u_i| \ge |\Sigma|$. The conclusion follows.   □

The following combinatorial result characterise universality by repetitions.

Theorem 11 (∗). *A word $w \in \Sigma^{\geq k}$ with $alph(w) = \Sigma$ is k-universal for $k \in \mathbb{N}_0$ iff $\mathrm{ScatFact}_k(w^n) = \mathrm{ScatFact}_k(w^{n+1})$ for an $n \in \mathbb{N}$. Moreover we have $\iota(w^n) \geq kn$ if $\iota(w) = k$.*

As witnessed by $w = \mathsf{aabb} \in \{\mathsf{a}, \mathsf{b}\}^*$, $\iota(w^n)$ can be greater than $n \cdot \iota(w)$: w is universal, not 2-universal but $w^2 = \mathsf{aab.ba.ab.b}$ is 3-universal. We study this phenomenon at the end of this section. Theorem 11 can also be used to compute an uncommon scattered factor of w and ww over arbitrary alphabets; note that the shortest such a factor has to have length $k + 1$ if $\iota(w) = k$.

Proposition 12 (∗). *Given a word $w \in \Sigma^*$ we can compute in linear time $O(|w|)$ one of the uncommon scattered factors of w und ww of minimal length.*

Remark 13. By Proposition 12, computing the shortest uncommon scattered factor of w and ww takes optimal $O(n)$ time, which is more efficient than running an algorithm computing the shortest uncommon scattered factor of two arbitrary words (see, e.g., [10,11], and note that we are not aware of any linear-time algorithm performing this task for integer alphabets). In particular, we can use Theorem 8 to find by binary search the smallest k for which two words have distinct k-spectra in $O(n \log n)$ time. In [17] a linear time algorithm solving this problem is given for binary alphabets; an extension seems non-trivial.

Continuing the idea of Theorem 11, we investigate even-length palindromes, i.e. appending w^R to w. The first result is similar to Theorem 11 for $n = 1$. Notice that $\iota(w) = \iota(w^R)$ follows immediately with the arch factorisation.

Corollary 14. *A word w is k-universal iff $\mathrm{ScatFact}_k(w) = \mathrm{ScatFact}_k(ww^R)$.*

In contrast to $\iota(w^2)$, $\iota(ww^R)$ is never greater than $2\iota(w)$.

Proposition 15 (∗). *Let $w \in \Sigma^*$ be a palindrome and $u = \mathrm{Pref}_{\lfloor \frac{|w|}{2} \rfloor}(w)$ with $\iota(u) = k \in \mathbb{N}$. For $|w|$ even we have $\iota(w) = 2k$ if $|w|$ even and for $|w|$ odd we get $\iota(w) = 2k + 1$ iff $w[\frac{n+1}{2}] \cup alph(r(u)) = \Sigma$.*

Remark 16. If we consider the universality of a word $w = w_1 \ldots w_m$ for $m \in \mathbb{N}$ with $w_i \in \{u, u^R\}$ for a given word $u \in \Sigma^*$, then a combination of the previous results can be applied. Each time either u^2 or $(u^R)^2$ occurs Theorem 11 can be applied (and the results about circular universality that finish this section). Whenever uu^R or u^Ru occur in w, the results of Proposition 15 are applicable.

Another generalisation of Theorem 11 is to investigate concatenations under permutations: for a morphic permutation π of Σ can we compute $\iota(w\pi(w))$?

Lemma 17 (∗). *Let $\pi : \Sigma^* \to \Sigma^*$ be a morphic permutation. Then $\iota(w) = \iota(\pi(w))$ for all $w \in \Sigma^*$ and especially the factors of the arch factorisation of w are mapped by π to the factors of the arch factorisation of $\pi(w)$.*

By Lemma 17 we have $2\iota(w) \leq \iota(w\pi(w)) \leq 2\iota(w)+1$. Consider the universal word $w = $ abcba. For $\pi(a) = $ c, $\pi(b) = $ b, and $\pi(c) = $ a we obtain $w\pi(w) = $ abc.bac.babc. which is 3-universal. However, for the identity id on Σ we get that $w\,\mathrm{id}(w)$ is 2-universal. We can show exactly the case when $\iota(w\pi(w)) = 2\iota(w)+1$.

Proposition 18 (∗). *Let* $\pi : \Sigma^* \to \Sigma^*$ *be a morphic permutation and* $w \in \Sigma^*$ *with the arch factorisation* $w = \mathrm{ar}_w(1)\ldots\mathrm{ar}_w(k)r(w)$ *and* $\pi(w)^R = \mathrm{ar}_{\pi(w)^R}(1)\ldots \mathrm{ar}_{\pi(w)^R}(k)r(\pi(w)^R)$ *for an appropriate* $k \in \mathbb{N}_0$. *Then* $\iota(w\pi(w)) = 2\iota(w)+1$ *iff* $\mathrm{alph}(r(w)r(\pi(w)^R)) = \Sigma$, *i.e. the both rests together are 1-universal.*

Proposition 18 ensures that, for a given word with a non-empty rest, we can raise the universality-index of $w\pi(w)$ by one if π is chosen accordingly.

Remark 19. Appending permutations of the word instead of its images under permutations of the alphabet, i.e. appending to w abelian equivalent words, does not lead to immediate results as the universality depends heavily on the permutation. If w is k-universal, a permutation π may arrange the letters in lexicographical order, so $\pi(w)$ would only be 1-universal. On the other hand, the universality can be increased by sorting the letters in 1-universal factors: $a_1^m a_2^m \ldots a_{|\Sigma|}^m$ for $\Sigma = \{a_1,\ldots,a_{|\Sigma|}\}$ is 1-universal but $(a_1 \ldots a_{|\Sigma|})^m$ is m-universal, for $m \in \mathbb{N}$.

In the rest of this section we present results regarding circular universality. Recall that a word w is k-circular universal if a conjugate of w is k-universal. Consider $\Sigma = \{a, b, c, d\}$ and $w = $ abbccdabacdbdc. Note that w is not 3-universal (dda $\notin \mathrm{ScatFact}_3(w)$) but 2-universal. Moreover, the conjugate bbccdabacdbdca of w is 3-universal; accordingly, w is 3-circular universal.

Lemma 20 (∗). *Let* $w \in \Sigma^*$. *If* $\iota(w) = k \in \mathbb{N}$ *then* $k \leq \zeta(w) \leq k+1$. *Moreover if* $\zeta(w) = k+1$ *then* $\iota(w) \geq k$.

Lemma 21 (∗). *Let* $w \in \Sigma^+$. *If* $\iota(w) = k$ *and* $\zeta(w) = k+1$ *then there exists* $v, z, u \in \Sigma^*$ *such that* $w = vzu$, *with* $u, v \neq \varepsilon$ *and* $\iota(z) = k$.

The following theorem connects the circular universality index of a word with the universality index of the repetitions of that word.

Theorem 22 (∗). *Let* $w \in \Sigma^*$. *If* $\iota(w) = k$ *and* $\zeta(w) = k+1$ *then* $\iota(w^s) = sk + s - 1$, *for all* $s \in \mathbb{N}$.

The other direction of Theorem 22 does not hold for arbitrary alphabets: Consider the 2-universal word $w = $ babccaabc. We have that w^2 is 5-universal but w is not 3-circular universal. Nevertheless, Lemma 21 helps us show that the converse of Theorem 22 holds for binary alphabets:

Theorem 23 (∗). *Let* $w \in \{a, b\}^*$ *with* $\iota(w) = k$ *and* $s \in \mathbb{N}$. *Then* $\iota(w^s) = sk + s - 1$ *if* $\zeta(w) = k+1$ *and* sk *otherwise.*

5 On Modifying the Universality Index

In this section we present algorithms answering the for us most natural questions regarding universality: is a specific factor v of $w \in \Sigma^*$ universal? what is the minimal $\ell \in \mathbb{N}$ such that w^ℓ is k-universal for a given $k \in \mathbb{N}$? how many (and which) words from a given set do we have to concatenate such that the resulting word is k-universal for a given $k \in \mathbb{N}$? what is the longest (shortest) prefix (suffix) of a word being k-universal for a given $k \in \mathbb{N}$? In the following lemma we establish some preliminary data structures.

Lemma 24 (∗). *Given a word $x \in \Sigma^n$ with $alph(x) = \Sigma$, we can compute in $O(n)$ and for all $j \in [n]$*

- *the shortest 1-universal prefix of $x[j..n]$: $u_x[j] = \min\{i \mid x[j..i]$ is universal$\}$,*
- *the value $\iota(x[j..n])$: $t_x[j] = \max\{t \mid \mathrm{ScatFact}_t(x[j..n]) = \Sigma^t\}$, and*
- *the minimal $\ell \in [n]$ with $\iota(x[j..\ell]) = \iota(x[j..|x|])$: $m_x[j] = \min\{i \mid \mathrm{ScatFact}_{t_x[j]}$ $(x[j..i]) = \Sigma^{t_x[j]}\}$.*

The data structures constructed in Lemma 24 allow us to test in $O(1)$ time the universality of factors $w[i..j]$ of a given word w, w.r.t. $alph(w) = \Sigma$: $w[i..j]$ is Σ-universal iff $j \geq u_w[i]$. The combinatorial results of Sect. 4 give us an initial idea on how the universality of repetitions of a word relates to the universality of that word: Theorem 22 shows that in order to compute the minimum s such that w^s is ℓ-universal, for a given *binary* word w and a number ℓ, can be reduced to computing the circular universality of w. Unfortunately, this is not the case for all alphabets, as also shown in Sect. 4. However, this number s can be computed efficiently, for input words over alphabets of all sizes. While the main idea for binary alphabets was to analyse the universality index of the conjugates of w (i.e., factors of length $|w|$ of ww), in the general case we can analyse the universality index of the suffixes of ww, by constructing the data structures of Lemma 24 for $x = ww$. The problem is then reduced to solving an equation over integers in order to identify the smallest ℓ such that w^ℓ is k-universal.

Proposition 25 (∗). *Given a word $w \in \Sigma^n$ with $alph(w) = \Sigma$ and $k \in \mathbb{N}$, we can compute the minimal ℓ such that w^ℓ is k-universal in $O(n + \frac{\log k}{\log n})$ time.*

We can extend the previous result to the more general (but less motivated) case of arbitrary concatenations of words from a given set, not just repetitions of the same word. The following preliminary results can be obtained. In all cases we give the number of steps of the algorithms, including arithmetic operations on $\log k$-bit numbers; the time complexities of these algorithms is obtained by multiplying these numbers by $O(\frac{\log k}{\log n})$.

1. Given the words $w_1, \ldots, w_p \in \Sigma^*$ with $|w_1 \cdots w_p| = n$ and $alph(w_1 \cdots w_p) = \Sigma$, and $k \in \mathbb{N}$, we can compute the minimal ℓ for which there exist $\{i_1, \ldots, i_\ell\} \subseteq [k]$ such that $w_{i_1} \cdots w_{i_\ell}$ is k-universal in $O(2^{3|\Sigma|}p^2 \log \ell + n)$ steps.

2. Given $k \in \mathbb{N}$ and $w_1, \ldots, w_p \in \{a, b\}^*$ with $\text{alph}(w_1 \cdots w_p) = \{a, b\}$ and $|w_1 \cdots w_p| = n$, we can compute the minimal ℓ for which there exist $\{i_1, \ldots, i_\ell\} \subseteq [k]$ such that $w_{i_1} \cdots w_{i_\ell}$ is k-universal in $O(n + \log \ell)$ steps.
3. Given $w_1, \ldots, w_p \in \Sigma^*$, with $\text{alph}(w_i) = \Sigma$ for all $i \in [p]$ and $|w_1 \cdots w_p| = n$, and $k \in \mathbb{N}$, we can compute in $O(n + p^3 |\Sigma| \log \ell)$ steps the minimal ℓ for which there exist $\{i_1, \ldots, i_\ell\} \subseteq [k]$ with $w_{i_1} \cdots w_{i_\ell}$ is k-universal.

Finally, we consider the case of decreasing the universality of a word by an operation opposed to concatenation, namely the deletion of a prefix or a suffix.

Theorem 26 (∗). *Given $w \in \Sigma^n$ with $\iota(w) = m$ and a number $\ell < m$, we can compute in linear time the shortest prefix (resp., suffix) $w[1..i]$ (resp., $w[i..n]$) such that $w[i+1..n]$ (resp., $w[1..i-1]$) has universality index ℓ.*

Theorem 26 allows us to compute which is the shortest prefix (suffix) we should delete so that we get a string of universality index ℓ. Its proof is based on the data structures of Lemma 24. For instance, to compute the longest prefix $w[1..i-1]$ of w which has universality index ℓ, we identify the first $\ell + 1$ factors of the decomposition of Theorem 10, assume that their concatenation is $w[1..i]$, and remove the last symbol of this string. A similar approach works for suffixes.

References

1. Bringman, K., Künnemann, M.: Multivariate fine-grained complexity of longest common subsequence. In: Proceedings of the SODA 2018, pp. 1216–1235. SIAM (2018)
2. Chen, H.Z.Q., Kitaev, S., Mütze, T., Sun, B.Y.: On universal partial words. Electron. Notes Discrete Math. **61**, 231–237 (2017)
3. Cormen, T.H., Leiserson, C.E., Rivest, R.L., Stein, C.: Introduction to Algorithms, 3rd edn. MIT Press, Cambridge (2009)
4. Crochemore, M., Hancart, C., Lecroq, T.: Algorithms on Strings. Cambridge University Press, Cambridge (2007)
5. Day, J.D., Fleischmann, P., Manea, F., Nowotka, D.: k-spectra of weakly-c-balanced words. In: Hofman, P., Skrzypczak, M. (eds.) DLT 2019. LNCS, vol. 11647, pp. 265–277. Springer, Cham (2019). https://doi.org/10.1007/978-3-030-24886-4_20
6. de Bruijn, N.G.: A combinatorial problem. Koninklijke Nederlandse Akademie v. Wetenschappen **49**, 758–764 (1946)
7. de Luca, A., Glen, A., Zamboni, L.Q.: Rich, Sturmian, and trapezoidal words. Theor. Comput. Sci. **407**(1–3), 569–573 (2008)
8. Dobkin, D.P., Lipton, R.J.: On the complexity of computations under varying sets of primitives. J. Comput. Syst. Sci. **18**(1), 86–91 (1979)
9. Droubay, X., Justin, J., Pirillo, G.: Episturmian words and some constructions of de Luca and Rauzy. Theor. Comput. Sci. **255**(1–2), 539–553 (2001)
10. Elzinga, C.H., Rahmann, S., Wang, H.: Algorithms for subsequence combinatorics. Theor. Comput. Sci. **409**(3), 394–404 (2008)
11. Fleischer, L., Kufleitner, M.: Testing Simon's congruence. In: Proceedings of the MFCS 2018, volume 117 of LIPIcs, pp. 62:1–62:13. Schloss Dagstuhl - Leibniz-Zentrum fuer Informatik (2018)

12. Freydenberger, D.D., Gawrychowski, P., Karhumäki, J., Manea, F., Rytter, W.: Testing k-binomial equivalence. CoRR, abs/1509.00622 (2015)
13. Gabow, H.N., Tarjan, R.E.: A linear-time algorithm for a special case of disjoint set union. In: Proceedings of the 15th STOC, pp. 246–251 (1983)
14. Gawrychowski, P., Lange, M., Rampersad, N., Shallit, J., Szykula, M.: Existential length universality. To appear at STACS, abs/1702.03961 (2020)
15. Goeckner, B., et al.: Universal partial words over non-binary alphabets. Theor. Comput. Sci. **713**, 56–65 (2018)
16. Halfon, S., Schnoebelen, P., Zetzsche, G.: Decidability, complexity, and expressiveness of first-order logic over the subword ordering. In: Proceedings of the LICS 2017, pp. 1–12 (2017)
17. Hebrard, J.-J.: An algorithm for distinguishing efficiently bit-strings by their subsequences. Theor. Comput. Sci. **82**(1), 35–49 (1991)
18. Holzer, M., Kutrib, M.: Descriptional and computational complexity of finite automata - a survey. Inf. Comput. **209**(3), 456–470 (2011)
19. Imai, H., Asano, T.: Dynamic segment intersection search with applications. In: Proceedings of the 25th Annual Symposium on Foundations of Computer Science, FOCS, pp. 393–402. IEEE Computer Society (1984)
20. Karandikar, P., Kufleitner, M., Schnoebelen, P.: On the index of Simon's congruence for piecewise testability. Inf. Process. Lett. **115**(4), 515–519 (2015)
21. Karandikar, P., Schnoebelen, P.: The height of piecewise-testable languages with applications in logical complexity. In: Proceedings of the CSL 2016, volume 62 of LIPIcs, pp. 37:1–37:22 (2016)
22. Karandikar, P., Schnoebelen, P.: The height of piecewise-testable languages and the complexity of the logic of subwords. Logic. Methods Comput. Sci. **15**(2) (2019)
23. Krötzsch, M., Masopust, T., Thomazo, M.: Complexity of universality and related problems for partially ordered NFAs. Inf. Comput. **255**, 177–192 (2017)
24. Kuske, D., Zetzsche, G.: Languages ordered by the subword order. In: Bojańczyk, M., Simpson, A. (eds.) FoSSaCS 2019. LNCS, vol. 11425, pp. 348–364. Springer, Cham (2019). https://doi.org/10.1007/978-3-030-17127-8_20
25. Lejeune, M., Leroy, J., Rigo, M.: Computing the k-binomial complexity of the Thue-Morse word. CoRR, abs/1812.07330 (2018)
26. Leroy, J., Rigo, M., Stipulanti, M.: Generalized Pascal triangle for binomial coefficients of words. CoRR, abs/1705.08270 (2017)
27. Lothaire, M.: Combinatorics on Words. Cambridge University Press, Cambridge (1997)
28. Maier, D.: The complexity of some problems on subsequences and supersequences. J. ACM **25**(2), 322–336 (1978)
29. Martin, M.H.: A problem in arrangements. Bull. Amer. Math. Soc. **40**(12), 859–864 (1934)
30. Mateescu, A., Salomaa, A., Sheng, Y.: Subword histories and Parikh matrices. J. Comput. Syst. Sci. **68**(1), 1–21 (2004)
31. Rampersad, N., Shallit, J., Zhi, X.: The computational complexity of universality problems for prefixes, suffixes, factors, and subwords of regular languages. Fundam. Inf. **116**(1–4), 223–236 (2012)
32. Rigo, M., Salimov, P.: Another generalization of abelian equivalence: binomial complexity of infinite words. Theor. Comput. Sci. **601**, 47–57 (2015)
33. Salomaa, A.: Connections between subwords and certain matrix mappings. Theor. Comput. Sci. **340**(2), 188–203 (2005)
34. Seki, S.: Absoluteness of subword inequality is undecidable. Theor. Comput. Sci. **418**, 116–120 (2012)

35. Simon, I.: Piecewise testable events. In: Brakhage, H. (ed.) GI-Fachtagung 1975. LNCS, vol. 33, pp. 214–222. Springer, Heidelberg (1975). https://doi.org/10.1007/3-540-07407-4_23
36. Wagner, R.A., Fischer, M.J.: The string-to-string correction problem. J. ACM **21**(1), 168–173 (1974)
37. Zetzsche, G.: The complexity of downward closure comparisons. In: Proceedings of the ICALP 2016, volume 55 of LIPIcs, pp. 123:1–123:14 (2016)

On Normalish Subgroups
of the R. Thompson Groups

Collin Bleak[(✉)] [iD]

University of St Andrews, St Andrews KY16 9SS, Scotland, UK
cb211@st-andrews.ac.uk
http://www-groups.mcs.st-and.ac.uk/~collin/

Abstract. Results in C^* algebras, of Matte Bon and Le Boudec, and of Haagerup and Olesen, apply to the R. Thompson groups $F \leq T \leq V$. These results together show that F is non-amenable if and only if T has a simple reduced C^*-algebra.

In further investigations into the structure of C^*-algebras, Breuillard, Kalantar, Kennedy, and Ozawa introduce the notion of a normalish subgroup of a group G. They show that if a group G admits no non-trivial finite normal subgroups and no *normalish* amenable subgroups then it has a simple reduced C^*-algebra. Our chief result concerns the R. Thompson groups $F < T < V$; we show that there is an elementary amenable group $E < F$ (where here, $E \cong \ldots) \wr \mathbb{Z}) \wr \mathbb{Z}) \wr \mathbb{Z})$ with E normalish in V.

The proof given uses a natural partial action of the group V on a regular language determined by a synchronizing automaton in order to verify a certain stability condition: once again highlighting the existence of interesting intersections of the theory of V with various forms of formal language theory.

Keywords: Thompson's group · Amenable · C^*-simplicity · Regular language · Synchronizing automata · Group actions · Normalish subgroups · Wreath product

1 Introduction

In this note we show that for the R. Thompson groups $F \leq T \leq V$ there is an elementary amenable group $E \leq F$ so that E is **normalish** in each of the groups F, T, and V.

1.1 General Motivating Background

Various weakenings of the notion of normal subgroup were introduced between 2014 and 2018 in order to obtain insight into the C^*-simplicity of the (reduced)

The author wishes to gratefully acknowledge support from the EPSRC grant EP/R032866/1.

© Springer Nature Switzerland AG 2020
N. Jonoska and D. Savchuk (Eds.): DLT 2020, LNCS 12086, pp. 29–42, 2020.
https://doi.org/10.1007/978-3-030-48516-0_3

group algebra $C_r^*(G)$ of a group G. This has had particular impact for infinite simple groups such as the R. Thompson groups T and V. The concept of a normalish subgroup of a group was introduced by in the seminal paper of Breuilliard, Kalantar, Kennedy, and Ozawa [5]. They show that a discrete group G with no non-trivial finite normal subgroups and no amenable normalish subgroups is C^*-simple. In that paper, they also obtain the just-previously-announced result of Haagerup and Olesen [8] that if the reduced group C^*-algebra $C_r^*(T)$ is simple, then F is non-amenable.

Meanwhile, Kennedy in [9] shows that a countable group G is C^*-simple (has simple reduced C^*-algebra) if and only if G admits no non-trivial amenable URS (uniformly recurrent subgroup). Using this, Le Boudec and Matte Bon in [10] show the converse of the stated Haagerup-Olesen result, if F is non-amenable, then the reduced C^*-algebra of T must be simple.

Indeed, for those interested in the question of the non-amenability of the R. Thompson group F, the focus has passed through the exploration of the uniformly recurrent subgroups of T to understanding the point stabilisers of the action of T on its Furstenberg boundary. Here, there are two possible cases, and F will be non-amenable precisely if these point stabilisers are trivial (see [10]). Despite this shift, we find the concept of normalish subgroups of simple groups like F and T to be of interest, and that is the focus of this note.

1.2   Core Results

Let $G \leq H$ be groups. The group G is **normalish in** H if for any finite set of elements $\{c_1, c_2, \ldots, c_k\}$ the intersection

$$\bigcap_{i=1}^{k} G^{c_i}$$

is infinite.

Our chief result is the following:

Theorem 1. *There is an embedding of the elementary amenable group*

$$^\infty(\mathbb{Z} \wr \mathbb{Z}) = \ldots \wr \mathbb{Z}) \wr \mathbb{Z}) \wr \mathbb{Z}$$

into R. Thompson's group F so that the image group E is normalish in V.

Observe the corollary that E is then an amenable normalish subgroup of both F and of T as well.

1.3   Specific History of the Core Result

We should mention some other history related to this result. In [1] we showed the existence of an infinite direct sum of copies of $\mathbb{Z}$ that could be found embedded as a normalish amenable subgroup of F, and discussed our conjecture (disproven here) that any normalish amenable subgroup of T should either contain an embedded subgroup isomorphic to R. Thompson's group F or to a non-abelian free subgroup. Meanwhile, the paper [10] shows that V contains an amenable normalish torsion group Λ: the subgroup of V consisting of those elements which are automorphisms of the infinite rooted binary tree $\mathcal{T}_2$. These automorphism arise as finite compositions of the tree automorphisms that swap the two child vertices of any particular vertex (copying the dependent trees identically). The group Λ is normalish for reasons that are very similar to why our own group E is normalish, and it is a limit of finite groups hence elementary amenable. However, the group Λ is not a subgroup of F nor of T.

1.4   An Unexpected Visitor: A Controlling Synchronizing Automaton

A note on the proof: for experts on R. Thompson groups, the embedded copy of E that we find will clearly be normalish in V after short inspection. However, the technical proof of this requires a bit of work in that the conjugation action on our generators needs to not introduce too many breakpoints into our group elements, and also in that we need to have enough group elements that the set is essentially closed under translations by arbitrary elements in V. The second task is the harder one if we are to avoid having further subgroups isomorphic to R. Thompson's group F. We approach this by introducing a partial action of V on a regular language which is determined by a synchronizing automaton. We link this to the action of V on an infinite specified subset of E. By considering our partial action on the regular language, we can show there is an infinite subset of E that is not moved off of itself too much under the action of finitely many elements of V.

Thanks:
We would like to thank Adrienne Le Boudec for kind and informative conversations where he has helped the author of the present note to understand some of the amazing events that have transpired in the field of C^*-algebras over the last six years.

2   The Interval and the Circle as Quotients of Cantor Space, and Some Related Language

Let $I := [0,1] \subset \mathbb{R}$ represent the unit interval in the real numbers. Let $\mathfrak{C} := \{0,1\}^\omega$ represent the Cantor space that arises as the infinite cartesian product of the discrete space $\{0,1\}$ with itself, with the product indexed by the ordinal ω. As we will act on our Cantor space from the right via prefix substitutions, we will express elements of Cantor space as left infinite strings, so a typical element $\overleftarrow{x}$ of $\mathfrak{C}$ will be written as $\overleftarrow{x} = \ldots x_2 x_1 x_0$ where each x_i is either a 0 or a 1. Note that in this usage, and for such left-infinite strings, we will refer to any finite rightmost contiguous substring as a **prefix** of the infinite string (and we will use the word prefix in this way as well when comparing finite strings, which we will formalise below). The monoid $\{0,1\}^*$ of finite strings under the concatenation operator "^" (e.g., $00110\hat{}1001 = 001101001$) will be central to our analysis and we might refer to an element of $\{0,1\}^*$ as an **address**, for reasons which will become clear.

We give the monoid of finite words $\{0,1\}^*$ the **prefix-based partial ordering** as follows: if $p_1, p_2 \in \{0,1\}^*$ with $p_1 = x_j x_{j-1} \ldots x_1 x_0$ and $p_2 = y_k y_{k-1} \ldots y_1 y_0$ (where each x_i and y_i is in the set $\{0,1\}$ for each valid index i), we say $p_1 \leq p_2$ if and only if $j \leq k$ and for all indices $0 \leq i \leq j$ we have $x_i = y_i$. Recall that with this partial ordering, a **complete antichain** $\mathcal{A}$ of $\{0,1\}^*$ is a finite set $\{p_1, p_2, \ldots, p_k\}$ so that for each pair of distinct indices i and j we have that p_i and p_j are **incomparable** (written $p_1 \perp p_2$, and meaning that both $p_1 \not\leq p_2$ and $p_2 \not\leq p_1$ are true) and for any $w \in \{0,1\}^*$ we have some index r so that either $w \leq p_r$ or $p_r \leq w$.

The monoid $\{0,1\}^*$ with the partial order above can be naturally drawn as a rooted infinite binary tree, with its vertices being the elements of $\{0,1\}^*$, and where we draw an edge from vertices r to s if $r \leq s$ and the length of s (denoted $|s|$) is one greater than the length of r. We will denote this tree as $\mathcal{T}_2$ and sketch a small neighbourhood of its root in the figure below (the tree $\mathcal{T}_2$ is often drawn so as to "open out" as one descends) (Fig. 1).

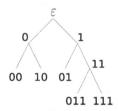

Fig. 1. A neighbourhood of the root ε of the tree $\mathcal{T}_2$

For any finite word $w = w_k w_{k-1} \ldots w_1 w_0 \in \{0,1\}^*$ we obtain the basic open set $\mathfrak{C}w$ for the topology of $\mathfrak{C}$. Specifically, $\mathfrak{C}w$ is the set of all points in Cantor space with prefix w:

$$\mathfrak{C}w = \{\overleftarrow{x}\hat{}w : \overleftarrow{x} \in \mathfrak{C}\}.$$

We will refer to such basic open sets as cones, and for a given finite word $w \in \{0,1\}^*$ the set $\mathfrak{C}w$ will be called the **cone at (address)** w. It is a standard fact that one can identify the Cantor space $\mathfrak{C}$ with the boundary of $\mathcal{T}_2$, or with the set of infinite descending paths in the tree (which correspond to infinite sequences of edge lables, if one labels each edge of $\mathcal{T}_2$ with a 0 or a 1, depending on the letter of the extension connecting the shorter address to the longer address).

Recall there is a standard quotient map $q : \mathfrak{C} \twoheadrightarrow [0, 1]$, which we define fully here in order to give some practice with our right-to-left indexing notation. Let

$$\bar{x} = \dots x_2 x_1 x_0 \in \mathfrak{C}.$$

We have

$$(\bar{x})q := \sum_{i=0}^{\infty} x_i \cdot \frac{1}{2^{i+1}},$$

which we can think of as the ordinary map which interprets a real number in $[0, 1]$ from its binary expansion.

We further recall that given any prefix $w = w_k w_{k-1} \dots w_1 w_0$ the map q identifies the two points $\bar{1}0 w_k w_{k-1} \dots w_1 w_0$ and $\bar{0}1 w_k w_{k-1} \dots w_1 w_0$. The resulting two-point equivalence classes map onto the dyadic rationals in $\mathbb{Z}[1/2] \cap (0, 1) \subset \mathbb{R}$, and further, the cone $\mathfrak{C}w$ at w maps to the closed interval $\mathcal{I}_w$ of radius $(1/2)^{k+2}$ centered at the diadic point d_w which is defined by the infinite sum

$$d_w := \sum_{i=0}^{\infty} w_i \cdot \frac{1}{2^{i+1}}.$$

where we set $w_{k+1} = 1$ and $w_m = 0$ for all $m > k + 1$. For example, if $w = 01$, then $k = 1$ and we have $w_0 = 1$, $w_1 = 0$, $w_2 = 1$ and $w_m = 0$ for all $m > 2$. Then, d_{01} is computed as

$$d_{01} = \left(w_0 \cdot \frac{1}{2^1} + w_1 \cdot \frac{1}{2^2} + w_2 \cdot \frac{1}{2^3} + \mathbf{0} \right) = \left(1 \cdot \frac{1}{2} + 0 \cdot \frac{1}{4} + 1 \cdot \frac{1}{8} \right) = \frac{5}{8}$$

and the interval $\mathcal{I}_{01}$ is of radius $1/(2^{1+2}) = 1/8$ centered at $d_{01} = 5/8$. In particular, we have $\mathcal{I}_{01} = [1/2, 3/4] = [5/8 - 1/8, 5/8 + 1/8]$.

For $w \in \{0, 1\}^*$, we call the interval $\mathcal{I}_w$ constructed as above **the standard dyadic interval at address** w (or "the standard dyadic interval centered at d_w"), noting that these intervals are naturally in a one-one correspondence with the words in the monoid $\{0, 1\}^*$ (we set $k = -1$ when $w = \varepsilon$, the empty word, so that we produce the interval $[0, 1]$, that is, the closed interval of radius $1/2$ centered at $1/2$).

To obtain the circle as a quotient of Cantor space we add one further identification, that is, we identify the point $\dots 000 = \bar{0}0$ with the point $\dots 111 = \bar{1}1$, noting that this simply identifies the real numbers 0 and 1 from the interval I.

When working in the unit interval, we will mostly use the real number parameterisation of points, but sometimes it is convenient to name a point by one of its names arising from the map q^{-1}. Similarly, for points on the circle, we will use either the parameterisation arising from the quotient map $I \to I/(0 \sim 1) = \mathbb{R}/\mathbb{Z}$ (this is equivalent to applying the map $p : I \to \mathbb{S}^1$ given by $t \mapsto e^{2\pi i t}$ where we consider $\mathbb{S}^1$ as the unit circle in the complex plane) or, we will use the parameterisation arising from the map $q \cdot p : \mathfrak{C} \to \mathbb{S}^1$, where a point on the circle is referred to by one of its preimage left-infinite strings under the map $q \cdot p$.

Our group elements will act on the right, and induce permutations of the underlying sets of the spaces under consideration. We establish some notation for our context. Let Y be a set. We will use the notation $\mathrm{Sym}(Y)$ for the group of bijections from Y to itself. For any element $g \in \mathrm{Sym}(Y)$ we define the **support of** g, written $\mathrm{supt}(g)$, as the set

$$\mathrm{supt}(g) := \{y \in Y : yg \neq y\},$$

that is, the set of points moved by g. In keeping with our right-actions notation, if $g, h \in \mathrm{Sym}(Y)$, then the conjugate of g by h, denoted g^h, is the map $h^{-1}gh$. That is, we apply h^{-1}, then g, and finally h again. We then obtain the following standard lemma from the theory of permutation groups.

Lemma 2. *Let Y be a set, and $g, h \in \mathrm{Sym}(Y)$. We have*

$$\mathrm{supt}(g^h) = \mathrm{supt}(g)h.$$

In particular, the support of g^h is the image of the support of g under the function h.

3   The R. Thompson Groups $F < T < V$

The Thompson groups $F < T < V$ are groups of homeomorphisms which have been well studied. In this note, we generally take F, T, and V as each being groups of homeomorphisms of the Cantor space $\mathfrak{C}$.

3.1   Describing Elements of F, T, and V

For two words $w_1, w_2 \in \{0,1\}^*$ with $|w_1| > 0$ and $|w_2| > 0$ we define the **cone map** $\phi_{w_1,w_2} : \mathfrak{C}w_1 \to \mathfrak{C}w_2$ by the rule $\bar{x}w_1 \mapsto \bar{x}w_2$, for each point $\bar{x}$ of $\mathfrak{C}$. It is immediate that this map is a homeomorphism from the Cantor space $\mathfrak{C}w_1$ to the Cantor space $\mathfrak{C}w_2$. Note that the map ϕ_{w_1,w_2} induces a map $\mathcal{I}_{w_1} \to \mathcal{I}_{w_2}$ which is a restriction of an affine map on the reals $\mathbb{R}$, and for this reason we might refer to ϕ_{w_1,w_2} as an "affine map" between the two subspaces of our larger Cantor space $\mathfrak{C}$. Note further that any such cone map ϕ_{w_1,w_2} is not just a homeomorphism from its domain to its range but also that it has many extensions to homeomorphisms from $\mathfrak{C} \to \mathfrak{C}$, and we can think of ϕ_{w_1,w_2} as being a subset of a larger (if $w_1 \neq \varepsilon \neq w_2$) function from $\mathfrak{C}$ to $\mathfrak{C}$ (which we in turn consider as a subset of $\mathfrak{C} \times \mathfrak{C}$).

We are now in a position to define the R. Thompson groups $F < T < V$.

An element $g \in \mathrm{Homeo}(\mathfrak{C})$ is an element of V if and only if we can write g as a **prefix replacement map**, as follows.

The element g is a prefix replacement map if and only if it admits some natural number $n > 1$, two complete antichains $D = \{a_1, a_2, \ldots, a_n\}$ and $R = \{r_1, r_2, \ldots, r_n\}$ for $\{0,1\}^*$, and a bijection $\sigma : D \to R$, so that when restricted to any cone $\mathfrak{C}a_i$ (for valid index i), the map g restricts and co-restricts to the cone map $\phi_{a_1, a_1 \cdot \sigma}$. In this context, we will write

$$g = (\{a_1, a_2, \ldots, a_n\}, \{r_1, r_2, \ldots, r_n\}, \sigma).$$

We observe in passing that any element of V admits infinitely many distinct prefix replacement maps representing it, but it is a standard exercise in R. Thompson theory that there is a unique, minimal prefix-map representation of v.

We observe that any complete antichain $\{a_1, a_2, \ldots, a_n\}$ for $\{0,1\}^*$ admits a natural left-to-right ordering $\prec$ induced from the arrangement of the addresses a_i on the tree (this is simply the dictionary order, when we take $0 \prec 1$ and read our strings from right to left). An element $g \in V$ is in the subgroup F if and only if, when expressed as a prefix replacement map, the permutation σ preserves the ordering $\prec$. An element $g \in V$ is in the subgroup T if and only if, when expressed as a prefix replacement map, the permution σ preserves the ordering $\prec$ up to some cyclic rotation. It is then a standard exercise that elements of F induce homeomorphisms of I through the quotient map q which are piecewise affine, respect the dyadic rationals, and where all slopes are powers of two and all breaks in slope occur over dyadic rationals. Similarly, it is a standard exercise that elements of T induce homeomorphisms of $\mathbb{S}^1$ through the quotient $q \cdot p$ which are piecewise affine, respect the dyadic rationals, and where all slopes are powers of two and all breaks in slope occur over dyadic rationals.

A standard introductory reference for the general theory of the R. Thompson groups F, T, and V is the paper [6].

3.2 The Element Family $\mathcal{X}$

We now single out a family

$$\mathcal{X} := \{x_w : w \in \{0,1\}^*\}$$

of elements of V of specific interest to our discussion.

Given a word $w \in \{0,1\}^*$, we specify the element x_w as the element of V which acts as the identity over the complement of the cone $\mathfrak{C}w$, and on the cone $\mathfrak{C}w$, acts according to the prefix map specified below (we only express the actual prefix substitutions here):

$$x_w := \begin{cases} 00\,\hat{}\,w \mapsto 0\,\hat{}\,w \\ 10\,\hat{}\,w \mapsto 01\,\hat{}\,w \\ 1\,\hat{}\,w \mapsto 11\,\hat{}\,w \end{cases}$$

In particular, the element x_w is the extension of the partial function

$$\phi_{00\,\hat{}\,w,0\,\hat{}\,w} \sqcup \phi_{10\,\hat{}\,w,01\,\hat{}\,w} \sqcup \phi_{1\,\hat{}\,w,11\,\hat{}\,w}$$

by the identity map away from the cone $\mathfrak{C}w$.

Note that it is easy to extend the set $\{00\,\hat{}\,w, 01\,\hat{}\,w, 1\,\hat{}\,w\}$ to a complete antichain $\{a_1, a_2, \ldots, a_{k-1}, 00\,\hat{}\,w, 10\,\hat{}\,w, 1\,\hat{}\,w\}$ for $\{0,1\}^*$ where $|w| = k$, and that in this case $\{a_1, a_2, \ldots, a_{k-1}, 0\,\hat{}\,w, 01\,\hat{}\,w, 11\,\hat{}\,w\}$ is also a complete antichain for $\{0,1\}^*$ (the set of addresses $\{a_i : 1 \leq i \leq k-1\}$ represents the minimal set of addresses one can use so that $\{a_i : i \in 1 \leq i \leq k-1\} \cup \{w\}$ is a complete antichain). Our map x_w acts as cone maps on each of the cones at the set of addresses $\{00\,\hat{}\,w, 01\,\hat{}\,w, 1\,\hat{}\,w\}$, and otherwise takes each cone $\mathfrak{C}a_i$ to itself with the identity map.

It is easy to see that $\mathcal{X} \subset F$ so also $\mathcal{X} \subset T$ and $\mathcal{X} \subset V$. The figure below depicts the graphs of x_ε and x_{10} as homeomorphisms of I as examples.

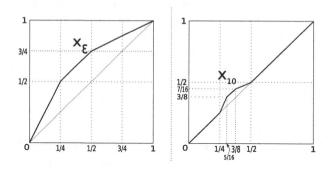

Fig. 2. The elements x_ε and x_{10}.

3.3   The Element Family $\mathcal{G}$

For each natural $n \in \mathbb{N}$, set $g_n := x_{(10)^n}$. The set $\mathcal{G} := \{g_i : i \in \mathbb{N}\}$ will be our second family of elements of F of interest. Note that we use $\mathbb{N}$ to represent the natural numbers, which we take to be the non-negative integers.

We observe that Fig. 2 also depicts g_0 and g_1 since $g_0 = x_\varepsilon$ and $g_1 = x_{10}$.

4   Realising $^{\infty}(\mathbb{Z} \wr \mathbb{Z})$ in F

Set $E := \langle \mathcal{G} \rangle \leq F \leq T \leq V$. For each index n, consider the group

$$W_n := \langle \{g_i : i \in \mathbb{N}, i < n\} \rangle,$$

where for clarity we specify $W_0 = \{1_V\}$. It is immediate that $W_m \leq W_n$ when $m < n$. Direct calculation shows that $\mathrm{supt}(g_n) \cap \mathrm{supt}(g_n^{g_m}) = \emptyset$ whenever $m < n$ (the content of the following lemma, which is not hard to prove). Specifically, following the arguments of [2,3] for natural index n we have $W_n \cong (\ldots((\mathbb{Z} \wr \mathbb{Z}) \wr \mathbb{Z}) \ldots \wr \mathbb{Z}) \wr \mathbb{Z}$ (with n appearances of $\mathbb{Z}$ in this expression), which is a solvable group of derived length n. We immediately obtain $E \cong {}^{\infty}(\mathbb{Z} \wr \mathbb{Z}) = \ldots \wr \mathbb{Z}) \wr \mathbb{Z}) \wr \mathbb{Z}$, as described in detail in [4]. As E admits a decomposition as a direct union of the solvable groups W_n, we obtain that E is elementary amenable (see Chou's paper [7] for details on the class of elementary amenable groups).

For what follows, set $G_{m,n} := \langle g_m, g_n \rangle$ for all natural numbers $m < n$.

Lemma 3. *Let $m < n$ be two natural numbers. We have*

1. *there is an isomorphism $G_{m,n} \cong G_{0,(n-m)}$ which is induced by a restriction map followed by a topological conjugacy,*

2. $\operatorname{supt}(g_n) \cap \operatorname{supt}(g_n^{g_m}) = \emptyset$, and therefore
3. $G_{m,n} \cong \mathbb{Z} \wr \mathbb{Z}$.

Proof. We first prove Point (1) that there is an isomorphism $G_{m,n} := \langle g_m, g_n \rangle \cong \langle g_0, g_{n-m} \rangle = G_{0,(n-m)}$ which is induced by a restriction map followed by a topological conjugacy.

To see this point, first observe that both of the elements g_n and g_m are supported wholly in the cone $\mathfrak{C}(10)^m$, so, the restriction of the maps g_n and g_m to the cone $\mathfrak{C}(10)^m$ results in an isomorphism of groups between the homeomorphism group $G_{m,n} = \langle g_m, g_n \rangle$, which acts on the Cantor space $\mathfrak{C}$, to a homeomorphism group $\widehat{G}_{m,n} = \langle \widehat{g}_m, \widehat{g}_n \rangle$ (these generators being the restrictions of the generators g_m and g_n respectively), so that the group $\widehat{G}_{m,n}$ is a group of homeomorphisms of the Cantor space $\mathfrak{C}(10)^m$. Now, the homeomorphism $\theta_m : \mathfrak{C}(10)^m \to \mathfrak{C}$ which is induced by deleting the prefix $(10)^m$ from all points in the Cantor space $\mathfrak{C}(10)^m$ provides a topological conjugacy which induces an isomorphism from the group $\widehat{G}_{m,n}$ to the group $G_{0,n-m} = \langle g_0, g_{(n-m)} \rangle$, as the reader can check that the image of the $\widehat{g}_m$ is the element g_0 and the image of $\widehat{g}_{(n)}$ is the element $g_{(n-m)}$ under this topological conjugacy.

For Point (2), we observe that the restrictions applied in the argument for Point (1) only removed areas from the domain of the elements g_m and g_n where these elements already acted as the identity. Therefore the support of g_n and of $g_n^{g_m}$ will be disjoint if and only if the supports of the elements $g_{(n-m)}$ and of $g_{(n-m)}^{g_0}$ are disjoint. In particular, we have our result if we prove that for any positive integer k, we have $\operatorname{supt}(g_k) \cap \operatorname{supt}(g_k^{g_0}) = \emptyset$.

However, $g_0 = x_\varepsilon$, which acts over the cone $\mathfrak{C}10$ as a cone map, affinely taking the cone $\mathfrak{C}10$ rightward to the cone $\mathfrak{C}01$ by the prefix substitution $10 \mapsto 01$. Now, the support of g_k is contained in the cone $\mathfrak{C}(10)^k$, a subset of the cone $\mathfrak{C}(10)$. Direct calculation now shows that the cone $\mathfrak{C}(10)^k$ is carried affinely to the cone $\mathfrak{C}(10)^{k-1}(01)$ by g_0, so Lemma 2 implies our result. We note in passing that we have shown that $g_k^{g_0} = x_{(10)^{k-1}01}$, or more specifically, that $x_{(01)^k}^{x_\varepsilon} = x_{(10)^{k-1}01}$, since our conjugator acted affinely.

For Point (3), recall Section 1.2.1 of [2], where an argument is given that two elements α_1 and α_2 of F generate a group isomorphic to $\mathbb{Z} \wr \mathbb{Z}$, with the element α_1 generating the top group of the wreath product. It happens that the element α_1 of that paper is the element we call x_ε here, while the element α_2 is the element we call x_{10} here. The proof of Section 1.2.1 essentially relies on only three facts: 1) the support of α_2 is contained in the support of α_1, 2) every point in the support of α_1 is on an infinite orbit under the action of $\langle \alpha_1 \rangle$, and 3) the support of α_2 is moved entirely off of itself by α_1. In our case with g_m and g_n, we again have these three conditions (with g_m playing the role of α_1), so we have our claimed Point (3).

The discussion above indicates the following lemma.

Lemma 4. *Let $v \in V$. There are $\{a_1, a_2, \ldots, a_n\}$ and $\{b_1, b_2, \ldots, b_n\}$, minimal cardinality (finite) antichains, together with a bijection σ between them, so that*

v can be described as the prefix replacement map

$$v = (\{a_1, a_2, \ldots, a_n\}, \{b_1, b_2, \ldots, b_n\}, \sigma).$$

If $w, u \in \{0,1\}^*$ so that $w = u \hat{\ } a_i$ for some i, then $x_w^v = x_{u \hat{\ }(a_i\sigma)}$.

Proof. By definition, we have $x_w^v = v^{-1}x_w v$. By our assumptions w has a_i as a prefix, and we note that the initial map v^{-1} restricts to a cone map from $\mathfrak{C}(a_i\sigma)$ to $\mathfrak{C}a_i$, that is, an affine map with image containing the support of x_w. The action of x_w off the cone $\mathfrak{C}a_i$ is as the identity, and in general is as described in the definition of x_w (it acts as a prefix replacement map, which modifies only the prefixes which begin with $w = u \hat{\ } a_i$, and these modifications appear in entries at indices larger than the length $|w|$). Finally, v acts on the cone $\mathfrak{C}a_i$ by affinely returning it to $\mathfrak{C}(a_i\sigma)$ as a cone map, (it simply transforms the prefix a_i to the prefix $a_i\sigma$, and preserves all later entries (with index offset of size $|a_i\sigma| - |a_i|$) at larger indices, for any point in the Cantor space $\mathfrak{C}(a_i)$). Therefore, $x_w^v = x_{u \hat{\ }(a_i\sigma)}$. □

5   On Partial Actions

The proof of Lemma 4 suggests the well-known fact that V has a natural partial action on the addresses in $\{0,1\}^*$. Let $v \in V$ and suppose there is a minimal natural number n and antichains $A = \{a_1, a_2, \ldots, a_n\}$ and $B = \{b_1, b_2, \ldots, b_n\}$ with a bijection σ between them so that v can be described as the prefix replacement map $v = (A, B, \sigma)$. The partial action of v on $\{0,1\}^*$ is defined precisely on the set of words in $\{0,1\}^*$ which admit one of the a_i as a prefix. Let us suppose $w \in \{0,1\}^*$ and $w = u \hat{\ } a_i$. We set $w \cdot v := u \hat{\ }(a_i\sigma)$.

We can now re-express the result of Lemma 4 in terms of the partial action of V on the set $\{0,1\}^*$.

Corollary 5. *Let $v \in V$ and $w_1, w_2 \in \{0,1\}^*$ so that $w_1 \cdot v = w_2$ under the partial action of V on $\{0,1\}^*$. If $u \in \{0,1\}^*$ then $x_{u \hat{\ } w_1}^v = x_{u \hat{\ } w_2}$.*

That is, we see that the group V admits a partial action on the set $\mathcal{X}$ which parallels its partial action on $\{0,1\}^*$. We now work to understand the action of V on elements of the group E.

Our first step in understanding this partial action is to analyse a formal language.

5.1   A Regular Language and An Action

Define the set $\mathcal{T} \subset \{0,1\}^*$ of **tokens** as follows:

$$\mathcal{T} := \{10^k, 01^k : k \in \mathbb{N}, k \neq 0\}.$$

We build a formal language $\mathcal{W}$ over the alphabet $\{0,1\}$ as follows. The language $\mathcal{W}$ is the set of all words which decompose as $w = w_j \hat{\ } w_{j-1} \hat{\ } \ldots \hat{\ } w_1$ for some natural j, where each w_i is a token. The language $\mathcal{W}$ is actually a regular

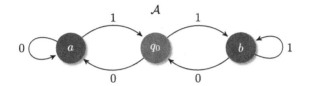

Fig. 3. The automaton $\mathcal{A}$ which accepts the language $\mathcal{W}$

language, which is recognised by the automaton $\mathcal{A}$ depicted in Fig. 3. The state q_0 of $\mathcal{A}$ is both the start and accept state of $\mathcal{A}$.

Recall that an automaton is a finite directed edge-labelled graph with a subset of its set of states called the start states of the automaton, and another subset of its set of states called the accept states of the automaton. Then, the language accepted by the automaton is precisely the set of all finite-length words which arise as the concatenated edge-labels of some finite path in the automaton from a start state to an end state.

One can see that our formal language $\mathcal{W}$ is indeed the language accepted by $\mathcal{A}$; the paths which leave the state q_0 and then eventually return (exactly once) have labels of the form 10^k or 01^k, for some non-zero natural number k. In particular, the language accepted by $\mathcal{A}$ is precisely the language of words built by concatenating tokens from $\mathcal{T}$.

We now set some terminology describing the structure of elements of the language $\mathcal{W}$. Observe firstly that the decomposition of any word in $\mathcal{W}$ into a concatenation of tokens is unique. Therefore, for each $w \in \mathcal{W}$ we can define the **token length of** w as the number of tokens in its decomposition as a concatenation of tokens. Note that we index these tokens from right to left: $w = w_k \,\hat{}\, w_{k-1} \,\hat{}\, \ldots \,\hat{}\, w_2 \,\hat{}\, w_1$.

We now observe that the partial action of V on $\{0,1\}^*$ restricts to an action of $\langle g_0 \rangle$ on the set $\mathcal{T}$. Below, the proofs of Lemmas 6 and 8 follow by simple inductions on a basic calculation.

Lemma 6. *For each integer k, the restriction of g_0^k to the cone $\mathfrak{C}10$ produces a cone map from the cone $\mathfrak{C}10$ to the cone $\mathfrak{C}w_k$, where w_k is given by the formula below:*

$$w_k = 10 \cdot g_0^k = \begin{cases} 10^{|k-1|} & k \leq 0 \\ 01^k & k > 0. \end{cases}$$

Proof. This proof is a simple induction. Recall that $g_0 = x_\varepsilon$.

If $k = 0$ we observe that our formula works as $g_0^0 = 1_V$, which maps the cone $\mathfrak{C}10$ to the cone $\mathfrak{C}10$ by the identity map, which is a cone map. If $k = 1$ then g_0 takes $\mathfrak{C}10$ to $\mathfrak{C}01$ as a cone map, in accordance with the definiton of x_ε. For all $k \geq 2$, g_0^k acts as $x_\varepsilon \cdot x_\varepsilon^{k-1}$, so first as a cone map from $\mathfrak{C}10$ to $\mathfrak{C}01$, and then it will continue to act as x_ε^{k-1} on this resulting cone. However, the cone $\mathfrak{C}01$ is contained in the cone $\mathfrak{C}1$, and so the prefix replacement of x_ε here replaces the initial prefix 1 with the prefix 11, and this process repeats so inductively we have our desired result for all integers $k \geq 0$. For negative integers k, the argument follows as x_ε^{-1} replaces the prefix 0 with the prefix 00, and so inductively, the cone at 10 is carried by a cone map to the cone at $10^{|k-1|}$ by x_ε^k.

A translation of the above result is as follows.

Corollary 7. *The partial action of V on the set $\{0,1\}^*$ restricts to a free, transitive action of $\langle g_0 \rangle$ on the set $\mathcal{T}$ of tokens.*

The following lemma simply extends the result of Lemma 6.

Lemma 8. *For each integer k and natural number i, the restriction of g_i^k to the cone $\mathfrak{C}(10)^{i+1}$ produces a cone map from the cone $\mathfrak{C}(10)^{i+1}$ to the cone $\mathfrak{C}w_{i,k}$, where $w_{i,k}$ is given by the formula below:*

$$w_{i,k} = (10)^{i+1} \cdot g_i^k = \begin{cases} 10^{|k-1|}(10)^i & k \leq 0 \\ 01^k(10)^i & k > 0. \end{cases}$$

Proof. The proof is similar to the proof of Lemma 6; $g_i = x_{(10)^i}$ acts as the identity off of the cone $\mathfrak{C}(10)^i$, and acts on the cone $\mathfrak{C}(10)^i$ in the same way that g_0 acts on the cone $\mathfrak{C}\varepsilon = \mathfrak{C}$ (this is essentially the content of the proof of Lemma 3 (1). That is, the prefix $(10)^i$ is fixed by all powers of g_i, but the word $(10)^{i+1}$ is changed by g_i on the final token "10" (the "0" at index $2i$ and the "1" at index $2i+1$).

Lemma 8 has the following related corollary.

Corollary 9. *Let i be a natural number. The partial action of V on the set $\{0,1\}^*$ restricts to a transitive and free action of $\langle g_i \rangle$ on the set of words $\{t\,\hat{}\,(10)^i : t \in \mathcal{T}\}$.*

6   Visiting the Family $\mathcal{X}$

We now discuss the intersection of the group $E = \langle \mathcal{G} \rangle$ with the family $\mathcal{X}$.

Lemma 10. *Let $w \in \mathcal{W}$, and $k \in \mathbb{N}$ so that w has token decomposition $w = w_k \hat{} w_{k-1} \hat{} \ldots \hat{} w_2 \hat{} w_1$. For each token w_i, let j_i be the integer so that $10 \cdot x_\varepsilon^{j_i} = w_i$ and also, recall that $x_\varepsilon = g_0$. If we set θ_w to be the product*

$$\theta_w := g_{k-1}^{j_k} g_{k-2}^{j_{k-1}} \cdots g_1^{j_2} g_0^{j_1}$$

then we have

$$x_w = g_k^{\theta_w}.$$

Proof. One constructs θ by modifying the prefix $(10)^k$ to the prefix w by acting on one token at a time, starting with the leftmost token (the kth token), and then working to the first token w_1. Progressively, each term in the product decomposition of θ acts on a cone containing the impact of the previous terms which have acted, the actions stack to create the following sequence of prefixes for the locations of the actions of the (partially) conjugated versions of $x_{(10)^k}$.

$$(10)^k \mapsto$$
$$w_k(10)^{k-1} \mapsto$$
$$w_k w_{k-1}(10)^{k-2} \mapsto$$
$$\cdots$$
$$w_k w_{k-1} \ldots w_2 10 \mapsto$$
$$w_k w_{k-1} \ldots w_2 w_1.$$

We therefore have the following corollary.

Corollary 11. *Let $\mathcal{X}_{\mathcal{W}} := \{x_w : w \in \mathcal{W}\}$. Then $\mathcal{X}_{\mathcal{W}} \subset E$.*

We now consider a special subset of $\mathcal{W}$. Set

$$\mathcal{S} := \{100\,\hat{}\,w, 011\,\hat{}\,w : w \in \{0,1\}^*\}.$$

Lemma 12. *The set $\mathcal{S}$ is a subset of $\mathcal{W}$.*

Proof. The automaton $\mathcal{A}$ of Fig. 3 has further properties not mentioned previously; it is highly connected and synchronizing. These properties together mean that given any particular state (let us say q_0), there is a non-empty set of **synchronizing words** W_{q_0} associated with q_0 so that, starting from any particular state s of the automaton and following a path labelled by any word in W_{q_0}, perforce, one will be lead to the state q_0.

Note that the words 100 and 011 are synchronizing words for the state q_0; no matter what state one starts in, after following the path labelled by the word 100 from that state, or the path labelled by the word 011 from that state, one arrives in the state q_0 (recall that we are reading these words from right-to-left!).

Thus, if we have some general word w and we append a suffix 011 or 100 to produce either $z = 011\,\hat{}\,w$ or $z = 100\,\hat{}\,w$ (that is, a general word z in $\mathcal{S}$), then upon reading this resulting word on the automaton $\mathcal{A}$ starting from the start state q_0, we will return to q_0; our word z is in the language $\mathcal{W}$ accepted by $\mathcal{A}$.

Below, we will actually be interested in the subset of $\mathcal{X}_{\mathcal{W}}$ where the words involved come from $\mathcal{S}$. Set

$$\mathcal{X}_S := \{x_w : w \in \mathcal{S}\}.$$

Corollary 13. *The set $\mathcal{X}_S$ is a subset of the group E.*

We now consider how V interacts with the set $\mathcal{X}_S$ under conjugation. The following lemma follows quickly by an application of Corollary 5.

Lemma 14. *Let $v \in V$. There is a natural number n so that for all $w \in \mathcal{S}$ with $|w| \geq n$ there is $z \in S$ so that $x_z^v = x_w$.*

Proof. Let us assume that we can represent v by some prefix replacement map

$$v = (\{a_1, a_2, \ldots, a_m\}, \{b_1, b_2, \ldots, b_m\}, \sigma)$$

where we assume that n is at least three larger than the length of the largest string in the range antichain $\{b_1, b_2, \ldots, b_m\}$. We then set b as the prefix of w appearing in the set $\{b_1, b_2, \ldots, b_m\}$, and set $a = b\sigma^{-1} \in \{a_1, a_2, \ldots, a_m\}$. Then, $w = c \,\hat{}\, b$ where c is some string of length at least three, and we have $(c\,\hat{}\,a) \cdot v = (c\,\hat{}\,b) = w$, so that, in particular, if we take $z := c\,\hat{}\,a$, then Corollary 5 assures us that $x_z^v = x_{c\,\hat{}\,b} = x_w$.

But now, as the word c has length at least three, we see that it must end with the string 100 or the string 011 (since w has one of these two length three suffixes), and in particular, $z \in \mathcal{S}$.

Thus, we have found that all sufficiently long strings w in $\mathcal{S}$ have that x_w is the conjugate image of x_z under v, for z another string in $\mathcal{S}$.

Proof of Theorem 1:

Proof. The group E of this note is infinite and amenable. We can further see that for any finite set $C := \{v_1, v_2, \ldots, v_k\} \subset V$, the elements x_w, for $w \in \mathcal{S}$ with w long enough (given by some particular integer dependent on the set C), all appear in all of the groups E^{v_i}. In particular, the intersection

$$\bigcap_{v_i \in C} E^{v_i}$$

is an infinite set, so that E is normalish in each of F, T, and V.

References

1. Bleak, C.: An exploration of normalish subgroups of R. Thompson's groups F and T, pp. 1–4 (2016, submitted) arXiv:1603.01726
2. Bleak, C.: An algebraic classification of some solvable groups of homeomorphisms. J. Algebra **319**(4), 1368–1397 (2008). MR 2383051
3. Bleak, C.: A geometric classification of some solvable groups of homeomorphisms. J. Lond. Math. Soc. (2) **78**(2), 352–372 (2008). MR 2439629
4. Bleak, C.: A minimal non-solvable group of homeomorphisms. Groups Geom. Dyn. **3**(1), 1–37 (2009). MR 2466019
5. Breuillard, E., Kalantar, M., Kennedy, M., Ozawa, N.: C^*-simplicity and the unique trace property for discrete groups. Publ. Math. Inst. Hautes Études Sci. **126**, 35–71 (2017). MR 3735864
6. Cannon, J.W., Floyd, W.J., Parry, W.R.: Introductory notes on Richard Thompson's groups. Enseign. Math. (2) **42**(3–4), 215–256 (1996)
7. Chou, C.: Elementary amenable groups. Illinois J. Math. **24**(3), 396–407 (1980)
8. Haagerup, U., Olesen, K.K.: Non-inner amenability of the Thompson groups T and V. J. Funct. Anal. **272**(11), 4838–4852 (2017). MR 3630641
9. Kennedy, M.: Characterisations of C^* simplicity, pp. 1–16 (2015, submitted). arxiv:1509.01870v3
10. Le Boudec, A., Matte Bon, N.: Subgroup dynamics and C^*-simplicity of groups of homeomorphisms. Ann. Sci. Éc. Norm. Supér. (4) **51**(3), 557–602 (2018). MR 3831032

Computing the Shortest String and the Edit-Distance for Parsing Expression Languages

Hyunjoon Cheon and Yo-Sub Han[(⊠)]

Department of Computer Science, Yonsei University, 50, Yonsei-ro, Seodaemun-gu,
Seoul 03722, Republic of Korea
{hyunjooncheon,emmous}@yonsei.ac.kr

Abstract. A distance between two languages is a useful tool to measure the language similarity, and is closely related to the intersection problem as well as the shortest string problem. A parsing expression grammar (PEG) is an unambiguous grammar such that the choice operator selects the first matching in PEG while it can be ambiguous in a context-free grammar. PEGs are also closely related to top-down parsing languages. We consider two problems on parsing expression languages (PELs). One is the r-shortest string problem that decides whether or not a given PEL contains a string of length shorter than r. The other problem is the edit-distance problem of PELs with respect to other language families such as finite languages or regular languages. We show that the r-shortest string problem and the edit-distance problem with respect to finite languages are NEXPTIME-complete, and the edit-distance problem with respect to regular languages is undecidable. In addition, we prove that it is impossible to compute a length bound $\mathcal{B}(G)$ of a PEG G such that $L(G)$ has a string w of length at most $\mathcal{B}(G)$.

Keywords: Formal languages · Parsing expression grammars · Edit-distance

1 Introduction

Perl-compatible regular expressions (PCREs) are popular tools for information retrieval and data processing. From a formal language viewpoint, the expressive power of PCREs is interesting. A simple PCRE can define a context-sensitive language. For example, the PCRE (a*)b\1b\1 represents a context-sensitive language $a^n b a^n b a^n$, which is not context-free. This implies that there are no simple matching algorithms for PCREs.

Ford [4] proposed parsing expression grammars as a recognition-based formal grammar that is a generalization of TMG recognition schema [2]. Parsing expression grammars (PEGs) are intuitive for pattern matching and have a simpler and efficient algorithm for matching than the algorithms for PCRE. Unlike other grammars such as regular expressions or context-free grammars (CFGs)

© Springer Nature Switzerland AG 2020
N. Jonoska and D. Savchuk (Eds.): DLT 2020, LNCS 12086, pp. 43–54, 2020.
https://doi.org/10.1007/978-3-030-48516-0_4

that match the input string if the grammars generate the whole input string, PEGs regard the string matching if PEGs recognize a prefix (not necessarily the whole string) of the input string deterministically. Thus, a PEG itself is unambiguous—each rule in the grammar has a strict order and the grammar tries to match the input according to its rule orders. This makes PEGs useful for parsing a string since it has a lookahead that can verify whether or not a prefix of the remaining input matches the given expression. Based on this property, the Packrat parsing algorithm [3] for PEG recognizes a prefix of its input string in polynomial time in the input size and the grammar size. This efficient matching algorithm makes PEGs to be alternatives of PCREs. IBM recently proposed Rosie pattern language (RPL) [1] based on PEG for pattern matching.

Medeiros et al. [11] designed a PEG construction from a PCRE and evaluated the pattern matching performance between PEGs and PCREs. Loff et al. [10] discussed a few computational aspects of PEGs. They showed that a PEG accepts a pair of an input and its output for any computable functions. They also proposed a new computational machine, scaffolding automata (SAs), that operates using a set of states and an auxiliary DAG structure, and proved the equivalence between SAs and the reversal of parsing expression languages (PELs). Koga [7] examined the context-freeness of a PEL, and showed that it is undecidable whether or not a PEL L belongs to a subfamily of context-free languages (CFLs).

For the edit-distance problems of languages, Mohri [12] showed that the edit-distance between two regular languages can be solved in polynomial-time while the same problem between two CFLs is undecidable. Konstantinidis [8] suggested an algorithm for computing the edit-distance of a given finite automaton (FA) and obtained an upper bound of the distance. Povarov [16] studied the neighborhood language according to the Hamming distance [5] that counts the number of different symbols between two strings of the same length. An r-neighborhood L_r of a language L according to a distance metric d is a set of strings whose distance from a string in L is at most r. From an FA with n states, we can construct an NFA with $n(r+1)$ states for r-Hamming-neighborhood language of $L(A)$. Under the similar construction, we can also construct an NFA for r-edit-distance neighborhood with the same bound. Han et al. [6] considered the edit-distance problem between a regular language and a CFL. They presented a construction that accepts an alignment between a pair of strings from each language, and designed an algorithm that computes the edit-distance between a regular language and a CFL in polynomial time based on the construction. Ng et al. [13] studied the edit-distance neighborhood of a regular languages. They showed that, for an n state FA A, there exists a DFA with at most $(r+2)^n - 1$ states that accepts an r-edit-distance neighborhood of $L(A)$.

We consider two decision problems on PEGs: the r-shortest string problem and the r-edit-distance problem. The r-shortest string problem determines whether or not a given PEL has a string whose length is at most length $r \geq 0$ string. The r-edit-distance problem decides whether or not the edit-distance between one PEL and another language is at most $r \geq 0$. We show that the r-shortest string problem and the edit-distance problem with respect to finite languages are NEXPTIME-complete. Moreover, we demonstrate that the r-edit-distance problem with respect to regular languages is undecidable. In addition, we prove that it is impossible to compute a length bound $\mathcal{B}(G)$ of a PEG G such that $L(G)$ has a string of length at most $\mathcal{B}(G)$.

2 Preliminaries

A Turing machine (TM) M is specified by a tuple $(Q, \Sigma, \Gamma, \delta, q_0, q_a, q_r)$, where Q is a set of states, Σ is an input alphabet, Γ is a tape alphabet, $\delta \subseteq Q \times \Gamma \times Q \times \Gamma \times \{L, R\}$ is a set of transitions, and three states $q_0, q_a, q_r \in Q$ are an initial, an accepting and a rejecting state, respectively. The blank tape symbol is denoted by $B \in \Gamma$.

A TM M accepts (rejects) a string $w \in \Sigma^*$ if, starting from the initial state q_0, M has a transition path that reaches an accepting state q_a (a rejecting state q_r, respectively) by following δ. We define the language $L(M)$ of M to be

$$L(M) = \{w \mid M \text{ accepts } w\}.$$

We assume that M halts on w if M reaches either q_a or q_r on w.

A configuration C of M is a string over $Q \cup \Gamma$, where C consists of one state symbol and tape symbols. The state symbol represents both the current state (by the symbol itself) and the head position (by its position). The initial configuration with the input string w is always $q_0 w$ and the head is at the first symbol of w. We say a configuration C is accepting (rejecting) if C is on the state q_a (q_r, respectively).

2.1 Parsing Expression Grammars

A parsing expression is an expression generated from the following grammars.

$$e \rightarrow \lambda \mid a \in \Sigma \mid A \in V \mid (e) \mid ee \mid e/e \mid \&e \mid !e,$$

where λ denotes the empty string, Σ is an alphabet and V is a set of variables. Given a parsing expression e and a string $w = xy$, we say that $(e, w) \rightarrow y$ if e recognizes a prefix x of w and $(e, w) \rightarrow f$ if e fails to recognize w. A PEG G is specified by a tuple (V, Σ, R, S), where V is a set of variables, Σ is an input alphabet, R is a set of rules and $S \in V$ is the starting variable. A rule $(A, e) \in R$ is a pair of a variable and a parsing expression—we sometimes denote it by $A \leftarrow e$. PEGs look similar to CFGs yet the main difference is that PEGs do not have the union operation that can nondeterministically choose either of productions. On the other hand, PEGs have a choice operator $(/)$ (e.g., A/B) that first tries to match its left side rule (A) and, if it fails, tries its right side rule (B) in order. This choice operator makes a PEG unambiguous. The expressions of the forms $\&e$ or $!e$ are called *and-* and *not-predicate* that recognizes the empty string λ if e recognizes (cannot recognize, respectively) the input string. In other words, these expressions work as unbounded lookahead.

Given a PEG $G = (V, \Sigma, R, S)$, we define the PEL $L(G)$ to be

$$L(G) = \{xy \mid (S, xy) \rightarrow y, y \in \Sigma^*\}.$$

This definition allows to contain any strings whose prefix is recognized by the grammar.

Example 1. The following PEG G over $\Sigma = \{a, b, c\}$ with a single rule $S \leftarrow \&a/!(bc)$ recognizes the language $\{(a + ba + bb + c)\Sigma^*\}$ by the following steps:

– The strings in $\{a\Sigma^*\}$ (e.g., abc) match the and-predicate. We do not try matching the second expression.
– The strings in $\{bc\Sigma^*\}$ (e.g., bca) *do not match* both expressions so G fails to recognize. (The second expression $!(bc)$ fails to match since bc matches those strings.)
– Other strings in $\{(ba + bb + c)\Sigma^*\}$ do not match the and-predicate but match the not-predicate.

Example 1 illustrates the role of & and ! in PEGs. Table 1 shows a few more PEG examples and languages (Σ matches any symbol).

Table 1. Example PEGs and their languages

PEG	Language
$S \leftarrow a$	$\{a\Sigma^*\}$
$S \leftarrow a/ab$	$\{a\Sigma^*\}$
$S \leftarrow \&(ab)$	$\{ab\Sigma^*\}$
$S \leftarrow a!\Sigma$	$\{a\}$

It is known that the emptiness, the universality of a PEL and the equivalence between two PELs are all undecidable [4]. It follows that the intersection emptiness between a regular language and a PEL is also undecidable: if the regular language is Σ^*, their intersection emptiness shows that the PEL is empty. Furthermore, PEGs have a linear-time parsing algorithm that uses Packrat parsing method. If the grammar G, however, is not fixed, the membership test $w \in L(G)$ can be done in $O(|G| \cdot |w|)$ time [3].

Note that PEGs may seem similar to conjunctive grammars [14]. However, a big difference is that PEGs are always unambiguous whereas conjunctive grammars can be ambiguous [4,15]. Also, in parsing, the conjunction operation in conjunctive grammars must consume the matching substring but the and-predicate in PEGs only verifies the matching string and consumes no symbols.

2.2   Edit-Distance

The edit-distance (or the Levenshtein distance) [9] between two strings x and y, denoted by $d(x, y)$, is the minimum number of edit operations—insertion, deletion and substitution—that transform x into y, where

- insertion adds a symbol into x
- deletion removes a symbol from x and
- substitution replaces a symbol from x with another symbol.

Then, we define the edit-distance between a string and a language, and between two languages as follows:

$$d(w, L) = \min_{x \in L} d(w, x) \qquad d(L_1, L_2) = \min_{x \in L_1} d(x, L_2).$$

3   The r-Shortest String Problem

The r-shortest string problem is to decide whether or not a language L contains a string of length of at most $r \geq 0$. For CFGs, this problem is straightforward. However, for PEGs, because of the ordered choice and the predicates in PEGs, the problem is not trivial. When recognizing a given input string, we need to check both the predicates and the corresponding rules that actually process the input string. Thus, we need to verify every rule and predicate, and find the common string for every rule. For instance, a PEG G of

$$S \leftarrow \&(10)A, \quad A \leftarrow 1(1/01)$$

cannot recognize the string 11 as the and-predicate does not match 11. Moreover, the r-shortest string problem is a special version of the r-edit-distance problem for two languages (the formal definition is in Definition 2), where one language is $\{\lambda\}$.

Definition 1 (r-shortest string (r-SS) problem). *Given a language L and an integer $r \geq 0$, the r-SS problem on L is to decide whether or not there exists a string $w \in L$ whose length is at most r.*

We prove that the r-SS problem on PELs is NEXPTIME-complete. We start with a simple NEXPTIME algorithm.

Lemma 1. *Given a PEG $G = (V, \Sigma, R, S)$ and an integer $r \geq 0$, The r-SS problem on $L(G)$ is in NEXPTIME.*

Proof. Consider the following algorithm.

1. Nondeterministically choose a string $w \in \Sigma^{\leq r}$,
2. Decide whether or not $w \in L(G)$.

We can guess the string w on the first step of the algorithm in at most r computation steps. The following membership test takes in quadratic time to the grammar size $|G|$ and the input size $|w| \leq r$ [3]. Thus, the entire algorithm is in NEXPTIME.   □

Next we show that the r-SS problem on PEL is NEXPTIME-hard using the bounded halting problem.

Theorem 1 (Bounded halting problem [17]). *Given a TM M, an input $w \in \Sigma^*$ and an integer $k \geq 0$, it is* NEXPTIME-complete *to decide whether or not M halts on w in at most k steps.*

Before the main proof, consider the following PEG that generates a string of an exponential length with respect to the grammar size.

Example 2. The following PEG G of $n + 1$ variables recognizes 0^{2^n} and, thus, $L(G) = \{0^{2^n} \Sigma^*\}$.

$$S \leftarrow A_{n-1} A_{n-1}$$
$$A_{n-1} \leftarrow A_{n-2} A_{n-2}$$
$$\vdots$$
$$A_1 \leftarrow A_0 A_0$$
$$A_0 \leftarrow 0$$

By substituting A_0 rule to recognize a set of symbols, say $A_0 \leftarrow \Sigma$, we can design a PEG for recognizing Σ^n using $O(\log n)$ rules.

Now, we are ready to show that the r-SS problem on PELs is NEXPTIME-hard.

Lemma 2. *Given a PEG G and an integer $r \geq 0$, the r-SS problem on $L(G)$ is* NEXPTIME-hard.

Proof. We prove the hardness by a poly-time reduction from the bounded halting problem to the r-SS problem. Given a TM $M = (Q, \Sigma, \Gamma, \delta, q_0, q_a, q_r)$, an input w and a nonnegative integer k, if M halts on w in at most k steps, we must have a finite computation of

$$C_1 \# C_2 \# C_3 \# \ldots \# C_n \#,$$

where each C_i is a configuration over $\Sigma \cup Q$ and $1 \leq n \leq k$. Such a halting computation is valid if and only if:

1. every configuration C_i is valid (has only one state and contains only tape symbols),
2. the initial configuration is $C_1 = q_0 w$,
3. the final configuration is on either the state q_a or q_r and
4. every computation step must follow the TM transitions δ.

We construct a PEG G that accepts every halting computation

$$C = C_1' \# C_2' \# C_3' \# \ldots \# C_n' \#$$

of M on input w, where $C_1' = B^k q_0 w B^k$ with the blank symbol B. Because C_1' has at least k symbols on both sides of its head, as long as we simulate at most k steps, we can assume that every configuration C_i has the same number $l = 2k + |C_1| = 2k + |w| + 1$ of symbols.

The following is a fragment of PEG G for M over the alphabet $\widetilde{\Sigma} = \Gamma \cup Q \cup \{\#\}$. We omit the polynomial size rules for the strings in the form of A^k, where A is a set of symbols for simplicity. (The construction is similar to Example 2)

$$S \leftarrow \&(B^k q_0 w B^k \#) D$$
$$F \leftarrow \&(Q_f \#)(\Gamma \cup Q)^l \# ! \widetilde{\Sigma}$$
$$D \leftarrow F$$

$$\begin{array}{ll} /cpa\&(\widetilde{\Sigma}^{l-2}qcb)D & (p,a,q,b,L) \in \delta, c \in \Gamma \\ /pa\&(\widetilde{\Sigma}^{l-1}bq)D & (p,a,q,b,R) \in \delta \\ /a\&(\widetilde{\Sigma}^l a)D & a \in \Gamma \cup \{\#\} \end{array}$$
$$Q_f \leftarrow (q_a/q_r)\Gamma^*/\Gamma Q_f$$

The constructed PEG G has three main rules: S, F and D. The rule S for the valid condition 2 checks whether or not the initial configuration is exactly $B^k q_0 w B^k$. Since we can design a PEG that accepts the prefix B^k in the size of $O(\log k)$, this rule does not violate the polynomial bound of reduction.

The rule F corresponding to the valid condition 3 determines whether or not a length l configuration is a halting configuration. F checks that the configuration has exactly one state, which is either q_a or q_r, and $l - 1$ tape symbol sequence ending with $\#$.

The rule D corresponding to the valid condition 4 represents the TM transitions. D first checks that the current configuration is a halting configuration by delegating checking to F. If the current configuration is not a halting configuration, it enumerates possible TM transitions between the current configuration and the next configuration.

The valid condition 1 holds because the initial configuration always has exactly one (q_0) state surrounded by $l - 1$ tape symbols and the rule D ensures that, if the previous configuration is valid for rule 1, then the next one is also valid. The rule Q_f decides whether or not the current configuration sequence, not necessarily l-length, has exactly one final state and a sequence of tape symbols.

The constructed grammar G can recognize a halting computation of a $(2k + |w| + 2)n$ length string, where n is the number of computation steps to halt. Thus, the $(r = (2k + |w| + 2)k)$-SS problem is equivalent to deciding the existence of a halting computation of at most k steps. This completes a polynomial time reduction.   □

By Lemmas 1 and 2, we establish the following statement.

Theorem 2. *Given a PEG G and an integer $r \geq 0$, the r-SS problem on $L(G)$ is* NEXPTIME-*complete.*

When r is in unary representation or a fixed constant, we can obtain a better result as follows:

Corollary 1. *When r is given in a unary representation, the r-SS problem is* NP-complete. *If r is a fixed constant, then we can solve the problem in polynomial time.*

Proof. The proof is similar to the proof for Lemma 2. The bounded halting problem is NP-complete when the input r is unary [17]. If we regard r to be a fixed constant, then the algorithm in the proof of Lemma 2 becomes polynomial. □

4   The r-Edit-Distance Problem

Next, we examine the r-edit-distance problem for PELs. This problem is crucial for verifying whether or not a string is in a given error bound from a PEL, or for measuring the similarity between a PEL and another language.

Definition 2 (r-edit-distance problem (r-ED)). *Given a language L, a string w and an integer $r \geq 0$, the r-ED problem between w and L is to decide whether or not $d(w, L) \leq r$.*

As the r-SS problem is a simple version of the r-ED problem, it is immediate that the r-ED problem is "harder" than r-SS problem.

Lemma 3. *Given a language L and an integer $r \geq 0$, the r-SS problem for L is mapping reducible to the r-ED between λ and L.*

Proof. If L has a string of length $n \leq r$, the edit-distance between the empty string λ and the string must be $n \leq r$. On the other hand, if L has no strings of length $n \leq r$, then the edit-distance between the empty string and L must be greater than r. □

Corollary 2. *Given a PEG G, a string w and an integer $r \geq 0$, the r-ED between w and $L(G)$ is* NEXPTIME-hard.

Now, we show that the r-ED problem on PELs is in NEXPTIME.

Lemma 4. *Given a PEG G, a string w and an integer $r \geq 0$, the r-ED between w and $L(G)$ is in* NEXPTIME.

Proof. Similar to the r-SS problem, the following algorithm shows that the r-ED problem is in NEXPTIME.

1. Choose $i \leq r$ nondeterministically.
2. Nondeterministically choose a valid edit operation on w and apply it.
3. Repeat the previous step i times to make the resulting string x has edit-distance of at most r.
4. Decide whether or not $x \in L(G)$.

Note that the length of x is between $\max(0, |w| - r)$ and $|w| + r$, which is exponential to the input size. □

Combining Corollary 2 and Lemma 4, we obtain the following statement.

Theorem 3. *Given a PEG G, a string w and an integer $r \geq 0$, the r-ED between w and $L(G)$ is* NEXPTIME-complete.

We can also solve the r-ED problem between a PEL and a finite language.

Theorem 4. *Given a PEG G, an acyclic DFA A and an integer $r \geq 0$, the r-ED problem between $L(G)$ and $L(A)$ is* NEXPTIME-complete.

Proof. It is easy to show that the problem is NEXPTIME-hard since the r-ED between a PEL and a single string is already NEXPTIME-complete.

The following is a naive NEXPTIME algorithm for the r-ED problem.

1. Nondeterministically choose a string $w \in L(A)$.
2. Apply the NEXPTIME algorithm for the r-ED problem between the given PEG G and the string w.

Since A is acyclic, $|w| \leq |A|$, we can guess w in polynomial time to the input size. Thus, the second step is in NEXPTIME bound with respect to the original input size. □

For the general case when we consider the r-ED problem for a PEG G and an arbitrary infinite language L, we should ensure that $L(G)$ is nonempty. If not, the problem becomes undecidable since $d(L(G), \Sigma^*) \leq 0$, which is equivalent to $L(G) \cap \Sigma^* \neq \emptyset$, decides whether or not $L(G)$ is empty [4]. Therefore, from now on, we assume that a PEL is a nonempty language. Now consider when L is regular. We show that this problem is undecidable even with a fixed r.

Theorem 5. *Given a nonempty PEG G, a DFA A and a fixed integer $r \geq 0$, it is undecidable that the r-ED problem between $L(G)$ and $L(A)$.*

Proof. It is easy to show that the case for $r = 0$ is undecidable as we cannot decide the emptiness of a PEL [4].

The case for $r > 0$, we will show a reduction from the Post Correspondence Problem (PCP), which is a well-known undecidable problem, to the r-ED problem by construct a PEG that recognizes possible solutions for the given PCP instance and a DFA for PCP solution encodings.

Consider a PCP instance $P = \{(x_1, y_1), (x_2, y_2), \ldots, (x_n, y_n)\}$, where every x_i's and y_i's are strings over Σ. We first construct a PEG G with the starting variable S for P and its alphabet $\widetilde{\Sigma} = \Sigma \cup \{\#, \$\}$:

$$S \leftarrow \&(X!\widetilde{\Sigma})\&(Y!\widetilde{\Sigma})\widetilde{\Sigma}\widetilde{\Sigma}/\$^{r+1}!\widetilde{\Sigma}$$
$$X \leftarrow x_1 X a_1 / x_2 X a_2 / \ldots / x_n X a_n / \#$$
$$Y \leftarrow y_1 Y a_1 / y_2 Y a_2 / \ldots / y_n Y a_n / \#$$

where \$, #, a_i's are new symbols not in Σ. Then $L(G)$ contains an encoded PCP solutions if P has a solution. Note that this grammar is similar to one on Ford's PCP to PEL emptiness reduction [4] but always nonempty by ensuring that $L(G)$ always contains the string $\$^{r+1}$. Also G cannot recognize the string # because S requires to recognize at least 2 symbols.

If P has no match, then the first choice rule on G cannot recognize any string, and thus G can recognize only $\$^{r+1}$. On the other hand, if P has at least one match, the first one can recognize the solution as well as the second rule.

Second, we construct a DFA A for the language

$$L(A) = \{w \# \alpha \mid w \in \Sigma^*, \alpha \in \{a_i\}^*\}$$

such that A represents a superset of valid PCP solution encodings.

Considering the PEG G and the DFA A, we can see that

- $d(L(G), L(A)) = 0 \le r$ if P has a match and
- $d(L(G), L(A)) = r + 1 \not\le r$ if P has no match. □

Furthermore, we cannot decide the r-ED problem between a PEL and a CFL since the classes of PELs or CFLs both contain regular languages.

Corollary 3. *Given a PEG G, a CFG (or a PEG) A and a fixed integer $r \ge 0$, the r-ED problem between $L(G)$ and $L(A)$ with bound r is undecidable.*

For other representations of r, the r-ED problems have similar results to the r-SS problems.

Corollary 4. *Given a PEG G, a finite language L and a unary (or a fixed) integer $r \ge 0$, the r-ED problem between L and $L(G)$ is NP-complete (done in polynomial time, respectively).*

5   Undecidability of Length Bound

Our next question is what is the length bound of shortest string in a PEG.

Definition 3. *We define $\mathcal{B}(G)$ to be a string length bound of a PEG G, where there is a nonempty string $w \in L(G)$ such that $|w| \le \mathcal{B}(G)$.*

If we can compute $\mathcal{B}(G)$, then we can use $\mathcal{B}(G)$ to find a bound, which gives rise to the shortest string as well as the edit-distance. For example, a regular language contains at least one string whose length is shorter than the number of its FA states, a context-free language contains at least one string whose length is $2^{|V|-1}$, where V is the set of variables of corresponding CFG in CNF. Unfortunately, for PEGs, we prove that we cannot find such bound.

Corollary 5. *On a PEG G, $\mathcal{B}(G)$ is not computable.*

Proof. We prove the statement by contradiction. Suppose that there exists a TM for $\mathcal{B}$. Then, by the definition of $\mathcal{B}$, for given a PEG G, $L(G)$ must contain a string w whose length is at most $\mathcal{B}(G)$. On the other hand, if $L(G)$ has at least one string, regardless of its length, $L(G)$ is not empty.

Then, we can decide whether or not $L(G)$ is nonempty by testing membership of every string in $\Sigma^{\mathcal{B}(G)}$ on G since $\mathcal{B}(G)$ is computable. If G can recognize any of those strings, G must be nonempty. This contradicts to the fact that deciding emptiness of G is not possible. □

6   Conclusions

Recently, PEGs became popular as a recognition-based formal grammar, which is always unambiguous and has a simple parsing algorithm. We have studied the shortest string problem and the edit-distance problem on PELs since we can use a language similarity metric, like edit-distance, to quantitatively verify errors between a grammar and a target string.

We have considered the r-SS problem about whether or not a PEL contains a string of length at most r and we have proved that the r-SS problem is NEXPTIME-complete. We have also examined the r-ED problem about whether or not the edit-distance between a PEL and a language is bounded up to r. The r-ED problem is decidable when we consider the edit-distance between a PEL and a finite language, and we prove that its complexity is NEXPTIME-complete. Finally, we have demonstrated that we cannot bound the length of strings in $L(G)$, where G is a PEG.

For future work, we plan to design a nontrivial algorithm that computes $\mathcal{B}(G)$ for a nonempty PEG G. We would also compute the edit-distance with the swap operation, which is another popular edit operation.

References

1. Rosie Pattern Language. https://rosie-lang.org. Accessed 7 Jan 2020
2. Birman, A., Ullman, J.D.: Parsing algorithms with backtrack. Inf. Control **23**(1), 1–34 (1973)
3. Ford, B.: Packrat parsing: simple, powerful, lazy, linear time, functional pearl. In: Proceedings of the Seventh ACM SIGPLAN International Conference on Functional Programming (ICFP 2002), pp. 36–47 (2002)
4. Ford, B.: Parsing expression grammars: a recognition-based syntactic foundation. In: Proceedings of the 31st ACM SIGPLAN-SIGACT Symposium on Principles of Programming Languages, POPL 2004, pp. 111–122 (2004)
5. Hamming, R.: Error detecting and error correcting codes. Bell Syst. Tech. J. **29**, 147–160 (1950)
6. Han, Y.S., Ko, S.K., Salomaa, K.: The edit-distance between a regular language and a context-free language. Int. J. Found. Comput. Sci. **24**(07), 1067–1082 (2013)
7. Koga, T.: Context-freeness of parsing expression languages is undecidable. Int. J. Found. Comput. Sci. **29**(7), 1203–1213 (2018)

8. Konstantinidis, S.: Computing the edit distance of a regular language. Inf. Comput. **205**(9), 1307–1316 (2007)

9. Levenshtein, V.I.: Binary codes capable of correcting deletions, insertions, and reversals. Soviet Phys. Dokl. **10**, 707–710 (1966)

10. Loff, B., Moreira, N., Reis, R.: The computational power of parsing expression grammars. In: Hoshi, M., Seki, S. (eds.) DLT 2018. LNCS, vol. 11088, pp. 491–502. Springer, Cham (2018). https://doi.org/10.1007/978-3-319-98654-8_40. Extended version is available on arXiv: https://arxiv.org/abs/1902.08272

11. Medeiros, S., Mascarenhas, F., Ierusalimschy, R.: From regexes to parsing expression grammars. Sci. Comput. Program. **93**, 3–18 (2014)

12. Mohri, M.: Edit-distance of weighted automata: general definitions and algorithms. Int. J. Found. Comput. Sci. **14**(06), 957–982 (2003)

13. Ng, T., Rappaport, D., Salomaa, K.: State complexity of neighbourhoods and approximate pattern matching. Int. J. Found. Comput. Sci. **29**(02), 315–329 (2018)

14. Okhotin, A.: Conjunctive grammars. J. Autom. Lang. Comb. **6**(4), 519–535 (2001)

15. Okhotin, A.: Unambiguous Boolean grammars. Inf. Comput. **206**(9–10), 1234–1247 (2008)

16. Povarov, G.: Descriptive complexity of the hamming neighborhood of a regular language. In: Proceedings of the 1st International Conference on Language and Automata Theory and Applications, pp. 509–520 (2007)

17. Sipser, M.: Introduction to the Theory of Computataion, 3rd edn. Cengage Learning, Boston (2013)

An Approach to the Herzog-Schönheim Conjecture Using Automata

Fabienne Chouraqui[✉]

University of Haifa, Campus Oranim, Kiryat Tiv'on, Israel
fabiennechouraqui@gmail.com

Abstract. Let G be a group and $H_1, \ldots, H_s$ be subgroups of G of indices $d_1, \ldots, d_s$ respectively. In 1974, M. Herzog and J. Schönheim conjectured that if $\{H_i \alpha_i\}_{i=1}^{i=s}$, $\alpha_i \in G$, is a coset partition of G, then $d_1, \ldots, d_s$ cannot be distinct. In this paper, we present a new approach to the Herzog-Schönheim conjecture based on automata and present a translation of the conjecture as a problem on automata.

Keywords: Free groups · Coset partitions · The Herzog-Schönheim conjecture · Automata

1 Introduction

Let G be a group and $H_1, \ldots, H_s$ be subgroups of G. If there exist $\alpha_i \in G$ such that $G = \bigcup_{i=1}^{i=s} H_i \alpha_i$, and the sets $H_i \alpha_i$, $1 \leq i \leq s$, are pairwise disjoint, then $\{H_i \alpha_i\}_{i=1}^{i=s}$ is *a coset partition of* G (or a *disjoint cover of* G). In this case, all the subgroups $H_1, \ldots, H_s$ can be assumed to be of finite index in G [17,21]. We denote by $d_1, \ldots, d_s$ the indices of $H_1, \ldots, H_s$ respectively [20]. The coset partition $\{H_i \alpha_i\}_{i=1}^{i=s}$ has *multiplicity* if $d_i = d_j$ for some $i \neq j$.

If G is the infinite cyclic group $\mathbb{Z}$, a coset partition of $\mathbb{Z}$ is $\{d_i \mathbb{Z} + r_i\}_{i=1}^{i=s}$, $r_i \in \mathbb{Z}$, with each $d_i \mathbb{Z} + r_i$ the residue class of r_i modulo d_i. These coset partitions of $\mathbb{Z}$ were first introduced by P. Erdös [11] and he conjectured that if $\{d_i \mathbb{Z} + r_i\}_{i=1}^{i=s}$, $1 < d_1 \leq \ldots \leq d_s$, $r_i \in \mathbb{Z}$, is a coset partition of $\mathbb{Z}$, then the largest index d_s appears at least twice. Erdös' conjecture was proved independently by H. Davenport with R.Rado and L. Mirsky with D. Newman using analysis of complex functions [12,21,22]. Furthermore, it was proved that the largest index d_s appears at least p times, where p is the smallest prime dividing d_s [21,22,34], that each index d_i divides another index d_j, $j \neq i$, and that each index d_k that does not properly divide any other index appears at least twice [22]. We refer also to [25–28,35] for more details on coset partitions of $\mathbb{Z}$ (also called covers of $\mathbb{Z}$ by arithmetic progressions) and to [13] for a proof of the Erdös' conjecture using group representations [16].

In 1974, M. Herzog and J. Schönheim extended Erdös' conjecture for arbitrary groups and conjectured that if $\{H_i \alpha_i\}_{i=1}^{i=s}$, $\alpha_i \in G$, is a coset partition

© Springer Nature Switzerland AG 2020
N. Jonoska and D. Savchuk (Eds.): DLT 2020, LNCS 12086, pp. 55–68, 2020.
https://doi.org/10.1007/978-3-030-48516-0_5

of G, then $d_1, \ldots, d_s$ cannot be distinct. In the 1980's, in a series of papers, M.A. Berger, A. Felzenbaum and A.S. Fraenkel studied the Herzog-Schönheim conjecture [2–4] and in [5] they proved the conjecture is true for the pyramidal groups [14], a subclass of the finite solvable groups. Coset partitions of finite groups with additional assumptions on the subgroups of the partition have been extensively studied. We refer to [6, 33, 36, 37]. In [18], the authors very recently proved that the conjecture is true for all groups of order less than 1440.

The common approach to the Herzog-Schönheim (HS) conjecture is to study it in finite groups. Indeed, given any group G, every coset partition of G induces a coset partition of a finite quotient group of G with the same indices [17]. In this paper, we present a completely different approach to the HS conjecture. The idea is to study it in free groups of finite rank and from there to provide answers for every group. This is possible since any finite or finitely generated group is a quotient group of a free group of finite rank and any coset partition of a quotient group F/N induces a coset partition of F with the same indices [7]. In order to study the Herzog-Schönheim conjecture in free groups of finite rank, we use the machinery of covering spaces. A pair $(\tilde{X}, p)$ is a *covering space* of a topological space X if $\tilde{X}$ is a path connected space, $p : \tilde{X} \to X$ is an open continuous surjection and every $x \in X$ has an open neighborhood U_x such that $p^{-1}(U_x)$ is a disjoint union of open sets in $\tilde{X}$, each of which is mapped homeomorphically onto U_x by p. For each $x \in X$, the non-empty set $Y_x = p^{-1}(x)$ is called *the fiber over x* and for all $x, x' \in X$, $\mid Y_x \mid = \mid Y_{x'} \mid$. If the cardinal of a fiber is m, one says that $(\tilde{X}, p)$ is a *m-sheeted covering* (m-fold cover) of X [15, 29].

The fundamental group of the bouquet with n leaves (or the wedge sum of n circles), X, is F_n, the free group of finite rank n and for any subgroup H of F_n of finite index d, there exists a d-sheeted covering space $(\tilde{X}_H, p)$ with a fixed basepoint. The underlying graph of $\tilde{X}_H$ is a directed labelled graph, with d vertices, called *the Schreier graph* and it can be seen as a finite complete bi-deterministic automaton; fixing the start and the end state at the basepoint, it recognises the set of elements in H. It is called *the Schreier coset diagram for F_n relative to the subgroup H* [32, p.107] or *the Schreier automaton for F_n relative to the subgroup H* [30, p.102]. The d vertices (or states) correspond to the d right cosets of H, any edge (or transition) has the form $Hg \xrightarrow{a} Hga$, $g \in F_n$, a a generator of F_n, and it describes the right action of F_n on the right cosets of H. If we fix the start state at the basepoint (H), and the end state at another vertex $H\alpha$, where α denotes the label of some path from the start state to the end state, then this automaton recognises the set of elements in $H\alpha$ and we call it *the Schreier automaton of $H\alpha$* and denote it by $\tilde{X}_{H\alpha}$.

In general, for any automaton M, with alphabet Σ, and d states, there exists a square matrix A of order $d \times d$, with a_{ij} equal to the number of directed edges from vertex i to vertex j, $1 \leq i, j \leq d$. This matrix is non-negative and it is called *the transition matrix of M* [10]. If for every $1 \leq i, j \leq d$, there exists $m \in \mathbb{Z}^+$ such that $(A^m)_{ij} > 0$, the matrix is said to be *irreducible*. For A an irreducible non-negative matrix, *the period of A* is the gcd of all $m \in \mathbb{Z}^+$ such that there is i with $(A^m)_{ii} > 0$. If M has a unique start state i and a unique end state j, then the number of words of length k (in the alphabet Σ) accepted by M is

$a_k = (A^k)_{ij}$. *The generating function of* M *is defined by* $p(z) = \sum\limits_{k=0}^{k=\infty} a_k z^k$. It is a rational function: the fraction of two polynomials in z with integer coefficients [10], [31, p. 575].

In [8], we study the properties of the transition matrices and generating functions of the Schreier automata in the context of coset partitions of the free group. Let $F_n = \langle \Sigma \rangle$, and Σ^* denote the free group and the free monoid generated by Σ, respectively. We will consider Σ^* as a subset of F_n. Let $\{H_i\alpha_i\}_{i=1}^{i=s}$ be a coset partition of F_n with $H_i < F_n$ of index $d_i > 1$, $\alpha_i \in F_n$, $1 \leq i \leq s$. Let $\tilde{X}_i$ denote the Schreier graph of H_i, with transition matrix A_i of period $h_i \geq 1$ and $\tilde{X}_{H_i\alpha_i}$ the Schreier automaton of $H_i\alpha_i$, with generating function $p_i(z)$, $1 \leq i \leq s$. For each $\tilde{X}_i$, A_i is a non-negative irreducible matrix and $a_{i,k} = (A_i^k)_{bf}$, $k \geq 0$, counts the number of words of length k that belong to $H_i\alpha_i \cap \Sigma^*$ (with b and f denoting the start and end state of $H_i\alpha_i$ respectively). Since F_n is the disjoint union of the sets $\{H_i\alpha_i\}_{i=1}^{i=s}$, each element in Σ^* belongs to one and exactly one such set, so n^k, the number of words of length k in Σ^*, satisfies $n^k = \sum\limits_{i=1}^{i=s} a_{i,k}$, for every $k \geq 0$, and moreover $\sum\limits_{k=0}^{k=\infty} n^k z^k = \sum\limits_{i=1}^{i=s} p_i(z)$. By using this kind of counting argument and studying the behaviour of the generating functions at their poles, we prove that if $h = max\{h_i \mid 1 \leq i \leq s\}$ is greater than 1, then there is a repetition of the maximal period $h > 1$ and that, under certain conditions, the coset partition has multiplicity. Furthermore, we recover the Davenport-Rado result (or Mirsky-Newman result) for the Erdős' conjecture and some of its consequences.

In this paper, we deepen further our study of the transition matrices of the Schreier automata in the context of coset partitions of F_n and give some new conditions that ensure a coset partition of F_n has multiplicity.

Theorem 1. *Let* F_n *be the free group on* $n \geq 2$ *generators. Let* $\{H_i\alpha_i\}_{i=1}^{i=s}$ *be a coset partition of* F_n *with* $H_i < F_n$ *of index* d_i, $\alpha_i \in F_n$, $1 \leq i \leq s$, *and* $1 < d_1 \leq \ldots \leq d_s$. *Let* $\tilde{X}_i$ *denote the Schreier graph of* H_i, *with transition matrix* A_i, *and period* $h_i \geq 1$, $1 \leq i \leq s$. *Let* $H = \{h_j \mid 1 \leq j \leq s, h_j > 1\}$. *Assume* $H \neq \emptyset$ *and different elements in* H *are pairwise coprime. Let* r_h *denote the number of repetitions of* h. *If for some* $h \in H$, $h \leq r_h \leq 2(h-1)$, *then* $\{H_i\alpha_i\}_{i=1}^{i=s}$ *has multiplicity.*

Furthermore, we show the Herzog-Schönheim conjecture in free groups can be translated into a conjecture on automata.

Conjecture 1. *Let* Σ *be a finite alphabet, and* Σ^* *be the free monoid generated by* Σ. *For every* $1 \leq i \leq s$, *let* M_i *be a finite, bi-deterministic and complete automaton with strongly-connected underlying graph. Let* d_i *be the number of states of* M_i ($d_i > 1$), *and* $L_i \subsetneq \Sigma^*$ *be the accepted language of* M_i. *If* Σ^* *is equal to the disjoint union of the* s *languages* $L_1, L_2, \ldots, L_s$, *then there are* $1 \leq j, k \leq s$, $j \neq k$, *such that* $d_j = d_k$.

Theorem 2. *If Conjecture 1 is true, then the Herzog-Schönheim conjecture is true.*

The paper is organized as follows. In Sect. 2, we give some preliminaries on automata and on irreducible non-negative matrices. In Sect. 3, we present a particular class of automata adapted to the study of the Herzog-Schönheim conjecture in free groups and describe some of their properties. In Sect. 4, we prove Theorem 1 and Theorem 2. We refer to [7] for more preliminaries and examples: Sect. 2, for free groups and covering spaces and Sect. 3.1, for graphs.

2   Automata, Non-negative Irreducible Matrices

2.1   Automata

We refer the reader to [30, p. 96], [9, p. 7], [23,24], [10]. A *finite state automaton* is a quintuple (S, Σ, μ, Y, s_0), where S is a finite set, called the *state set*, Σ is a finite set, called the *alphabet*, $\mu : S \times \Sigma \to S$ is a function, called the *transition function*, Y is a (possibly empty) subset of S called the *accept (or end) states*, and s_0 is called the *start state*. It can be represented by a directed graph with vertices the states and each transition $\mu(s, a) = s'$ is represented by a labelled edge $s \xrightarrow{a} s'$ from s to s' with label $a \in \Sigma$. The *label of a path* p of length n is the product $a_1 a_2 \ldots a_n$ of the labels of the edges of p. The finite state automaton $M = (S, \Sigma, \mu, Y, s_0)$ is *deterministic* if there is only one initial state and each state is the source of exactly one arrow with any given label from Σ. In a deterministic automaton, a path is determined by its starting point and its label [30, p. 105]. It is *co-deterministic* if there is only one final state and each state is the target of exactly one arrow with any given label from Σ. The automaton $M = (S, \Sigma, \mu, Y, s_0)$ is *bi-deterministic* if it is both deterministic and co-deterministic. An automaton M is *complete* if for each state $s \in S$ and for each $a \in \Sigma$, there is exactly one edge from s labelled a. We say that an automaton or a graph is *strongly-connected* if there is a directed path from any state to any other state.

Definition 2.1. Let $M = (S, \Sigma, \mu, Y, s_0)$ be a finite state automaton. Let Σ^* be the free monoid generated by Σ. Let $\mathrm{Map}(S, S)$ be the monoid consisting of all maps from S to S. The map $\phi : \Sigma \to \mathrm{Map}(S, S)$ given by μ (i.e. $\phi(a) \colon s \mapsto \mu(s, a)$), can be extended in a unique way to a monoid homomorphism $\phi : \Sigma^* \to \mathrm{Map}(S, S)$. The range of this map is a monoid called *the transition monoid of* M, which is generated by $\{\phi(a) \mid a \in \Sigma\}$. An element $w \in \Sigma^*$ is *accepted* by M if the corresponding element of $\mathrm{Map}(S, S)$, $\phi(w)$, takes s_0 to an element of the accept states set Y. The set $L \subseteq \Sigma^*$ recognized by M is called *the language accepted by* M, denoted by $L(M)$.

For any directed graph with d vertices or any finite state automaton M, with alphabet Σ, and d states, there exists a square matrix A of order $d \times d$, with $a_{ij} = A_{ij}$ equal to the number of directed edges from vertex i to vertex j, $1 \leq i, j \leq d$. This matrix is non-negative (i.e $a_{ij} \geq 0$) and it is called *the transition matrix* (as in [10]) or *the adjacency matrix* (as in [31, p. 575]). For any $k \geq 1$, $(A^k)_{ij}$ is equal to the number of directed paths of length k from vertex i to vertex j. So, if M is a bi-deterministic automaton with alphabet Σ, d states, start state i, accept state j and transition matrix A, then $(A^k)_{ij}$ is the number of words of length k in the free monoid Σ^* accepted by M.

2.2   Irreducible Non-negative Matrices

We refer to [1, Ch. 16], [19, Ch. 8]. There is a vast literature on the topic. Let A be a transition matrix of order $d \times d$ of a directed graph or an automaton with d states, as defined in Sect. 2.1. If for every $1 \leq i, j \leq d$, there exists $m \in \mathbb{Z}^+$ such that $(A^m)_{ij} > 0$, the matrix is said to be *irreducible*. For A an irreducible non-negative matrix, *the period of A* is the gcd of all $m \in \mathbb{Z}^+$ such that there is i with $(A^m)_{ii} > 0$. If the period is 1, A is called *aperiodic*. In [19], an irreducible and aperiodic matrix A is called *primitive* and the period h is called the *index of imprimitivity*.

Let A be an irreducible non-negative matrix of order $d \times d$ with period $h \geq 1$ and spectral radius r. Then the Perron-Frobenius theorem states that r is a positive real number and it is a simple eigenvalue of A, λ_{PF}, called the *Perron-Frobenius (PF) eigenvalue*. It satisfies $\min_j \sum_i a_{ij} \leq \lambda_{PF} \leq \max_i \sum_j a_{ij}$.

The matrix A has a right eigenvector v_R with eigenvalue λ_{PF} whose components are all positive and likewise, a left eigenvector v_L with eigenvalue λ_{PF} whose components are all positive. Both right and left eigenspaces associated with λ_{PF} are one-dimensional. The behaviour of irreducible non-negative matrices depends strongly on whether the matrix is aperiodic or not.

Theorem 2.2. *[19, Ch. 8] Let A be a $d \times d$ irreducible non-negative matrix of period $h \geq 1$, with PF eigenvalue λ_{PF}. Let v_L and v_R be left and right eigenvectors of λ_{PF} whose components are all positive, with $v_L v_R = 1$.*

$$\text{If } h = 1, \ \lim_{k \to \infty} \frac{A^k}{\lambda_{PF}^k} = P, \text{ and if } h > 1, \ \lim_{k \to \infty} \frac{1}{k} \sum_{m=0}^{m=k-1} \frac{A^m}{\lambda_{PF}^m} = P; \ P = v_R v_L.$$

3   A Particular Class of Automaton Adapted to the Study of the HS Conjecture

3.1   The Schreier Automaton of a Coset of a Subgroup

We now introduce the particular class of automata we are interested in, that is a slightly modified version of *the Schreier automaton for F_n relative to the subgroup H* [30, p. 102], [32, p. 107]. We refer to [7] for concrete examples.

Definition 3.1. Let $F_n = \langle \Sigma \rangle$, and Σ^* denote the free group and the free monoid generated by Σ, respectively. Let $H < F_n$ be of index d. Let $(\tilde{X}_H, p)$ be the covering of the n-leaves bouquet with basepoint $\tilde{x}_1$ and vertices $\tilde{x}_1, \tilde{x}_2, ..., \tilde{x}_d$. Let $t_i \in \Sigma^*$ denote the label of a directed path of minimal length from $\tilde{x}_1$ to $\tilde{x}_i$. Let $\tilde{X}_H$ be the Schreier coset diagram for F_n relative to the subgroup H, with $\tilde{x}_1$ representing the subgroup H and the other vertices $\tilde{x}_2, ..., \tilde{x}_d$ representing the cosets Ht_i accordingly. We call $\tilde{X}_H$ the *Schreier graph of H*, with this correspondence between the vertices $\tilde{x}_1, \tilde{x}_2, ..., \tilde{x}_d$ and the cosets Ht_i accordingly.

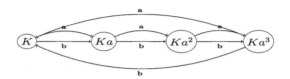

From the correspondence between the vertices and the cosets as described in Definition 3.1, there exists a directed path from any vertex $\tilde{x}_i$ to any other vertex $\tilde{x}_j$ in $\tilde{X}_H$, that is $\tilde{X}_H$ is a strongly-connected graph. Furthermore, it is n-regular. So, its transition matrix A is non-negative and irreducible, with PF eigenvalue n (the sum of the elements at each row and at each column is equal to n).

Definition 3.2. Let $F_n = \langle \Sigma \rangle$, and Σ^* denote the free group and the free monoid generated by Σ, respectively. Let $H < F_n$ be of index d. Let $\tilde{X}_H$ be the Schreier graph of H. Using the notation from Definition 3.1, let $\tilde{x}_1$ be the start state and $\tilde{x}_f$ be the end state for some $1 \leq f \leq d$. We call the automaton obtained *the Schreier automaton of Ht_f* and denote it by $\tilde{X}_{Ht_f}$. The language accepted by $\tilde{X}_{Ht_f}$ is the set of elements in Σ^* that belong to Ht_f. We call the elements in $\Sigma^* \cap Ht_f$, *the positive words in Ht_f*. The identity may belong to this set (and in fact, it does, for $f = 1$).

Example 3.3. Let $\Sigma = \{a, b\}$; $F_2 = \langle a, b \rangle$. Let $K \leq F_2$, of index 4.

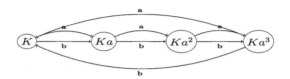

Fig. 1. The Schreier graph $\tilde{X}_K$ of $K = \langle a^4, b^4, ab^{-1}, a^2b^{-2}, a^3b^{-3} \rangle$.

The transition matrix of $\tilde{X}_K$ is $\begin{pmatrix} 0 & 2 & 0 & 0 \\ 0 & 0 & 2 & 0 \\ 0 & 0 & 0 & 2 \\ 2 & 0 & 0 & 0 \end{pmatrix}$ with period 4. If K and Ka are the start and end states, L is the set of positive words in Ka.

3.2 Properties of the Schreier Automata in Coset Partitions

We recall here some results proved in [8].

Theorem 3.4. *[8] Let F_n be the free group on $n \geq 1$ generators. Let $\{H_i\alpha_i\}_{i=1}^{i=s}$ be a coset partition of F_n with $H_i < F_n$ of index d_i, $\alpha_i \in F_n$, $1 \leq i \leq s$, and $1 < d_1 \leq \ldots \leq d_s$. Let $\tilde{X}_i$ denote the Schreier graph of H_i, with transition matrix A_i, and period $h_i \geq 1$, $1 \leq i \leq s$.*

(i) *Assume $h_k = max\{h_i \mid 1 \leq i \leq s\} > 1$. Then there exists $j \neq k$ such that $h_j = h_k$.*

(ii) *Let $h_\ell > 1$, such that h_ℓ does not properly divide any other period h_i, $1 \leq i \leq s$. Then there exists $j \neq \ell$ such that $h_j = h_\ell$.*

(iii) For every h_i, there exists $j \neq i$ such that either $h_i = h_j$ or $h_i \mid h_j$.

□

If $n = 1$ in Theorem 3.4, $\{H_i \alpha_i\}_{i=1}^{i=s}$ is a coset partition of $\mathbb{Z}$ and we recover the Davenport-Rado result (or Mirsky-Newman result) for the Erdős' conjecture and some of its consequences. Indeed, for every index d, the Schreier graph of $d\mathbb{Z}$ has a transition matrix with period equal to d, so a repetition of the period is equivalent to a repetition of the index. For the unique subgroup H of $\mathbb{Z}$ of index d, its Schreier graph $\tilde{X}_H$ is a closed directed path of length d (with each edge labelled 1). So, its transition matrix A is the permutation matrix corresponding to the $d-$cycle $(1, 2, ..., d)$, and it has period d. In particular, the period of A_s is d_s, and there exists $j \neq s$ such that $d_j = d_s$. Also, if the period (index) d_k of A_k does not properly divide any other period (index), then there exists $j \neq k$ such that $d_j = d_k$. For the free groups in general, we prove that in some cases, the repetition of the period implies the repetition of the index (see [8]).

4   Proof of the Main Results

4.1   Properties of the Transition Matrix of the Schreier Graph

We study the properties of the transition matrix of a Schreier graph.

Lemma 4.1. *Let $H < F_n$ of index d, with Schreier graph $\tilde{X}_H$ and transition matrix A with period $h \geq 1$. Then the following properties hold:*

(i) The vectors $v_L = \frac{1}{d}(1, 1, ..., 1)$, $v_R = (1, 1, ..., 1)^T$ are left and right eigenvectors of n whose components are all positive, with $v_L v_R = 1$.

(ii) The matrix $P = v_R v_L$ is of order $d \times d$ with all entries equal to $\frac{1}{d}$.

(iii) If $h = 1$, then $\lim\limits_{k \to \infty} \frac{A^k}{n^k} = P$ and if $h > 1$, then $\lim\limits_{k \to \infty} \frac{1}{k} \sum\limits_{j=0}^{j=k-1} \frac{A^j}{n^j} = P$.

Proof. *(i), (ii), (iii)* As the sum of every row and every column in A is equal to n, $\lambda_{PF} = n$ with right eigenvector $v_R = (1, 1, ..., 1)^T$ and left eigenvector $(1, 1, ..., 1)$. Since $(1, 1, ..., 1)v_R = d$, $v_L = \frac{1}{d}(1, 1, ..., 1)$ is a left eigenvector that satisfies $v_L v_R = 1$. Computing $v_R v_L$ gives the matrix P of order $d \times d$ with all entries equal to $\frac{1}{d}$. *(iii)* results from Theorem 2.2.   □

The behaviour of exponents of an aperiodic $d \times d$ matrix of a Schreier graph $\tilde{X}_H$ is well known: for every $1 \leq i, j \leq d$, $\lim\limits_{k \to \infty} \frac{(A^k)_{ij}}{n^k} = \frac{1}{d}$, from Lemma 4.1. It means that the proportion of positive words of every length k (k large enough) that belong to any coset of H tends to the fixed value $\frac{1}{d}$. We turn now to the study of $\lim\limits_{k \to \infty} \frac{(A^k)_{ij}}{n^k}$, where A is the transition matrix of a Schreier graph $\tilde{X}_H$ of period $h > 1$.

Definition 4.2. For $1 \leq i, j \leq d$, we define m_{ij}, $0 \leq m_{ij} \leq d$, to be the minimal natural number such that $(A^{m_{ij}})_{ij} \neq 0$.

By definition, if $i \neq j$, then m_{ij} is the minimal length of a directed path from i to j in $\tilde{X}_H$ and if $i = j$, then $m_{ij} = 0$. Whenever $h > 1$, only for the exponents $m_{ij} + kh$, $k \geq 0$, $(A^{m_{ij}+kh})_{ij} \neq 0$, that is only positive words of length $m_{ij} + kh$ are accepted by the Schreier automaton, with i and j the start and end states respectively. Note that if H is a subgroup of $\mathbb{Z} = \langle 1 \rangle$ of index d, its transition matrix A is a permutation matrix with period d and $m_{ij} = r$, where $d\mathbb{Z} + r$ is the coset with i and j the start and end states respectively.

Lemma 4.3. *Let $H < F_n$ be of index d, with Schreier graph $\tilde{X}_H$ and transition matrix A with period $h > 1$. Then, the following properties hold:*

(i) $\frac{(A^k)_{ij}}{n^k} = 0$, *whenever* $k \not\equiv m_{ij} (mod\, h)$, $1 \leq i, j \leq d$.

(ii) $\lim_{k \to \infty} \frac{(A^{m_{ij}+kh})_{ij}}{n^k} = \frac{h}{d}$, $1 \leq i, j \leq d$.

(iii) *for every* $0 \leq m \leq h - 1$, *there is i such that* $m_{1i} \equiv m (mod\, h)$.

(iv) *h divides d.*

Proof. (i) By definition, whenever $k \not\equiv m_{ij}(mod\, h)$, $(A^k)_{ij} = 0$.

(ii), (iii), (iv) We define an $d \times h$ matrix B in the following way. Each row i is labelled by a right coset of H in the same order as they appear in the rows and columns of A and each column by $m = 0, 1, 2, ..., h - 1$, and $(B)_{ij} = \lim_{k \to \infty} \frac{(A^{j+kh})_{1i}}{n^{j+kh}}$. Roughly, $(B)_{ij}$ is the proportion of positive words of very large length (congruent to $j\,(mod\, h)$) that belong to the corresponding coset of H. From (i):

$$(B)_{ij} = \begin{cases} 0 & \text{if } m_{1i} \not\equiv j(mod\, h), \\ \lim_{k \to \infty} \frac{(A^{m_{1i}+kh})_{1i}}{n^{m_{1i}+kh}} & \text{if } m_{1i} \equiv j(mod\, h). \end{cases}$$

So, at each row of B, there is a single non-zero entry. As F_n is partitioned by the d cosets of H, all the non-zero elements in B are equal and for every $k \geq 0$ and every $1 \leq i \leq d$, $\sum_{f=1}^{f=d}(A^k)_{if} = n^k$, in particular $\sum_{f=1}^{f=d} \frac{(A^k)_{1f}}{n^k} = 1$.

So, $\sum_{i=1}^{i=d}(B)_{ij} = \sum_{i=1}^{i=d} \lim_{k \to \infty} \frac{(A^{j+kh})_{1i}}{n^{j+kh}} = \lim_{k \to \infty} \sum_{i=1}^{i=d} \frac{(A^{j+kh})_{1i}}{n^{j+kh}} = 1$, that is the sum of elements in each column of B is equal to 1. If $h = d$, B is a square matrix and the right cosets can be arranged such that their labels m are in growing order and we have necessarily a diagonal matrix (otherwise there would be a column of zeroes). So, $\lim_{k \to \infty} \frac{(A^{m_{1i}+kh})_{1i}}{n^{m_{1i}+kh}} = 1$ and (ii), (iii), (iv) hold. Now, assume $d > h$. At each column, there is at least one non-zero entry, so (iv) holds. Furthermore, the number of non-zero entries in each column needs to be the same, so h divides d and for any i, $\frac{d}{h} * (\lim_{k \to \infty} \frac{(A^{m_{1i}+kh})_{1i}}{n^{m_{1i}+kh}}) = 1$. That is, $\lim_{k \to \infty} \frac{(A^{m_{1i}+kh})_{1i}}{n^{m_{1i}+kh}} = \frac{h}{d}$.

Furthermore, $\lim_{k \to \infty} \frac{1}{h} \sum_{j=0}^{j=h-1} \frac{(A^{j+kh})_{1i}}{n^{j+kh}} = \frac{1}{d}$. $\qquad \square$

4.2 Conditions that Ensure Multiplicity in a Coset Partition

Let F_n be the free group on $n \geq 1$ generators. Let $\{H_i\alpha_i\}_{i=1}^{i=s}$ be a coset partition of F_n with $H_i < F_n$ of index d_i, $\alpha_i \in F_n$, $1 \leq i \leq s$, and $1 < d_1 \leq \ldots \leq d_s$. Let $\tilde{X}_i$ denote the Schreier graph of H_i, with transition matrix A_i, and period $h_i \geq 1$, $1 \leq i \leq s$. In the following lemmas, we prove, under additional assumptions, that there exist conditions that ensure multiplicity.

Lemma 4.4. *Assume that among all periods $h_1, \ldots h_s$, there exists a unique value $h > 1$. Let r denote the number of repetitions of h. Then, $r \geq h$. Furthermore, if $h \leq r \leq 2(h-1)$, then $\{H_i\alpha_i\}_{i=1}^{i=s}$ has multiplicity.*

Proof. For every $1 \leq i \leq s$, $(A_i^k)_{1f_i}$ denotes the number of positive words of length k that belong to the coset $H_i\alpha_i$. Let $I = \{1 \leq i \leq s \mid h_i = h\}$. For every $i \in I$, we denote by m_i the minimal natural number such that $(A_i^{m_i})_{1f_i} \neq 0$. We define an $r \times h$ matrix C in the following way. Each row i is labelled by a right coset $H_i\alpha_i$, where $i \in I$ and each column by $m = 0, 1, 2, \ldots, h-1$, and:

$$(C)_{ij} = \begin{cases} 0 & \text{if } m_i \not\equiv j \pmod{h}, \\ \lim\limits_{k\to\infty} \frac{(A_i^{m_i+hk})_{1f_i}}{n^{m_i+hk}} & \text{if } m_i \equiv j \pmod{h}. \end{cases}$$

Roughly, $(C)_{ij}$ is the proportion of positive words of very large length (congruent to $j \pmod{h}$) that belong to $H_i\alpha_i$, where $i \in I$. At each row of C there is a unique non-zero entry. Since $\{H_i\alpha_i\}_{i=1}^{i=s}$ is a coset partition of F_n, for every k, $\sum_{i=1}^{i=s}(A_i^k)_{1f_i} = n^k$, that is $\sum_{i=1}^{i=s}\frac{(A_i^k)_{1f_i}}{n^k} = 1$. If A_i is aperiodic, then

$$\lim_{k\to\infty} \frac{(A_i^k)_{1f_i}}{n^k} = \frac{1}{d_i} \text{ from Lemma 4.1. So, } 1 = \lim_{k\to\infty}\sum_{i=1}^{i=s}\frac{(A_i^k)_{1f_i}}{n^k} = \sum_{i=1}^{i=s}\lim_{k\to\infty}\frac{(A_i^k)_{1f_i}}{n^k} =$$

$\sum_{i\notin I}\frac{1}{d_i} + \sum_{i\in I}\lim_{k\to\infty}\frac{(A_i^k)_{1f_i}}{n^k}$. That is, $\sum_{i\in I}\lim_{k\to\infty}\frac{(A_i^k)_{1f_i}}{n^k} = 1 - \sum_{i\notin I}\frac{1}{d_i} = \sum_{i\in I}\frac{1}{d_i}$, since $\sum_{i=1}^{i=s}\frac{1}{d_i} = 1$. So, the sum of elements in each column of C is equal to $\sum_{i\in I}\frac{1}{d_i}$ and from Lemma 4.3, the non-zero entries in C have the form $\frac{h}{d_i}$. If $r < h$, then there is necessarily a column of zeroes, so $r \geq h$. If $r = h$, then C is a square matrix and the right cosets can be arranged such that their labels m are in growing order and we have necessarily a diagonal matrix (otherwise there would be a column of zeroes). So, for every $i \in I$, $\frac{h}{d_i} = \sum_{i\in I}\frac{1}{d_i}$. That is, the coset partition $\{H_i\alpha_i\}_{i=1}^{i=s}$ has multiplicity with all the d_i equal for $i \in I$. Now, assume $r > h$. At each column, there is at least one non-zero entry and there are necessarily columns with several non-zero entries. By a simple combinatorial argument, the number n_0 of columns with a single non-zero entry satisfies $h - (r - h) \leq n_0 \leq h - 1$, that is $2h - r \leq n_0 \leq h - 1$. If we assume $r \leq 2(h-1)$, then $n_0 \geq 2h - r - 2(h-1) \geq 2$, that is the number of columns with a single non-zero entry is at least 2, so there are at least two $i \in I$, such that $\frac{h}{d_i} = \sum_{i\in I}\frac{1}{d_i}$, and the coset partition $\{H_i\alpha_i\}_{i=1}^{i=s}$ has multiplicity. Note that for every $0 \leq m \leq h-1$, there is i such that $m_i \equiv m \pmod{h}$. $\qquad\square$

Lemma 4.5. *Assume there exist at least two coprime values $h, h' > 1$ of periods $h_1, \ldots, h_s$. Let r and r' denote the number of repetitions of h and h' respectively. If $r = h$ or $r \leq 2(h-1)$ or $r' = h'$ or $r' \leq 2(h'-1)$, then $\{H_i \alpha_i\}_{i=1}^{i=s}$ has multiplicity.*

Proof. Let $I = \{1 \leq i \leq s \mid h_i = h\}$ and $I' = \{1 \leq i \leq s \mid h_i = h'\}$. Assume with no loss of generality that $h' < h$. From the same argument as in the proof of Lemma 4.4, $\sum_{i \in I \cup I'} \lim_{k \to \infty} \frac{(A_i^k)_{1f_i}}{n^k} = \sum_{i \in I \cup I'} \frac{1}{d_i}$. We show that each period can be considered independently, that is each period has its own matrix C as defined in the proof of Lemma 4.4. We define an $(r' + r) \times L$ matrix D, where $L = 2hh'$, in the following way. The first r' rows are labelled by right cosets $H_i \alpha_i$, where $i \in I'$, the last r rows are labelled by right cosets $H_i \alpha_i$, where $i \in I$ and each column by $m = 0, 1, 2, \ldots, h' - 1, \ldots, h - 1, h, \ldots, L - 1$, and:

$$(D)_{ij} = \begin{cases} 0 & \text{if } i \in I', \ m_i \not\equiv j \pmod{h'}, \\ \lim_{k \to \infty} \frac{(A_i^{m_i + h'k})_{1f_i}}{n^{m_i + h'k}} & \text{if } i \in I', \ m_i \equiv j \pmod{h'} \\ 0 & \text{if } i \in I, \ m_i \not\equiv j \pmod{h}, \\ \lim_{k \to \infty} \frac{(A_i^{m_i + hk})_{1f_i}}{n^{m_i + hk}} & \text{if } i \in I, \ m_i \equiv j \pmod{h}. \end{cases}$$

So, the sum of elements in each column of D is equal to $\sum_{i \in I \cup I'} \frac{1}{d_i}$ and from Lemma 4.3, the non-zero entries in D have the form $\frac{h}{d_i}$ for $i \in I$ and $\frac{h'}{d_i}$, for $i \in I'$. Let $0 \leq m \leq h' - 1$, be the minimal number such that the sum of entries of the m-th column is $\sum_{i \in J_0} \frac{h}{d_i} + \sum_{i \in J_0'} \frac{h'}{d_i}$, where $\emptyset \neq J_0 \subset I$ and $\emptyset \neq J_0' \subset I'$. So, for every $0 \leq k \leq h' - 1$, the sum of entries of the $(m + kh)$-th column is $\sum_{i \in J_0} \frac{h}{d_i} + \sum_{i \in J_k} \frac{h'}{d_i}$, and this implies necessarily $\sum_{i \in J_0'} \frac{h'}{d_i} = \sum_{i \in J_1'} \frac{h'}{d_i} = \ldots = \sum_{i \in J_{h'-1}'} \frac{h'}{d_i}$.

We show that $\{\frac{h'}{d_i} \mid i \in J_0'\}$, $\{\frac{h'}{d_i} \mid i \in J_1'\}$,..., $\{\frac{h'}{d_i} \mid i \in J_{h'-1}'\}$ appear in the first h' columns of D (not necessarily in this order). Let $0 \leq k, l \leq h' - 1$, $k \neq l$. Assume by contradiction that $m + kh \equiv m + lh \pmod{h'}$. So, h' divides $h(k - l)$. As h and h' are coprime, h' divides $k - l$, a contradiction. So, for every $0 \leq k, l \leq h' - 1$, $k \neq l$, $m + kh \not\equiv m + lh \pmod{h'}$. As there are exactly h' values, these correspond to $0, 1, \ldots, h' - 1 \pmod{h'}$, and $\{\frac{h'}{d_i} \mid i \in J_0'\}$, $\{\frac{h'}{d_i} \mid i \in J_1'\}$,..., $\{\frac{h'}{d_i} \mid i \in J_{h'-1}'\}$ appear in the first h' columns of D with $\sum_{i \in J_0'} \frac{h'}{d_i} = \ldots = \sum_{i \in J_{h'-1}'} \frac{h'}{d_i}$.

Furthermore, $\sum_{i \in J_0'} \frac{h'}{d_i} = \ldots = \sum_{i \in J_{h'-1}'} \frac{h'}{d_i} = \sum_{i \in I'} \frac{1}{d_i}$. Indeed, on one hand, the sum of elements in the first r' rows and h' columns is equal to $h' \sum_{i \in J_0'} \frac{h'}{d_i}$ and on the second hand, it is equal to $\sum_{i \in I'} \frac{h'}{d_i}$. Using the same argument, for every $0 \leq k \leq h - 1$, the sum of entries of the $(m + kh')$-th column is $\sum_{i \in J_k} \frac{h}{d_i} + \sum_{i \in J_0'} \frac{h'}{d_i}$, and this implies necessarily $\sum_{i \in J_0} \frac{h}{d_i} = \sum_{i \in J_1} \frac{h}{d_i} = \ldots = \sum_{i \in J_{h-1}} \frac{h}{d_i}$. We show that $\{\frac{h}{d_i} \mid i \in J_0\}, \{\frac{h}{d_i} \mid i \in J_1\}, \ldots, \{\frac{h}{d_i} \mid i \in J_{h-1}\}$ appear in the first h columns of D (not necessarily in this order). Let $0 \leq k, l \leq h - 1$, $k \neq l$. Assume by contradiction that $m + kh' \equiv m + lh' \pmod{h}$. So, h divides $h'(k - l)$. As h and h' are coprime, h divides $k - l$, a contradiction. So, for every $0 \leq k, l \leq h - 1$, $k \neq l$, $m + kh' \not\equiv m + lh' \pmod{h}$. As there are exactly h values, these correspond to $0, 1, \ldots, h - 1 \pmod{h}$, and $\{\frac{h}{d_i} \mid i \in J_0\}, \{\frac{h}{d_i} \mid i \in J_1\}, \ldots, \{\frac{h}{d_i} \mid i \in J_{h-1}\}$ appear in the first h columns of D, with $\sum_{i \in J_0} \frac{h}{d_i} = \sum_{i \in J_1} \frac{h}{d_i} = \ldots = \sum_{i \in J_{h-1}} \frac{h}{d_i}$. Furthermore,

$$\sum_{i \in J_0} \frac{h}{d_i} = \ldots = \sum_{i \in J_{h-1}} \frac{h}{d_i} = \sum_{i \in I} \frac{1}{d_i}.$$ So, each period has its own matrix C and we apply the results of Lemma 4.3. □

From Lemma 4.5, coprime periods h and h' can be considered independently. But, if h and h' are not coprime, then the situation is different. Indeed, consider the following coset partition of F_2: $F_2 = H \cup Ka \cup Ka^3$, where K is the subgroup described in Example 3.3 and $H = \langle a^2, b^2, ab \rangle < F_2$ of index 2. The period of the transition matrix of $\tilde{X}_H$ is $h' = 2$ and the period of the transition matrix of $\tilde{X}_K$ is $h = 4$ and the corresponding matrix D as defined in the proof of Lemma 4.5 is $D = \begin{pmatrix} 1 & 0 & 1 & 0 \\ 0 & 1 & 0 & 0 \\ 0 & 0 & 0 & 1 \end{pmatrix}$, with the first row labelled H, the second row Ka, the third row Ka^3 and at each column $0 \leq m \leq 3$. So, if h' divides h, each period cannot have its own matrix C. Yet, using the same kind of arguments as before, it is not difficult to prove that $r \geq h - \frac{h}{h'}r'$ and that if $r \leq 2(h - \frac{h}{h'}r' - 1)$ then the coset partition has multiplicity. We now turn to the proof of Theorem 1.

Proof of Theorem 1. We assume that H, the set of periods greater than 1, is not empty and that different elements in H are pairwise coprime. Let r_h denote the number of repetitions of h. From the proof of Lemma 4.5, each period has its own matrix C and we apply the results of Lemma 4.3. That is, if for some $h \in H$, $h \leq r_h \leq 2(h - 1)$, then $\{H_i \alpha_i\}_{i=1}^{i=s}$ has multiplicity. □

4.3 Translation of the HS Conjecture in Terms of Automata

Let $F_n = \langle \Sigma \rangle$, and Σ^* denote the free group and the free monoid generated by Σ, respectively. Let $\{H_i \alpha_i\}_{i=1}^{i=s}$ be a coset partition of F_n with $H_i < F_n$ of index $d_i > 1$, $\alpha_i \in F_n$, $1 \leq i \leq s$. Let $\tilde{X}_i$ be the Schreier automaton of $H_i \alpha_i$, with language $L_i = \Sigma^* \cap H_i \alpha_i$.

Proof of Theorem 2. Assume Conjecture 1 is true. For every $1 \leq i \leq s$, the Schreier automaton $\tilde{X}_i$ is a finite, bi-deterministic and complete automaton with strongly-connected underlying graph and alphabet Σ. Since F_n is the disjoint union of the sets $\{H_i \alpha_i\}_{i=1}^{i=s}$, each word in Σ^* belongs to one and exactly one such language, so Σ^* is the disjoint union of the s languages $L_1, L_2, \ldots, L_s$. Since Conjecture 1 is true, there is a repetition of the number of states and this implies the coset partition $\{H_i \alpha_i\}_{i=1}^{i=s}$ has multiplicity, that is the HS conjecture in free groups of finite rank is true. From [7, Thm. 6], this implies the HS conjecture is true for all finitely generated groups, in particular for all finite groups. So, the HS conjecture is true for all groups.   □

A question that arises naturally is whether the HS Conjecture implies Conjecture 1, that is do the conditions of Conjecture 1 imply necessarily the existence of a coset partition of a free group. First, we note that any finite, bi-deterministic, complete automaton M with d states, finite alphabet Σ, and a strongly-connected underlying graph can represent an automaton with accepted language all the words that belong to some coset of a subgroup H of index d in F_Σ, the free group generated by Σ. Indeed, the start state is replaced by H and each state is replaced by a right coset $H\alpha$, where $\alpha \in \Sigma^*$ is the label of a directed path from the start to it. As M is complete and bi-deterministic, at each vertex v, there are $|\Sigma|$ directed edges into v with each such edge labelled by a different label $a \in \Sigma$, and $|\Sigma|$ directed edges out of v, with each such edge labelled by a different label $a \in \Sigma$. For each $a \in \Sigma$, there exists $a^{-1} \in F_\Sigma$ and for each directed edge $H\alpha \xrightarrow{a} H\alpha a$, $\alpha \in \Sigma^*$, in the underlying graph of M, there exists another directed edge $H\alpha \xleftarrow{a^{-1}} H\alpha a$, which is implicit and not drawn. This fact is crucial for the construction of an automaton with accepted language all the words that belong to some coset $H\alpha$ and not only the positive words that belong to this coset. In fact, this is how the Schreier automaton for a free group relative to a subgroup H is defined initially (see [30, p. 102], [32, p. 107]).

So, the existence of the s automata $M_1, \ldots, M_s$, satisfying the conditions of Conjecture 1, with accepted languages $L_1, \ldots, L_s$ respectively, leads to the existence of s automata $M_1', \ldots, M_s'$ with accepted language $L_1', \ldots, L_s'$, where L_i' denotes the set of words that belong to the coset $H_i\alpha_i$, and $H_i < F_\Sigma$, $1 \le i \le s$. The question is now: does the assumption that Σ^* is equal to the disjoint union of the s languages $L_1, L_2, \ldots, L_s$ imply necessarily that $\{H_i\alpha_i\}_{i=1}^{i=s}$ is a coset partition of F_Σ. If all the s automata $M_1, \ldots, M_s$ satisfy the following additional conditions: Σ is equal to the disjoint union of two sets S and S^-, where $S^- = \{a^- \mid a \in S\}$ and a^- is such that for each directed edge $H\alpha \xrightarrow{a} H\alpha a$, $\alpha \in S^*$, in the underlying graph, there exists another directed edge $H\alpha \xleftarrow{a^-} H\alpha a$ in the underlying graph (and it is drawn), then clearly $\Sigma^* = \bigsqcup_{i=1}^{i=s} L_i$ implies $F_\Sigma = \bigsqcup_{i=1}^{i=s} H_i\alpha_i$ (since $F_\Sigma = \Sigma^*$ and L_i is the set of words that belong to $H_i\alpha_i$). But, for automata that do not satisfy these additional conditions, it is not clear at all if this is still true. Indeed, it does not seem that if the languages $L_1, \ldots, L_s$ are mutually disjoint, then the languages $L_1', \ldots, L_s'$ are also mutually disjoint and moreover, that $\Sigma^* = \bigsqcup_{i=1}^{i=s} L_i$ implies necessarily $F_\Sigma = \bigsqcup_{i=1}^{i=s} L_i'$.

References

1. Bellman, R.: Matrix Analysis. S.I.A.M. Press, Philadelphia (1997)
2. Berger, M.A., Felzenbaum, A., Fraenkel, A.S.: Improvements to two results concerning systems of residue sets. Ars. Combin. **20**, 69–82 (1985)
3. Berger, M.A., Felzenbaum, A., Fraenkel, A.S.: The Herzog-Schönheim conjecture for finite nilpotent groups. Can. Math. Bull. **29**, 329–333 (1986)
4. Berger, M.A., Felzenbaum, A., Fraenkel, A.S.: Lattice parallelotopes and disjoint covering systems. Discrete Math. **65**, 23–44 (1987)
5. Berger, M.A., Felzenbaum, A., Fraenkel, A.S.: Remark on the multiplicity of a partition of a group into cosets. Fund. Math. **128**, 139–144 (1987)
6. Brodie, M.A., Chamberlain, R.F., Kappe, L.C.: Finite coverings by normal subgroups. Proc. Am. Math. Soc. **104**, 669–674 (1988)
7. Chouraqui, F.: The Herzog-Schönheim conjecture for finitely generated groups. Int. J. Algebra Comput. **29**(6), 1083–1112 (2019)
8. Chouraqui, F.: About an extension of the Davenport-Rado result to the Herzog-Schönheim conjecture for free groups, arXiv:1901.09898
9. Epstein, D.B.A., Cannon, J.W., Holt, D.F., Levy, S.V.F., Paterson, M.S., Thurston, W.P.: Word Processing in Groups. Jones and Bartlett Publishers, Boston (1992)
10. Epstein, D.B.A., Iano-Fletcher, A.R., Zwick, U.: Growth functions and automatic groups. Exp. Math. **5**(4), 297–315 (1996)
11. Erdös, P.: On integers of the form $2^k + p$ and some related problems. Summa Brasil. Math. **2**, 113–123 (1950)
12. Erdös, P.: Problems and results in Number theory, Recent Progress in Analytic Number Theory, vol. 1, pp. 1–13. Academic Press, London (1981)
13. Ginosar, Y.: Tile the group. Elem. Math. **72** (2018). Swiss Mathematical Society
14. Ginosar, Y., Schnabel, O.: Prime factorization conditions providing multiplicities in coset partitions of groups. J. Comb. Number Theory **3**(2), 75–86 (2011)
15. Hatcher, A.: Algebraic Topology. Cambridge University Press, Cambridge (2002)
16. Herzog, M., Schönheim, J.: Research problem no. 9. Can. Math. Bull. **17**, 150 (1974)
17. Korec, I., Znám, Š.: On disjoint covering of groups by their cosets. Math. Slovaca **27**, 3–7 (1977)
18. Margolis, L., Schnabel, O.: The Herzog-Schonheim conjecture for small groups and harmonic subgroups, arXiv:1803.03569
19. Meier, C.D.: Matrix Analysis and Applied Linear Algebra. SIAM: Society for Industrial and Applied Mathematics, Philadelphia (2010)
20. Neumann, B.H.: Groups covered by finitely many cosets. Publ. Math. Debrecen **3**, 227–242 (1954)
21. Newman, M.: Roots of unity and covering sets. Math. Ann. **191**, 279–282 (1971). https://doi.org/10.1007/BF01350330
22. Novák, B., Znám, Š.: Disjoint covering systems. Am. Math. Monthly **81**, 42–45 (1974)
23. Pin, Jean-Eric: On reversible automata. In: Simon, Imre (ed.) LATIN 1992. LNCS, vol. 583, pp. 401–416. Springer, Heidelberg (1992). https://doi.org/10.1007/BFb0023844
24. Pin, J.E.: Mathematical foundations of automata theory. https://www.irif.fr/~jep/PDF/MPRI/MPRI.pdf

25. Porubský, Š.: Natural exactly covering systems of congruences. Czechoslovak Math. J. **24**, 598–606 (1974)
26. Porubský, Š.: Covering systems and generating functions. Acta Arithmetica **26**(3), 223–231 (1975)
27. Porubský, Š.: Results and problems on covering systems of residue classes. Mitt. Math. Sem. Giessen **150**, 1–85 (1981)
28. Porubský, Š., Schönheim, J.: Covering systems of Paul Erdös. Past, present and future. In: Paul Erdös and His Mathematics, vol. 11, pp. 581–627. János Bolyai Math. Soc. (2002)
29. Rotman, J.J.: An Introduction to Algebraic Topology. Graduate Texts in Mathematics, vol. 119. Springer, Heidelberg (1988). https://doi.org/10.1007/978-1-4612-4576-6
30. Sims, C.C.: Computation with Finitely Presented Groups. Encyclopedia of Mathematics and Its Applications, vol. 48. Cambridge University Press, Cambridge (1994)
31. Stanley, R.P.: Enumerative Combinatorics. Wadsworth and Brooks/Cole, Monterey (1986)
32. Stillwell, J.: Classical Topology and Combinatorial Group Theory. Graduate Texts in Mathematics, vol. 72. Springer, New York (1980). https://doi.org/10.1007/978-1-4684-0110-3
33. Sun, Z.W.: Finite covers of groups by cosets or subgroups. Int. J. Math. **17**(9), 1047–1064 (2006)
34. Sun, Z.W.: An improvement of the Znám-Newman result. Chin. Q. J. Math. **6**(3), 90–96 (1991)
35. Sun, Z.W.: Covering the integers by arithmetic sequences II. Trans. Am. Math. Soc. **348**, 4279–4320 (1996)
36. Tomkinson, M.J.: Groups covered by abelian subgroups. London Mathematical Society Lecture Note Series, vol. 121. Cambridge University Press, Cambridge (1986)
37. Tomkinson, M.J.: Groups covered by finitely many cosets or subgroups. Comm. Algebra **15**, 845–859 (1987)

On the Fine Grained Complexity of Finite Automata Non-emptiness of Intersection

Mateus de Oliveira Oliveira[1] and Michael Wehar[2]([⊠])

[1] University of Bergen, Bergen, Norway
mateus.oliveira@uib.no
[2] Swarthmore College, Swarthmore, PA, USA
mwehar1@swarthmore.edu

Abstract. We study the fine grained complexity of the DFA non-emptiness of intersection problem parameterized by the number k of input automata (k-DFA-NEI). More specifically, we are given a list $\langle \mathcal{A}_1, ..., \mathcal{A}_k \rangle$ of DFA's over a common alphabet Σ, and the goal is to determine whether $\bigcap_{i=1}^{k} \mathcal{L}(\mathcal{A}_i) \neq \emptyset$. This problem can be solved in time $O(n^k)$ by applying the classic Rabin-Scott product construction. In this work, we show that the existence of algorithms solving k-DFA-NEI in time slightly faster than $O(n^k)$ would imply the existence of deterministic sub-exponential time algorithms for the simulation of nondeterministic linear space bounded computations. This consequence strengthens the existing conditional lower bounds for k-DFA-NEI and implies new non-uniform circuit lower bounds.

Keywords: Finite automata · Intersection non-emptiness · Fine grained complexity · Parameterized complexity

1 Introduction

1.1 History

In the DFA non-emptiness of intersection problem (DFA-NEI), the input consists of a list $\langle \mathcal{A}_1, ..., \mathcal{A}_k \rangle$ of DFA's over a common alphabet Σ, and the goal is to determine whether the intersection of the languages $\mathcal{L}(\mathcal{A}_1), ..., \mathcal{L}(\mathcal{A}_k)$ is non-empty. When no restriction is imposed on the input, DFA-NEI is a PSPACE-complete problem [18]. Nevertheless, the classic Rabin-Scott product construction for finite automata yields a simple algorithm that solves DFA-NEI in time $O(n^k)$ where n is the number of states and k is the number of input automata. Therefore, for a fixed number of input automata, the problem can be solved in polynomial time.

In this work, we study the fine grained complexity of DFA-NEI parameterized by the number of input automata k. For clarity, we refer to this parameterized version as k-DFA-NEI. Interestingly, Rabin and Scott's six-decades-old

Mateus de Oliveira Oliveira acknowledges support from the Bergen Research Foundation and from the Research Council of Norway (project number 288761).

N. Jonoska and D. Savchuk (Eds.): DLT 2020, LNCS 12086, pp. 69–82, 2020.
https://doi.org/10.1007/978-3-030-48516-0_6

time $O(n^k)$ algorithm for k-DFA-NEI remains unimproved, and in particular, time $O(n^2)$ is still the best we can get for deciding non-emptiness of intersection for two DFA's.

Kasai and Iwata [17] are believed to be the first to provide conditional lower bounds for k-DFA-NEI. They showed that k-DFA-NEI requires deterministic time $\Omega(n^{(k-2)/2})$ under the conjecture that NSPACE$[k \cdot \log n] \not\subseteq$ DTIME$[n^{k-\varepsilon}]$ for all $\varepsilon > 0$.

Almost two decades later, Karakostas, Lipton, and Viglas showed that faster algorithms for certain variants of k-DFA-NEI would imply both faster algorithms for certain NP-hard problems and new complexity class separations [16]. In particular, they showed that an algorithm solving k-DFA-NEI in time $n^{o(k)}$ would have two consequences. First, this would imply that the well studied SUB-SET SUM problem can be solved in time $O(2^{\varepsilon \cdot n})$ for every $\varepsilon > 0$. Second, this would imply that NTIME$[n] \subseteq$ DTIME$[2^{o(n)}]$. They also showed some remarkable consequences of the existence of algorithms solving k-DFA-NEI in time $s \cdot r^{o(k)}$ where s is the number of states in the largest input automaton and r is the number of states in the second largest input automaton. In particular, such an algorithm would imply that NSPACE$[O(\log s)] \subset$ DTIME$[s^{1+\epsilon}]$ for all $\epsilon > 0$, which would further imply that $P \neq NL$. Additionally, by padding, we would also have NSPACE$[s] \subseteq$ DTIME$[2^{o(s)}]$. It is worth noting that this last result strongly requires that the runtime has only a marginal dependence on the size s of the largest automaton. Further, this last result is in a similar spirit as conditional lower bounds for weighted satisfiability problems from [8–10].

It was shown by Fernau and Krebs [12], and independently in [30], that an algorithm solving k-DFA-NEI in time $n^{o(k)}$ would contradict the celebrated exponential time hypothesis (ETH). Using a refinement of the proof technique introduced in [16], it was shown in [29,30] that if k-DFA-NEI can be solved in time $n^{o(k)}$, then $P \neq NL$. Additional results on the parameterized complexity of non-emptiness of intersection for DFA's are presented in [17,19,26] and results on the fine grained complexity of non-emptiness of intersection specifically for two and three DFA's are presented in [23].

1.2   Our Results

Finer Simulations for Nondeterministic Linear Space. Our first result (Theorem 1) provides a finer reduction from the problem of simulating a nondeterministic space bounded Turing machine to k-DFA-NEI. The following two corollaries of Theorem 1 fill in some gaps in the literature related to non-emptiness of intersection. In this work, NSPACE$[n]$ denotes the class of functions computable by 2-tape Turing Machines over a binary alphabet using at most n bits on its work tape.

(Corollary 1.1) If we can solve k-DFA-NEI in time $n^{o(k)}$, then NSPACE$[n] \subseteq$ DTIME$[2^{o(n)}]$ [28].[1]

---

[1] This work was not formally published.

(Corollary 1.2) If there exists $k \geq 2$ and $\varepsilon > 0$ such that k-DFA-NEI can be solved in time $O(n^{k-\varepsilon})$, then $\mathrm{NSPACE}[n + o(n)] \subseteq \mathrm{DTIME}[2^{(1-\delta)n}]$ for some $\delta > 0$ [30].

As mentioned in the first part of the introduction, the conclusion that $\mathrm{NSPACE}[n] \subseteq \mathrm{DTIME}[2^{o(n)}]$ can be obtained from the results in [16] under the assumption that there exists an algorithm for k-DFA-NEI running in time $s \cdot r^{o(k)}$ where s is the size of the largest automaton and r is the size of the second largest automaton. Corollary 1.1 relaxes this assumption to the existence of an algorithm running in time $n^{o(k)}$, with no regard to the way in which the sizes of the input automata compare with each other. We observe that the same assumption as ours was shown in [16] to imply that $\mathrm{NTIME}[n] \subseteq \mathrm{DTIME}[2^{o(n)}]$. Therefore, we improve the consequence in [16] from $\mathrm{NTIME}[n] \subseteq \mathrm{DTIME}[2^{o(n)}]$ to $\mathrm{NSPACE}[n] \subseteq \mathrm{DTIME}[2^{o(n)}]$.

Corollary 1.2 states that for each $k > 1$, any additive constant improvement on the running time of the Rabin-Scott algorithm for k-DFA-NEI would imply the existence of faster than state-of-the art algorithms for the simulation of nondeterministic linear space bounded computations. In particular, an algorithm solving non-emptiness of intersection for two DFA's in time $O(n^{2-\varepsilon})$, for some $\varepsilon > 0$, would imply that $\mathrm{NSPACE}[n + o(n)] \subseteq \mathrm{DTIME}[2^{(1-\delta)n}]$ for some $\delta > 0$.

Contradicting Stronger Versions of ETH and SETH. In the satisfiability problem for Boolean formulas (SAT), we are given a Boolean formula. The goal is to determine if there exists an assignment that satisfies the formula. It is common to restrict the inputs for SAT to formulas in conjunctive normal form (CNF-SAT). Further, it is common to restrict the inputs for SAT to formulas in conjunctive normal form with clause width at most k (k-CNF-SAT) for some fixed number k.

The *Exponential Time Hypothesis* (ETH) asserts that for some $\varepsilon > 0$, 3-CNF-SAT cannot be solved in time $(1 + \varepsilon)^n$ [14]. The *strong exponential time hypothesis* (SETH) asserts that for every $\varepsilon > 0$, there is a large enough integer k such that k-CNF-SAT cannot be solved in time $(2-\varepsilon)^n$ [7,14,15]. ETH has been used to rule out the existence of subexponential algorithms for many decision problems [14], parameterized problems [8,20], approximation problems [22], and counting problems [11]. On the other hand, SETH has been useful in establishing tight lower bounds for many problems in P such as EDIT DISTANCE [3], k-DOMINATING SET [24], k-DFA-NEI [30], and many other problems [2,27,33].

Our next results state that slightly faster algorithms for k-DFA-NEI would contradict much stronger versions of ETH and SETH. First, we show that if there exists $k \geq 2$ such that k-DFA-NEI can be solved in time $O(n^{k-\varepsilon})$ for some $\varepsilon > 0$, then satisfiability for n-variable Boolean formulas of size $2^{o(n)}$ can be solved in time $O(2^{(1-\delta)n})$ for some $\delta > 0$ (Corollary 2). The inexistence of such fast algorithms for satisfiability for n-variable Boolean formulas of sub-exponential size is a safer assumption than SETH. Going further, we show that if k-DFA-NEI can be solved in time $n^{o(k)}$, then satisfiability for n-input fan-in-2 Boolean circuits of depth $O(n)$ and size $2^{o(n)}$ can be solved in time $2^{o(n)}$

(Corollary 3). We note that this consequence is stronger than the existence of algorithms solving CNF-SAT in sub-exponential time. Indeed, CNF formulas of polynomial size are a very weak model of computation, which are unable, for instance, to compute the parity of their input bits [13]. On the other hand, circuits of linear depth can already simulate complicated cryptographic primitives. Therefore, the inexistence of satisfiability algorithms for such circuits is a safer assumption than ETH.

Non Uniform Circuit Lower Bounds. Finally, from the results mentioned above together with results obtained within the context of Williams' *algorithms versus lower bounds* framework [1,31,32] (as well as [4]), we infer that faster algorithms for k-DFA-NEI would imply non-uniform circuit lower bounds that are sharper than what is currently known. In particular, an algorithm running in time $n^{o(k)}$ for k-DFA-NEI would imply that there are problems in E^{NP} that cannot be solved by non-uniform fan-in-2 Boolean circuits of linear depth and sub-exponential size (Corollary 4). We note that currently it is still open whether every problem in E^{NP} can be solved by non-uniform fan-in-2 Boolean circuits of linear size. Additionally, we show that an algorithm running in time $O(n^{2-\varepsilon})$ for 2-DFA-NEI would imply that there are problems in E^{NP} that cannot be solved by non-uniform Boolean formulas of sub-exponential size (Corollary 5).

Further, we have that even polylogarithmic improvements for the running time of algorithms solving 2-DFA-NEI would imply interesting lower bounds. More specifically, if 2-DFA-NEI can be solved in time $O(n^2/\log^c n)$ for every $c > 0$, then there are functions that can be computed in NTIME$[2^{O(n)}]$ but not by non-uniform NC^1 circuits.

Analogous conditional non-uniform circuit lower bounds have been obtained in [1] under the assumptions that the EDIT DISTANCE problem can be computed in time $O(n^{2-\varepsilon})$ for some $\varepsilon > 0$ and in time $O(n^2/\log^c n)$ for every $c \geq 1$. It is worth noting that Theorem 5 which establishes conditional lower bounds for fan-in-2 Boolean circuits of linear depth and sub-exponential size is not explicitly stated in [1] and no parallel to the associated conditional lower bound for k-DFA-NEI is given for EDIT DISTANCE.

2   Reducing Acceptance in NSPACE[n] to DFA-NEI

In this section we provide a reduction from the problem of simulating 2-tape Turing machines to DFA-NEI. For any k, the reduction in Theorem 1 outputs k DFA's each with at most $O(m^2 \cdot n \cdot \sigma^{1+c} \cdot 2^{\frac{\sigma}{k}})$ states where m denotes the number of states in the Turing machine, n denotes the input string length, σ denotes the amount of space on the binary work tape, and c denotes the maximum number of occurrences of a special delimiter symbol # that can simultaneously appear on the work tape during the computation. The parameter c is a constant associated with the Turing machine and is independent of the parameters n and σ.

2-tape Turing Machines: A 2-tape Turing machine with binary alphabet is a machine with a two-way read-only input tape and a two-way binary work tape.

More formally, it is a tuple $M = (Q, \{0,1\}, c, q_0, F, \delta)$ where Q is a set of states, c is the maximum number of occurrences of special delimiter symbol $\#$, $q_0 \in Q$ is an initial state, F is a set of final states, and

$$\delta : Q \times (\{0,1\} \cup \{\#\})^2 \to \mathcal{P}(Q \times \{-1,0,1\}^2 \times (\{0,1\} \cup \{\#\}))$$

is a partial transition function that assigns to each triple

$$(q, b_1, b_2) \in Q \times (\{0,1\} \cup \{\#\})^2,$$

a set of tuples

$$\delta(q, b_1, b_2) \subseteq Q \times \{-1,0,1\}^2 \times (\{0,1\} \cup \{\#\}).$$

We say that a tuple

$$(q, d, d', w) \in Q \times \{-1,0,1\}^2 \times (\{0,1\} \cup \{\#\})$$

is an instruction that sets the machine to state q, moves the input head from position p to position $p + d$, moves the work head from position p' to position $p' + d'$, and writes symbol w at position p' on the work tape. The transition function δ specifies that if the machine M is currently at state q, reading symbol b_1 on the input tape and symbol b_2 on the work tape, then the next instruction of the machine must be an element of the set $\delta(q, b_1, b_2)$.

Configurations: A *space-σ configuration* for M on input $x \in \{0,1\}^*$ is a tuple

$$(q, h, h', y) \in Q \times [|x|] \times [\sigma] \times (\{0,1\} \cup \{\#\})^\sigma$$

where intuitively, $q \in Q$ is the current state of M, $h \in [|x|]$ is the position of M's input tape head, $h' \in [\sigma]$ is the position of M's work tape head, and $y \in (\{0,1\} \cup \{\#\})^\sigma$ is the binary string (containing at most c special delimiter symbols) corresponding to the first σ symbols on the work tape of M.

Configuration Sequences: A *space-σ configuration sequence* for M on input $x \in \{0,1\}^*$ is a sequence of the form

$$S \equiv (q_0, h_0, h_0', y_0) \xrightarrow{(q_1, d_1, d_1', r_1, r_1', w_1)} (q_1, h_1, h_1', y_1)$$

$$\xrightarrow{(q_2, d_2, d_2', r_2, r_2', w_2)} (q_2, h_2, h_2', y_2)$$

$$\dots$$

$$\xrightarrow{(q_k, d_k, d_k', r_k, r_k', w_k)} (q_k, h_k, h_k', y_k)$$

satisfying the following conditions.

1. For each $i \in \{0, 1, ..., k\}$, (q_i, h_i, h_i', y_i) is a space-σ configuration for M on x.
2. q_0 is the initial state of M, $y_0 = 0^\sigma$, meaning that the work tape is initialized with zeros, and $h_0 = h_0' = 1$, meaning that the input tape head and work tape head are in the first position of their respective tapes.

3. For each $i \in \{1, ..., k\}$, $(q_i, d_i, d'_i, w_i) \in \delta(q_{i-1}, x[h_{i-1}], y_{i-1}[h'_{i-1}])$, meaning the state of the machine at time i, the directions taken by both heads at time i, and the symbol written on the work tape at time i are compatible with the transition function δ, and depend only on the state at time $i-1$ and on the symbols that are read at time $i-1$.

4. For each $i \in \{1, ..., k\}$, $h_i = h_{i-1} + d_i$, $h'_i = h'_{i-1} + d'_i$, $r_i = x[h_{i-1}]$, $r'_i = y_{i-1}[h'_{i-1}]$, and y_i is obtained from y_{i-1} by substituting w_i for the symbol $y_{i-1}[h'_{i-1}]$, and by leaving all other symbols untouched. Intuitively, this means that the configuration at time i is obtained from the configuration at time $i-1$ by the application of the transition (q_i, d_i, d'_i, w_i).

5. For each $i \in \{0, 1, ..., k\}$, y_i contains at most c occurrences of the special delimiter symbol $\#$.

We say that the sequence

$$I \equiv (q_1, d_1, d'_1, r_1, r'_1, w_1)(q_2, d_2, d'_2, r_2, r'_2, w_2)...(q_k, d_k, d'_k, r_k, r'_k, w_k)$$

that induces a configuration sequence S as above is a *space-σ instruction sequence* for M on input x. We say that I is accepting if $q_k \in F$.

Remark 1. As suggested in [17], the technique provided in the proof of Proposition 3 from [25] can be applied to remove the special delimiter symbol from the work tape by increasing the Turing machine's state and space complexities. However, without a formal proof in the literature of the stated result from [17] and because it is not required in our work, we decided to refrain from using it.

Theorem 1. *Let a nondeterministic m-state 2-tape Turing machine M with binary tape alphabet (other than at most c occurrences of symbol $\#$) and an input string x of length n be given. If M uses at most σ symbols on the work tape, then for every k, we can efficiently compute k DFA's $\langle \mathcal{A}_1, \mathcal{A}_2, ..., \mathcal{A}_k \rangle$ each with a binary alphabet and $O(m^2 \cdot n \cdot \sigma^{1+c} \cdot 2^{\frac{\sigma}{k}})$ states such that M accepts x if and only if $\bigcap_{i=1}^k \mathcal{L}(\mathcal{A}_i) \neq \emptyset$.*

Proof. The Turing machine M accepts x if and only if there exists an accepting space-σ instruction sequence for M on x. We build k DFA's that read in a binary string and collectively determine whether the string encodes an accepting space-σ instruction sequence for M on x.

Consider splitting the work tape of M into k equal sized blocks each consisting of $\frac{\sigma}{k}$ work tape cells. A block-i space-σ configuration for M on input x consists of the state, input tape head, work tape head, the contents of the work tape from position $lbound_i := (i-1) \cdot \frac{\sigma}{k} + 1$ to position $rbound_i := i \cdot \frac{\sigma}{k}$, and all c positions of the occurrences of special delimiter symbol $\#$. We construct k DFA's $\langle \mathcal{A}_1, \mathcal{A}_2, ..., \mathcal{A}_k \rangle$ where each DFA $\mathcal{A}_i$ keeps track of the current block-i space-σ configuration for M on input x. The DFA's read in space-σ instructions one at a time and transition accordingly where each instruction is encoded as a unique bit string of length $O(\log(m))$.

The start state of DFA $\mathcal{A}_i$ represents the block-i space-σ configuration $(q_0, 1, 1, 0^{\frac{\sigma}{k}})$ where q_0 is the start state of M. Further, a state representing a block-i space-σ configuration $(q_j, h_j, h'_j, contents_j)$ is accepting if q_j is a final state of M. Suppose that the DFA $\mathcal{A}_i$ is currently at a state representing a block-i space-σ configuration $(q_j, h_j, h'_j, contents_j)$ and reads in a space-σ instruction (q, d, d', r, r', w). The DFA $\mathcal{A}_i$ transitions to a state representing a block-i space-σ configuration $(q_{j+1}, h_{j+1}, h'_{j+1}, contents_{j+1})$ if:

1. $(q, d, d', w) \in \delta(q_j, r, r')$ and $q = q_{j+1}$
2. $h_{j+1} = h_j + d$ and $h'_{j+1} = h'_j + d'$
3. $1 \le h_j, h_{j+1} \le n$ and $1 \le h'_j, h'_{j+1} \le \sigma$
4. $r = x[h_j]$
5. if $lbound_i \le h'_j \le rbound_i$, then $r' = contents_j[h'_j - lbound_i + 1]$ and
   $w = contents_{j+1}[h'_j - lbound_i + 1]$

Collectively the DFA's determine whether the input string encodes an accepting space-σ instruction sequence for M on x. Therefore, the Turing machine M accepts x if and only if there exists an accepting space-σ instruction sequence for M on x if and only if $\bigcap_{i=1}^{k} \mathcal{L}(\mathcal{A}_i) \ne \emptyset$. Further, the DFA's each have at most $O(m^2 \cdot n \cdot \sigma^{1+c} \cdot 2^{\frac{\sigma}{k}})$ states because there are $O(m)$ space-σ instructions and $O(m \cdot n \cdot \sigma^{1+c} \cdot 2^{\frac{\sigma}{k}})$ block-i space-σ configurations. $\square$

Remark 2. The preceding simulation is sufficient for our purposes. However, we suggest that it could be refined by having only one DFA keep track of the Turing machine's tape heads. When $\sigma = O(\log(n))$, such a refinement could be used to obtain a tighter connection between k-DFA-NEI and nondeterministic logspace.

Corollary 1. *We obtain the following directly from the preceding theorem:*

1. *If we can solve k-DFA-NEI in time $n^{o(k)}$, then $\mathrm{NSPACE}[n] \subseteq \mathrm{DTIME}[2^{o(n)}]$.*
2. *If there exists $k \ge 2$ and $\varepsilon > 0$ such that k-DFA-NEI can be solved in time $O(n^{k-\varepsilon})$, then $\mathrm{NSPACE}[n + o(n)] \subseteq \mathrm{DTIME}[2^{(1-\delta)n}]$ for some $\delta > 0$.*

3 Non-emptiness of Intersection and Conditional Lower Bounds

In this section we apply results obtained in Theorem 1 to show that even a slight improvement in running time of the classic algorithm for non-emptiness of intersection for finite automata would yield faster than state of the art algorithms for satisfiability for Boolean formulas and Boolean circuits. Therefore, this result implies that the impossibility of obtaining better algorithms for non-emptiness of intersection for k finite automata can be based on assumptions that are safer than the exponential time hypothesis (ETH). An analogous result is proven with respect to non-emptiness of intersection for a constant number of finite automata (say two). We will show that the existence of algorithms that are faster than time $O(n^{2-\varepsilon})$ for non-emptiness of intersection for 2 DFA's would contradict assumptions that are safer than the strong exponential time hypothesis (SETH). We note that the endeavour of basing lower bounds for algorithms in P on assumptions that are safer than ETH or SETH has been pursued before [1]. In this work, we obtain lower bounds using assumptions like those used in [1], with the advantage that our reductions are simpler. Therefore, we believe that the techniques employed here provide a cleaner framework that can potentially be used to strengthen the analysis of the fine grained complexity of other algorithmic problems in P.

Finally, by applying Williams' *algorithms versus lower bounds* framework, we are able to show that faster algorithms for non-emptiness of intersection for finite automata would also imply non-uniform circuit lower bounds that are much better than those that are currently known.

3.1   Satisfiability for Boolean Formulas

Lemma 1. *Satisfiability for n-variable Boolean formulas of size s is solvable by a nondeterministic 2-tape Turing machine with binary alphabet using at most $n + O(\log(s))$ bits and a fixed number of delimiter symbol $\#$ occurrences on the work tape.*[2]

Proof. The machine uses n tape cells to guess an assignment $x \in \{0,1\}^n$ to the input variables. Subsequently, using $O(\log s)$ work tape cells, the machine evaluates the Boolean formula from the input tape on the guessed assignment from the work tape. This evaluation problem is referred to as the Boolean formula value problem (BFVP) and has been shown to be solvable in space $O(\log s)$ on formulas of size s in [6,21]. Storing both the assignment and the formula evaluation on the same work tape, it will be necessary to use a fixed number of occurrences of the delimiter symbol $\#$ as left/right tape markers, a delimiter between assignment and formula evaluation, and markers for remembering the position in the formula evaluation and the assignment when the work tape head moves from formula evaluation to assignment or vice versa.   □

By combining Theorem 1 with Lemma 1, we obtain the following.

Theorem 2. *If there exists $k \geq 2$ and $\varepsilon > 0$ such that k-DFA-NEI can be solved in time $O(n^{k-\varepsilon})$, then satisfiability for n-variable Boolean formulas of size s is solvable in time $\mathrm{poly}(s) \cdot 2^{n(1-\delta)}$ for some $\delta > 0$.*

Proof. Suppose that there exists $k \geq 2$ and $\varepsilon > 0$ such that k-DFA-NEI can be solved in time $O(n^{k-\varepsilon})$. Let a n-variable Boolean formula ϕ of size s be given. By Theorem 1 and Lemma 1, we can reduce satisfiability for ϕ to non-emptiness of intersection for k DFA's each with $\mathrm{poly}(s) \cdot 2^{\frac{n}{k}}$ states. Therefore, by assumption, we can determine whether ϕ has a satisfying assignment in time $s^{O(k)} \cdot 2^{\frac{k-\varepsilon}{k} \cdot n}$. It follows that satisfiability for n-variable Boolean formulas of size s is solvable in time $\mathrm{poly}(s) \cdot 2^{n(1-\delta)}$ for some $\delta > 0$.   □

---

[2] In addition, both QBF and satisfiability for nondeterministic branching programs are solvable by nondeterministic 2-tape Turing machines with binary alphabet using at most $n + O(\log(s))$ bits and a fixed number of delimiter symbol $\#$ occurrences.

It was shown in [30] that if there exists some $k \geq 2$ and $\varepsilon > 0$ such that k-DFA-NEI can be solved in time $O(n^{k-\varepsilon})$, then SETH is false. The following corollary improves this result by showing a much stronger consequence. The corollary follows directly from Theorem 2.

Corollary 2. *If there exists $k \geq 2$ and $\varepsilon > 0$ such that k-DFA-NEI can be solved in time $O(n^{k-\varepsilon})$, then satisfiability for n-variable Boolean formulas of size $2^{o(n)}$ can be solved in time $O(2^{n(1-\delta)})$ for some $\delta > 0$.*

Note that while CNF's of bounded width and polynomial size are a very weak computational model, Boolean formulas of sub-exponential size can already simulate any circuit in the class NC. Therefore, the consequence of Corollary 2 would contradict the NC-SETH hypothesis, a more robust version of SETH which states that satisfiability for circuits of polynomial size and polylogarithmic depth cannot be solved in time $O(2^{n(1-\delta)})$ for any $\delta > 0$ [1]. In the next subsection we show that the existence of an algorithm running in time $n^{o(k)}$ for k-DFA-NEI would imply faster satisfiability algorithms for even larger classes of circuits.

3.2   Satisfiability for Boolean Circuits

In the CIRCUIT VALUE problem (CV), we are given a n-input fan-in-2 Boolean circuit C and a string $x \in \{0,1\}^n$. The goal is to determine whether the circuit $C(x)$, obtained by initializing the input variables of C according to x, evaluates to 1. Let the size of C denote the number of gates of C. The next lemma, which is a classic result in complexity theory [5], states that the circuit value problem for circuits of depth d and size s can be solved in space $O(d) + O(\log s)$ on a 2-tape Turing machine.

Lemma 2 (Borodin [5]). *There is a deterministic 2-tape Turing machine M with binary alphabet, that takes as input a pair $\langle C, x \rangle$ where x is a string in $\{0,1\}^n$ and C is a n-input fan-in-2 Boolean circuit of depth d and size s, and determines, using at most $O(d) + O(\log s)$ work tape cells, whether $C(x)$ evaluates to 1.*

In the satisfiability problem for Boolean circuits, we are given a n-input fan-in-2 Boolean circuit C. The goal is to determine whether there exists a string $x \in \{0,1\}^n$ such that $C(x)$ evaluates to 1. As a consequence of Lemma 2, we have that satisfiability for Boolean circuits can be decided by a nondeterministic 2-tape Turing machine using at most $n + O(d) + O(\log s)$ tape cells.

Lemma 3. *There is a nondeterministic 2-tape Turing machine M with binary alphabet and a fixed number of delimiter symbol $\#$ occurrences, that takes as input a n-input fan-in-2 Boolean circuit C of depth d and size s, and determines, using at most $n + O(d) + O(\log s)$ tape cells, whether C is satisfiable.*

By combining Theorem 1 with Lemma 3, we obtain the following.

Theorem 3. *If k-DFA-NEI can be solved in time $O(n^{f(k)})$, then satisfiability for n-input fan-in-2 Boolean circuits of depth d and size s can be solved in time $s^{O(f(k))} \cdot 2^{\frac{f(k)}{k} \cdot (n + O(d))}$ where k is allowed to depend on n, d, and s.*

Proof. Suppose that k-DFA-NEI can be solved in time $O(n^{f(k)})$. Let a n-input fan-in-2 Boolean circuit C of depth d and size s be given. By Theorem 1 and Lemma 3, for any k, we can reduce satisfiability for C to non-emptiness of intersection for k DFA's each with $\text{poly}(s) \cdot 2^{\frac{n + O(d)}{k}}$ states. Therefore, by assumption, we can determine whether C has a satisfying assignment in time $s^{O(f(k))} \cdot 2^{\frac{f(k)}{k} \cdot (n + O(d))}$. It follows that satisfiability for n-input fan-in-2 Boolean circuits of depth d and size s is solvable in the desired time.   □

It was shown in [12,30] that if k-DFA-NEI can be solved in time $n^{o(k)}$, then ETH is false. The following corollary improves this result by showing a much stronger consequence. The corollary follows directly from Theorem 3.

Corollary 3. *If k-DFA-NEI can be solved in time $n^{o(k)}$, then satisfiability for fan-in-2 Boolean circuits of depth $O(n)$ and size $2^{o(n)}$ can be solved in time $2^{o(n)}$.*

Note that the consequence in the preceding corollary is stronger than ETH because circuits of linear depth and sub-exponential size are a stronger computational model than formulas in conjunctive normal form.

3.3   Circuit Lower Bounds

Corollaries 2 and 3 lead us to the following questions.

- What are the barriers (beyond ETH) to solving satisfiability for fan-in-2 Boolean circuits of depth $O(n)$ and size $2^{o(n)}$ more efficiently?
- What are the barriers (beyond SETH) to solving satisfiability for Boolean formulas of size $2^{o(n)}$ more efficiently?
- What are the barriers to solving satisfiability only slightly faster for Boolean formulas of polynomial size?

Below, we investigate the preceding questions and reference recent works [1,4,31,32] that connect satisfiability problems to non-uniform circuit complexity lower bounds. From these connections, we observe that faster algorithms for k-DFA-NEI would imply new non-uniform circuit complexity lower bounds.

Barriers Beyond ETH. The following is a slightly modified restatement of a technical theorem from [1] (with related results in [4,31,32]) that connects circuit satisfiability to non-uniform circuit complexity lower bounds for E^{NP}. Recall that E^{NP} is the class of functions that can be computed by Turing machines that operate in time $2^{O(n)}$ with the help of an NP oracle. Note that the strings that are passed to each call of the oracle may have size up to $2^{O(n)}$.

Theorem 4 (Theorem 8 in [1]). *Let $S(n)$ be a time constructible and monotone non-decreasing function such that $n \leq S(n) \leq 2^{o(n)}$. Let $\mathcal{C}$ be a class of circuits. Suppose there is a SAT algorithm for n-input circuits which are arbitrary functions of three $O(S(n))$-size circuits from $\mathcal{C}$, that runs in time $O(2^n/(n^{10} \cdot S(n)))$. Then E^{NP} does not have $S(n)$-size circuits from $\mathcal{C}$.*

Notice that if we take three circuits of linear depth and sub-exponential size and combine their outputs using a constant number of additional gates, then the resulting circuit still has linear depth and sub-exponential size. Therefore, a satisfiability algorithm running in time $2^{o(n)}$ for n-input fan-in-2 Boolean circuits of linear depth and sub-exponential size would imply that there are functions in E^{NP} that cannot be computed by such circuits. This would be a significant consequence in complexity theory since to date it is not even known whether all functions in E^{NP} can be computed by non-uniform circuits of linear size. In view of our discussion, Theorem 5 directly follows from Theorem 4, but is not explicitly stated in [1].

Theorem 5. *If satisfiability for n-input fan-in-2 Boolean circuits of depth $O(n)$ and size $2^{o(n)}$ is solvable in time $O(2^{(1-\delta)n})$ for some $\delta > 0$, then E^{NP} does not have non-uniform fan-in-2 Boolean circuits of $O(n)$ depth and $2^{o(n)}$ size.*

By combining the preceding theorem with Corollary 3, we obtain the following.

Corollary 4. *If k-DFA-NEI can be solved in time $n^{o(k)}$, then E^{NP} does not have non-uniform fan-in-2 Boolean circuits of $O(n)$ depth and $2^{o(n)}$ size.*

Barriers Beyond SETH. Next, we look at another known result that connects formula satisfiability to non-uniform formula complexity lower bounds for E^{NP}. The following theorem is similar to Corollary 1.1 in [1] and directly follows from Theorem 4.

Theorem 6. *If satisfiability for n-variable Boolean formulas of size $2^{o(n)}$ is solvable in time $O(2^{(1-\delta)n})$ for some $\delta > 0$, then E^{NP} does not have non-uniform Boolean formulas of size $2^{o(n)}$.*

By combining the preceding theorem with Corollary 2, we obtain the following.

Corollary 5. *If there exists $k \geq 2$ and $\varepsilon > 0$ such that k-DFA-NEI can be solved in time $O(n^{k-\varepsilon})$, then E^{NP} does not have non-uniform Boolean formulas of size $2^{o(n)}$.*

Barriers to Slightly Faster Algorithms. Finally, we investigate the possible consequences of polylogarithmic improvements to the running time of algorithms for k-DFA-NEI, and in particular for 2-DFA-NEI. The following is a restatement of Theorem 7 in [1] (related to Theorem 1.3 in [31]).

Theorem 7 (Theorem 7 in [1]). *Suppose that there is a satisfiability algorithm for bounded fan-in formulas of size n^r running in time $O(2^n/n^r)$, for each constant $r > 0$. Then $\mathrm{NTIME}[2^{O(n)}]$ is not contained in non-uniform NC^1.*

By combining the preceding theorem with Theorem 1 and Lemma 1, we obtain the following.

Corollary 6. *If k-DFA-NEI can be solved in time $O(n^k/(\log n)^c)$ for every $c > 0$, then $\mathrm{NTIME}[2^{O(n)}]$ is not contained in non-uniform NC^1.*

Proof. Suppose that there exists $k \geq 2$ such that k-DFA-NEI can be solved in time $O(n^k/(\log n)^c)$ for every $c > 0$. By combining Theorem 1 and Lemma 1, it follows that for all $r > 0$, satisfiability for n-variable Boolean formulas of size n^r can be reduced to intersecting k DFA's with at most $poly(n) \cdot 2^{\frac{n}{k}}$ states where k and r are treated as constants. Therefore, satisfiability for n-variable Boolean formulas of size n^r can be solved in time

$$\frac{(poly(n) \cdot 2^{\frac{n}{k}})^k}{\log^c(poly(n) \cdot 2^{\frac{n}{k}})} \leq \frac{poly(n) \cdot 2^n}{(O(\log n) + \frac{n}{k})^c} \leq O(\frac{n^d \cdot 2^n}{n^c})$$

for all $c > 0$ and some constant d that is independent of c. If we set $c = d \cdot r$, then we have that satisfiability for Boolean formulas of size n^r can be solved in time $O(2^n/n^r)$. It follows that satisfiability for Boolean formulas of size n^r can be solved in time $O(2^n/n^r)$ for all $r > 0$. Moreover, by Theorem 7, it follows that $\mathrm{NTIME}[2^{O(n)}]$ does not have non-uniform NC^1 circuits. $\square$

Notice that if we set $k = 2$ in the preceding corollary, then it follows that if 2-DFA-NEI can be solved in time $O(n^2/(\log n)^c)$ for every $c > 0$, then $\mathrm{NTIME}[2^{O(n)}]$ is not contained in non-uniform NC^1.

4 Conclusion

We analyzed the fine grained complexity of the non-emptiness of intersection problem parameterized by the number of input DFA's (k-DFA-NEI). Despite the fact that this problem has been studied for at least six decades, the fastest known algorithm for k-DFA-NEI is still the $O(n^k)$ time algorithm obtained by a direct application of the classic Rabin-Scott product construction.

The lack of progress in the task of developing a faster algorithm for k-DFA-NEI motivated the search for evidence supporting the possibility that substantially faster algorithms for this problem do not exist. In this work, we have simplified and strengthened several previous conditional lower bounds for k-DFA-NEI under a unified perspective. In particular, we have shown that if k-DFA-NEI can be solved in time $n^{o(k)}$ then $\mathrm{NSPACE}[n] \subseteq \mathrm{DTIME}[2^{o(n)}]$. Additionally, we have shown that solving non-emptiness of intersection for two DFA's in time $O(n^{2-\varepsilon})$ for some $\varepsilon > 0$ would imply that $\mathrm{NSPACE}[n + o(n)] \subseteq \mathrm{DTIME}[2^{(1-\delta)n}]$ for some $\delta > 0$. Further, we have unveiled several connections between k-DFA-NEI and non-uniform circuit complexity theory. In particular, we have shown that even improving non-emptiness of intersection for two DFA's by a $\log^c n$ factor for every $c > 0$ would imply non-uniform NC^1 circuit lower bounds.

References

1. Abboud, A., Hansen, T.D., Williams, V.V., Williams, R.: Simulating branching programs with edit distance and friends: or: a polylog shaved is a lower bound made. In: Proceedings of the 48th Annual ACM Symposium on Theory of Computing (STOC 2016), pp. 375–388. ACM (2016)
2. Abboud, A., Williams, V.V.: Popular conjectures imply strong lower bounds for dynamic problems. In: Proceedings of the 55th Annual Symposium on Foundations of Computer Science (FOCS 2014), pp. 434–443. IEEE (2014)
3. Backurs, A., Indyk, P.: Edit distance cannot be computed in strongly subquadratic time (unless SETH is false). In: Proceedigs of the 47th Annual ACM Symposium on Theory of Computing (STOC 2015), pp. 51–58. ACM (2015)
4. Ben-Sasson, E., Viola, E.: Short PCPs with projection queries. In: Esparza, J., Fraigniaud, P., Husfeldt, T., Koutsoupias, E. (eds.) ICALP 2014. LNCS, vol. 8572, pp. 163–173. Springer, Heidelberg (2014). https://doi.org/10.1007/978-3-662-43948-7_14
5. Borodin, A.: On relating time and space to size and depth. SIAM J. Comput. **6**(4), 733–744 (1977)
6. Buss, S.R.: Algorithms for boolean formula evaluation and for tree contraction. In: Arithmetic, Proof Theory and Computational Complexity, pp. 96–115. Oxford University Press (1993)
7. Calabro, C., Impagliazzo, R., Paturi, R.: The complexity of satisfiability of small depth circuits. In: Chen, J., Fomin, F.V. (eds.) IWPEC 2009. LNCS, vol. 5917, pp. 75–85. Springer, Heidelberg (2009). https://doi.org/10.1007/978-3-642-11269-0_6
8. Chen, J., et al.: Tight lower bounds for certain parameterized NP-hard problems. Inf. Comput. **201**(2), 216–231 (2005)
9. Chen, J., Huang, X., Kanj, I.A., Xia, G.: W-Hardness under linear FPT-reductions: structural properties and further applications. In: Wang, L. (ed.) COCOON 2005. LNCS, vol. 3595, pp. 975–984. Springer, Heidelberg (2005). https://doi.org/10.1007/11533719_98
10. Chen, J., Huang, X., Kanj, I.A., Xia, G.: Strong computational lower bounds via parameterized complexity. J. Comput. Syst. Sci. **72**(8), 1346–1367 (2006)
11. Dell, H., Husfeldt, T., Marx, D., Taslaman, N., Wahlén, M.: Exponential time complexity of the permanent and the Tutte polynomial. ACM Trans. Algorithms (TALG) **10**(4), 1–32 (2014)
12. Fernau, H., Krebs, A.: Problems on finite automata and the exponential time hypothesis. Algorithms **10**(1), 24 (2017)
13. Furst, M., Saxe, J.B., Sipser, M.: Parity, circuits, and the polynomial-time hierarchy. Math. Syst. Theory **17**(1), 13–27 (1984)
14. Impagliazzo, R., Paturi, R.: On the complexity of k-SAT. J. Comput. Syst. Sci. **62**, 367–375 (2001)
15. Impagliazzo, R., Paturi, R., Zane, F.: Which problems have strongly exponential complexity? J. Comput. Syst. Sci. **63**, 512–530 (2001)
16. Karakostas, G., Lipton, R.J., Viglas, A.: On the complexity of intersecting finite state automata and NL versus NP. Theor. Comput. Sci. **302**(1), 257–274 (2003)
17. Kasai, T., Iwata, S.: Gradually intractable problems and nondeterministic logspace lower bounds. Math. Syst. Theory **18**(1), 153–170 (1985). https://doi.org/10.1007/BF01699467
18. Kozen, D.: Lower bounds for natural proof systems. In: Proceedings of the 8th Annual Symposium on Foundations of Computer Science (FOCS 1977), vol. 77, pp. 254–266 (1977)

19. Lange, K.-J., Rossmanith, P.: The emptiness problem for intersections of regular languages. In: Havel, I.M., Koubek, V. (eds.) MFCS 1992. LNCS, vol. 629, pp. 346–354. Springer, Heidelberg (1992). https://doi.org/10.1007/3-540-55808-X_33

20. Lokshtanov, D., Marx, D., Saurabh, S.: Slightly superexponential parameterized problems. In: Proceedings of the 22nd Annual ACM-SIAM Symposium on Discrete Algorithms (SODA 2011), pp. 760–776 (2011)

21. Lynch, N.: Log space recognition and translation of parenthesis languages. J. ACM **24**(4), 583–590 (1977)

22. Marx, D.: On the optimality of planar and geometric approximation schemes. In: Proceedings of the 48th Annual Symposium on Foundations of Computer Science (FOCS 2007), pp. 338–348. IEEE (2007)

23. de Oliveira Oliveira, M., Wehar, M.: Intersection non-emptiness and hardness within polynomial time. In: Hoshi, M., Seki, S. (eds.) DLT 2018. LNCS, vol. 11088, pp. 282–290. Springer, Cham (2018). https://doi.org/10.1007/978-3-319-98654-8_23

24. Pătraşcu, M., Williams, R.: On the possibility of faster SAT algorithms. In: Proceedings of the 21st Annual Symposium on Discrete Algorithms (SODA 2010), pp. 1065–1075. SIAM (2010)

25. Seiferas, J.I.: Relating refined space complexity classes. J. Comput. Syst. Sci. **14**(1), 100–129 (1977)

26. Todd Wareham, H.: The parameterized complexity of intersection and composition operations on sets of finite-state automata. In: Yu, S., Păun, A. (eds.) CIAA 2000. LNCS, vol. 2088, pp. 302–310. Springer, Heidelberg (2001). https://doi.org/10.1007/3-540-44674-5_26

27. Wang, J.R., Williams, R.R.: Exact algorithms and strong exponential time hypothesis. In: Encyclopedia of Algorithms, pp. 657–661 (2016)

28. Wehar, M.: Intersection emptiness for finite automata. Honors thesis, Carnegie Mellon University (2012)

29. Wehar, M.: Hardness results for intersection non-emptiness. In: Esparza, J., Fraigniaud, P., Husfeldt, T., Koutsoupias, E. (eds.) ICALP 2014. LNCS, vol. 8573, pp. 354–362. Springer, Heidelberg (2014). https://doi.org/10.1007/978-3-662-43951-7_30

30. Wehar, M.: On the complexity of intersection non-emptiness problems. Ph.D. thesis, University at Buffalo (2016)

31. Williams, R.: Non-uniform ACC circuit lower bounds. In: IEEE 26th Annual Conference on Computational Complexity (CCC 2011), pp. 115–125, June 2011

32. Williams, R.: Improving exhaustive search implies superpolynomial lower bounds. SIAM J. Comput. **42**(3), 1218–1244 (2013)

33. Williams, V.V.: Hardness of easy problems: basing hardness on popular conjectures such as the strong exponential time hypothesis (invited talk). In: Proceedings of the 10th International Symposium on Parameterized and Exact Computation (IPEC 2015), LIPICs, vol. 43, pp. 17–29 (2015)

The State Complexity
of Lexicographically Smallest Words
and Computing Successors

Lukas Fleischer$^{(\boxtimes)}$ and Jeffrey Shallit

School of Computer Science, University of Waterloo, 200 University Avenue West,
Waterloo, ON N2L 3G1, Canada
{lukas.fleischer,shallit}@uwaterloo.ca

Abstract. Given a regular language L over an ordered alphabet Σ, the
set of lexicographically smallest (resp., largest) words of each length is
itself regular. Moreover, there exists an unambiguous finite-state transducer that, on a given word $w \in \Sigma^*$, outputs the length-lexicographically
smallest word larger than w (henceforth called the *L-successor* of w). In
both cases, naïve constructions result in an exponential blowup in the
number of states. We prove that if L is recognized by a DFA with n
states, then $2^{\Theta(\sqrt{n \log n})}$ states are sufficient for a DFA to recognize the
subset $S(L)$ of L composed of its lexicographically smallest words. We
give a matching lower bound that holds even if $S(L)$ is represented as
an NFA. We then show that the same upper and lower bounds hold for
an unambiguous finite-state transducer that computes L-successors.

1 Introduction

One of the most basic problems in formal language theory is the problem of
enumerating the words of a language L. Since, in general, L is infinite, language
enumeration is often formalized in one of the following two ways:

1. A function that maps an integer $n \in \mathbb{N}$ to the n-th word of L.
2. A function that takes a word and maps it to the next word in L.

Both descriptions require some linear ordering of the words in order for them
to be well-defined. Usually, *radix order* (also known as length-lexicographical
order) is used. Throughout this work, we focus on the second formalization.

   While enumeration is non-computable in general, there are many interesting special cases. In this paper, we investigate the case of fixed regular languages, where successors can be computed in linear time [1,2,9]. Moreover,
Frougny [7] showed that for every regular language L, the mapping of words to
their successors in L can be realized by a finite-state transducer. Later, Angrand
and Sakarovitch refined this result [3], showing that the successor function of
any regular language is a finite union of functions computed by sequential

© Springer Nature Switzerland AG 2020
N. Jonoska and D. Savchuk (Eds.): DLT 2020, LNCS 12086, pp. 83–95, 2020.
https://doi.org/10.1007/978-3-030-48516-0_7

transducers that operate from right to left. However, to the best of our knowledge, no upper bound on the size of smallest transducer computing the successor function was known.

In this work, we consider transducers operating from left to right, and prove that the optimal upper bound for the size of transducers computing successors in L is in $2^{\Theta(\sqrt{n \log n})}$, where n is the size of the smallest DFA for L.

The construction used to prove the upper bound relies heavily on another closely related result. Many years before Frougny published her proof, it had already been shown that if L is a regular language, the set of all lexicographically smallest (resp., largest) words of each length is itself regular; see, e.g., [11,12]. This fact is used both in [3] and in our construction. In [12], it was shown that if L is recognized by a DFA with n states, then the set of all lexicographically smallest words is recognized by a DFA with 2^{n^2} states. While it is easy to improve this upper bound to $n2^n$, the exact state complexity of this operation remained open. We prove that $2^{\Theta(\sqrt{n \log n})}$ states are sufficient and that this upper bound is optimal. We also prove that nondeterminism does not help with recognizing lexicographically smallest words, i.e., the corresponding lower bound still holds if the constructed automaton is allowed to be nondeterministic.

The key component to our results is a careful investigation of the structure of lexicographically smallest words. This is broken down into a series of technical lemmas in Sect. 3, which are interesting in their own right. Some of the other techniques are similar to those already found in [3], but need to be carried out more carefully to achieve the desired upper bound.

For some related results, see [5,10].

2   Preliminaries

We assume familiarity with basic concepts of formal language theory and automata theory; see [8,13] for a comprehensive introduction. Below, we introduce concepts and notation specific to this work.

Ordered Words and Languages. Let Σ be a finite ordered alphabet. Throughout the paper, we consider words ordered by *radix order*, which is defined by $u < v$ if either $|u| < |v|$ or there exist factorizations $u = xay$, $v = xbz$ with $|y| = |z|$ and $a, b \in \Sigma$ such that $a < b$. We write $u \leqslant v$ if $u = v$ or $u < v$. In this case, the word u is *smaller* than v and the word v is *larger* than u.

For a language $L \subseteq \Sigma^*$ and two words $u, v \in \Sigma^*$, we say that v is the L-*successor of* u if $v \in L$ and $w \notin L$ for all $w \in \Sigma^*$ with $u < w < v$. Similarly, u is the L-*predecessor of* v if $u \in L$ and $w \notin L$ for all $w \in \Sigma^*$ with $u < w < v$. A word is L-*minimal* if it has no L-predecessor. A word is L-*maximal* if it has no L-successor. Note that every nonempty language contains exactly one L-minimal word. It contains a (unique) L-maximal word if and only if L is finite. A word $u \in \Sigma^*$ is L-*length-preserving* if it is not L-maximal and the L-successor of u has length $|u|$. Words that are not L-length-preserving are called L-*length-increasing*.

Note that by definition, an L-maximal word is always L-length-increasing. For convenience, we sometimes use the terms *successor* (resp., *predecessor*) instead of Σ^*-*successor* (resp., Σ^*-*predecessor*).

For a given language $L \subseteq \Sigma^*$, the set of all smallest words of each length in L is denoted by $S(L)$. It is formally defined as follows:

$$S(L) = \{u \in L \mid \forall v \in L \colon v < u \implies |v| < |u|\}.$$

Similarly, we define $B(L)$ to be the set of all L-length-increasing words:

$$B(L) = \{u \in L \mid \forall v \in L \colon v > u \implies |v| > |u|\}.$$

A language $L \subseteq \Sigma^*$ is *thin* if it contains at most one word of each length, i.e., $|L \cap \Sigma^n| \in \{0,1\}$ for all $n \geqslant 1$. It is easy to see that for every language $L \subseteq \Sigma^*$, the languages $S(L)$ and $B(L)$ are thin.

Finite Automata and Transducers. A *nondeterministic finite automaton* (NFA for short) is a 5-tuple $(Q, \Sigma, \cdot, q_0, F)$ where Q is a finite set of *states*, Σ is a finite alphabet, $q_0 \in Q$ is the *initial state*, $F \subseteq Q$ is the set of *accepting states* and $\cdot : Q \times \Sigma \to 2^Q$ is the *transition function*. We usually use the notation $q \cdot a$ instead of $\cdot(q, a)$, and we extend the transition function to $2^Q \times \Sigma^*$ by letting $X \cdot \varepsilon = X$ and $X \cdot wa = \bigcup_{q \in X \cdot w} q \cdot a$ for all $X \subseteq Q$, $w \in \Sigma^*$, and $a \in \Sigma$. For a state $q \in Q$ and a word $w \in \Sigma^*$, we also use the notation $q \cdot w$ instead of $\{q\} \cdot w$ for convenience. A word $w \in \Sigma^*$ is *accepted* by the NFA if $q_0 \cdot w \cap F \neq \emptyset$. We sometimes use the notation $p \xrightarrow{a} q$ to indicate that $q \in p \cdot a$. An NFA is *unambiguous* if for every input, there exists at most one accepting run. Unambiguous NFA are also called unambiguous finite state automata (UFA). A *deterministic finite automaton* (DFA for short) is an NFA $(Q, \Sigma, \cdot, q_0, F)$ with $|q \cdot a| = 1$ for all $q \in Q$ and $a \in \Sigma$. Since this implies $|q \cdot w| = 1$ for all $w \in \Sigma^*$, we sometimes identify the singleton $q \cdot w$ with the only element it contains.

A *finite-state transducer* is a nondeterministic finite automaton that additionally produces some output that depends on the current state, the current letter and the successor state. For each transition, we allow both the input and the output letter to be empty. Formally, it is a 6-tuple $(Q, \Sigma, \Gamma, \cdot, q_0, F)$ where Q is a finite set of states, Σ and Γ are finite alphabets, $q_0 \in Q$ is the initial state and $F \subseteq Q$ is the set of accepting states, and $\cdot : Q \times (\Sigma \cup \{\varepsilon\}) \to 2^{Q \times (\Gamma \cup \{\varepsilon\})}$ is the *transition function*. One can extend this transition function to the product $2^Q \times \Sigma^*$. To this end, we first define the ε-*closure* of a set $T \subseteq Q \times \Sigma^*$ as the smallest superset C of T with $\{(q \cdot \varepsilon, w) \mid (q, w) \in C\} \subseteq C$. We then define $X \cdot \varepsilon$ to be the ε-closure of $\{(q, \varepsilon) \mid q \in X\}$ and $X \cdot wa$ to be the ε-closure of $\{(q', ub) \mid (q, u) \in X \cdot w, (q', b) \in q \cdot a\}$ for all $X \subseteq Q$, $w \in \Sigma^*$ and $a \in \Sigma$. We sometimes use the notation $p \xrightarrow{a|b} q$ to indicate that $(q, b) \in p \cdot a$. A finite-state transducer is *unambiguous* if, for every input, there exists at most one accepting run.

3 The State Complexity of $S(L)$

It is known that if L is a regular language, then both $S(L)$ and $B(L)$ are also regular [11,12]. In this section, we investigate the state complexity of the operations $L \mapsto S(L)$ and $L \mapsto B(L)$ for regular languages. Since the operations are symmetric, we focus on the former. To this end, we first prove some technical lemmas. The first lemma is a simple observation that helps us investigate the structure of words in $S(L)$.

Lemma 1. *Let $x, u, y, v, z \in \Sigma^*$ with $|u| = |v|$. Then $xuuyz < xuyvz$ or $xyvvz < xuyvz$ or $xuuyz = xuyvz = xyvvz$.*

Proof. Note that uy and yv are words of the same length. If $uy < yv$, then $xuuyz < xuyvz$. Similarly, $uy > yv$ immediately yields $xuyvz > xyvvz$. The last case is $uy = yv$, which implies $xuuyz = xuyvz = xyvvz$. □

Using this observation, we can generalize a well-known factorization technique for regular languages to minimal words. For a DFA with state set Q, a state $q \in Q$ and a word $w = a_1 \cdots a_n \in \Sigma^*$, we define

$$\mathrm{tr}(q, w) = (q, q \cdot a_1, \dots, q \cdot a_1 \cdots a_n)$$

to be the sequence of all states that are visited when starting in state q and following the transitions labeled by the letters from w.

Lemma 2. *Let $\mathcal{A}$ be a DFA over Σ with n states and with initial state q_0. Then for every word $w \in \Sigma^*$, there exists a factorization $w = u_1 v_1^{i_1} \cdots u_k v_k^{i_k}$ with $u_1, v_1, \dots, u_k, v_k \in \Sigma^*$ and $i_1, \dots, i_k \geqslant 1$ such that, for all $j \in \{1, \dots, k\}$, the following hold:*

(a) $q_0 \cdot u_1 v_1^{i_1} \cdots u_{j-1} v_{j-1}^{i_{j-1}} u_j = q_0 \cdot u_1 v_1^{i_1} \cdots u_{j-1} v_{j-1}^{i_{j-1}} u_j v_j$,
(b) $|u_j v_j| \leqslant n$, *and*
(c) v_j *is not a prefix of* $u_{j+1} v_{j+1}^{i_{j+1}} \cdots u_k v_k^{i_k}$.

Additionally, if $w \in S(L(\mathcal{A}))$, this factorization can be chosen such that

(d) *the lengths $|v_j|$ are pairwise disjoint (i.e., $|\{|v_1|, \dots, |v_k|\}| = k$) and*
(e) *there exists at most one $j \in \{1, \dots, k\}$ with $i_j > n$.*

Proof. To construct the desired factorization, initialize $j := 1$ and $q := q_0$ and follow these steps:

1. If $w = \varepsilon$, we are done. If $w \neq \varepsilon$ and the states in $\mathrm{tr}(q, w)$ are pairwise distinct, let $u_j = w$ and $v_j = \varepsilon$ and we are done. Otherwise, factorize $w = xy$ with $|x|$ minimal such that $\mathrm{tr}(q_0, x)$ contains exactly one state twice, i.e., $|x|$ distinct states in total.
2. Choose the unique factorization $x = uv$ such that $q \cdot u = q \cdot uv$ and $v \neq \varepsilon$.
3. Let $q := q \cdot x$ and $w := y$.

4. If $j > 1$ and $u = \varepsilon$ and $v = v_{j-1}$, increment i_{j-1} and go back to step 1. Otherwise, let $u_j := u$, $v_j := v$ and $j := j + 1$; then go back to step 1.

This factorization satisfies the first three properties by construction. It remains to show that if $w \in S(L(\mathcal{A}))$, then Properties (d) and (e) are satisfied as well.

Let us begin with Property (d). For the sake of contradiction, assume that there exist two indices a, b with $a < b$ and $|v_a| = |v_b|$. Note that by construction, v_a and v_b must be nonempty. Moreover, by Property (a), the words

$$w' := u_1 v_1^{i_1} \cdots u_a v_a^{i_a+1} \cdots u_b v_b^{i_b-1} \cdots u_k v_k^{i_k} \text{ and}$$
$$w'' := u_1 v_1^{i_1} \cdots u_a v_a^{i_a-1} \cdots u_b v_b^{i_b+1} \cdots u_k v_k^{i_k}$$

both belong to $L(\mathcal{A})$. However, since $w \in S(L(\mathcal{A}))$, neither w' nor w'' can be strictly smaller than w. Using Lemma 1, we obtain that $w' = w$. This contradicts Property (c).

Property (e) can be proved by using the same argument: Assume that there exist indices a, b with $a < b$ and $i_a, i_b > n$. The words $v_a^{|v_b|}$ and $v_b^{|v_a|}$ have the same lengths. We define

$$w' := u_1 v_1^{i_1} \cdots u_a v_a^{i_a+|v_b|} \cdots u_b v_b^{i_b-|v_a|} \cdots u_k v_k^{i_k},$$
$$w'' := u_1 v_1^{i_1} \cdots u_a v_a^{i_a-|v_b|} \cdots u_b v_b^{i_b+|v_a|} \cdots u_k v_k^{i_k},$$

and obtain $w = w'$, which is a contradiction as above.   □

The existence of such a factorization almost immediately yields our next technical ingredient.

Lemma 3. *Let $\mathcal{A}$ be a DFA with $n \geqslant 3$ states. Let q_0 be the initial state of $\mathcal{A}$ and let $w \in S(L(\mathcal{A}))$. Then there exists a factorization $w = xy^i z$ with $i \in \mathbb{N}$, $|xz| \leqslant n^3$ and $|y| \leqslant n$ such that $q_0 \cdot xy = q_0 \cdot x$. In particular, $xy^* z \subseteq L(\mathcal{A})$.*

Proof. Let $w = u_1 v_1^{i_1} \cdots u_k v_k^{i_k}$ be a factorization that satisfies all properties in the statement of Lemma 2. Suppose first that all exponents i_j are at most n. Using Properties (b) and (d), we obtain $k \leqslant n+1$ and the maximum length of w is achieved when all lengths $\ell \in \{0, \ldots, n\}$ are present among the factors v_j and the corresponding u_j have lengths $n - |v_j|$. This yields

$$|w| \leqslant \sum_{\ell=0}^{n} (n-\ell+n\ell) = n(n+1)+(n-1)\sum_{\ell=1}^{n} \ell = n^2+n+\frac{(n-1)n(n+1)}{2} \leqslant n^3$$

where the last inequality uses $n \geqslant 3$. Therefore, we may set $x := w$, $y := \varepsilon$ and $z := \varepsilon$.

If not all exponents are at most n, by Property (e), there exists a unique index j with $i_j > n$. In this case, let $x := u_1 v_1^{i_1} \cdots u_{j-1} v_{j-1}^{i_{j-1}} u_j$, $y := v_j$ and $z := u_{j+1} v_{j+1}^{i_{j+1}} \cdots u_k v_k^{i_k}$. The upper bound $|xz| \leqslant n^3$ still follows by the argument above, and $|y| \leqslant n$ is a direct consequence of Property (b). Moreover, $w \in L(\mathcal{A})$ and Property (a) together imply that $xy^* z \subseteq L(\mathcal{A})$.   □

For the next lemma, we need one more definition. Let $\mathcal{A}$ be a DFA with initial state q_0. Two tuples (x, y, z) and (x', y', z') are *cycle-disjoint with respect to $\mathcal{A}$* if the sets of states in $\mathrm{tr}(q_0 \cdot x, y)$ and $\mathrm{tr}(q_0 \cdot x', y')$ are either equal or disjoint.

Lemma 4. *Let $\mathcal{A}$ be a DFA with $n \geqslant 3$ states and initial state q_0. Let (x, y, z) and (x', y', z') be tuples that are not cycle-disjoint with respect to $\mathcal{A}$ such that*

$$q_0 \cdot x = q_0 \cdot xy, \quad q_0 \cdot x' = q_0 \cdot x'y', \quad |xz|, |x'z'| \leqslant n^3 \text{ and } |y|, |y'| \leqslant n.$$

Then either $xy^*z \cap S(L(\mathcal{A}))$ or $x'(y')^*z' \cap S(L(\mathcal{A}))$ only contains words of length at most $n^3 + n^2$.

Proof. Since the tuples are not cycle-disjoint with respect to $\mathcal{A}$, we can factorize $y = uv$ and $y' = u'v'$ such that $q_0 \cdot xu = q_0 \cdot x'u'$.

Note that since $q_0 \cdot xuv = q_0 \cdot x$, the sets of states in $\mathrm{tr}(q_0 \cdot x, uv)$ and $\mathrm{tr}(q_0 \cdot xu, (vu)^i)$ coincide for all $i \geqslant 1$. By the same argument, the sets of states in $\mathrm{tr}(q_0 \cdot x', u'v')$ and $\mathrm{tr}(q_0 \cdot x'u', (v'u')^i)$ coincide for all $i \geqslant 1$.

If the powers $(vu)^{|y'|}$ and $(v'u')^{|y|}$ were equal, then $\mathrm{tr}(q_0 \cdot xu, (vu)^{|y'|})$ and $\mathrm{tr}(q_0 \cdot x'u', (v'u')^{|y|})$ coincide. By the previous observation, this would imply that the tuples (x, y, z) and (x', y', z') are cycle-disjoint, a contradiction. We conclude $(vu)^{|y'|} \neq (v'u')^{|y|}$.

By symmetry, we may assume that $(vu)^{|y'|} < (v'u')^{|y|}$. But then, for every word of the form $x'(y')^i z' \in L(\mathcal{A})$ with $i > |y|$, there exists a strictly smaller word $x'u'(vu)^{|y'|}(v'u')^{i-|y|-1}v'z'$ in $L(\mathcal{A})$. To see that this word indeed belongs to $L(\mathcal{A})$, note that $q_0 \cdot x'u'vu = q_0 \cdot xuvu = q_0 \cdot xu = q_0 \cdot x'u'$. This means that all words in $x'(y')^*z' \cap S(L)$ are of the form $x'(y')^i z'$ with $i \leqslant |y|$.   □

The previous lemmas now allow us to replace any language L by another language that has a simple structure and approximates L with respect to $S(L)$.

Lemma 5. *Let $\mathcal{A}$ be a DFA over Σ with $n \geqslant 3$ states. Then there exist an integer $k \leqslant n^4 + n^3$ and tuples $(x_1, y_1, z_1), \ldots, (x_k, y_k, z_k) \in (\Sigma^*)^3$ such that the following properties hold:*

(i) $S(L(\mathcal{A})) \subseteq \bigcup_{i=1}^{k} x_i y_i^* z_i \subseteq L(\mathcal{A})$,
(ii) $|x_i z_i| \leqslant n^3 + n^2$ *for all* $i \in \{1, \ldots k\}$, *and*
(iii) $\sum_{\ell \in Y} \ell \leqslant n$ *where* $Y = \{|y_1|, \ldots, |y_k|\}$.

Proof. If we ignore the required upper bound $k \leqslant n^4 + n^3$ and Property (iii) for now, the statement follows immediately from Lemma 3 and the fact that there are only finitely many different tuples (x, y, z) with $|xz| \leqslant n^3$ and $|y| \leqslant n$. We start with such a finite set of tuples $(x_1, y_1, z_1), \ldots, (x_k, y_k, z_k)$ and show that we can repeatedly eliminate tuples until at most $n^4 + n^3$ cycle-disjoint tuples remain. The desired upper bound $\sum_{\ell \in Y} \ell \leqslant n$ then follows automatically.

In each step of this elimination process, we handle one of the following cases:

- If there are two distinct tuples (x_i, y_i, z_i) and (x_j, y_j, z_j) with $|x_i z_i| = |x_j z_j|$ and $y_i = y_j$, there are two possible scenarios. If $x_i z_i < x_j z_j$, then for every word in $x_j y_j^* z_j$ there exists a smaller word in $x_i y_i^* z_i$ and we can remove (x_j, y_j, z_j) from the set of tuples. By the same argument, we can remove the tuple (x_i, y_i, z_i) if $y_i = y_j$ and $x_i z_i > x_j z_j$.
- Now consider the case that there are two distinct tuples (x_i, y_i, z_i) and (x_j, y_j, z_j) with $|x_i z_i| = |x_j z_j|$ and $|y_i| = |y_j|$ but $y_i \neq y_j$. We first check whether $x_i z_i < x_j z_j$. If true, we add the tuple (x_i, ε, z_i), otherwise we add (x_j, ε, z_j). If $x_i y_i < x_j y_j$, we know that each word in $x_j y_j^+ z_j$ has a smaller word in $x_i y_i^+ z_i$, and we remove the tuple (x_j, y_j, z_j). Otherwise, we can remove (x_i, y_i, z_i) by the same argument.
- The last case is that there exist two tuples (x_i, y_i, z_i) and (x_j, y_j, z_j) that are not cycle-disjoint. By Lemma 4, we can remove at least one of these tuples and replace it by multiple tuples of the form (x, ε, z). Note that the newly introduced tuples might be of the form (x, ε, z) with $|xz| > n^3$ but Lemma 4 asserts that they still satisfy $|xz| \leqslant n^3 + n^2$.

Note that we introduce new tuples of the form (x, ε, z) during this elimination process. These new tuples are readily eliminated using the first rule.

After iterating this elimination process, the remaining tuples are pairwise cycle-disjoint and the pairs $(|x_i z_i|, |y_i|)$ assigned to these tuples (x_i, y_i, z_i) are pairwise disjoint. Properties (ii) and (iii) yield the desired upper bound on k. □

Remark 1. While $S(L)$ can be approximated by a language of the simple form given in Lemma 5, the language $S(L)$ itself does not necessarily have such a simple description. An example of a regular language L where $S(L)$ does not have such a simple form is given in the proof of Theorem 2.

The last step is to investigate languages L of the simple structure described in the previous lemma and show how to construct a small DFA for $S(L)$.

Lemma 6. *Let $n \in \mathbb{N}$. Let $L = \bigcup_{i=1}^k x_i y_i^* z_i$ with $k \leqslant n^4 + n^3$ and $|x_i z_i| \leqslant n^3 + n^2$ for all $i \in \{1, \ldots k\}$ and $\sum_{\ell \in Y} \ell \leqslant n$ where $Y = \{|y_1|, \ldots, |y_k|\}$. Then $S(L)$ is recognized by a DFA with $2^{\mathcal{O}(\sqrt{n \log n})}$ states.*

Proof. We describe how to construct a DFA of the desired size that recognizes the language $S(L)$. This DFA is the product automaton of multiple components.

In one component (henceforth called the *counter component*), we keep track of the length of the processed input as long as at most $n^3 + n^2$ letters have been consumed. If more than $n^3 + n^2$ letters have been consumed, we only keep track of the length of the processed input modulo all numbers $|y_i|$ for $i \in \{1, \ldots, k\}$.

For each $i \in \{1, \ldots k\}$, there is an additional component (henceforth called the *i-th activity component*). In this component, we keep track of whether the currently processed prefix u of the input is a prefix of a word in $x_i y_i^*$, whether u is a prefix of a word in $x_i y_i^* z_i$ and whether $u \in x_i y_i^* z_i$. Note that if some prefix of the input is not a prefix of a word in $x_i y_i^* z_i$, no longer prefix of the input can be a prefix of a word in $x_i y_i^* z_i$. The information stored in the counter component suffices to compute the possible letters of $x_i y_i^* z_i$ allowed to be read in each step to maintain the prefix invariants.

It remains to describe how to determine whether a state is final. To this end, we use the following procedure. First, we determine which sets of the form $x_i y_i^* z_i$ the input word leading to the considered state belongs to. These languages are called the *active languages* of the state. They can be obtained from the activity components of the state. If there are no active languages, the state is immediately marked as not final. If the length of the input word w leading to the considered state is $n^3 + n^2$ or less, we can obtain $|w|$ from the counter component and reconstruct w from the set of active languages. If the length of the input is larger than $n^3 + n^2$, we cannot fully recover the input from the information stored in the state. However, we can determine the shortest word w with $|w| > n^3 + n^2$ such that $|w|$ is consistent with the length information stored in the counter component and w itself is consistent with the set of active languages. In either case, we then compute the set A of all words of length $|w|$ that belong to any (possibly not active) language $x_i y_i^* z_i$ with $1 \leqslant i \leqslant k$. If w is the smallest word in A, the state is final, otherwise it is not final.

The desired upper bound on the number of states follows from known estimates on the least common multiple of a set of natural numbers with a given sum; see e.g., [6]. □

We can now combine the previous lemmas to obtain an upper bound on the state complexity of $S(L)$.

Theorem 1. *Let L be a regular language that is recognized by a DFA with n states. Then $S(L)$ is recognized by a DFA with $2^{O(\sqrt{n \log n})}$ states.*

Proof. By Lemma 5, we know that there exists a language L' of the form described in the statement of Lemma 6 with $S(L) \subseteq L' \subseteq L$. Since $L' \subseteq L$ implies $S(L') \subseteq S(L)$ and since $S(S(L)) = S(L)$, this also means that $S(L') = S(L)$. Lemma 6 now shows that there exists a DFA of the desired size. □

To show that the result is optimal, we provide a matching lower bound.

Theorem 2. *There exists a family of DFA $(A_n)_{n \in \mathbb{N}}$ over a binary alphabet such that A_n has n states and every NFA for $S(L(A_n))$ has $2^{\Omega(\sqrt{n \log n})}$ states.*

Proof. For $i \in \{1, \ldots k\}$, let p_i be the i-th prime number and let $p = p_1 \cdots p_k$. We define a language

$$L = 1^* \cup \bigcup_{1 \leqslant i \leqslant k} 1^i 0^{k-i+1} \{1, 1^2, \ldots, 1^{p_i-1}\} (1^{p_i})^*.$$

It is easy to see that L is recognized by a DFA with $k^2 + p_1 + \cdots + p_k$ states. We show that $S(L)$ is not recognized by any NFA with less than p states. From known estimates on the prime numbers (e.g., [4, Sec. 2.7]), this suffices to prove our claim.

Let $\mathcal{A}$ be a NFA for $S(L)$ and assume, for the sake of contradiction, that $\mathcal{A}$ has less than p states. Note that since for each $i \in \{1, \ldots, k\}$, the integer p is a multiple of p_i, the language L does not contain any word of the form $1^i 0^{k-i+1} 1^p$. Therefore, the word 1^{k+1+p} belongs to $S(L)$ and by assumption, an accepting path for this word in $\mathcal{A}$ must contain a loop of some length $\ell \in \{1, \ldots, p-1\}$. But then $1^{k+1+p+\ell}$ is accepted by $\mathcal{A}$, too. However, since $1 \leqslant \ell < p$, there exists some $i \in \{1, \ldots, k\}$ such that p_i does not divide ℓ. This means that p_i also does not divide $p + \ell$. Thus, $1^i 0^{k-i+1} 1^{p+\ell} \in L$, contradicting the fact that $1^{k+1+p+\ell}$ belongs to $S(L)$. $\qquad\square$

Combining the previous two theorems, we obtain the following corollary.

Corollary 1. *Let L be a language that is recognized by a DFA with n states. Then, in general, $2^{\Theta(\sqrt{n \log n})}$ states are necessary and sufficient for a DFA or NFA to recognize $S(L)$.*

By reversing the alphabet ordering, we immediately obtain similar results for largest words.

Corollary 2. *Let L be a language that is recognized by a DFA with n states. Then, in general, $2^{\Theta(\sqrt{n \log n})}$ states are necessary and sufficient for a DFA or NFA to recognize $B(L)$.*

4   The State Complexity of Computing Successors

One approach to efficient enumeration of a regular language L is constructing a transducer that reads a word and outputs its L-successor [3,7]. We consider transducers that operate from left to right. Since the output letter in each step might depend on letters that have not yet been read, this transducer needs to be nondeterministic. However, the construction can be made *unambiguous*, meaning that for any given input, at most one computation path is accepting and yields the desired output word. In this paper, we prove that, in general, $2^{\Theta(\sqrt{n \log n})}$ states are necessary and sufficient for a transducer that performs this computation.

Our proof is split into two parts. First, we construct a transducer that only maps L-length-preserving words to their corresponding L-successors. All other words are rejected. This construction heavily relies on results from the previous section. Then we extend this transducer to L-length-increasing words by using a technique called *padding*. For the first part, we also need the following result.

Theorem 3. *Let $L \subseteq \Sigma^*$ be a thin language that is recognized by a DFA with n states. Then the languages*

$$L_\leqslant = \{v \in \Sigma^* \mid \exists u \in L \colon |u| = |v| \text{ and } v \leqslant u\} \text{ and}$$
$$L_\geqslant = \{v \in \Sigma^* \mid \exists u \in L \colon |u| = |v| \text{ and } v \geqslant u\}$$

are recognized by UFA with $2n$ states.

Proof. Let $\mathcal{A} = (Q, \Sigma, \cdot, q_0, F)$ be a DFA for L and let $n = |Q|$. We construct a UFA with $2n$ states for $L_\leqslant$. The statement for $L_\geqslant$ follows by symmetry.

The state set of the UFA is $Q \times \{0, 1\}$, the initial state is $(q_0, 0)$ and the set of final states is $F \times \{0, 1\}$. The transitions are

$$(q, 0) \xrightarrow{a} (q \cdot a, 0) \qquad \text{for all } q \in Q \text{ and } a \in \Sigma,$$

$$(q, 0) \xrightarrow{a} (q \cdot b, 1) \qquad \text{for all } q \in Q \text{ and } a, b \in \Sigma \text{ with } a < b,$$

$$(q, 1) \xrightarrow{a} (q \cdot b, 1) \qquad \text{for all } q \in Q \text{ and } a, b \in \Sigma.$$

It is easy to verify that this automaton indeed recognizes $L_\leqslant$. To see that this automaton is unambiguous, consider an accepting run of a word w of length ℓ. Note that the sequence of first components of the states in this run yield an accepting path of length ℓ in $\mathcal{A}$. Since $L(\mathcal{A})$ is thin, this path is unique. Therefore, the sequence of first components is uniquely defined. The second components are then uniquely defined, too: they are 0 up to the first position where w differs from the unique word of length ℓ in L, and 1 afterwards. □

For a language $L \subseteq \Sigma^*$, we denote by $B_\geqslant(L)$ the language of all words from Σ^* such that there exists no strictly larger word of the same length in L. Combining Theorem 1 and Theorem 3, the following corollary is immediate.

Corollary 3. *Let L be a language that is recognized by a DFA with n states. Then there exists a UFA with $2^{\mathcal{O}(\sqrt{n \log n})}$ states that recognizes the language $B_\geqslant(L)$.*

For a language $L \subseteq \Sigma^*$, we define

$$X(L) = \{u \in \Sigma^* \mid \forall v \in L \colon |u| \neq |v|\} .$$

If L is regular, it is easy to construct an NFA for the complement of $X(L)$, henceforth denoted as $\overline{X(L)}$. To this end, we take a DFA for L and replace the label of each transition with all letters from Σ. This NFA can also be viewed as an NFA over the unary alphabet $\{\Sigma\}$; here, Σ is interpreted as a letter, not a set. It can be converted to a DFA for $\overline{X(L)}$ by using Chrobak's efficient determinization procedure for unary NFA [6]. The resulting DFA can then be complemented to obtain a DFA for $X(L)$:

Corollary 4. *Let L be a language that is recognized by a DFA with n states. Then there exists a DFA with $2^{\mathcal{O}(\sqrt{n \log n})}$ states that recognizes the language $X(L)$.*

We now use the previous results to prove an upper bound on the size of a transducer performing a variant of the L-successor computation that only works for L-length-preserving words.

Theorem 4. *Let L be a language that is recognized by a DFA with n states. Then there exists an unambiguous finite-state transducer with $2^{\mathcal{O}(\sqrt{n \log n})}$ states that rejects all L-length-increasing words and maps every L-length-preserving word to its L-successor.*

Proof. Let $\mathcal{A} = (Q, \Sigma, \cdot, q_0, F)$ be a DFA for L and let $n = |Q|$. For every $q \in Q$, we denote by $\mathcal{A}_q$ the DFA that is obtained by making q the new initial state of $\mathcal{A}$. We use $\mathcal{A}_q^S$ to denote DFA with $2^{\mathcal{O}(\sqrt{n \log n})}$ states that recognizes the language $S(L(\mathcal{A}_q))$. These DFA exist by Theorem 1. Moreover, by Corollary 3, there exist UFA with $2^{\mathcal{O}(\sqrt{n \log n})}$ states that recognize the languages $B_{\geqslant}(L(\mathcal{A}_q))$. We denote these UFA by $\mathcal{A}_q^B$. Similarly, we use $\mathcal{A}_q^X$ to denote DFA with $2^{\mathcal{O}(\sqrt{n \log n})}$ states that recognize $X(L(\mathcal{A}_q))$. These DFA exist by Corollary 4.

In the finite-state transducer, we first simulate $\mathcal{A}$ on a prefix u of the input, copying the input letters in each step, i.e., producing the output u. At some position, after having read a prefix u leading up to the state $q := q_0 \cdot u$, we nondeterministically decide to output a letter b that is strictly larger than the current input letter a. From then on, we guess an output letter in each step and start simulating multiple automata in different components. In one component, we simulate $\mathcal{A}_{q \cdot a}^B$ on the remaining input. In another component, we simulate $\mathcal{A}_{q \cdot b}^S$ on the guessed output. In additional components, for each $c \in \Sigma$ with $a < c < b$, we simulate $\mathcal{A}_{q \cdot c}^X$ on the input. The automata in all components must accept in order for the transducer to accept the input.

The automaton $\mathcal{A}_{q \cdot a}^B$ verifies that there is no word in L that starts with the prefix ua, has the same length as the input word and is strictly larger than the input word. The automaton $\mathcal{A}_{q \cdot b}^S$ verifies that there is no word in L that starts with the prefix ub, has the same length as the input word and is strictly smaller than the output word. It also certifies that the output word belongs to L. For each letter c, the automaton $\mathcal{A}_{q \cdot c}^X$ verifies that there is no word in L that starts with the prefix uc and has the same length as the input word.

Together, the components ensure that the guessed output is the unique successor of the input word, given that it is L-length-preserving. It is also clear that L-length-increasing words are rejected, since the $\mathcal{A}_{q \cdot b}^S$-component does not accept for any sequence of nondeterministic choices. $\qquad \square$

The construction given in the previous proof can be extended to also compute L-successors of L-length-increasing words. However, this requires some quite technical adjustments to the transducer. Instead, we use a technique called *padding*. A very similar approach appears in [3, Prop. 5.1].

We call the smallest letter of an ordered alphabet Σ the *padding symbol* of Σ. A language $L \subseteq \Sigma^*$ is $\diamond$-*padded* if $\diamond$ is the padding symbol of Σ and $L = \diamond^* K$ for some $K \subseteq (\Sigma \setminus \{\diamond\})^*$. The key property of padded languages is that all words prefixed by a sufficiently long block of padding symbols are L-length-preserving.

Lemma 7. *Let $\mathcal{A}$ be a DFA over Σ with n states such that $L(\mathcal{A})$ is a $\diamond$-padded language. Let $\Gamma = \Sigma \setminus \{\diamond\}$ and let $K = L(\mathcal{A}) \cap \Gamma^*$. Let $u \in \Gamma^*$ be a word that is not K-maximal. Then the $L(\mathcal{A})$-successor of $\diamond^n u$ has length $|\diamond^n u|$.*

Proof. Let v be the K-successor of u. By a standard pumping argument, we have $|u| \leqslant |v| \leqslant |u| + n$. This means that $\diamond^{n+|u|-|v|} v$ is well-defined and belongs to $L(\mathcal{A})$. Note that this word is strictly greater than $\diamond^n u$ and has length $|\diamond^n u|$. Thus, the $L(\mathcal{A})$-successor of $\diamond^n u$ has length $|\diamond^n u|$, too. $\qquad \square$

We now state the main result of this section.

Theorem 5. *Let $\mathcal{A}$ be a deterministic finite automaton over Σ with n states. Then there exists an unambiguous finite-state transducer with $2^{O(\sqrt{n \log n})}$ states that maps every word to its $L(\mathcal{A})$-successor.*

Proof. We extend the alphabet by adding a new padding symbol $\diamond$ and convert $\mathcal{A}$ to a DFA for $\diamond^* L$ by adding a new initial state. The language L' accepted by this new DFA is $\diamond$-padded. By Theorem 4 and Lemma 7, there exists an unambiguous transducer of the desired size that maps every word from $\diamond^{n+1} \Sigma^*$ to its successor in L'. It is easy to modify this transducer such that all words that do not belong to $\diamond^{n+1} \Sigma^*$ are rejected. We then replace every transition that reads a $\diamond$ by a corresponding transition that reads the empty word instead. Similarly, we replace every transition that outputs a $\diamond$ by a transition that outputs the empty word instead. Clearly, this yields the desired construction for the original language L. A careful analysis of the construction shows that the transducer remains unambiguous after each step.   □

We now show that this construction is optimal up to constants in the exponent. The idea is similar to the construction used in Theorem 2.

Theorem 6. *There exists a family of deterministic finite automata $(\mathcal{A}_n)_{n \in \mathbb{N}}$ such that $\mathcal{A}_n$ has n states whereas the smallest unambiguous transducer that maps every word to its $L(\mathcal{A}_n)$-successor has $2^{\Omega(\sqrt{n \log n})}$ states.*

Proof. Let $k \in \mathbb{N}$. Let $p_1, \ldots, p_k$ be the k smallest prime numbers such that $p_1 < \cdots < p_k$ and let $p = p_1 \cdots p_k$. We construct a deterministic finite automaton $\mathcal{A}$ with $2 + p_1 + \cdots + p_k$ states such that the smallest transducer computing the desired mapping has at least p states. From known estimates on the prime numbers (e.g., [4, Sec. 2.7]), this suffices to prove our claim.

The automaton is defined over the alphabet $\Sigma = \{1, \ldots, k\} \cup \{\#\}$. It consists of an initial state q_0, an error state q_{err}, and states (i, j) for $i \in \{1, \ldots, k\}$ and $j \in \{0, \ldots, p_i - 1\}$ with transitions defined as follows:

$$q_0 \cdot a = \begin{cases} (a, 0), & \text{for } a \in \{1, \ldots, k\}; \\ q_{\mathsf{err}}, & \text{if } a = \#; \end{cases}$$

$$(i, j) \cdot a = \begin{cases} (i, j+1 \bmod p_i), & \text{if } a = \#; \\ q_{\mathsf{err}}, & \text{for } a \in \{1, \ldots, k\}. \end{cases}$$

The set of accepting states is $\{(i, 0) \mid 1 \leqslant i \leqslant k\}$. The language $L(\mathcal{A})$ is the set of all words of the form $i \#^j$ with $1 \leqslant i \leqslant k$ such that j is a multiple of p_i.

Assume, to get a contradiction, that there exists an unambiguous transducer with less than p states that maps w to the smallest word in $L(\mathcal{A})$ strictly greater than w. Consider an accepting run of this transducer on some input of the form $2 \#^{\ell p}$ with $\ell \in \mathbb{N}$ large enough such that the run contains a cycle. Clearly, since $\ell p + 1$ and p are coprime, the output of the transducer has to be $2 \#^{\ell p + 2}$. We fix one cycle in this run.

If the number of # read in this cycle does not equal the number of # output in this cycle, by using a pumping argument, we can construct a word of the form $2\#^j$ that is mapped to a word or the form $i\#^{j'}$ with $|j' - j| > 2$. This contradicts the fact that $2\#^{2\mathbb{N}}$ is a subset of $L(\mathcal{A})$. Therefore, we may assume that both the number of letters read and output on the cycle is $r \in \{1, \ldots, p-1\}$.

Again, by a pumping argument, this implies that $2\#^{\ell p + jr}$ is mapped to $2\#^{\ell p + jr + 2}$ for every $j \in \mathbb{N}$. Since $r < p$, at least one of the prime numbers p_i is coprime to r. Therefore, we can choose j such that $jr + 1 \equiv 0 \pmod{p_i}$. However, this means that $p_i\#^{\ell p + jr + 1}$ belongs to $L(\mathcal{A})$, contradicting the fact that the transducer maps $2\#^{\ell p + jr}$ to $2\#^{\ell p + jr + 2}$.   □

Combining the two previous theorems, we obtain the following corollary.

Corollary 5. *Let L be a language that is recognized by a DFA with n states. Then, in general, $2^{\Theta(\sqrt{n \log n})}$ states are necessary and sufficient for an unambiguous finite-state transducer that maps words to their L-successors.*

References

1. Ackerman, M., Mäkinen, E.: Three new algorithms for regular language enumeration. In: Ngo, H.Q. (ed.) COCOON 2009. LNCS, vol. 5609, pp. 178–191. Springer, Heidelberg (2009). https://doi.org/10.1007/978-3-642-02882-3_19
2. Ackerman, M., Shallit, J.: Efficient enumeration of words in regular languages. Theoret. Comput. Sci. **410**(37), 3461–3470 (2009)
3. Angrand, P.-Y., Sakarovitch, J.: Radix enumeration of rational languages. RAIRO - Theoret. Inform. Appl. **44**(1), 19–36 (2010)
4. Bach, E., Shallit, J.: Algorithmic Number Theory. MIT Press, Cambridge (1996)
5. Berthé, V., Frougny, C., Rigo, M., Sakarovitch, J.: On the cost and complexity of the successor function. In: Arnoux, P., Bédaride, N., Cassaigne, J. (eds.) Proceedings of WORDS 2007, Technical Report, Institut de mathématiques de Luminy, pp. 43–56 (2007)
6. Chrobak, M.: Finite automata and unary languages. Theoret. Comput. Sci. **47**, 149–158 (1986). Erratum, 302:497–498, 2003
7. Frougny, C.: On the sequentiality of the successor function. Inf. Comput. **139**(1), 17–38 (1997)
8. Hopcroft, J.E., Motwani, R., Ullman, J.D.: Introduction to Automata Theory, Languages, and Computation, 3rd edn. Addison-Wesley Longman Publishing Co., Inc., New York (2006)
9. Mäkinen, E.: On lexicographic enumeration of regular and context-free languages. Acta Cybern. **13**(1), 55–61 (1997)
10. Okhotin, A.S.: On the complexity of the string generation problem. Discret. Math. Appl. **13**, 467–482 (2003)
11. Sakarovitch, J.: Deux remarques sur un théorème de S. Eilenberg. RAIRO - Theoret. Inform. Appl. **17**(1), 23–48 (1983)
12. Shallit, J.: Numeration systems, linear recurrences, and regular sets. Inf. Comput. **113**(2), 331–347 (1994)
13. Shallit, J.: A Second Course in Formal Languages and Automata Theory. Cambridge University Press, Cambridge (2008)

Reconstructing Words from Right-Bounded-Block Words

Pamela Fleischmann[1]([✉]), Marie Lejeune[2], Florin Manea[3], Dirk Nowotka[1], and Michel Rigo[2]

[1] Kiel University, Kiel, Germany
{fpa,dn}@informatik.uni-kiel.de
[2] University of Liège, Liège, Belgium
{m.lejeune,m.rigo}@uliege.be
[3] University of Göttingen, Göttingen, Germany
florin.manea@informatik.uni-goettingen.de

Abstract. A reconstruction problem of words from scattered factors asks for the minimal information, like multisets of scattered factors of a given length or the number of occurrences of scattered factors from a given set, necessary to uniquely determine a word. We show that a word $w \in \{a, b\}^*$ can be reconstructed from the number of occurrences of at most $\min(|w|_a, |w|_b) + 1$ scattered factors of the form $a^i b$, where $|w|_a$ is the number of occurrences of the letter a in w. Moreover, we generalize the result to alphabets of the form $\{1, \ldots, q\}$ by showing that at most $\sum_{i=1}^{q-1} |w|_i (q - i + 1)$ scattered factors suffices to reconstruct w. Both results improve on the upper bounds known so far. Complexity time bounds on reconstruction algorithms are also considered here.

1 Introduction

The general scheme for a so-called *reconstruction problem* is the following one: given a sufficient amount of information about substructures of a hidden discrete structure, can one uniquely determine this structure? In particular, what are the fragments about the structure needed to recover it all. For instance, a square matrix of size at least 5 can be reconstructed from its principal minors given in any order [20].

In graph theory, given some subgraphs of a graph (these subgraphs may share some common vertices and edges), can one uniquely rebuild the original graph? Given a finite undirected graph $G = (V, E)$ with n vertices, consider the multiset made of the n induced subgraphs of G obtained by deleting exactly one vertex from G. In particular, one knows how many isomorphic subgraphs of a given class appear. Two graphs leading to the same multiset (generally called a *deck*) are said to be *hypomorphic*. A conjecture due to Kelly and Ulam states that two hypomorphic graphs with at least three vertices are isomorphic [14,21]. A similar

M. Lejeune—Supported by a FNRS fellowship.
F. Manea—Supported by the DFG grant MA 5725/2-1.

© Springer Nature Switzerland AG 2020
N. Jonoska and D. Savchuk (Eds.): DLT 2020, LNCS 12086, pp. 96–109, 2020.
https://doi.org/10.1007/978-3-030-48516-0_8

conjecture in terms of edge-deleted subgraphs has been proposed by Harary [11]. These conjectures are known to hold true for several families of graphs.

A *finite word*, i.e., a finite sequence of letters of some given alphabet, can be seen as an edge- or vertex-labeled linear tree. So variants of the graph reconstruction problem can be considered and are of independent interest. Participants of the Oberwolfach meeting on Combinatorics on Words in 2010 [2] gave a list of 18 important open problems in the field. Amongst them, the twelfth problem is stated as *reconstruction from subwords of given length*. In the following statement and all along the paper, a *subword* of a word is understood as a subsequence of not necessarily contiguous letters from this word, i.e., subwords can be obtained by deleting letters from the given word. To highlight this latter property, they are often called *scattered subwords* or *scattered factors*, which is the notion we are going to use.

Definition 1. *Let k, n be natural numbers. Words of length n over a given alphabet are said to be k-reconstructible whenever the multiset of scattered factors of length k (or k-deck) uniquely determines any word of length n.*

Notice that the definition requires multisets to store the information how often a scattered factor occurs in the words. For instance, the scattered factor ba occurs three times in baba which provides more information for the reconstruction than the mere fact that ba is a scattered factor.

The challenge is to determine the function $f(n) = k$ where k is the least integer for which words of length n are k-reconstructible. This problem has been studied by several authors and one of the first trace goes back to 1973 [13]. Results in that direction have been obtained by M.-P. Schützenberger (with the so-called *Schützenberger's Guessing game*) and L. Simon [25]. They show that words of length n sharing the same multiset of scattered factors of length up to $\lfloor n/2 \rfloor + 1$ are the same. Consequently, words of length n are $(\lfloor n/2 \rfloor + 1)$-reconstructible. In [15], this upper bound has been improved: Krasikov and Roditty have shown that words of length n are k-reconstructible for $k \geq \lfloor 16\sqrt{n}/7 \rfloor + 5$. On the other hand Dudik and Schulmann [6] provide a lower bound: if words of length n are k-reconstructible, then $k \geq 3^{(\sqrt{2/3} - o(1)) \log_3^{1/2} n}$. Bounds were also considered in [19]. Algorithmic complexity of the reconstruction problem is discussed, for instance, in [5]. Note that the different types of reconstruction problems have application in philogenetic networks, see, e.g., [12], or in the context of molecular genetics [7] and coding theory [16].

Another motivation, close to combinatorics on words, stems from the study of k-binomial equivalence of finite words and k-binomial complexity of infinite words (see [23] for more details). Given two words of the same length, they are k-binomially equivalent if they have the same multiset of scattered factors of length k, also known as k-*spectrum* ([1, 18, 24]). Given two words x and y of the same length, one can address the following problem: decide whether or not x and y are k-binomially equivalent? A polynomial time decision algorithm based on automata and a probabilistic algorithm have been addressed in [10]. A variation of our work would be to find, given k and n, a minimal set of scattered factors for which the knowledge of the number of occurrences in x and y permits to decide k-binomial equivalence.

Over an alphabet of size q, there are q^k pairwise distinct length-k factors. If we relax the requirement of only considering scattered factors of the same length, another interesting question is to look for a minimal (in terms of cardinality) multiset of scattered factors to reconstruct entirely a word. Let the *binomial coefficient* $\binom{u}{x}$ be the number of occurrences of x as a scattered factor of u. The general problem addressed in this paper is therefore the following one.

Problem 2. Let Σ be a given alphabet and n a natural number. We want to reconstruct a hidden word $w \in \Sigma^n$. To that aim, we are allowed to pick a word u_i and ask questions of the type "What is the value of $\binom{w}{u_i}$?". Based on the answers to questions related to $\binom{w}{u_1}, \ldots, \binom{w}{u_i}$, we can decide which will be the next question (i.e. decide which word will be u_{i+1}). We want to have the shortest sequence $(u_1, \ldots, u_k)$ uniquely determining w by knowing the values of $\binom{w}{u_1}, \ldots, \binom{w}{u_k}$.

We naturally look for a value of k less than the upper bound for k-reconstructibility.

In this paper, we firstly recall the use of Lyndon words in the context of reconstructibility. A word w over a totally ordered alphabet is called *Lyndon word* if it is the lexicographically smallest amongst all its rotations, i.e., $w = xy$ is smaller than yx for all non trivial factorisations $w = xy$. Every binomial coefficient $\binom{w}{x}$ for arbitrary words w and x over the same alphabet can be deduced from the values of the coefficients $\binom{w}{u}$ for Lyndon words u that are lexicographically less than or equal to x. This result is presented in Sect. 2 along with the basic definitions. We consider an alphabet equipped with a total order on the letters. Words of the form $\mathsf{a}^n\mathsf{b}$ with letters $\mathsf{a} < \mathsf{b}$ and a natural number n are a special form of Lyndon words, the so-called *right-bounded-block* words.

We consider the reconstruction problem from the information given by the occurrences of right-bounded-block words as scattered factors of a word of length n. In Sect. 3 we show how to reconstruct a word uniquely from $m + 1$ binomial coefficients of right-bounded-block words where m is the minimum number of occurrences of a and b in the word. We also prove that this is less than the upper bound given in [15]. In Sect. 4 we reduce the problem for arbitrary finite alphabets $\{1, \ldots, q\}$ to the binary case. Here we show that at most $\sum_{i=1}^{q-1} |w|_i \, (q - i + 1) \leq q|w|$ binomial coefficients suffice to uniquely reconstruct w with $|w|_i$ being the number of occurrences of letter i in w. Again, we compare this bound to the best known one for the classical reconstruction problem (from words of a given length). In the last section of the paper we also propose several results of algorithmic nature regarding the efficient reconstruction of words from given scattered factors.

Due to space restrictions some proofs (marked with ∗) can be found in [9].

2   Preliminaries

Let $\mathbb{N}$ be the set of natural numbers, $\mathbb{N}_0 = \mathbb{N} \cup \{0\}$, and let $\mathbb{N}_{\geq k}$ be the set of all natural numbers greater than or equal to k. Let $[n]$ denote the set $\{1, \ldots, n\}$ and $[n]_0 = [n] \cup \{0\}$ for an $n \in \mathbb{N}$.

An *alphabet* $\Sigma = \{a, b, c, \ldots\}$ is a finite set of letters and a *word* is a finite sequence of letters. We let Σ^* denote the set of all finite words over Σ. The *empty word* is denoted by ε and Σ^+ is the free semigroup $\Sigma^* \backslash \{\varepsilon\}$. The length of a word w is denoted by $|w|$. Let $\Sigma^{\leq k} := \{w \in \Sigma^* \mid |w| \leq k\}$ and Σ^k be the set of all words of length exactly $k \in \mathbb{N}$. The number of occurrences of a letter $a \in \Sigma$ in a word $w \in \Sigma^*$ is denoted by $|w|_a$. The ith letter of a word w is given by $w[i]$ for $i \in [|w|]$. The powers of $w \in \Sigma^*$ are defined recursively by $w^0 = \varepsilon$, $w^n = ww^{n-1}$ for $n \in \mathbb{N}$. A word $u \in \Sigma^*$ is a *factor* of $w \in \Sigma^*$, if $w = xuy$ holds for some words $x, y \in \Sigma^*$. Moreover, u is a *prefix* of w if $x = \varepsilon$ holds and a *suffix* if $y = \varepsilon$ holds. The factor of w from the ith to the jth letter will be denoted by $w[i..j]$ for $1 \leq i \leq j \leq |w|$. Two words $u, v \in \Sigma^*$ are called *conjugates or rotations* of each other if there exist $x, y \in \Sigma^*$ with $u = xy$ and $v = yx$. Additional basic information about combinatorics on words can be found in [17].

Definition 3. *Let $<$ be a total ordering on Σ. A word $w \in \Sigma^*$ is called* right-bounded-block word *if there exist $x, y \in \Sigma$ with $x < y$ and $\ell \in \mathbb{N}_0$ with $w = x^\ell y$.*

Definition 4. *A word $u = a_1 \cdots a_n \in \Sigma^n$, for $n \in \mathbb{N}$, is a* scattered factor *of a word $w \in \Sigma^+$ if there exist $v_0, \ldots, v_n \in \Sigma^*$ with $w = v_0 a_1 v_1 \cdots v_{n-1} a_n v_n$. For words $w, u \in \Sigma^*$, define $\binom{w}{u}$ as the number of occurrences of u as a scattered factor of w.*

Remark 5. Notice that $|w|_x = \binom{w}{x}$ for all $x \in \Sigma$.

The following definition addresses Problem 2.

Definition 6. *A word $w \in \Sigma^n$ is called* uniquely reconstructible/determined *by the set $S \subset \Sigma^*$ if for all words $v \in \Sigma^n \backslash \{w\}$ there exists a word $u \in S$ with $\binom{w}{u} \neq \binom{v}{u}$.*

Consider $S = \{ab, ba\}$. Then $w = abba$ is not uniquely reconstructible by S since $\left[\binom{w}{ab}, \binom{w}{ba}\right] = [2, 2]$ is also the 2-vector of binomial coefficients of $baab$. On the other hand $S = \{a, ab, ab^2\}$ reconstructs w uniquely. The following remark gives immediate results for binary alphabets.

Remark 7. Let $\Sigma = \{a, b\}$ and $w \in \Sigma^n$. If $|w|_a \in \{0, n\}$ then w contains either only b or a and by the given length n of w, w is uniquely determined by $S = \{a\}$. This fact is in particular an equivalence: $w \in \Sigma^n$ can be uniquely determined by $\{a\}$ iff $|w|_a \in \{0, n\}$. If $|w|_a \in \{1, n-1\}$, w is not uniquely determined by $\{a\}$ as witnessed by ab and ba for $n = 2$. It is immediately clear that the additional information $\binom{w}{ab}$ leads to unique determinism of w.

Lyndon words play an important role regarding the reconstruction problem. As shown in [22] only scattered factors which are Lyndon words are necessary to determine a word uniquely, i.e., S can always be assumed to be a set of Lyndon words.

Definition 8. *Let $<$ be a total ordering on Σ. A word $w \in \Sigma^*$ is a Lyndon word iff for all $u, v \in \Sigma^+$ with $w = uv$, we have $w <_{lex} vu$ where $<_{lex}$ is the lexicographical ordering on words induced by $<$.*

Proposition 9 ([22]). *Let w and u be two words. The binomial coefficient $\binom{w}{u}$ can be computed using only binomial coefficients of the type $\binom{w}{v}$ where v is a Lyndon word of length up to $|u|$ such that $v \leq_{lex} u$.*

To obtain a formula to compute the binomial coefficient $\binom{w}{u}$ for $w, u \in \Sigma^*$ by binomial coefficients $\binom{w}{v_i}$ for Lyndon words $v_1, \ldots, v_k$ with $v_i \in \Sigma^{\leq |u|}$, $i \in [k]$, and $k \in \mathbb{N}$ the definitions of shuffle and infiltration are necessary [17].

Definition 10. *Let $n_1, n_2 \in \mathbb{N}$, $u_1 \in \Sigma^{n_1}$, and $u_2 \in \Sigma^{n_2}$. Set $n = n_1 + n_2$. The shuffle of u_1 and u_2 is the polynomial $u_1 \sqcup\!\sqcup u_2 = \sum_{I_1, I_2} w(I_1, I_2)$ where the sum has to be taken over all pairs (I_1, I_2) of sets that are partitions of $[n]$ such that $|I_1| = n_1$ and $|I_2| = n_2$. If $I_1 = \{i_{1,1} < \ldots < i_{1,n_1}\}$ and $I_2 = \{i_{2,1} < \ldots < i_{2,n_2}\}$, then the word $w(I_1, I_2)$ is defined such that $w[i_{1,1}]w[i_{1,2}] \cdots w[i_{1,n_1}] = u_1$ and $w[i_{2,1}]w[i_{2,2}] \cdots w[i_{2,n_2}] = u_2$ hold.*

The infiltration is a variant of the shuffle in which equal letters can be merged.

Definition 11. *Let $n_1, n_2 \in \mathbb{N}$, $u_1 \in \Sigma^{n_1}$, and $u_2 \in \Sigma^{n_2}$. Set $n = n_1 + n_2$. The infiltration of u_1 and u_2 is the polynomial $u_1 \downarrow u_2 = \sum_{I_1, I_2} w(I_1, I_2)$, where the sum has to be taken over all pairs (I_1, I_2) of sets of cardinality n_1 and n_2 respectively, for which the union is equal to the set $[n']$ for some $n' \leq n$. Words $w(I_1, I_2)$ are defined as in the previous definition. Note that some $w(I_1, I_2)$ are not well defined if $i_{1,j} = i_{2,k}$ but $u_1[j] \neq u_2[k]$. In that case they do not appear in the previous sum.*

Considering for instance $u_1 = \mathsf{aba}$ and $u_2 = \mathsf{ab}$ gives the polynomials

$$u_1 \sqcup\!\sqcup u_2 = 2\mathsf{ababa} + 4\mathsf{aabba} + 2\mathsf{aabab} + 2\mathsf{abaab},$$

$$u_1 \downarrow u_2 = \mathsf{aba} \sqcup\!\sqcup \mathsf{ab} + \mathsf{aba} + 2\mathsf{abba} + 2\mathsf{aaba} + 2\mathsf{abab}.$$

Based on Definitions 10 and 11, we are able to give a formula to compute a binomial coefficient from the ones making use of Lyndon words. This formula is given implicitly in [22, Theorem 6.4]: Let $u \in \Sigma^*$ be a non-Lyndon word. By [22, Corollary 6.2] there exist non-empty words $x, y \in \Sigma^*$ and with $u = xy$ and such that every word appearing in the polynomial $x \sqcup\!\sqcup y$ is lexicographically less than or equal to u. Then, for all word $w \in \Sigma^*$, we have

$$\binom{w}{u} = \frac{1}{(x \sqcup\!\sqcup y, u)}\left[\binom{w}{x}\binom{w}{y} - \sum_{v \in \Sigma^*, v \neq u}(x \downarrow y, v)\binom{w}{v}\right],$$

where (P, v) is a notation giving the coefficient of the word v in the polynomial P. One may apply recursively this formula until only Lyndon factors are considered. Some examples can be found in [9].

3 Reconstruction from Binary Right-Bounded-Block Words

In this section we present a method to reconstruct a binary word uniquely from binomial coefficients of right-bounded-block words. Let $n \in \mathbb{N}$ be a natural number and $w \in \{a, b\}^n$ a word. Since the word length n is assumed to be known, $|w|_a$ is known if $|w|_b$ is given and vice versa. Set for abbreviation $k_u = \binom{w}{u}$ for $u \in \Sigma^*$. Moreover we assume w.l.o.g. $k_a \leq k_b$ and that k_a is known (otherwise substitute each a by b and each b by a, apply the following reconstruction method and revert the substitution). This implies that w is of the form

$$b^{s_1} a b^{s_2} \ldots b^{s_{k_a}} a b^{s_{k_a}+1} \tag{1}$$

for $s_i \in \mathbb{N}_0$ and $i \in [|w|_a + 1]$ with $\sum_{i \in [k_a+1]} s_i = n - k_a = k_b$ and thus we get for $\ell \in [k_a]_0$

$$k_{a^\ell b} = \binom{w}{a^\ell b} = \sum_{i=\ell+1}^{k_a+1} \binom{i-1}{\ell} s_i. \tag{2}$$

Remark 12. Notice that for fixed $\ell \in [k_a]_0$ and $c_i = \binom{i-1}{\ell}$ for $i \in [k_a + 1] \setminus [\ell]$, we have $c_i < c_{i+1}$ and especially $c_{\ell+1} = 1$ and $c_{\ell+2} = \ell + 1$.

Equation (2) shows that reconstructing a word uniquely from binomial coefficients of right-bounded-block words equates to solve a system of Diophantine equations. The knowledge of $k_b, \ldots, k_{a^\ell b}$ provides $\ell + 1$ equations. If the equation of $k_{a^\ell b}$ has a unique solution for $\{s_{\ell+1}, \ldots, s_{k_a+1}\}$ (in this case we say, by language abuse, that $k_{a^\ell b}$ is *unique*), then the system in row echelon form has a unique solution and thus the binary word is uniquely reconstructible. Notice that $k_{a^{k_a}b}$ is always unique since $k_{a^{k_a}b} = s_{k_a+1}$.

Consider $n = 10$ and $k_a = 4$. This leads to $w = b^{s_1} a b^{s_2} a b^{s_3} a b^{s_4} a b^{s_5}$ with $\sum_{i \in [5]} s_i = 6$. Given $k_{ab} = 4$ we get $4 = s_2 + 2s_3 + 3s_4 + 4s_5$. The s_i are not uniquely determined. If $k_{a^2b} = 2$ is also given, we obtain the equation $2 = s_3 + 3s_4 + 6s_5$ and thus $s_3 = 2$ and $s_4 = s_5 = 0$ is the only solution. Substituting these results in the previous equation leads to $s_2 = 0$ and since we only have six b, we get $s_1 = 4$. Hence $w = b^4 a^2 b^2 a^2$ is uniquely reconstructed by $S = \{a, ab, a^2b\}$.

The following definition captures all solutions for the equation defined by $k_{a^\ell b}$ for $\ell \in [k_a]_0$.

Definition 13. Set $M(k_{a^\ell b}) = \{(r_{\ell+1}, \ldots, r_{k_a+1}) \mid k_{a^\ell b} = \sum_{i=\ell+1}^{k_a+1} \binom{i-1}{\ell} r_i\}$ for fixed $\ell \in [k_a]_0$. We call $k_{a^\ell b}$ unique if $|M(k_{a^\ell b})| = 1$.

By Remark 12 the coefficients of each equation of the form (2) are strictly increasing. The next lemma provides the range each $k_{a^\ell b}$ may take under the constraint $\sum_{i=1}^{k_a+1} s_i = n - k_a$.

Lemma 14. *Let $n \in \mathbb{N}$, $k \in [n]_0$, $j \in [k+1]$ and $c_1, \ldots, c_{k+1}, s_1, \ldots, s_{k+1} \in \mathbb{N}_0$ with $c_i < c_{i+1}$, for $i \in [k]$, and $\sum_{i=1}^{k+1} s_i = n - k$. The sum $\sum_{i=j}^{k+1} c_i s_i$ is maximal iff $s_{k+1} = n - k$ (and consequently $s_i = 0$ for all $i \in [k]$).*

Proof. The case $k = 0$ is trivial. Consider the case $n = k$, i.e., $\sum_{i=1}^{k+1} s_i = 0$. This implies immediately $s_i = 0$ for all $i \in [k+1]$ and the equivalence holds. Assume for the rest of the proof $k < n$. If $s_{k+1} = n - k$, then $s_i = 0$ for all $i \leq k$ and $\sum_{i=j}^{k+1} c_i s_i = c_{k+1}(n - k)$. Let us assume that the maximal value for $\sum_{i=j}^{k+1} c_i s_i$ can be obtained in another way and that there exist $s'_1, \ldots, s'_{k+1} \in \mathbb{N}_0$, $\ell \in [n-k]$ such that $\sum_{i=1}^{k+1} s'_i = n - k$ and $s'_{k+1} = n - k - \ell$. Thus

$$c_{k+1}(n - k) \leq \sum_{i=j}^{k+1} c_i s'_i = \left(\sum_{i=j}^{k} c_i s'_i \right) + c_{k+1}(n - k - \ell).$$

This implies $\sum_{i=j}^{k} c_i s'_i \geq c_{k+1}\ell$. Since the coefficients are strictly increasing we get $\sum_{i=j}^{k} c_i s'_i \leq c_k \sum_{i=j}^{k} s'_i < c_{k+1}\ell$, hence the contradiction.   □

Corollary 15. *Let $k_a \in [n]_0$, $\ell \in [k_a]_0$, and $s_1, \ldots, s_{k_a+1} \in \mathbb{N}_0$ with $\sum_{i=1}^{k_a+1} s_i = n - k_a$. Then $\binom{w}{a^\ell b} \in \left[\binom{k_a}{\ell}(n - k_a) \right]_0$.*

Proof. It follows directly from Eq. (2) and Lemma 14.   □

The following lemma shows some cases in which $k_{a^\ell b}$ is unique.

Lemma 16. *Let $k_a \in [n]$, $\ell \in [k_a]_0$ and $s_1, \ldots, s_{k_a+1} \in \mathbb{N}_0$ with $\sum_{i=1}^{k_a+1} s_i = n - k_a$. If $k_{a^\ell b} \in [\ell]_0 \cup \{\binom{k_a}{\ell}(n - k_a)\}$ or $k_{a^\ell b} = \binom{k_a-1}{\ell}r + \binom{k_a}{\ell}(n - k_a - r)$ for $r \in [k_b]_0$ then $k_{a^\ell b}$ is unique.*

Proof. Consider firstly $k_{a^\ell b} \in [\ell]_0$. By Remark 12 we have $c_{\ell+1} = 1$ and $c_{\ell+2} = \ell + 1$. By $c_i < c_{i+1}$ we obtain immediately $s_i = 0$ for $i \in [k_a + 1]\setminus[\ell + 1]$. By setting $s_{\ell+1} = k_{a^\ell b}$ the claim is proven. If $k_{a^\ell b} = \binom{k_a}{\ell}(n - k_a)$, $s_{k_a+1} = (n - k_a)$ and $s_i = 0$ for $i \in [k_a]_0$ is the only possibility. Let secondly be $r \in [k_b]_0$ and $k_{a^\ell b} = \binom{k_a-1}{\ell}r + \binom{k_a}{\ell}(n - k_a - r)$ and suppose that $k_{a^\ell b}$ is not unique. This implies $s_{k_a+1} < n - k_a - r$. Assume that $s_{k_a+1} = n - k_a - r'$ for $r' \in [k_b]_{>r}$. Thus there exists $x \in \mathbb{N}$ with $\binom{k_a}{\ell}(n - k_a - r') + x = \frac{(k_a-1)!(k_a(n-k_a)-\ell r)}{\ell!(k_a-\ell)!}$, i.e., $x = \frac{(k_a-1)!(k_a r'-\ell r)}{\ell!(k_a-\ell)!}$. By $k_b = n - k_a$ we have $x \leq \binom{k_a-1}{\ell}r' = \frac{(k_a-1)!(k_a r'-\ell r')}{\ell!(k_a-\ell)!}$ (we only have r' occurrences of **b** left to distribute). By $r' > r$ we have $\frac{(k_a-1)!(k_a r'-\ell r)}{\ell!(k_a-\ell)!} = x < \frac{(k_a-1)!(k_a r'-\ell r)}{\ell!(k_a-\ell)!}$ - a contradiction.   □

Since we are not able to fully characterise the uniquely determined values for each $k_{a^\ell b}$ for arbitrary n and ℓ, the following proposition gives the characterisation for $\ell \in \{0, 1\}$. Notice that we use k_a immediately since it is determinable by n and $k_{a^0 b} = k_b$.

Proposition 17 (∗). *The word $w \in \Sigma^n$ is uniquely determined by k_a and k_{ab} iff one of the following occurs*

- *$k_a = 0$ or $k_a = n$ (and obviously $k_{ab} = 0$),*
- *$k_a = 1$ or $k_a = n - 1$ and k_{ab} is arbitrary,*
- *$k_a \in [n - 2]_{\geq 2}$ and $k_{ab} \in \{0, 1, k_a(n - k_a) - 1, k_a(n - k_a)\}$.*

In all cases not covered by Proposition 17 the word cannot be uniquely determined by $\binom{w}{a}$ and $\binom{w}{ab}$. The following theorem combines the reconstruction of a word with the binomial coefficients of right-bounded-block words.

Theorem 18. *Let $j \in [k_a]_0$. If $k_{a^j b}$ is unique, then the word $w \in \Sigma^n$ is uniquely determined by $\{b, ab, a^2b, \ldots, a^jb\}$.*

Proof. If $k_{a^j b}$ is unique, the coefficients $s_{j+1}, \ldots, s_{k_a + 1}$ are uniquely determined. Substituting backwards the known values in the first $j - 1$ Eq. (2) (for $\ell = 1, \ldots, j - 1$) we can now obtain successively the values for $s_j, \ldots, s_1$. □

Corollary 19. *Let ℓ be minimal such that $k_{a^\ell b}$ is unique. Then w is uniquely determined by $\{a, ab, a^2b, \ldots, a^\ell b\}$ and not uniquely determined by any $\{a, ab, a^2b, \ldots, a^ib\}$ for $i < \ell$.*

Proof. It follows directly from Theorem 18. □

By [15] an upper bound on the number of binomial coefficients to uniquely reconstruct the word $w \in \Sigma^n$ is given by the amount of the binomial coefficients of the $(\lfloor \frac{16}{7}\sqrt{n}\rfloor + 5)$-spectrum. Notice that implicitly the full spectrum is assumed to be known. As proven in Sect. 2, Lyndon words up to this length suffice. Since there are $\frac{1}{n}\sum_{d|n}\mu(d) \cdot 2^{\frac{n}{d}}$ Lyndon words of length n, the combination of both results presented in [15, 22] states that, for $n > 6$,

$$\sum_{i=1}^{\lfloor \frac{16}{7}\sqrt{n}\rfloor + 5} \frac{1}{i}\sum_{d|i}\mu(d) \cdot 2^{\frac{i}{d}} \tag{3}$$

binomial coefficients are sufficient for a unique reconstruction with the Möbius function μ. Up to now, it was the best known upper bound.

Theorem 18 shows that $\min\{k_a, k_b\} + 1$ binomial coefficients are enough for reconstructing a binary word uniquely. By Proposition 17 we need exactly one binomial coefficient if $n \in [3]$ and at most two if $n = 4$. For $n \in \{5, 6\}$ we need at most $n - 2$ different binomial coefficients. The following theorem shows that by Theorem 18 we need strictly less binomial coefficients for $n > 6$.

Theorem 20 (∗). *Let $w \in \Sigma^n$. We have that $\min\{k_a, k_b\} + 1$ binomial coefficients suffice to uniquely reconstruct w. If $k_a \leq k_b$, then the set of sufficient binomial coefficients is $S = \{b, ab, a^2b, ..., a^hb\}$ where $h = \lfloor \frac{n}{2}\rfloor$. If $k_a > k_b$, then the set is $S = \{a, ba, b^2a, ..., b^ha\}$. This bound is strictly smaller than (3).*

Remark 21. By Lemma 16 we know that $k_{a^\ell b}$ is unique if it is in $[\ell]_0$ or exactly $\binom{k_a}{\ell}(n-k_a)$. The probability for the latter is $\frac{1}{2^n}$ for $w \in \{a, b\}^n$. If $k_{a^\ell b} = m \in [\ell]_0$ we get by (2) immediately $s_{\ell+1} = m$ and $s_i = 0$ for $\ell+2 \leq i \leq k_a+1$. Hence, the values for s_j for $j \in [\ell]$ are not determined. By $\sum_{i \in [\ell]} s_i = n - k_a - m$ there are $d = \sum_{i \in [\ell]_0} \binom{\ell}{\ell-i}\binom{n-k_a-m-1}{i-1}$ possibilities to fulfill the constraints, i.e., we have a probability of $\frac{d}{2^n}$ to have such a word.

4   Reconstruction for Arbitrary Alphabets

In this section we address the problem of reconstructing words over arbitrary alphabets from their scattered factors. We begin with a series of results of algorithmic nature. Let $\Sigma = \{a_1, \ldots, a_q\}$ be an alphabet equipped with the ordering $a_i < a_j$ for $1 \leq i < j \leq q \in \mathbb{N}$.

Definition 22. *Let $w_1, \ldots, w_k \in \Sigma^*$ for $k \in \mathbb{N}$, and $K = (k_a)_{a \in \Sigma}$ a sequence of $|\Sigma|$ natural numbers. A K-valid marking of $w_1, \ldots, w_k$ is a mapping $\psi : [k] \times \mathbb{N} \to \mathbb{N}$ such that for all $j \in [k]$, $i, \ell \in [|w_j|]$, and $a \in \Sigma$ there holds*

- *if $w_j[i] = a$ then $\psi(j, i) \leq k_a$,*
- *if $i < \ell \leq |w_j|$ and $w_j[i] = w_j[\ell] = a$ then $\psi(j, i) < \psi(j, \ell)$.*

A K-valid marking of $w_1, \ldots, w_k$ is represented as the string $w_1^\psi, w_2^\psi, \ldots, w_k^\psi$, where $w_j^\psi[i] = (w_j[i])_{\psi(j,i)}$ for fresh letters $(w_j[i])_{\psi(j,i)}$.

For instance, let $k = 2$, $\Sigma = \{a, b\}$, and $w_1 = aab$, $w_2 = abb$. Let $k_a = 3, k_b = 2$ define the sequence K. A K-valid marking of w_1, w_2 would be $w_1^\psi = (a)_1(a)_3(b)_1, w_2^\psi = (a)_2(b)_1(b)_2$ defining ψ implicitly by the indices. We used parentheses in the marking of the letters in order to avoid confusions.

We recall that a topological sorting of a directed graph $G = (V, E)$, with $V = \{v_1, \ldots, v_n\}$, is a linear ordering $v_{\sigma(1)} < v_{\sigma(2)} < \ldots < v_{\sigma(n)}$ of the nodes, defined by the permutation $\sigma : [n] \to [n]$, such that there exists no edge in E from $v_{\sigma(i)}$ to $v_{\sigma(j)}$ for any $i > j$ (i.e., if v_a comes after v_b in the linear ordering, for some $a = \sigma(i)$ and $b = \sigma(j)$, then we have $i > j$ and there should be no edge between v_a and v_b). It is a folklore result that any directed graph G has a topological sorting if and only if G is acyclic.

Definition 23. *Let $w_1, \ldots, w_k \in \Sigma^*$ for $k \in \mathbb{N}$, $K = (k_a)_{a \in \Sigma}$ a sequence of $|\Sigma|$ natural numbers, and ψ a K-valid marking of $w_1, \ldots, w_k$. Let G_ψ be the graph that has $\sum_{a \in \Sigma} k_a$ nodes, labelled with the letters $(a)_1, \ldots, (a)_{k_a}$, for all $a \in \Sigma$, and the directed edges $((w_j[i])_{\psi(j,i)}, (w_j[i+1])_{\psi(j,i+1)})$, for all $j \in [k]$, $i \in [|w_j|]$, and $((a)_i, (a)_{i+1})$, for all occuring i and $a \in \Sigma$. We say that there exists a valid topological sorting of the ψ-marked letters of the words $w_1, \ldots, w_k$ if there exists a topological sorting of the nodes of G_ψ, i.e., G_ψ is a directed acyclic graph.*

The graph associated with the K-valid marking of w_1, w_2 from above would have the five nodes $(a)_1, (a)_2, (a)_3, (b)_1, (b)_2$ and the six directed edges

$((a)_1, (a)_3)$, $((a)_3, (b)_1)$, $((a)_2, (b)_1)$, $((b)_1, (b)_2)$, $((a)_1, (a)_2)$, $((a)_2, (a)_3)$ (where the direction of the edge is from the left node to the right node of the pair defining it). This graph has the topological sorting $(a)_1(a)_2(a)_3(b)_1(b)_2$.

Theorem 24 (∗). *For $w_1, \dots, w_k \in \Sigma^*$ and a sequence $K = (k_a)_{a \in \Sigma}$ of $|\Sigma|$ natural numbers, there exists a word w such that w_i is a scattered factor of w with $|w|_a = k_a$, for all $i \in [k]$ and all $a \in \Sigma$, if and only if there exist a K-valid marking ψ of the words $w_1, \dots, w_k$ and a valid topological sorting of the ψ-marked letters of the words $w_1, \dots, w_k$.*

Next we show that in Theorem 24 uniqueness propagates in the $\Leftarrow$-direction.

Corollary 25. *Let $w_1, \dots, w_k \in \Sigma^*$ and $K = (k_a)_{a \in \Sigma}$ a sequence of $|\Sigma|$ natural numbers. If the following hold*

- *there exists a unique K-valid marking ψ of the words $w_1, \dots, w_k$,*
- *in the unique K-valid marking ψ we have that for each $a \in \Sigma$ and $\ell \in [k_a]$ there exists $i \in [k]$ and $j \in [|w_i|]$ with $\psi(i, j) = \ell$, and*
- *there exists a unique valid topological sorting of the ψ-marked letters of the words $w_1, \dots, w_k$*

then there exists a unique word w such that w_i is a scattered factor of w, for all $i \in [k]$ and $|w|_a = k_a$ for all $a \in \Sigma$.

Proof. Let w be the word obtained by writing in order the letters of the unique valid topological sorting of the ψ-marked letters of the words $w_1, \dots, w_k$ and removing their markings. It is clear that w' has w_i as a scattered factor, for all $i \in [k]$, and that $|w|_a = k_a$, for all $a \in \Sigma$. The word w is uniquely defined (as there is no other K-valid marking nor valid topological sorting of the ψ-marked letters), and $|w|_a = k_a$, for all $a \in \Sigma$. □

In order to state the second result, we need the projection $\pi_S(w)$ of a word $w \in \Sigma^*$ on $S \subseteq \Sigma$: $\pi_S(w)$ is obtained from w by removing all letters from $\Sigma \setminus S$.

Theorem 26. *Set $W = \{w_{a,b} \mid a < b \in \Sigma\}$ such that*

- *$w_{a,b} \in \{a, b\}^*$ for all $a, b \in \Sigma$,*
- *for all $w, w' \in W$ and all $a \in \Sigma$, if $|w|_a \cdot |w'|_a > 0$, then $|w|_a = |w'|_a$.*

Then there exists at most one $w \in \Sigma^*$ such that $w_{a,b}$ is $\pi_{\{a,b\}}(w)$ for all $a, b \in \Sigma$.

Proof. Notice firstly $|W| = \frac{q(q-1)}{2}$. Let $k_a = |w_{a,b}|_a$, for $a < b \in \Sigma$. These numbers are clearly well defined, by the second item in our hypothesis. Let $K = (k_a)_{a \in \Sigma}$. It is immediate that there exists a unique K-valid marking ψ of the words $(w_{a,b})_{a < b \in \Sigma}$. As each two marked letters $(a)_i$ and $(b)_j$ (i.e., each two nodes $(a)_i$ and $(b)_j$ of G_ψ) appear in the marked word $w_{a,b}^\psi$, we know the order in which these two nodes should occur in a topological sorting of G_ψ. This means that, if G_ψ is acyclic, then it has a unique topological sorting. Our statement follows now from Corollary 25. □

Remark 27. Given the set $W = \{w_{\mathsf{a},\mathsf{b}} \mid \mathsf{a} < \mathsf{b} \in \Sigma\}$ as in the statement of Theorem 26, with $k_{\mathsf{a}} = |w_{\mathsf{a},\mathsf{b}}|_{\mathsf{a}}$, for $\mathsf{a} < \mathsf{b} \in \Sigma$, and $K = (k_{\mathsf{a}})_{\mathsf{a} \in \Sigma}$, we can produce the unique K-valid marking ψ of the words $(w_{\mathsf{a},\mathsf{b}})_{\mathsf{a} < \mathsf{b} \in \Sigma}$ in linear time $O(\sum_{\mathsf{a} < \mathsf{b} \in \Sigma} |w_{\mathsf{a},\mathsf{b}}|) = O((q-1) \sum_{\mathsf{a} \in \Sigma} k_{\mathsf{a}})$: just replace the ith letter a of $w_{\mathsf{a},\mathsf{b}}$ by $(\mathsf{a})_i$, for all a and i. The graph G_ψ has $O((q-1) \sum k_{\mathsf{a}})$ edges and $O(\sum k_{\mathsf{a}})$ vertices and can be constructed in linear time $O((q-1) \sum k_{\mathsf{a}})$. Sorting G_ψ topologically takes $O((q-1) \sum k_{\mathsf{a}})$ time (see, e.g., the handbook [4]). As such, we conclude that reconstructing a word $w \in \Sigma^*$ from its projections over all two-letter-subsets of Σ can be done in linear time w.r.t. the total length of the respective projections.

Theorem 26 is in a sense optimal: in order to reconstruct a word over Σ uniquely, we need all its projections on two-letter-subsets of Σ. That is, it is always the case that for a strict subset U of $\{\{\mathsf{a},\mathsf{b}\} \mid \mathsf{a} < \mathsf{b} \in \Sigma\}$, with $|U| = \frac{q(q-1)}{2} - 1$, there exist two words $w' \neq w$ such that $\{\pi_p(w') \mid p \in U\} = \{\pi_p(w) \mid p \in U\}$. We can, in fact, show the following results:

Theorem 28. *Let $S_1, \ldots, S_k$ be subsets of Σ. The following hold:*

1. *If each pair $\{\mathsf{a}, \mathsf{b}\} \subseteq \Sigma$ is included in at least one of the sets S_i, then we can reconstruct any word uniquely from its projections $\pi_{S_1}(\cdot), \ldots, \pi_{S_k}(\cdot)$.*
2. *If there exists a pair $\{\mathsf{a}, \mathsf{b}\}$ that is not contained in any of the sets $S_1, \ldots, S_k$, then there exist two words w and w' such that $w \neq w'$ and $\pi_{S_1}(w) = \pi_{S_1}(w'), \ldots, \pi_{S_k}(w) = \pi_{S_k}(w')$.*

Proof. The first part is, once again, a consequence of Corollary 25. The second part can be shown by assuming that $\Sigma = \{\mathsf{a}_1, \ldots, \mathsf{a}_q\}$ and the pair $\{\mathsf{a}_1, \mathsf{a}_2\}$ is not contained in any of the sets $S_1, \ldots, S_k$. Then, for $w = \mathsf{a}_1 \mathsf{a}_3 \mathsf{a}_4 \ldots \mathsf{a}_q$ and $w' = \mathsf{a}_2 \mathsf{a}_3 \mathsf{a}_4 \ldots \mathsf{a}_q$, we have that $\pi_{S_1}(w) = \pi_{S_1}(w'), \ldots, \pi_{S_k}(w) = \pi_{S_k}(w')$. ☐

In this context, we can ask how efficiently can we decide if a word is uniquely reconstructible from the projections $\pi_{S_1}(\cdot), \ldots, \pi_{S_k}(\cdot)$ for $S_1, \ldots, S_k \subset \Sigma$.

Theorem 29 (∗). *Given the sets $S_1, \ldots, S_k \subset \Sigma$, we decide whether we can reconstruct any word uniquely from its projections $\pi_{S_1}(\cdot), \ldots, \pi_{S_k}(\cdot)$ in $O(q^2 k)$ time. Moreover, under the* Strong Exponential Time Hypothesis *(see the survey [3] and the references therein), there is no $O(q^{2-d} k^c)$ algorithm for solving the above decision problem, for any $d, c > 0$.*

Coming now back to combinatorial results, we use the method developed in Sect. 3 to reconstruct a word over an arbitrary alphabet. We show that we need at most $\sum_{i \in [q]} |w|_i (q + 1 - i)$ different binomial coefficients to reconstruct w uniquely for the alphabet $\Sigma = \{1, \ldots, q\}$. In fact, following the results from the first part of this section, we apply this method on all combinations of two letters. Consider for an example that for $w \in \{\mathsf{a}, \mathsf{b}, \mathsf{n}\}^6$ the following binomial coefficients $\binom{w}{\mathsf{a}^0 \mathsf{b}} = 1$, $\binom{w}{\mathsf{a}^0 \mathsf{n}} = 2$, $\binom{w}{\mathsf{a}^1 \mathsf{n}} = 0$, $\binom{w}{\mathsf{a}^1 \mathsf{b}} = 3$, $\binom{w}{\mathsf{b}^1 \mathsf{n}} = 2$, and $\binom{w}{\mathsf{a}^2 \mathsf{n}} = 1$ are given. By $|w| = 6$, $|w|_{\mathsf{b}} = 1$, and $|w|_{\mathsf{n}} = 2$, we get $|w|_{\mathsf{a}} = 3$. Applying the method from Sect. 3 for $\{\mathsf{a}, \mathsf{b}\}$, $\{\mathsf{a}, \mathsf{n}\}$, and $\{\mathsf{b}, \mathsf{n}\}$ we obtain the scattered factors ba^3, anana, and bn^2. Combining all these three scattered factors gives us uniquely banana. Notice that in this example we only needed six binomial coefficients instead of ten, which is the worst case.

Remark 30. As seen in the example we have not only the word length but also $\binom{w}{x}$ for all $x \in \Sigma$ but one. Both information give us the remaining single letter binomial coefficient and hence we will assume that we know all of them.

For convenience in the following theorem consider $\Sigma = \{1, \ldots, q\}$ for $q > 2$ and set $\alpha := \lfloor \frac{16}{7} \sqrt{n} \rfloor + 5$. In the general case the results by [22] and [15] yield that

$$\sum_{i \in [\alpha]} \frac{1}{i} \frac{(q+1)^{\frac{i}{2}} - 1}{q} \tag{4}$$

is smaller than the best known upper bound on the number of binomial coefficients sufficient to reconstruct a word uniquely.

The following theorem generalises Theorem 20 on an arbitrary alphabet.

Theorem 31 (∗). *For uniquely reconstructing a word $w \in \Sigma^*$ of length at least $q - 1$, $\sum_{i \in [q]} |w|_i (q + 1 - i)$ binomial coefficients suffice, which is strictly smaller than (4).*

Remark 32. Since the estimation in Theorem 31 depends on the distribution of the letters in contrast to the method of reconstruction, it is wise to choose an order $<$ on Σ such that $x < y$ if $|w|_x \leq |w|_y$. In the example we have chosen the *natural* order $a < b < n$ which leads in the worst case to fourteen binomial coefficients that has to be taken into consideration. If we chose the order $b < n < a$ the formula from Theorem 31 provides that ten binomial coefficients suffice. This observation leads also to the fact that less binomial coefficients suffice for a unique determinism if the letters are not distributed equally but some letters occur very often and some only a few times.

Remark 33. Let's note that the number of binomial coefficients we need is at most qn. Indeed, we will prove that $\sum_{i \in [q]} |w|_i (q + 1 - i) \leq qn$. We have $qn = qn + n - n = q \sum_{i \in [q]} |w|_i + \sum_{i \in [q]} |w|_i - \sum_{i \in [q]} |w|_i \geq q \sum_{i \in [q]} |w|_i + \sum_{i \in [q]} |w|_i - \sum_{i \in [q]} (|w|_i i) = \sum_{i \in [q]} |w|_i (q + 1 - i)$.

5   Conclusion

In this paper we have proven that a relaxation of the so far investigated reconstruction problem from scattered factors from k-spectra to arbitrary sets yields that less scattered factors than the best known upper bound are sufficient to reconstruct a word uniquely. Not only in the binary but also in the general case the distribution of the letters plays an important role: in the binary case the amount of necessary binomial coefficients is smaller the larger $|w|_a - |w|_b$ is. The same observation results from the general case - if all letters are equally distributed in w then we need more binomial coefficients than in the case where some letters rarely occur and others occur much more often. Nevertheless the restriction to right-bounded-block words (that are intrinsically Lyndon words) shows that a word can be reconstructed by fewer binomial coefficients if scattered factors from different spectra are taken. Further investigations may lead into two directions: firstly a better characterisation of the uniqueness of the $k_{a^{\ell}b}$ would be helpful to understand better in which cases less than the worst case amount of binomial coefficients suffices and secondly other sets than the right-bounded-block words could be investigated for the reconstruction problem.

References

1. Berstel, J., Karhumäki, J.: Combinatorics on words - a tutorial. Bull. Eur. Assoc. Theor. Comput. Sci. EATCS **79**, 178–228 (2003)
2. Berthé, V., Karhumäki, J., Nowotka, D., Shallit, J.: Mini-workshop: combinatorics on words. Oberwolfach Rep. **7**, 2195–2244 (2010). https://doi.org/10.4171/OWR/2010/37
3. Bringmann, K.: Fine-grained complexity theory (tutorial). In: Niedermeier, R., Paul, C. (eds.) 36th International Symposium on Theoretical Aspects of Computer Science (STACS 2019), Leibniz International Proceedings in Informatics, vol. 4, pp. 1–7 (2019)
4. Cormen, T.H., Leiserson, C.E., Rivest, R.L., Stein, C.: Introduction to Algorithms, 3rd edn. MIT Press, Cambridge (2009)
5. Dress, A.W.M., Erdős, P.L.: Reconstructing words from subwords in linear time. Ann. Comb. **8**, 457–462 (2004)
6. Dudik, M., Schulman, L.J.: Reconstruction from subsequences. J. Combin. Theory, Ser. A **103**, 337–348 (2003)
7. Erdős, P.L., Ligeti, P., Sziklai, P., Torney, D.C.: Subwords in reverse-complement order. Ann. Comb. **10**, 415–430 (2006)
8. Ferov, M.: Irreducible polynomial modulo p, Bachelor thesis at Charles University Prague (2008)
9. Fleischmann, P., Lejeune, M., Manea, F., Nowotka, D., Rigo, M.: Reconstructing words from right-bounded-block words, 21 p. (2020). arXiv:2001.11218
10. Freydenberger, D.D., Gawrychowski, P., Karhumäki, J., Manea, F., Rytter, W.: Testing k-binomial equivalence. In: Multidisciplinary Creativity: homage to G. Păun on his 65th birthday, pp. 239–248, Ed. Spandugino, Bucharest (2015). arXiv:1509.00622
11. Harary, F.: On the reconstruction of a graph from a collection of subgraphs. In: Theory of Graphs and its Applications (Proceedings of Symposium Smolenice, 1963), pp. 47–52. Publ. House Czechoslovak Acad. Sci., Prague (1964)
12. van Iersel, L., Moulton, V.: Leaf-reconstructibility of phylogenetic networks. SIAM J. Discrete Math. **32**, 2047–2066 (2018)
13. Kalashnik, L.I.: The reconstruction of a word from fragments. In: Numerical Mathematics and Computer Technology, pp. 56–57. Akad. Nauk Ukrain. SSR Inst. Mat, Preprint IV (1973)
14. Kelly, P.J.: A congruence theorem for trees. Pac. J. Math. **7**, 961–968 (1957)
15. Krasikov, I., Roditty, Y.: On a reconstruction problem for sequences. J. Combin. Theory Ser. A **77**, 344–348 (1997)
16. Levenshtein, V.I.: On perfect codes in deletion and insertion metric. Discret. Math. Appl. **2**, 241–258 (1992)
17. Lothaire, M.: Combinatorics on Words. Cambridge University Press, Cambridge (1997)

18. Maňuch, J.: Characterization of a word by its subwords. In: Developments in Language Theory, pp. 210–219. World Scientific (1999)
19. Manvel, B., Meyerowitz, A., Schwenk, A., Smith, K., Stockmeyer, P.: Reconstruction of sequences. Discret. Math. **94**, 209–219 (1991)
20. Manvel, B., Stockmeyer, P.K.: On reconstruction of matrices. Math. Mag. **44**, 218–221 (1971)
21. O'Neil, P.V.: Ulam's conjecture and graph reconstructions. Am. Math. Mon. **77**, 35–43 (1970)
22. Reutenauer, C.: Free lie algebras. In: Cohn, P.M., Dales, H.G. (eds.) London Mathematical Society Monographs New Series (1993)
23. Rigo, M., Salimov, P.: Another generalization of abelian equivalence: binomial complexity of infinite words. Theoret. Comput. Sci. **601**, 47–57 (2015)
24. Rozenberg, G., Salomaa, A.: Handbook of Formal Languages (3 volumes). Springer, Heidelberg (1997). https://doi.org/10.1007/978-3-642-59126-6
25. Simon, I.: Piecewise testable events. In: Brakhage, H. (ed.) GI-Fachtagung 1975. LNCS, vol. 33, pp. 214–222. Springer, Heidelberg (1975). https://doi.org/10.1007/3-540-07407-4_23

A Study of a Simple Class of Modifiers: Product Modifiers

Pascal Caron, Edwin Hamel-de-le-court[(✉)], and Jean-Gabriel Luque

LITIS, Université de Rouen, Avenue de l'Université,
76801 Saint-Étienne du Rouvray Cedex, France
{Pascal.Caron,Jean-Gabriel.Luque}@univ-rouen.fr,
Edwin.Hamel-de-le-court@etu.univ-rouen.fr

Abstract. A modifier is a k-ary operator acting on DFAs and producing a DFA. Modifiers are involved in the theory of state complexity. We define and study a class of simple modifiers, called product modifiers, and we link closely the regular operations they encode to boolean operations.

1 Introduction

State complexity is a measure of complexity defined on regular operations. It allows to write the size of the minimal automaton recognizing the output as a function of the sizes of the minimal automata recognizing the inputs. The topic dates back to the 70s, from the seminal paper of Maslov [14] describing, explicitly but without any proof, the state complexities of several operations. Since the 90s, this area of research became very active and the state complexity of numerous operations has been computed. See, for example, [6,11–13,15] and [8] for a survey of the subject.

However, a few general methods are commonly used in order to compute state complexities. The most common method consists in providing a witness, which is a specific example reaching what is proven to be an upper bound. The witness itself is, in general, found by trial and error, sometimes using a witness that worked for a number of other operations and modifying it to fit the specific needs of the operation considered. In many cases, for example [1,7] or [4], the witness is constructed by considering, explicitly or implicitly, the whole monoid of the transformations acting on the states of the minimal automata recognizing the input languages. This method has been theorized in two independently written papers [2,5]. More precisely, the approach consists, on the one hand, in describing states as combinatorial objects and finding upper bounds using combinatorial tools, and, on the other hand, in building a huge witness, called a *monster*, chosen in a set of automata having as many transition functions as possible. This method can be applied to obtain the state complexity to the wide range of 1-*uniform* operations that are associated to operators, called *modifiers*, that act on automata to produce an automaton in a certain restrictive way. In this paper, we examine the regular operations described by the class of some very simple modifiers called *product modifiers*. These modifiers are characterized

© Springer Nature Switzerland AG 2020
N. Jonoska and D. Savchuk (Eds.): DLT 2020, LNCS 12086, pp. 110–121, 2020.
https://doi.org/10.1007/978-3-030-48516-0_9

by the fact that they build the Cartesian product automaton with the transitions took from the input automata. We investigate many properties of this class and in particular we completely describe the set of the regular operations that can be encoded by product modifiers. The paper is organized as follows. Section 2 gives definitions and notations about automata. In Sect. 3, we partially recall the monster approach. Finally, in Sect. 4, we define product modifiers and characterize the regular operations they encode in Sect. 5.

2   Preliminaries

2.1   Operations over Sets

The *set of subsets* of E is denoted by 2^E and the *set of mappings* of E into itself is denoted by E^E. The *symmetric difference* of two sets E_1 and E_2 is denoted by $\oplus$ and defined by $E_1 \oplus E_2 = (E_1 \cup E_2) \backslash (E_1 \cap E_2)$.

Let $(E_1, \ldots, E_k)$ be a k-tuple of finite sets, and let $(\delta_1, \ldots, \delta_k)$ be a k-tuple such that δ_i is a function from E_i to E_i for every $i \in \{1, \ldots, k\}$. For any k-tuple $(e_1, \ldots, e_k)$ such that $e_i \in E_i$ for all $i \in \{1, \ldots, k\}$, we denote by $(\delta_1, \ldots, \delta_k)(e_1, \ldots, e_k)$ the k-tuple $(\delta_1(e_1), \ldots, \delta_k(e_k))$.

Let E be a set, $f : E^j \to E$ and $g : E^k \to E$ for some $j, k \in \mathbb{N} \setminus \{0\}$. A *composition* is a function $f \circ_p g : E^{j+k-1} \to E$ defined for some $1 \leq p \leq j$ by

$$f \circ_p g(e_1, \ldots, e_{j+k-1}) = f(e_1, \ldots, e_{p-1}, g(e_p, \ldots, e_{p+k-1}), e_{p+k}, \ldots, e_{j+k-1}),$$

for any $e_1, \ldots, e_{j+k-1} \in E$.

2.2   Languages and Automata

Let Σ be a finite alphabet. A *word* w over Σ is a finite sequence of symbols of Σ. The set of all finite words over Σ is denoted by Σ^*. A *language* over Σ is a subset of Σ^*. We define the *complement* of a language $L \subseteq \Sigma^*$ by $L^c = \Sigma^* \setminus L$.

A *complete and deterministic finite automaton* (DFA) is a 5-tuple $A = (\Sigma, Q, i, F, \delta)$ where Σ is the input alphabet, Q is a finite set of states, $i \in Q$ is the initial state, $F \subset Q$ is the set of final states and δ is the transition function from $Q \times \Sigma$ to Q that is defined for every $q \in Q$ and every $a \in \Sigma$. We can extend transition functions in a natural way to functions from $Q \times \Sigma^*$ to Q, and again to functions from $2^Q \times \Sigma^*$ to Q. For any word w, we denote by δ^w the function $q \to \delta(q, w)$.

Let $A = (\Sigma, Q, i, F, \delta)$ be a DFA. A word $w \in \Sigma^*$ is *recognized* by the DFA A if $\delta(i, w) \in F$. The *language recognized* by a DFA A is the set $\mathrm{L}(A)$ of words recognized by A. By Kleene's theorem, a language is regular if and only if it is recognized by a DFA. It is well known that for any DFA, there exists a unique minimal one (up to isomorphism) among all DFAs recognizing the same language ([10]).

2.3 State Complexity

A *unary regular operation* is a function from regular languages of Σ into regular languages of Σ. A *k-ary regular operation* is a function from the set of k-tuples of regular languages over Σ into regular languages over Σ.

The state complexity of a regular language L denoted by $\mathrm{sc}(L)$ is the number of states of its minimal DFA. This notion extends to regular operations: the state complexity of a unary regular operation $\otimes$ is the function $\mathrm{sc}_\otimes$ such that, for all $n \in \mathbb{N}$, $\mathrm{sc}_\otimes(n)$ is the maximum of all the state complexities of $\otimes(L)$ when L is of state complexity n, *i.e.*

$$\mathrm{sc}_\otimes(n) = \max\{\mathrm{sc}(\otimes(L)) | \mathrm{sc}(L) = n\}.$$

This can be generalized, and the state complexity of a k-ary operation $\otimes$ is the k-ary function $\mathrm{sc}_\otimes$ such that, for all $(n_1, \ldots, n_k) \in (\mathbb{N})^k$,

$$\mathrm{sc}_\otimes(n_1, \ldots, n_k) = \max\{\mathrm{sc}(\otimes(L_1, \ldots, L_k)) \mid \text{ for all } i \in \{1, \ldots, k\}, \mathrm{sc}(L_i) = n_i\}.$$

Then, a witness for $\otimes$ is a way to assign to each $(n_1, \ldots, n_k)$, where each n_i is assumed sufficiently big, a k-tuple of languages $(L_1, \ldots, L_k)$ with $\mathrm{sc}(L_i) = n_i$, for all $i \in \{1, \ldots, k\}$, satisfying $\mathrm{sc}_\otimes(n_1, \ldots, n_k) = \mathrm{sc}(\otimes(L_1, \ldots, L_k))$.

3 Modifiers and 1-uniform Operations

We describe a class of regular operations, called 1-uniform which are interesting for the study of state complexity [3,5]. We then define operations on DFA called modifiers, and describe a subset of these operations that correspond to the set of 1-uniform regular operations.

3.1 Definition and First Properties

Definition 1. *Let Σ and Γ be two alphabets. A morphism is a function ϕ from Σ^* to Γ^* such that, for all $w, v \in \Sigma^*$, $\phi(wv) = \phi(w)\phi(v)$. Notice that ϕ is completely defined by its value on letters. A morphism ϕ is 1-uniform if the image by ϕ of any letter is a letter.*

The preimage $\phi^{-1}(L)$ of a regular language L by a morphism ϕ is regular, see, *e.g.*, [9]. This allows us to introduce the notion of 1-uniform regular operation.

Definition 2. *A k-ary regular operation $\otimes$ is 1-uniform if, for any k-tuple of regular languages $(L_1, \ldots, L_k)$, for any 1-uniform morphism ϕ, we have $\otimes(\phi^{-1}(L_1), \ldots, \phi^{-1}(L_k)) = \phi^{-1}(\otimes(L_1, \ldots, L_k))$.*

Obviously, 1-uniformity is stable by composition. Many well-known regular operations are 1-uniform. See [5] for a non-exhaustive list of examples like the complement, the Kleene star, the reverse, the cyclic shift, and the mirror, all boolean operations and catenation among others.

Each 1-uniform regular k-ary operation corresponds to a construction over DFAs, which is handy when we need to compute the state complexity of its elements. Such a construction on DFAs has some constraints that are described in the following definitions.

Definition 3. *The* state configuration *of a DFA* $A = (\Sigma, Q, i, F, \delta)$ *is the triplet* (Q, i, F).

Definition 4. *A* k-modifier *is a* k-ary *operation acting on a* k-tuple of DFAs $(A_1, \ldots, A_k)$, *on the same alphabet* Σ, *and producing a DFA* $\mathfrak{m}(A_1, ..., A_k)$ *such that*

- *its alphabet is* Σ,
- *its state configuration depends only on the state configurations of the DFAs* $A_1, \ldots, A_k$,
- *for any letter* $a \in \Sigma$, *the transition function of a in* $\mathfrak{m}(A_1, \ldots, A_k)$ *depends only on the state configurations of the DFAs* $A_1, \ldots, A_k$ *and on the transition functions of a in each of the DFAs* $A_1, ..., A_k$.

Example 1. For any DFA $A = (\Sigma, Q, i, F, \delta)$, define $\mathfrak{Star}(A) = (\Sigma, 2^Q, \emptyset, \{E | E \cap F \neq \emptyset\} \cup \{\emptyset\}, \delta_1)$, where for any $a \in \Sigma$, $\delta_1^a(\emptyset) = \delta^a(i)$ if $\delta^a(i) \notin F$ and $\delta_1^a(\emptyset) = \delta^a(i)$ otherwise, and, for all $E \neq \emptyset$, $\delta_1^a(E) = \delta^a(E)$ if $\delta^a(E) \cap F = \emptyset$ and $\delta_1^a(E) = \delta^a(E) \cup \{i\}$ otherwise. The modifier $\mathfrak{Star}$ describes a construcion on DFA associated to the Star operation on languages, *i.e.* for all DFA A, $L(A)^* = L(\mathfrak{Star}(A))$.

Example 2. For any DFAs $A = (\Sigma, Q_1, i_1, F_1, \delta_1)$ and $B = (\Sigma, Q_2, i_2, F_2, \delta_2)$, let $\mathfrak{Xor}(A, B) = (\Sigma, Q_1 \times Q_2, (i_1, i_2), (F_1 \times (Q_2 \setminus F_2)) \cup (Q_1 \setminus F_1) \times F_2), (\delta_1, \delta_2))$. The modifier $\mathfrak{Xor}$ describes the classical construction associated to the symmetrical difference, *i.e* for all DFAs A and B, $L(A) \oplus L(B) = L(\mathfrak{Xor}(A, B))$.

Definition 5. *A* k-modifier $\mathfrak{m}$ *is* 1-uniform *if, for every pair of* k-tuples of DFAs $(A_1, \ldots, A_k)$ *and* $(B_1, \ldots, B_k)$ *such that* $L(A_j) = L(B_j)$ *for all* $j \in \{1, \ldots, k\}$, *we have* $L(\mathfrak{m}(A_1, \ldots, A_k)) = L(\mathfrak{m}(B_1, \ldots, B_k))$. *In that case, there exists a regular operation* $\otimes_{\mathfrak{m}}$ *such that, for all* k-tuples $(A_1, \ldots, A_k)$ *of DFAs, we have* $\otimes_{\mathfrak{m}}(L(A_1), \ldots, L(A_k)) = L(\mathfrak{m}(A_1, \ldots, A_k))$. *We say that* $\mathfrak{m}$ *describes the operation* $\otimes_{\mathfrak{m}}$.

We easily check that, for modifiers, the 1-uniformity is stable by composition.

Claim. Let $\mathfrak{m}_1$ and $\mathfrak{m}_2$ be respectively a j-modifier and a k-modifier describing, respectively, operations $\otimes_1$ and $\otimes_2$. The modifier $\mathfrak{m}_1 \circ_p \mathfrak{m}_2$ describes $\otimes_1 \circ_p \otimes_2$.

The correspondence between 1-uniform modifiers and 1-uniform operations is stated in the following Theorem proved in [3].

Theorem 1. *A* k-ary *operation* $\otimes$ *is* 1-uniform *if and only if there exists a* k-modifier $\mathfrak{m}$ *such that* $\otimes = \otimes_{\mathfrak{m}}$.

Modifiers have been defined, for the first time, in [2] as a tool to compute state complexity of 1-uniform operations.

3.2   Functional Notations

When there is no ambiguity, for any character X and any integer k given by the context, we write $\underline{X}$ for $(X_1, \cdots, X_k)$. The number k will often be the arity of the regular operation or of the modifier we are considering.

From Definition 4, any k-modifier $\mathfrak{m}$ can be seen as a 4-tuple of mappings $(\mathfrak{Q}, \mathfrak{i}, \mathfrak{f}, \mathfrak{d})$ acting on k DFAs $\underline{A}$ with $A_j = (\Sigma, Q_j, i_j, F_j, \delta_j)$ to build a DFA $\mathfrak{m}\underline{A} = (\Sigma, Q, i, F, \delta)$, where $Q = \mathfrak{Q}(\underline{Q}, \underline{i}, \underline{F})$, $i = \mathfrak{i}(\underline{Q}, \underline{i}, \underline{F})$, $F = \mathfrak{f}(\underline{Q}, \underline{i}, \underline{F})$ and $\forall a \in \Sigma$, $\delta^a = \mathfrak{d}(\underline{i}, \underline{F}, \underline{\delta}^a)$. For the sake of clarity, we do not write explicitly the domains of the 4-tuple of mappings but the reader can derive them easily from the above equalities. Notice that we do not need to point out explicitly the dependency of $\mathfrak{d}$ on $\underline{Q}$ because the information is already contained in $\underline{\delta}^a$. We identify modifiers and such 4-tuples of mappings with each other. Below we revisit the definition of $\mathfrak{Xor}$ according to this formalism.

Example 3. $\mathfrak{Xor} = (\mathfrak{Q}, \mathfrak{i}, \mathfrak{f}, \mathfrak{d})$ where

$$\mathfrak{Q}((Q_1, Q_2), (i_1, i_2), (F_1, F_2)) = Q_1 \times Q_2, \quad \mathfrak{i}((Q_1, Q_2), (i_1, i_2), (F_1, F_2)) = (i_1, i_2),$$
$$\mathfrak{f}((Q_1, Q_2), (i_1, i_2), (F_1, F_2)) = F_1 \times (Q_2 \setminus F_2) \cup (Q_1 \setminus F_1) \times F_2,$$
$$\mathfrak{d}((i_1, i_2), (F_1, F_2), (\delta_1, \delta_2)) = (\delta_1, \delta_2).$$

4   Product Modifiers

In this section, we study a kind of simple modifier called *product modifiers* and show that they are closely linked to boolean operations.

Definition 6. *A k-modifier* $\mathfrak{m} = (\mathfrak{Q}, \mathfrak{i}, \mathfrak{f}, \mathfrak{d})$ *is a* product modifier *if, for any k-tuple of finite sets $\underline{Q}$, for any k-tuple of finite sets $\underline{F}$ such that $F_j \subseteq Q_j$ for all j, and for any $\underline{i} \in Q_1 \times \cdots \times Q_k$*

1. $\mathfrak{Q}(\underline{Q}, \underline{i}, \underline{F}) = Q_1 \times \cdots \times Q_k$.
2. $\forall a \in \Sigma$, $\mathfrak{d}(\underline{i}, \underline{F}, \underline{\delta}^a) = \underline{\delta}^a$, *with* $\underline{\delta}^a(\underline{q}) = (\delta_1^a(q_1), \delta_2^a(q_2), ..., \delta_k^a(q_k))$.

In other words, if $\mathfrak{m}$ is a product modifier, then $\mathfrak{m}\underline{A}$ is the product automaton of the A_j, but with final states $\mathfrak{f}(\underline{Q}, \underline{i}, \underline{F})$ and initial state $\mathfrak{i}(\underline{Q}, \underline{i}, \underline{F})$. Intuitively, product modifiers do not change the transition functions of the automata they act on, but seek only to change their final and initial states. We can easily check that the class of product modifiers is stable by composition.

For the sake of simplicity, in this section, $\mathfrak{m}$ denotes any k-ary product (but not necessarily 1-uniform) modifier and $\underline{A} = (A_1, \ldots, A_k)$ any sequence of k DFAs, with $A_j = (\Sigma, Q_j, i_j, F_j, \delta_j)$. Recall that $\underline{i} = (i_1, \ldots, i_k)$, $\underline{Q} = (Q_1, \ldots, Q_k)$ and $\underline{F} = (F_1, \ldots, F_k)$. We also denote $\mathfrak{m}\underline{A} = (\Sigma, Q', i', F', \underline{\delta})$.

We define the complementary product to get an easier access to the intersection of languages and their complement.

Definition 7. *For any k-tuple $\underline{P}$ of finite sets, for any k-tuple $\underline{G}$ of finite sets such that $G_j \subseteq P_j$ for all j, and for any $d \subseteq \{1, 2, ..., k\}$, we define* $\mathrm{cp}(d, \underline{G}, \underline{P}) = X_1 \times \cdots \times X_k$, *where $X_i = P_i \backslash G_i$ if $i \in d$ and $X_i = G_i$ otherwise.*

Example 4. $\mathrm{cp}(\{1,3\}, (\{1\}, \{2,3\}, \{2\}), (\{1,2\}, \{1,2,3,4\}, \{1,2,3\})) = \{2\} \times \{2,3\} \times \{1,3\}$.

Lemma 1. *The set* $\{\mathrm{cp}(d, \underline{F}, \underline{Q}) \mid d \subseteq \{1, \ldots, k\}\}$ *is a partition of* Q'.

Proof. Let $d \neq d'$ and suppose that there exists $j \in d \setminus d'$. For any element $\underline{q} \in \mathrm{cp}(d, \underline{F}, \underline{Q})$, we have $q_j \notin F_j$ and, for any element $\underline{q}' \in \mathrm{cp}(d', \underline{F}, \underline{Q})$, we have $q'_j \in F_j$. It follows that $\mathrm{cp}(d, \underline{F}, \underline{Q}) \cap \mathrm{cp}(d', \underline{F}, \underline{Q}) = \emptyset$.

Furthermore, consider an element $\underline{q} \in Q'$ and set $d = \{j \mid q_j \notin F_j\}$. Obviously, $\underline{q} \in \mathrm{cp}(d, \underline{F}, \underline{Q})$. This proves our result. $\square$

The following lemma sets a restriction on the form of $\mathfrak{f}$ on each of its entries, given that $\mathfrak{i}$ does not change the initial states in its entries.

Lemma 2. *Assume* $\underline{i}' = \underline{i}$. *If* $\mathfrak{m}$ *is 1-uniform then there exists* $E \subseteq 2^{\{1,2,\ldots,k\}}$ *such that* $F' = \bigcup_{d \in E} \mathrm{cp}(d, \underline{F}, \underline{Q})$.

Proof. Let us prove the contrapositive statement and assume that there is no set $E \subseteq 2^{\{1,2,\ldots,k\}}$ such that $F' = \bigcup_{d \in E} \mathrm{cp}(d, \underline{F}, \underline{Q})$. From Lemma 1, there exists $d \subseteq \{1, 2, \ldots, k\}$ such that $F' \cap \mathrm{cp}(d, \underline{F}, \underline{Q}) \notin \{\emptyset, \mathrm{cp}(d, \underline{F}, \underline{Q})\}$. Let d be such a set and let $G = \mathrm{cp}(d, \underline{F}, \underline{Q})$. The idea of the proof is to construct, with the states in G, two k-tuple of automata $\underline{B}$ and $\underline{C}$ that recognize the same languages, and such that $\mathrm{L}(\mathfrak{m}\underline{B})$ and $\mathrm{L}(\mathfrak{m}\underline{C})$ are different.

We distinguish two cases :

- First, suppose that $\underline{i} \in G$. If $\underline{i} \in F'$ then we choose $\underline{j} \in G \backslash F'$, otherwise we choose $\underline{j} \in G \cap F'$. Consider the two k-tuples of DFAs $\underline{B}$ and $\underline{C}$ such that $B_l = (\{a\}, Q_l, i_l, F_l, \beta_l)$ and $C_l = (\{a\}, Q_l, i_l, F_l, \gamma_l)$, where, for all positive integer $l \leq k$, $\beta_l^a(i_l) = j_l$ if $x = i_l$, $\beta_l^a(x) = x$ if $x \in Q_{i_l} \setminus \{i_l\}$, and $\gamma_l^a(x) = x$, for any $x \in Q_{i_l}$. Let us remark that, as $\underline{i}, \underline{j} \in G = \mathrm{cp}(d, \underline{F}, \underline{Q})$, i_l and j_l are either both in F_l (if $l \notin d$), or both not in F_l (if $l \in d$) by definition of cp. Therefore, i_l and j_l have the same finality in B_l, which is also their finality in C_l, and either B_l and C_l recognize a^*, or B_l and C_l recognize $\emptyset$.
  As described in Fig. 1, the transition functions β of $\mathfrak{m}\underline{B}$ and γ of $\mathfrak{m}\underline{C}$ satisfy $\beta^a(\underline{i}) = \underline{j}$ and $\gamma^a(\underline{i}) = \underline{i}$.
  The finality of $\underline{i}$ is the same in $\mathfrak{m}\underline{B}$ and $\mathfrak{m}\underline{C}$. However, it is not the same finality as $\underline{j}$ in $\mathfrak{m}\underline{B}$ and $\mathfrak{m}\underline{C}$. Therefore, we have $(a \in \mathrm{L}(\mathfrak{m}\underline{B}) \wedge a \notin \mathrm{L}(\mathfrak{m}\underline{C}))$ or $(a \notin \mathrm{L}(\mathfrak{m}\underline{B}) \wedge a \in \mathrm{L}(\mathfrak{m}\underline{C}))$. As a consequence, $\mathrm{L}(\mathfrak{m}\underline{B}) \neq \mathrm{L}(\mathfrak{m}\underline{C})$ and this implies that $\mathfrak{m}$ is not 1-uniform.
- Suppose now that $\underline{i} \notin G$. Let $\underline{j} \in G \backslash F'$, and let $\underline{j}' \in G \cap F'$. Consider the two k-tuple of DFAs $\underline{B}$ and $\underline{C}$ such that $B_l = (\{a, b\}, Q_l, i_l, F_l, \beta_l)$ and $C_l = (\{a, b\}, Q_l, i_l, F_l, \gamma_l)$, where, for all letters $u \in \{a, b\}$, for all positive integer $l \leq k$ and all $x \in Q_l$,

$$\beta_l^u(x) = \begin{cases} j_l & \text{if } x = i_l \wedge u = a \\ j'_l & \text{if } x = j_l \wedge u = b \\ x & \text{otherwise.} \end{cases} \quad \text{and} \quad \gamma_l^u(x) = \begin{cases} j'_l & \text{if } x = i_l \wedge u = a \\ j_l & \text{if } x = j'_l \wedge u = b \\ x & \text{otherwise.} \end{cases}$$

Fig. 1. Part of $\mathrm{m}\underline{B}$ and $\mathrm{m}\underline{C}$.

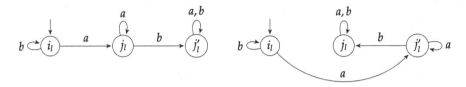

Fig. 2. Parts of B_l and C_l.

For any positive integer $l \leq k$, B_l and C_l recognize the same language. Indeed, from Fig. 2, as $\underline{j}, \underline{j}' \in G = \mathrm{cp}(d, \underline{F}, Q)$, j_l and j_l' have the same finality in B_l and C_l by definition of cp, we distinguish the cases :

 - $i_l \in F_l$ and $j_l \in F_l$. $\mathrm{L}(B_l) = \mathrm{L}(C_l) = (a + b)^*$
 - $i_l \in F_l$ and $j_l \notin F_l$. $\mathrm{L}(B_l) = \mathrm{L}(C_l) = b^*$
 - $i_l \notin F_l$ and $j_l \in F_l$. $\mathrm{L}(B_l) = \mathrm{L}(C_l) = b^* a(a + b)^*$
 - $i_l \notin F_l$ and $j_l \notin F_l$. $\mathrm{L}(B_l) = \mathrm{L}(C_l) = \emptyset$

As $\mathrm{m}\underline{B}$ and $\mathrm{m}\underline{C}$ are cartesian products of the B_l and the C_l respectively, if we call β the transition function of $\mathrm{m}\underline{B}$ and γ the transition function of $\mathrm{m}\underline{C}$, we have $\beta^a(\underline{i}) = \underline{j}$, $\beta^b(\underline{j}) = \underline{j}'$, $\gamma^a(\underline{i}) = \underline{j}'$, and $\gamma^b(\underline{j}') = \underline{j}$.

The finality of $\underline{j}$ is the same in $\mathrm{m}\underline{B}$ and $\mathrm{m}\underline{C}$. However, it is different from the finality of $\underline{j}'$ in $\mathrm{m}\underline{B}$ and $\mathrm{m}\underline{C}$. Therefore, we have $(ab \in \mathrm{L}(\mathrm{m}\underline{B}) \wedge ab \notin \mathrm{L}(\mathrm{m}\underline{C}))$ or $(ab \notin \mathrm{L}(\mathrm{m}\underline{B}) \wedge ab \in \mathrm{L}(\mathrm{m}\underline{C}))$. As a consequence, $\mathrm{L}(\mathrm{m}\underline{B}) \neq \mathrm{L}(\mathrm{m}\underline{C})$ which implies that m is not 1-uniform. □

The following two lemmas state that, for product modifiers, we can set $\underline{i}' = \underline{i}$ without changing the regular operation associated to m.

Lemma 3. *If* m *is* 1-*uniform then* $\underline{i}'$ *and* $\underline{i}$ *have the same finality.*

Proof. Let us prove the contrapositive of our statement. Assume that $\underline{i}'$ and $\underline{i}$ do not have the same finality, *i.e.* $(\underline{i} \notin F' \wedge \underline{i}' \in F')$ or $(\underline{i} \in F' \wedge \underline{i}' \notin F')$. Consider the two k-tuples of DFAs $\underline{B}$ and $\underline{C}$ such that $B_l = (\{a\}, Q_l, i_l, F_l, \beta_l)$ and $C_l = (\{a\}, Q_l, i_l, F_l, \gamma_l)$, where, for any $l \in \{1, \ldots, k\}$, $\beta_l^a(i_l') = i_l$, $\beta_l^a(q) = q$ when $q \neq i_l'$ and $\gamma_l^a(q) = q$. Let us remark that B_l and C_l recognize $\{a\}^*$ if $i_l \in F_l$, and $\emptyset$ otherwise. In any case, they recognize the same language.

If we denote by β the transition function of $\mathrm{m}\underline{B}$ and by γ the transition function of $\mathrm{m}\underline{C}$, we have $\beta^a(\underline{i}') = \underline{i}$ and $\gamma^a(\underline{i}') = \underline{i}'$. Recall that $\underline{i}'$ is the initial state of $\mathrm{m}\underline{B}$ and $\mathrm{m}\underline{C}$. Since $\underline{i}$ and $\underline{i}'$ do not have the same finality, the word a belongs to one of the languages $\mathrm{L}(\mathrm{m}\underline{B})$ or $\mathrm{L}(\mathrm{m}\underline{C})$ but not both (see Fig. 3). Hence the two automata do not recognize the same language and, as a consequence, m is not 1-uniform. □

Fig. 3. Part of $\mathfrak{m}\underline{B}$ and $\mathfrak{m}\underline{C}$.

We define an equivalence relation on states of the output of product modifiers whose relationship with the finality of states is examined in Lemma 4.

Definition 8. *Let $\underline{j}$ and $\underline{j}'$ be two k-tuples. We define the equivalence relation $\sim_{\underline{j},\underline{j}'}$ on k-tuples by $(x_1,\ldots,x_k) \sim_{\underline{j},\underline{j}'} (y_1,\ldots,y_k)$ if and only if for all $l \in \{1,\ldots,k\}$, $j_l = j'_l$ implies $x_l = y_l$.*

Example 5. We have $(3,3,2,5,1) \sim_{(1,4,3,2,3),(2,4,2,2,6)} (1,3,5,5,2)$.
We do not have $(3,3,2,5,1) \sim_{(1,4,3,2,3),(2,4,2,2,6)} (1,3,5,1,2)$.

Lemma 4. *If $\mathfrak{m}$ is 1-uniform then $L(\mathfrak{m}\underline{A}) = L((\Sigma, Q', \underline{i}, F', \underline{\delta}))$.*

Proof. One has to investigate the two complementary cases:

- *There exists two states $\underline{q} \in F', \underline{q}' \in Q' \setminus F'$ such that $\underline{q} \sim_{\underline{i},\underline{i}'} \underline{q}'$.*
  In this case we prove that $\underline{i} = \underline{i}'$, in other words $\mathfrak{m}A = (\Sigma, Q', \underline{i}, F', \delta)$. Let us show the contrapositive of the property. Suppose $\underline{i} \neq \underline{i}'$. We have to show that $\mathfrak{m}$ is not 1-uniform. By Lemma 3, $\underline{i} \in F' \wedge \underline{i}' \in F'$ or $\underline{i} \notin F' \wedge \underline{i}' \notin F'$. Consider the two k-tuples of DFAs $\underline{B}$ and $\underline{C}$ such that $B_l = (\{a\}, Q_l, i_l, F_l, \beta_l)$ and $C_l = (\{a\}, Q_l, i_l, F_l, \gamma_l)$, where for all $l \in \{1,\ldots,k\}$ and all $q \in Q_l$,

$$\beta_l^a(q) = \begin{cases} q_l \text{ if } q = i'_l \\ q \text{ otherwise} \end{cases} \quad \text{and} \quad \gamma_l^a(q) = \begin{cases} q'_l \text{ if } q = i'_l \\ q \text{ otherwise.} \end{cases}$$

  Let us remark that either $i'_l = i_l$, which implies $q_l = q'_l$, and $B_l = C_l$, or $i'_l \neq i_l$, and B_l and C_l recognize $\{a\}^*$ if $i_l \in F_l$ and $\emptyset$ otherwise. In any case, they recognize the same language. Recall that β is the transition function of $\mathfrak{m}\underline{B}$ and γ is the transition function of $\mathfrak{m}\underline{C}$. We have $\beta^a(\underline{i}') = \underline{q}$ and $\gamma^a(\underline{i}') = \underline{q}'$. Thus we have $a \in L(\mathfrak{m}\underline{B})$ and $a \notin L(\mathfrak{m}\underline{C})$. Therefore, $L(\mathfrak{m}\underline{B}) \neq L(\mathfrak{m}\underline{C})$ and $\mathfrak{m}$ is not 1-uniform.

- *For any two states $\underline{q}, \underline{q}' \in Q', \underline{q} \sim_{\underline{i},\underline{i}'} \underline{q}'$ implies that $\underline{q}$ and $\underline{q}'$ have the same finality.*

  First, for any letter $a \in \Sigma$, any two states $\underline{q}, \underline{q}' \in Q'$, the equivalence $\underline{q} \sim_{\underline{i},\underline{i}'} \underline{q}'$ implies

$$\underline{\delta}^a(\underline{q}) = (\delta_1^a(q_1), \delta_2^a(q_2), \ldots, \delta_k^a(q_k)) \sim_{\underline{i},\underline{i}'} (\delta_1^a(q'_1), \delta_2^a(q'_2), \ldots, \delta_k^a(q'_k)) = \underline{\delta}^a(\underline{q}').$$

  This property extends inductively to any word $w \in \Sigma^*$, i.e. $\underline{q} \sim_{\underline{i},\underline{i}'} \underline{q}'$ implies $\underline{\delta}^w(\underline{q}) \sim_{\underline{i},\underline{i}'} \underline{\delta}^w(\underline{q}')$. In particular, applying this to $\underline{q} = \underline{i}$ and $\underline{q}' = \underline{i}'$, we have $\delta^w(\underline{i}') \in F'$ if and only if $\delta^w(\underline{i}) \in F'$. As a direct consequence, the languages recognized by the two automata are the same. □

From Lemma 4, one can assume without loss of generality that $\underline{i} = \underline{i}'$. Hence, applying Lemma 2, we obtain

Corollary 1. *If* $\mathfrak{m}$ *is 1-uniform then there exists* $E \subseteq 2^{\{1,2,\ldots,k\}}$ *such that* $F' = \bigcup_{d \in E} \mathrm{cp}(d, \underline{F}, \underline{Q})$.

5 Quasi-boolean Operations

Before stating our main result, we need to clarify what is meant by a *boolean operation*. A boolean operation is an operation associated to an expression involving only the operators union, intersection and complement. It is well known that such an expression is equivalent to one written as a union of intersection of languages or their complement. More formally,

Definition 9. *A* k-*ary boolean operation* $\otimes$ *over regular languages* $L_1, \ldots, L_k$ *is defined as*

$$\otimes \underline{L} = \bigcup_{d \in E} \left(\bigcap_{i \in d} L_i \cap \bigcap_{i \notin d} L_i{}^c \right),$$

for some $E \subseteq 2^{\{1,\ldots,k\}}$. *Notice that there is a one-to-one correspondence between the boolean* k-*ary operations and the sets* $E \subseteq 2^{\{1,\ldots,k\}}$. *So we denote* $E_\otimes = E$.

Example 6. The classical boolean operation union can be written this way: for any two regular languages L_1 and L_2,

$$L_1 \cup L_2 = (L_1 \cap L_2{}^c) \cup (L_1 \cap L_2) \cup (L_1{}^c \cap L_2) = \bigcup_{d \in E} \left(\bigcap_{i \in d} L_i \cap \bigcap_{i \notin d} L_i{}^c \right),$$

with $E = \{\{1\}, \{2\}, \{1,2\}\}$.

We easily check that boolean operations are 1-uniform and can be associated to some product modifiers. More formally,

Lemma 5. *Assume that* $\otimes$ *is a* k-*ary boolean operation. Then* $\otimes = \otimes_\mathfrak{m}$, *where* $\mathfrak{m} = (\mathfrak{Q}, \mathfrak{i}, \mathfrak{f}, \mathfrak{d})$ *is a product modifier such that* $\mathfrak{i}(\underline{Q}, \underline{i}, \underline{F}) = \underline{i}$ *and*

$$\mathfrak{f}(\underline{Q}, \underline{i}, \underline{F}) = \bigcup_{d \in E_\otimes} \mathrm{cp}(d, \underline{F}, \underline{Q}).$$

From Definition 9, we construct a wider class of operators that we prove to be in correspondence with product modifiers.

Definition 10. *For any* k-*ary regular operation* $\otimes$, *for any* $\underline{v} \in \{0,1\}^k$, *we denote by* $\otimes^{\underline{v}}$ *the restriction of* $\otimes$ *to the set*

$$\mathcal{L}^{\underline{v}} = \{(L_1, \ldots, L_k) \mid \forall i \in \{1, \ldots, k\}, L_i \text{ is regular and } v_i = 0 \Leftrightarrow \epsilon \in L_i\}.$$

We say that $\otimes$ *is a* k-*ary quasi-boolean operation if for all* $\underline{v} \in \{0,1\}^k$, $\otimes^{\underline{v}}$ *is a boolean operation, i.e. for any* $\underline{v}$, *there exists a boolean operation* $\otimes_1$ *such that for any* $\underline{L} \in \mathcal{L}^{\underline{v}}$ *we have* $\otimes_1 \underline{L} = \otimes^{\underline{v}} \underline{L}$.

Example 7. Consider the unary operator defined by $\otimes L = L$ if $\epsilon \in L$ and L^c otherwise. This operation is clearly not boolean. Nevertheless, since for each $L \in \mathcal{L}^{(0)}$ we have $\otimes L = L$ and for each $L \in \mathcal{L}^{(1)}$ we have $\otimes L = L^c$, the operation $\otimes$ is quasi-boolean.

These operations do not have a higher state complexity than boolean operations, as we show in the following statement.

Proposition 1. *For any quasi-boolean k-ary operation $\otimes$, we have*

$$\mathrm{sc}_\otimes(n_1, \ldots, n_k) \le n_1 \cdots n_k.$$

Proof. Lemma 5 implies that $\mathrm{sc}_\otimes(n_1, \ldots, n_k) \le n_1 \cdots n_k$ for any boolean operation $\otimes$. We we prove our statement, by remarking that, for any quasi-boolean operation $\otimes$, we have $\mathrm{sc}_\otimes(n_1, \ldots, n_k) \le \max\{\mathrm{sc}_{\otimes^{\underline{v}}}(n_1, \ldots, n_k) \mid \underline{v} \in \{0,1\}^k\}$. $\square$

We now introduce our main result that characterizes the operations encoded by 1-uniform product modifiers.

Theorem 2. *An operation $\otimes$ is quasi-boolean if and only if there exists a 1-uniform product modifier $\mathfrak{m}$ such that $\otimes = \otimes_\mathfrak{m}$.*

Proof. Let $\otimes$ be a k-ary quasi-boolean operation. We construct a modifier $\mathfrak{m}$ such that $\otimes = \otimes_\mathfrak{m}$ as follows. We consider the product modifier $\mathfrak{m} = (\mathfrak{Q}, \mathfrak{i}, \mathfrak{f}, \mathfrak{d})$ such that $\mathfrak{i}(\underline{Q}, \underline{i}, \underline{F}) = \underline{i}$ and,

$$\mathfrak{f}(\underline{Q}, \underline{i}, \underline{F}) = \bigcup_{d \in E_{\otimes \underline{v}}} \mathrm{cp}(d, \underline{F}, \underline{Q}),$$

where $\underline{v} \in \{0,1\}^k$ is such that $v_j = 0$ if and only if $i_j \in F_j$.

Let $\underline{L} \in \mathcal{L}^{\underline{v}}$ for some $\underline{v} \in \{0,1\}^k$. For any k-tuple of DFAs $\underline{A}$ such that $A_j = (\Sigma, Q_j, i_j, F_j, \delta_j)$ recognizes L_j, we have $i_j \in F_j$ if and only if $v_j = 0$. From Lemma 5, one has $\mathrm{L}(\mathfrak{m}\underline{A}) = \otimes^{\underline{v}}\underline{L}$. Hence, $\mathfrak{m}$ is $1 - uniform$ and $\otimes = \otimes_\mathfrak{m}$.

Now, we prove the converse. Let $\otimes$ be a regular operation such that there exists a 1-uniform product modifier $\mathfrak{m}$ satisfying $\otimes_\mathfrak{m} = \otimes$. We use a *reductio ad absurdum* argument by assuming that $\otimes$ is not quasi-boolean. Let $\underline{v} \in \{0,1\}^k$ be such that $\otimes^{\underline{v}}$ is not a boolean operation. Let $\underline{A}$ be a k-tuple of DFAs with $A_l = (\Sigma, Q_l, i_l, F_l, \alpha_l)$ such that $(\mathrm{L}(A_1), \ldots, \mathrm{L}(A_k)) \in \mathcal{L}^{\underline{v}}$. Furthermore, we assume that for all $l \in \{1, \ldots, k\}$, $F_l \notin \{\emptyset, Q_l\}$. By Corollary 1, there exists $E \subseteq 2^{\{1,2,\ldots,k\}}$ such that $F = \mathfrak{f}(\underline{Q}, \underline{i}, \underline{F}) = \bigcup_{d \in E} \mathrm{cp}(d, \underline{F}, \underline{Q})$. There-

fore, $\mathfrak{m}\underline{A} = \bigcup_{d \in E} \left(\bigcap_{l \in d} \mathrm{L}(A_l) \cap \bigcap_{l \in \{1,2,\ldots,k\}\setminus d} \mathrm{L}(A_l)^c \right)$ which is obviously a boolean operation applied to $(\mathrm{L}(A_1), \ldots, \mathrm{L}(A_k))$. Since $\otimes^{\underline{v}}$ is not a boolean operation, there exists $\underline{A}'$, with $A_l' = (\Sigma', Q_l', i_l', F_l', \alpha_l')$, a k-tuple of DFAs such that

$(\mathrm{L}(A_1'), \ldots, \mathrm{L}(A_k')) \in \mathcal{L}^{\underline{v}}$ and $\mathfrak{m}\underline{A}' \ne \bigcup_{d \in E} \left(\bigcap_{l \in d} \mathrm{L}(A_l') \cap \bigcap_{l \in \{1,2,\ldots,k\}\setminus d} \mathrm{L}(A_l')^c \right)$. We construct new k-tuples of DFAs $\underline{B}$ and $\underline{B}'$ such that $L(B_l) = L(B_l')$ but such that $L(\mathfrak{m}\underline{B}) \neq L(\mathfrak{m}\underline{B}')$, contradicting the 1-uniformity of $\mathfrak{m}$. By Corollary 1, there exists $H \subseteq 2^{\{1,\ldots,k\}}$ such that $F' = \mathfrak{f}(\underline{Q}', \underline{i}', \underline{F}') = \bigcup_{d \in H} \mathrm{cp}(d, \underline{F}', \underline{Q}')$.

We have to examine two cases. Either there exists $\underline{p}' \in F'$ such that $\underline{p}' \notin \bigcup_{d \in E} \mathrm{cp}(d, \underline{F}', \underline{Q}')$, or there exists $\underline{p}' \in \bigcup_{d \in E} \mathrm{cp}(d, \underline{F}', \underline{Q}')$ such that $\underline{p}' \notin F'$. We only describe the first case, as the other one is treated symmetrically. Therefore, Lemma 1 implies that there exists $d \in H \setminus E$ such that $\underline{p}' \in \mathrm{cp}(d, \underline{F}', \underline{Q}')$. Let $\underline{p} \in \mathrm{cp}(d, \underline{F}, \underline{Q})$. Notice that $\underline{p} \notin F$ while each p_l has the same finality in B_l as p_l' in B_l'. Also remark that, as $(\mathrm{L}(A_1), \ldots, \mathrm{L}(A_k))$ and $(\mathrm{L}(A_1'), \ldots, \mathrm{L}(A_k'))$ are in $\mathcal{L}^{\underline{v}}$, for all $l \in \{1, \ldots, k\}$, $v_l = 0$ implies $i_l \in F_l$ and $i_l' \in F_l'$, and $v_l = 1$ implies $i_l \notin F_l$ and $i_l' \notin F_l'$.

Now consider the two k-tuples of DFAs $\underline{B}$ and $\underline{B}'$ such that $B_l = (\{a\}, Q_l, i_l, F_l, \beta_l)$ and $B_l' = (\{a\}, Q_l', i_l', F_l', \beta_l')$, where β_l and β_l' are defined, for all positive integer $l \leq k$ and all $(q, q') \in Q_l \times Q_l'$, by :

$$\beta_l^a(q) = \begin{cases} p_l & \text{if } q = i_l \\ q & \text{otherwise.} \end{cases} \quad \text{and} \quad \beta_l'^a(q') = \begin{cases} p_l' & \text{if } q' = i_l' \\ q' & \text{otherwise.} \end{cases}$$

We notice that, for all $l \in \{1, \ldots, k\}$, B_l and B_l' recognize the same language L_l. Indeed, since i_l and i_l' have the same finality and p_l and p_l' have the same finality, one has to examine four cases which are summarized in Table 1.

Furthermore, we have $\underline{\beta}^a(\underline{i}) = \underline{p} \notin F$, and $\underline{\beta}'^a(\underline{i}') = \underline{p}' \in F'$, which means that $a \notin L(\mathfrak{m}\underline{B})$ and $a \in L(\mathfrak{m}\underline{B}')$, which contradicts the 1-uniformity of $\mathfrak{m}$. □

Table 1. Common values of $L(B_l)$ and $L(B_l')$.

$L(B_l) = L(B_l')$	$i_l \in F_l$	$i_l \notin F_l$
$p_l \in F_l$	$\{a_\emptyset^*\}$	$\{a\}^+$
$p_l \notin F_l$	$\{\epsilon\}$	$\emptyset$

6   Conclusion

We have shown that some very simple modifiers, namely product modifiers, encode a class of very low state complexity operations. This is a non-trivial example of a set of modifiers closed by composition whose associated regular operations are completely described. The proof techniques open perspectives to explore other classes of modifiers closed by composition. The aim for our future works is to establish a kind of atlas, as complete as possible, of the set of modifiers in relation to the theory of state complexity.

References

1. Brzozowski, J., Jirásková, G., Liu, B., Rajasekaran, A., Szykuła, M.: On the state complexity of the shuffle of regular languages. In: Câmpeanu, C., Manea, F., Shallit, J. (eds.) DCFS 2016. LNCS, vol. 9777, pp. 73–86. Springer, Cham (2016). https://doi.org/10.1007/978-3-319-41114-9_6
2. Caron, P., Hame-De-Le-Court, E., Luque, J.G., Patrou, B.: New tools for state complexity. Discret. Math. Theor. Comput. Sci. **22**(1) (2020)
3. Caron, P., Hame-De-Le-Court, E., Luque, J.G.: Algebraic and combinatorial tools for state complexity : application to the star-xor problem. In: Leroux, J., Raskin, J.F. (eds.) Proceedings Tenth International Symposium on Games, Automata, Logics, and Formal Verification, GandALF 2019, Bordeaux, France, 2–3 September 2019, vol. 305 of EPTCS, pp. 154–168 (2019)
4. Caron, P., Luque, J.-G., Mignot, L., Patrou, B.: State complexity of catenation combined with a boolean operation: a unified approach. Int. J. Found. Comput. Sci. **27**(6), 675–704 (2016)
5. Davies, S.: A general approach to state complexity of operations: formalization and limitations. In: Hoshi, M., Seki, S. (eds.) DLT 2018. LNCS, vol. 11088, pp. 256–268. Springer, Cham (2018). https://doi.org/10.1007/978-3-319-98654-8_21
6. Domaratzki, M.: State complexity of proportional removals. J. Automata Lang. Comb. **7**(4), 455–468 (2002)
7. Domaratzki, M., Okhotin, A.: State complexity of power. Theor. Comput. Sci. **410**(24–25), 2377–2392 (2009)
8. Gao, Y., Moreira, N., Reis, R., Sheng, Y.: A survey on operational state complexity. J. Automata Lang. Comb. **21**(4), 251–310 (2017)
9. Hopcroft, J.E., Motwani, R., Ullman, J.D.: Introduction to Automata Theory, Languages, and Computation, 3rd edn. Addison-Wesley, Boston (2007). Pearson international edition
10. Hopcroft, J.E., Ullman, J.D.: Introduction to Automata Theory, Languages and Computation. Addison-Wesley, Boston (1979)
11. Jirásek, J., Jirásková, G., Szabari, A.: State complexity of concatenation and complementation. Int. J. Found. Comput. Sci. **16**(3), 511–529 (2005)
12. Jirásková, G.: State complexity of some operations on binary regular languages. Theor. Comput. Sci. **330**(2), 287–298 (2005)
13. Jirásková, G., Okhotin, A.: State complexity of cyclic shift. ITA **42**(2), 335–360 (2008)
14. Maslov, A.N.: Estimates of the number of states of finite automata. Soviet Math. Dokl. **11**, 1373–1375 (1970)
15. Sheng, Y.: State complexity of regular languages. J. Automata Lang. Comb. **6**(2), 221 (2001)

Operations on Permutation Automata

Michal Hospodár[1][(✉)] [iD] and Peter Mlynárčik[1,2]

[1] Mathematical Institute, Slovak Academy of Sciences, Grešákova 6,
040 01 Košice, Slovakia
hosmich@gmail.com, mlynarcik1972@gmail.com
[2] Faculty of Humanities and Natural Sciences, University of Prešov,
Ul. 17. novembra 1, 081 16 Prešov, Slovakia

Abstract. We investigate the class of languages recognized by permutation deterministic finite automata. Using automata constructions and some properties of permutation automata, we show that this class is closed under Boolean operations, reversal, and quotients, and it is not closed under concatenation, power, Kleene closure, positive closure, cut, shuffle, cyclic shift, and permutation. We prove that the state complexity of Boolean operations, Kleene closure, positive closure, and right quotient on permutation languages is the same as in the general case of regular languages. Next, we get the tight upper bounds on the state complexity of concatenation ($m2^n - 2^{n-1} - m + 1$), square ($n2^{n-1} - 2^{n-2}$), reversal ($\binom{n}{\lceil n/2 \rceil}$), and left quotient ($\binom{m}{\lceil m/2 \rceil}$; tight if $m \le n$). All our witnesses are unary or binary, and the binary alphabet is always optimal, except for Boolean operations in the case of $\gcd(m, n) = 1$. In the unary case, the state complexity of all considered operations is the same as for regular languages, except for quotients and cut. In case of quotients, it is $\min\{m, n\}$, and in case of cut, it is either $2m - 1$ or $2m - 2$, depending on whether there exists an integer ℓ with $2 \le \ell \le n$ such that $m \bmod \ell \ne 0$.

1 Introduction

A deterministic finite automaton (DFA) is said to be a permutation DFA if every input symbol induces a permutation on the state set. A language recognized by a permutation DFA is called a permutation language. The class of permutation languages has been introduced by Thierrin [22] who also proved some closure properties of this class using cancellative congruences. The aim of this paper is to study this class in detail, including the operational state complexity.

The state complexity of a regular operation is the number of states sufficient and necessary in the worst case for a DFA to accept the language resulting from the operation, considered as a function of the numbers of states in DFAs for operands. The state complexity of basic regular operations was given by Maslov [15] and Yu et al. [23].

Research supported by VEGA grant 2/0132/19 and grant APVV-15-0091.

If the operands of an operation belong to some subclass of regular languages, then the complexity of this operation may be significantly smaller than in the general case. The operational state complexity in several subclasses of regular languages has been studied in the literature. Câmpeanu et al. [6] considered finite languages, Han and Salomaa [8,9] investigated prefix-free and suffix-free languages, and Brzozowski et al. [3,4] examined ideal and closed languages. The classes of co-finite, star-free, union-free, and unary languages have been studied as well [1,5,13,18].

In this paper, we first study closure properties of the class of permutation languages. We provide alternative proofs to those in [22]; our proofs use automata constructions and properties of permutation DFAs. Moreover, we prove that permutation languages are not closed under power, positive closure, cut, shuffle, cyclic shift, and permutation. Then we study the operational state complexity in the class of permutation languages. For each considered operation, we obtain tight upper bounds on its state complexity. Our witnesses are defined over unary or binary alphabets. In the case of left and right quotients, we need $m \leq n$ to prove tightness.

2   Preliminaries

We assume that the reader is familiar with the backgrounds in formal languages and automata theory. For details, the reader may refer to [11,21].

For a rational number r, we denote the set $\{i \in \mathbb{Z} \mid 0 \leq i < r\}$ by $\mathbf{r}$. For example, we have $\mathbf{3} = \{0, 1, 2\}$, and for an integer n we have $\mathbf{n} = \{0, 1, \ldots, n-1\}$. Next, we have $\frac{7}{2} = \{0, 1, 2, 3\}$ and $\frac{n}{2} = \{0, 1, \ldots, \lceil \frac{n}{2} \rceil - 1\}$. For a finite set S, its size is denoted by $|S|$, and its power-set by 2^S.

Let Σ be a non-empty alphabet of symbols. Then Σ^* denotes the set of all strings over Σ including the empty string ε. A language is any subset of Σ^*.

Given two languages K and L over Σ, the *complement* of L is the language $L^c = \Sigma^* \setminus L$, and the operations of *intersection*, *union*, *difference*, and *symmetric difference* are defined as standard set operations. The *concatenation* of K and L is $KL = \{uv \mid u \in K \text{ and } v \in L\}$. The k-*th power* of L is defined as $L^0 = \{\varepsilon\}$ and $L^k = LL^{k-1}$ if $k \geq 1$. The second power is called *square*. The *Kleene closure* of L is $L^* = \bigcup_{i \geq 0} L^i$. The *positive closure* of L is $L^+ = \bigcup_{i \geq 1} L^i$. The *right quotient* of K by L is $KL^{-1} = \{w \mid wx \in K \text{ for some } x \text{ in } L\}$. The *left quotient* of K by L is $L^{-1}K = \{w \mid xw \in K \text{ for some } x \text{ in } L\}$. The *cut* of K and L is $K \,!\, L = \{uv \mid u \in K, v \in L, \text{ and } uv' \notin K \text{ for every non-empty prefix } v' \text{ of } v\}$. The *shuffle* of K and L is $K \sqcup L = \{u_0 v_1 u_1 \cdots v_k u_k \mid u_0 u_1 \cdots u_k \in K \text{ and } v_1 v_2 \cdots v_k \in L\}$. The *cyclic shift* of L is $\text{shift}(L) = \{uv \mid vu \in L\}$. The *permutation* of L is $\text{per}(L) = \{w \mid \psi(w) = \psi(v) \text{ for some } v \text{ in } L\}$ where $\psi(w)$ is the Parikh vector of w.

A *nondeterministic finite automaton with multiple initial states* (MNFA) is a quintuple $M = (Q, \Sigma, \cdot, I, F)$ where Q is a finite non-empty set of states, Σ is a non-empty alphabet of input symbols, $I \subseteq Q$ is the set of initial states, $F \subseteq Q$ is the set of final states, and $\cdot : Q \times \Sigma \to 2^Q$ is the transition function which is naturally extended to the domain $2^Q \times \Sigma^*$. For a state q, a set S, and a string w, we write qw and Sw instead of $q \cdot w$ and $S \cdot w$ if it does not cause any confusion. The language accepted by M is the set $L(M) = \{w \in \Sigma^* \mid Iw \cap F \neq \emptyset\}$.

For states p and q and a symbol a, we write that M has a transition (p, a, q) if $q \in pa$. We also say that p has an *out-transition* on a and q has an *in-transition* on a. A state q is called a *sink state* if it has the transition (q, a, q) for each input symbol a and no other out-transitions. To omit a state means to remove it from the set of states and to remove all its in-transitions and out-transitions from the transition function.

The *reverse* of an MNFA M is the MNFA M^R obtained from M by reverting all transitions and swapping the role of initial and final states. A subset S of Q is *reachable* in M if $S = Iw$ for some string w, and it is *co-reachable* in M if it is reachable in M^R.

If $|I| = 1$, we say that M is a *nondeterministic finite automaton* (NFA) and write $(Q, \Sigma, \cdot, s, F)$ instead of $(Q, \Sigma, \cdot, \{s\}, F)$.

An NFA $(Q, \Sigma, \cdot, s, F)$ is called *deterministic* (DFA) if $|qa| = 1$ for each state q and each symbol a. Similarly, an MNFA with this property is said to be an MDFA. We write $pa = q$ instead of $pa = \{q\}$ and use the notation $p \xrightarrow{a} q$. A DFA is *minimal* if its language cannot be accepted by any smaller DFA (with respect to number of states). It is well known that a DFA is minimal if and only if all its states are reachable and pairwise distinguishable. The state complexity of a language L, $\mathrm{sc}(L)$, is the number of states in a minimal DFA accepting L.

Every MNFA $M = (Q, \Sigma, \cdot, I, F)$ can be converted to an equivalent deterministic finite automaton $\mathcal{D}(M) = (2^Q, \Sigma, \cdot, I, \{S \mid S \cap F \neq \emptyset\})$ [19]. This DFA is called the *subset automaton* of M. The DFA $\mathcal{D}(M)$ may not be minimal since some of its states may be unreachable or equivalent to other states. To prove distinguishability of states in subset automata, we use the following observation.

Lemma 1 ([12, Lemma 1]). *If for each state q of an MNFA M, the singleton set $\{q\}$ is co-reachable in M, then all states of the subset automaton $\mathcal{D}(M)$ are pairwise distinguishable.*

To prove minimality of unary DFAs, we use the following lemma.

Lemma 2 ([16, Lemma 1]). *A unary DFA $(\mathbf{n}, \{a\}, \cdot, 0, F)$ with $i \cdot a = i + 1$ for $i = 0, 1, \ldots, n - 2$ and $(n - 1) \cdot a = k$ for some k in $\mathbf{n}$ is minimal if and only if the two following conditions hold:*

(1) its loop (the DFA obtained by omitting states in $\mathbf{k}$ and making k the initial state) is minimal,

(2) if $k \neq 0$, then the states $k - 1$ and $n - 1$ do not have the same finality.

A DFA $A = (Q, \Sigma, \cdot, s, F)$ is said to be a *permutation DFA* if for each $p, q \in Q$ and each $a \in \Sigma$, $p \cdot a = q \cdot a$ implies that $p = q$. A language is said to be a *permutation language* if it is recognized by a permutation DFA.

In a permutation DFA, each input symbol a induces a permutation on Q, namely $q \mapsto q \cdot a$. To describe transitions in permutation automata, we use the following notation. The formula $a: (p_1, p_2, \ldots, p_k)$ denotes that $p_i \cdot a = p_{i+1}$

for $i = 1, 2, \ldots, k - 1$, $p_k \cdot a = p_1$, and $q \cdot a = q$ if $q \neq p_i$ for $i = 1, 2, \ldots, k$. For example, we denote the maximal cyclic permutation on $\mathbf{n}$ by $(0, 1, \ldots, n-1)$, the transposition of 0 and 1 by $(0, 1)$, and the identity by (0). The next observation shows that a language is a permutation language if and only if its minimal DFA is a permutation DFA.

Proposition 3. *Let A be a permutation DFA. Then the minimal DFA equivalent to A is a permutation DFA.*

Proof. Assume that a permutation DFA $A = (Q_A, \Sigma, \cdot, s_A, F_A)$ is equivalent to a minimal DFA $B = (Q_B, \Sigma, \circ, s_B, F_B)$ which is not a permutation DFA. Then there exist two distinct states p and q in B and an input symbol a in Σ such that $p \circ a = q \circ a$. Since B is minimal, the states p and q are reachable from s_B by some strings u and v, respectively. Let $p' = s_A \cdot u$ and $q' = s_A \cdot v$ be the corresponding states in A. Since B is minimal, the states p and q are not equivalent, and therefore also p' and q' are not equivalent. However, $p' \cdot a$ and $q' \cdot a$ are equivalent since $p \circ a = q \circ a$. Moreover, for every positive i, the states $p' \cdot a^i$ and $q' \cdot a^i$ are equivalent as well. Since A is a permutation DFA, there exist positive integers k and ℓ such that $p' \cdot a^k = p'$ and $q' \cdot a^\ell = q'$. Then $p' \cdot a^{k\ell} = p'$ and $q' \cdot a^{k\ell} = q'$. It follows that p' and q' are equivalent, a contradiction. □

Notice that the converse does not hold; for example, the minimal unary DFA for a^* is a permutation DFA, however, this language is accepted by the two-state DFA $(\mathbf{2}, \{a\}, \cdot, 0, \mathbf{2})$ with $0 \cdot a = 1$ and $1 \cdot a = 1$ which is not a permutation DFA. We use the following corollary of Proposition 3 without citing it.

Corollary 4. *A language L is a permutation language if and only if the minimal DFA accepting L is a permutation DFA.*

Lemma 5. *If a DFA A has a reachable sink state and $L(A) \notin \{\emptyset, \Sigma^*\}$, then $L(A)$ is not a permutation language.*

Proof. Let A have a reachable sink state. Let B be the minimal DFA for $L(A)$. Then B must have a reachable sink state d. Since $L(A) \notin \{\emptyset, \Sigma^*\}$, the state d is reached from some other reachable state q by a symbol a. Then $q \cdot a = d \cdot a$, hence B is not a permutation automaton, and the lemma follows. □

A MDFA $(Q, \Sigma, \cdot, I, F)$ is said to be a *permutation MDFA* if each input symbol a in Σ induces a permutation on Q.

Lemma 6. *Every language recognized by a permutation MDFA is a permutation language.*

Proof. Let $M = (Q, \Sigma, \cdot, I, F)$ be a permutation MDFA. Let us show that the subset automaton $\mathcal{D}(M)$ is a permutation DFA. Let S and T be subsets of Q and assume that $Sa = Ta$. Since M is a permutation automaton, we have $|S| =$

$|Sa| = |Ta| = |T|$. If Sa is empty, then S and T are empty as well. Otherwise, let $|Sa| = k$ and let $Sa = \{p_1, p_2, \ldots, p_k\}$. For each p_i in Sa there is a state s_i in S and a state t_i in T such that $s_i \cdot a = p_i = t_i \cdot a$. This means that $s_i = t_i$. Moreover, if $i \neq j$, then $p_i \neq p_j$, and therefore $s_i \neq s_j$ and $t_i \neq t_j$ since M is a permutation MDFA. It follows that $S = \{s_1, s_2, \ldots, s_k\} = \{t_1, t_2, \ldots, t_k\} = T$, so the subset automaton $\mathcal{D}(M)$ is a permutation DFA. $\qquad\square$

Lemma 7. *Let M be an n-state permutation MDFA. Then $L(M)$ is accepted by a permutation DFA with at most $\binom{n}{\lceil n/2 \rceil}$ states. This bound is met by permutation MDFA $M = (\mathbf{n}, \{a, b\}, \cdot, \frac{\mathbf{n}}{2}, \{0\})$ with $a\colon (0, 1, \ldots, n-1)$ and $b\colon (0, 1)$.*

Proof. Let $M = (Q, \Sigma, \cdot, I, F)$ be a permutation MDFA. Let $|I| = k$. Then each reachable set in the subset automaton $\mathcal{D}(M)$ is of size k. Hence $\mathcal{D}(M)$ has at most $\binom{n}{k}$ states. Since $\binom{n}{k} \leq \binom{n}{\lceil n/2 \rceil}$, we get our upper bound. Let us show that this upper bound is met by the MDFA M from the statement of the lemma. Since a and b are generators of the symmetric group, and the symmetric group acts transitively on all subsets of given fixed size, each subset of $\mathbf{n}$ of size $\lceil n/2 \rceil$ is reachable in the subset automaton $\mathcal{D}(M)$ from the initial subset. Since each singleton set $\{i\}$ is co-reachable in M via a string in a^*, all states in $\mathcal{D}(M)$ are pairwise distinguishable by Lemma 1. This gives our lower bound. $\qquad\square$

3   Closure Properties of Permutation Languages

In this section, we study the closure properties of the class of permutation languages. The results for Boolean operations, concatenation, Kleene closure, reversal, and quotients have been already obtained by Thierrin [22] using cancellative congruences; here we provide alternative proofs using automata constructions and properties of permutation automata. Moreover, we consider the operations of power, positive closure, cut, shuffle, cyclic shift, and permutation.

Theorem 8 (Closure Properties). *The class of permutation languages is closed under complementation, intersection, union, difference, symmetric difference, reversal, right and left quotient, and it is not closed under concatenation, cut, shuffle, power, Kleene and positive closure, cyclic shift, and permutation.*

Proof. (a) Let K and L be accepted by permutation DFAs A and B, respectively. The language L^c is accepted by a DFA with the same transitions as in B, hence it is a permutation language. Next, intersection, union, difference, and symmetric difference are accepted by a product automaton with appropriately defined final states. The resulting product automaton is a permutation DFA. The language L^R is accepted by a permutation MDFA obtained from B by reversing all transitions and by swapping the roles of initial and final states, and it is a permutation language by Lemma 6. If $A = (Q, \Sigma, \cdot, s, F)$, then the right quotient KL^{-1} is accepted by a permutation DFA obtained from A by making all states in the set $\{q \mid q \cdot w \in F$ for some w in $L\}$ final. The left quotient $L^{-1}K$ is accepted by a permutation MDFA obtained from A by making all states in $\{s \cdot w \mid w \in L\}$ initial, so it is a permutation language by Lemma 6.

(b) Let $k \geq 2$ and $K = L = a(aa)^*$. Then K and L are unary permutation languages and we have $KL = K \, ! \, L = K \sqcup L = aa(aa)^*$ and $L^k = a^k(aa)^*$, which are not permutation languages.

(c) Let $L = aa(aaa)^*$. Then L is a unary permutation language and we have $L^* = \{\varepsilon, aa\} \cup a^4a^*$ and $L^+ = \{aa\} \cup a^4a^*$, which are co-finite and different from a^*. By Lemma 5, they are not permutation languages.

(d) Let $A = (\mathbf{3}, \{a, b\}, \cdot, 0, \{0, 2\})$ be a permutation DFA with $a\colon (0, 1, 2)$ and $b\colon (0, 1)$. Construct an MNFA with ε-transitions for shift$(L(A))$ as described in [14, p. 340]. In the corresponding subset automaton, the initial subset is $I = \{q_{00}, q_{11}, q_{22}, p_{00}, p_{20}\}$, and we have $Ib = \{q_{01}, q_{10}, q_{22}, p_{01}, p_{10}, p_{20}, p_{21}\}$ and $Iba = \{q_{02}, q_{11}, q_{20}, p_{00}, p_{02}, p_{11}, p_{20}, p_{21}, p_{22}\}$. Notice that the set Ib is not final and $Iba \supseteq \{p_{20}, p_{21}, p_{22}\}$. The transitions on states p_{20}, p_{21}, p_{22} are the same as in A, so a and b perform permutations on these three states. This means that $Iba \cdot w$ is always a superset of $\{p_{20}, p_{21}, p_{22}\}$ and therefore each string is accepted from the final state Iba. Hence Iba is equivalent to a final sink state, which means that the minimal DFA for shift$(L(A))$ has a reachable final sink state. By Lemma 5, shift$(L(A))$ is not a permutation language.

(e) Let $A = (\mathbf{3}, \{a, b\}, \cdot, 0, \{2\})$ be a permutation DFA with $a\colon (0, 1)$, $b\colon (1, 2)$. The language per$(L(A))$ consists of strings which have at least one a and at least one b; notice that A has loops $(0, b, 0)$ and $(2, a, 2)$. It is accepted by the DFA $(4, \{a, b\}, \circ, 0, \{3\})$ with the transitions $(0, a, 1)$, $(0, b, 2)$, $(1, a, 1)$, $(1, b, 3)$, $(2, a, 3)$, $(2, b, 2)$, $(3, a, 3)$, and $(3, b, 3)$ which has a reachable sink state 3. Hence per$(L(A))$ is not a permutation language by Lemma 5. □

4   State Complexity of Operations on Permutation Languages

In this section, we study the state complexity of operations on permutation languages. Let us start with Boolean operations.

Theorem 9 (Boolean Operations). *Let $m, n \geq 3$. Let K and L be languages accepted by permutation DFAs with m and n states, respectively. Then we have* $\mathrm{sc}(K \cup L), \mathrm{sc}(K \cap L), \mathrm{sc}(K \setminus L), \mathrm{sc}(K \oplus L) \leq mn$, *and these upper bounds are met by binary permutation languages. Moreover, if* $\gcd(m, n) = 1$, *then these bounds are met by unary permutation languages.*

Proof. The upper bound mn is the same as for regular languages and it is met by binary permutation languages $K = \{w \in \{a, b\}^* \mid |w|_a = 0 \bmod m\}$ and $L = \{w \in \{a, b\}^* \mid |w|_b = 0 \bmod n\}$ by [13, Theorem 8(1-4)]. In the unary case, the permutation languages $(a^m)^*$ and $(a^n)^*$ meet the upper bound for all four operations whenever $\gcd(m, n) = 1$; distinguishability in the mn-state cyclic DFA can be proved by using the Chinese Remainder Theorem. □

We continue with concatenation, power, Kleene and positive closure, and reversal. We first consider the unary case, then we deal with a general alphabet.

Theorem 10 (Basic Operations: Unary Case). *Let* $m, n \geq 2$. *Let* K *and* L *be unary languages accepted by permutation DFAs with* m *and* n *states, respectively. Then*

(a) $\mathrm{sc}(KL) = \mathrm{sc}(K \sqcup L) \leq mn$, *and this bound is tight if* $\gcd(m, n) = 1$;
(b) $\mathrm{sc}(L^k) \leq k(n-1) + 1$, *and this bound is tight;*
(c) $\mathrm{sc}(L^+), \mathrm{sc}(L^*) \leq (n-1)^2 + 1$, *and this bound is tight;*
(d) $\mathrm{sc}(L^R) \leq n$, *and this bound is tight.*

Proof. (a) In the case of unary languages, we have $KL = K \sqcup L$, and the upper bound mn follows from [23, Theorem 5.5]. For tightness in the case when $\gcd(m, n) = 1$, consider the unary permutation languages $K = a^{m-1}(a^m)^*$ and $L = a^{n-1}(a^n)^*$. Then $\mathrm{sc}(KL) = mn$ by [23, Theorem 5.4].

(b) The upper bound is the same as for regular languages [20, Theorem 3], and it is met by the permutation language $a^{n-1}(a^n)^*$ [20, Proof of Theorem 4].

(c) Let $A = (Q, \{a\}, \cdot, s, F)$ be a permutation DFA for L. If $F = \{s\}$, then $L^+ = L^* = L$, and $\mathrm{sc}(L^+) = \mathrm{sc}(L^*) \leq n$. Otherwise, if L is empty, then the upper bound follows since $n \geq 2$. Let L be non-empty and $F \neq \{s\}$. We construct an n-state NFA for L^+ by adding the transitions (q, a, s) whenever $q \cdot a \in F$. In the corresponding subset automaton, the initial subset is $\{s\}$ and there exists an integer t such that $t \leq n-1$ and $|\{s\} \cdot a^t| \geq 2$. Since A is a cyclic unary DFA, we cannot reach the singleton set $\{s\}$ from any other state of the subset automaton. It follows that to get a DFA for L^*, we only need to mark the initial state $\{s\}$ as final. It is shown in [23, Proof of Theorem 5.3] that the subset automaton for L^+ has at most $(n-1)^2 + 1$ reachable states, and that this upper bound is met by the permutation language $(aa)^*$ if $n = 2$ and $a^{n-1}(a^n)^*$ if $n \geq 3$.

(d) The upper bound n is met by the permutation language $(a^n)^*$. $\qquad\square$

Theorem 11 (Basic Operations: General Case). *Let* $m, n \geq 2$. *Let* K *and* L *be languages accepted by permutation DFAs with* m *and* n *states. Then*

(a) $\mathrm{sc}(KL) \leq m2^n - 2^{n-1} - m + 1$,
(b) $\mathrm{sc}(L^2) \leq n2^{n-1} - 2^{n-2}$,
(c) $\mathrm{sc}(L^+) \leq 2^{n-1} + 2^{n-2} - 1$ *and* $\mathrm{sc}(L^*) \leq 2^{n-1} + 2^{n-2}$,
(d) $\mathrm{sc}(L^R) \leq \binom{n}{\lceil n/2 \rceil}$,

and these bounds are met by binary witnesses and the binary alphabet is optimal.

Proof. (a) If $K = \emptyset$ or $L = \emptyset$, then $KL = \emptyset$. Otherwise, let K and L be accepted by permutation DFAs $A = (Q_A, \Sigma, \cdot_A, s_A, F_A)$ and $B = (Q_B, \Sigma, \cdot_B, s_B, F_B)$ with $Q_A \cap Q_B = \emptyset$, $|Q_A| = m$, $|Q_B| = n$, $|F_A| = k \geq 1$, and $|F_B| \geq 1$. Then KL is accepted by an MNFA $(Q_A \cup Q_B, \Sigma, \cdot, I, F_B)$ where $I = \{s_A\}$ if $s_A \notin F_A$ and $I = \{s_A, s_B\}$ otherwise, and the transition function $\cdot$ contains all transitions of $\cdot_A$ and $\cdot_B$ and moreover the transitions (q, a, s_B) whenever $q \cdot_A a \in F_A$. In the corresponding subset automaton, only the states of the form $\{q\} \cup S$ with $q \in Q_A$ and $S \subseteq Q_B$ are reachable; denote such a state by (q, S). Moreover, if $q \in F_A$, then $s_B \in S$. Next, since B is a permutation automaton, we have $Q_B \cdot w = Q_B$ for every string w in Σ^*. It follows that every string is accepted from (p, Q_B) for each p in Q_A, and all these m states are equivalent to each other. This gives the upper bound $(m - k)2^n + k2^{n-1} - m + 1 \leq m2^n - 2^{n-1} - m + 1$.

For tightness, consider the permutation DFAs $A = (\mathbf{m}, \{a, b\}, \cdot_A, 0, \{m-1\})$ with a: $(0, 1, \ldots, m-1)$ and b: (0) and $B = (\mathbf{n}, \{a, b\}, \cdot_B, 0, \{n-1\})$ with a: (0) and b: $(0, 1, \ldots, n-1)$. Notice that in A, the symbol a induces the maximal cyclic permutation while b is an identity, and the transitions in B are symmetric. Construct an NFA for $L(A)L(B)$ as described above. Let us show by induction on $|S|$ that each state (i, S) with $i = 0, 1, \ldots, m-2$ and each state $(m-1, S)$ with $0 \in S$ is reachable in the corresponding subset automaton. The base case, $|S| = 0$, holds true since for every i with $1 \leq i \leq m-2$, the state $(i, \emptyset)$ is reached from the initial state $(0, \emptyset)$ by a^i. Now assume that our claim holds for each S' with $|S'| = k$. Let $|S| = k + 1$. Consider three cases:

(1) $i = m - 1$. Then we must have $0 \in S$. Take $S' = S \setminus \{0\}$. Then $(m - 2, S')$ is reachable by the induction hypothesis and it is sent to $(m - 1, S)$ by a.
(2) $0 \leq i \leq m - 2$ and $0 \in S$. Then the set $(m - 1, S)$ is reachable as shown in case (1) and it is sent to (i, S) by a^{i+1}.
(3) $0 \leq i \leq m-2$ and $0 \notin S$. Take $S' = \{s - \min S \mid s \in S\}$. Then $0 \in S'$, so the set (i, S') is reachable as shown in case (2) and it is sent to (i, S) by $b^{\min S}$.

Now we prove distinguishability. Each state $(i, \mathbf{n})$ is equivalent to a final sink state. Consider a state (i, S) with $S \neq \mathbf{n}$. Let $s \notin S$. Then b^{n-1-s} is rejected from (i, S), so (i, S) is not equivalent to a final sink state. Let (i, S) and (j, T) be two distinct states of the subset automaton with $S, T \neq \mathbf{n}$. First let $S \neq T$, and without loss of generality, let $s \in S \setminus T$. Then the string b^{n-1-s} is accepted from S and it is rejected from T. Now let $S = T$, and without loss of generality, let $i < j$. Since $S \neq \mathbf{n}$, there exists a state s with $s \notin S$. Let $S' = \{(p - s) \bmod n \mid p \in S\}$. Then $0 \notin S'$. Consider the string $b^{n-s}a^{m-1-j}$. It leads (i, S) to $(i+m-1-j, S')$ and it leads (j, S) to $(m-1, \{0\} \cup S')$. Since the second components of the resulting states differ in the state 0, the states (i, S) and (j, S) are distinguishable. This completes our proof for concatenation.

(b) If the DFA for L has a unique final state which is initial, then $L^2 = L$. Otherwise, let L be accepted by a permutation DFA A with the initial state s and at least one final state f with $f \neq s$. Construct the NFA N for $L^2 = LL$ as described in case (a). If the initial state s of A is final, then the initial state of the subset automaton $\mathcal{D}(N)$ is $(s, \{s\})$. It follows that for each reachable state (q, S) of $\mathcal{D}(N)$, we have $q \in S$. Moreover, if q is final, then $s \in S$. In total, we get at most $2^{n-1} + 2^{n-2} + (n - 2)2^{n-1} = n2^{n-1} - 2^{n-2}$ reachable states. Now assume that s is not final. Let us show that no state (q, S) with $q \in S$ is reachable in $\mathcal{D}(N)$. Since s is not final, the initial state of $\mathcal{D}(N)$ is $(s, \emptyset)$ with $s \notin \emptyset$. Now let (q, S) be any reachable state with $q \notin S$. Consider the state $(p, T) = (q, S)a$ for any input symbol a. Assume for a contradiction that $p \in T$. It follows that $qa = p$ and either there is a state r in S such that $ra = p$, or $p = s$ and p is final. In the former case, we get a contradiction with the fact that A is a permutation DFA. The latter case cannot occur since s is not final. It follows by induction that for every reachable state (q, S) we have $q \notin S$. In total, taking into account that $s \in S$ whenever q is final, we get at most $(n - 1)2^{n-1} + 2^{n-2}$ reachable states. This gives the upper bound.

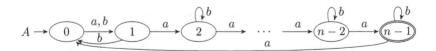

Fig. 1. The witness DFA for square meeting the upper bound $n2^{n-1} - 2^{n-2}$.

To prove tightness, consider the permutation DFA $A = (\mathbf{n}, \{a, b\}, \cdot, 0, \{n-1\})$ with transitions $a \colon (0, 1, \ldots, n - 1)$ and $b \colon (0, 1)$ shown in Fig. 1. Construct an NFA for $L(A)^2$ as described above. Let us show by induction on $|S|$ that each state (i, S) with $i = 0, 1, \ldots, n - 2$ and $i \notin S$ and each state $(n - 1, S)$ with $0 \in S$ and $n - 1 \notin S$ is reachable. The base case, $|S| = 0$, holds true since for every i with $1 \le i \le n - 2$, the state $(i, \emptyset)$ is reached from the initial state $(0, \emptyset)$ by a^i. Now assume that our claim holds for each S' with $|S'| = k$. Let $|S| = k + 1$. Consider four cases:

(1) $i = n - 1$. Then $0 \in S$ and $n - 1 \notin S$. Take $S' = (S \setminus \{0\}) \cdot a^{n-1}$. Then $n - 2 \notin S'$, so $(n - 2, S')$ is reachable by induction and it is sent to $(n - 1, S)$ by a.
(2) $i = 0$ and $1 \in S$. Take $S' = S \cdot a^{n-1}$. Since $0 \notin S$, we have $n - 1 \notin S'$, so $(n - 1, S')$ is reachable as shown in case (1) and it is sent to $(0, S)$ by a.
(3) $i = 0$ and $\min S > 1$. Take $S' = S \cdot a^{n - \min S + 1}$. Then $0 \notin S'$ and $1 \in S'$, so $(0, S')$ is reachable as shown in case (2) and it is sent to $(0, S)$ by $(ab)^{\min S - 1}$.
(4) $1 \le i \le n - 2$. Take $S' = S \cdot a^{n-i}$. Since $i \notin S$, we have $0 \notin S'$, so the set $(0, S')$ is reachable as shown in case (1) or (2) and it is sent to (i, S) by a^i.

Now we prove distinguishability. Let (i, S) and (j, T) be two distinct states of the subset automaton with $i \notin S$ and $j \notin T$. First let $S \neq T$, and without loss of generality, assume that $s \in S \setminus T$. Then the string a^{n-1-s} is accepted from S and rejected from T. Now let $S = T$, and without loss of generality, let $i < j$. We have $i, j \notin S$. Consider the string $a^{n-j}(ab)^{j-i-1}$. Let $S' = S \cdot a^{n-j}$. Since $j \notin S$, we have $0 \notin S'$. Next,

$$(i, S) \xrightarrow{a^{n-j}} (n - j + i, S') \xrightarrow{(ab)^{j-i-1}} (n - 1, S' \cdot (ab)^{j-i-1} \cup \{0\}),$$

$$(j, S) \xrightarrow{a^{n-j}} (0, S' \cup \{1\}) \xrightarrow{(ab)^{j-i-1}} (j - i - 1, (S' \cup \{1\}) \cdot (ab)^{j-i-1}).$$

Since $0 \notin S'$, we have $0 \notin (S' \cup \{1\}) \cdot (ab)^{j-i-1}$; notice that the only state which is set to 0 by a string in $(ab)^*$ is 0. This means that the resulting states differ in the second component, therefore (i, S) and (j, S) are distinguishable. This completes our proof for square.

(c) The upper bound for Kleene closure is the same as for regular languages. For tightness, consider the n-state DFA $A = (\mathbf{n}, \{a, b\}, \cdot, 0, \{n - 1\})$ with $a\colon (0, 1, \ldots, n - 1)$ and $b\colon (1, 2, \ldots, n - 2)$. Notice that this DFA, shown in Fig. 2, differs from the DFA in [17, Figure 4] just by having the unique final state $n - 1$ and by having the transition $(n - 2, b, 1)$ instead of $(n - 2, b, 0)$. The language $L(A)^*$ is accepted by MNFA $(\{q_0\} \cup \mathbf{n}, \{a, b\}, \circ, \{q_0, 0\}, \{q_0, n - 1\})$ where $\circ$ has the same transitions as $\cdot$ and moreover we have $0 \in (n - 2) \circ a$ and $0 \in (n - 1) \circ b$. Let $\mathcal{R} = \{q_0, 0\} \cup \{S \subseteq \mathbf{n} \mid S \neq \emptyset \text{ and if } n - 1 \in S, \text{ then } 0 \in S\}$. We can show by induction on the size of sets that each set in $\mathcal{R}$ is reachable in the corresponding subset automaton. We also can prove that the sets in $\mathcal{R}$ are pairwise distinguishable.

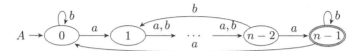

Fig. 2. The witness DFA for Kleene closure meeting the upper bound $(3/4)2^n$.

To get an NFA for L^+, we only remove the state q_0 from the MNFA for L^* described above. The corresponding subset automaton has at most $(3/4)2^n - 1$ reachable states, and this upper bound is met by our witness for Kleene closure.

(d) If L is accepted by an n-state permutation DFA A, then L^R is accepted by a permutation MDFA A^R obtained from A by reversing all transitions and by swapping the roles of initial and final states. We have $\mathrm{sc}(L^R) \leq \binom{n}{\lceil n/2 \rceil}$ and this upper bound is met by the language accepted by the permutation DFA $(\mathbf{n}, \{a, b\}, \cdot, 0, \frac{\mathbf{n}}{2})$ with $a\colon (n - 1, n - 2, \ldots, 0)$ and $b\colon (0, 1)$ by Lemma 7. $\square$

Notice that for concatenation, the upper bound is almost the same as for regular languages, while for square, the bound is exactly one half of the upper bound in the general case. In the next theorem, we consider quotients.

Theorem 12 (Left and Right Quotient). *Let $m, n \geq 2$. Let K and L be languages accepted by permutation DFAs with m and n states, respectively. Then $\mathrm{sc}(KL^{-1}) \leq m$, and this bound is met by unary languages if $m \leq n$. Next, $\mathrm{sc}(L^{-1}K) \leq \binom{m}{\lceil m/2 \rceil}$, and this bound is met by binary languages if $m \leq n$. In the unary case, the tight upper bound for both quotients is $\min\{m, n\}$.*

Proof. We have $\mathrm{sc}(KL^{-1}) \leq m$ by the construction of a DFA for the right quotient. To get tightness in the case of $m \leq n$, let $K = L = (a^m)^*$. Then $KL^{-1} = K$, so $\mathrm{sc}(KL^{-1}) = m$.

The language $L^{-1}K$ is accepted by an m-state permutation MDFA. We have $\mathrm{sc}(L^{-1}K) \le \binom{m}{\lceil m/2 \rceil}$ by Lemma 7. For tightness, let $m \le n$. Let K be the language accepted by DFA $A = (\mathbf{m}, \{a, b\}, \cdot, 0, \{0\})$ with $a\colon (0, 1, \ldots, m-1)$ and $b\colon (0, 1)$. Let L be the language accepted by DFA $B = (\mathbf{n}, \{a, b\}, \cdot, 0, \frac{m}{2})$ with $a\colon (0, 1, \ldots, m-1)$ and $b\colon (0, 1)$; notice that states $m, m+1, \ldots, n-1$ are unreachable and the transitions on a and b in states $0, 1, \ldots, m-1$ are the same as in DFA A. The language $L^{-1}K$ is accepted by an MDFA M obtained from A by making all states in $\frac{m}{2}$ initial; notice that after reading every input string, both A and B are in the same state. Then $\mathrm{sc}(L^{-1}K) = \binom{m}{\lceil m/2 \rceil}$ by Lemma 7.

In the unary case, we have $L^{-1}K = KL^{-1}$. Therefore $\mathrm{sc}(L^{-1}K) \le m$. Let us show that $\mathrm{sc}(KL^{-1}) \le n$. The languages K and L are periodic; let their periods be k and ℓ, respectively. By construction, the quotient KL^{-1} has period k. If $a^{i+\ell} \in KL^{-1}$, then $a^i \in KL^{-1}$ since L has period ℓ. Let $a^i \in KL^{-1}$. Then $a^i w \in K$ for some $w \in L$. It follows that $a^i a^\ell a^{(k-1)\ell} w \in K$ since K has period k. Next, $a^{(k-1)\ell} w \in L$ since L has period ℓ. It follows that $a^{i+\ell} \in KL^{-1}$. Thus KL^{-1} has period ℓ, so $\mathrm{sc}(KL^{-1}) \le \ell \le n$. This gives the upper bound $\min\{m, n\}$ which is met by $K = L = (a^{\min\{m,n\}})^*$.   □

Finally, we consider the cut operation on permutation languages. We briefly recall the construction of a DFA for the cut operation, cf. [2, p. 74], [7, p. 91] or [10, p. 193]. For two languages accepted by DFAs A and B with m and n states, respectively, we can construct the *cut automaton* $A\,!\,B$ as follows. The cut automaton has states in a grid with a row for every state of A and a column for every state of B, and one additional column which we denote by $\perp$. The column $\perp$ corresponds to the situation that we have not read a string in $L(A)$ yet. When we reach a final state in A for the first time, we leave the column $\perp$ and enter the product part of $A\,!\,B$ in the corresponding state. Unless we reach a final state of A again, the transitions in $A\,!\,B$ are the same as in the product automaton $A \times B$. When we reach a final state in A again, we reset to a state in the column corresponding to the initial state of B. The final states of $A\,!\,B$ are all states in columns corresponding to the final states of B.

Formally, let $A = (Q_A, \Sigma, \cdot_A, s_A, F_A)$ and $B = (Q_B, \Sigma, \cdot_B, s_B, F_B)$ be two DFAs. Let $\perp \notin Q_B$. Define the cut automaton $A\,!\,B = (Q, \Sigma, \cdot, s, Q_A \times F_B)$ where $Q = (Q_A \times \{\perp\}) \cup (Q_A \times Q_B)$, $s = (s_A, \perp)$ if $\varepsilon \notin L(A)$ and $s = (s_A, s_B)$ otherwise, and for each state (p, q) in Q and each input symbol a in Σ, we have

$$(p, \perp) \cdot a = \begin{cases} (p \cdot_A a, \perp), & \text{if } p \cdot_A a \notin F_A\,; \\ (p \cdot_A a, s_B), & \text{otherwise;} \end{cases}$$

and

$$(p, q) \cdot a = \begin{cases} (p \cdot_A a, q \cdot_B a), & \text{if } p \cdot_A a \notin F_A\,; \\ (p \cdot_A a, s_B), & \text{otherwise.} \end{cases}$$

The state complexity of cut in the general case is already solved in [7] where binary permutation languages were used as witnesses. The unary case is more interesting, as shown in the next theorem.

Theorem 13 (Cut Operation). *Let $m \geq 1$ and $n \geq 3$. Let K and L be languages accepted by permutation DFAs with m and n states, respectively. Then $\mathrm{sc}(K \,!\, L) \leq (m - 1)n + m$, and this upper bound is met by binary languages. If K and L are unary, then $\mathrm{sc}(K \,!\, L) \leq 2m - 1$. This bound is tight if $m \bmod \ell \neq 0$ for some ℓ with $2 \leq \ell \leq n$, otherwise the tight upper bound is $2m - 2$.*

Proof. If $K = \emptyset$ or $L = \emptyset$, then $K \,!\, L = \emptyset$, so $\mathrm{sc}(K \,!\, L) = 1$. Let K and L be non-empty. In the general case, the witness languages from [7, Proof of Theorem 3.1] are accepted by permutation DFAs and they meet the upper bound $(m-1)n+m$.

In the unary case, since each non-empty permutation language is infinite, the upper bound is $2m - 1$ by [7, Proof of Theorem 3.2].

First assume that there exists ℓ with $2 \leq \ell \leq n$ such that $m \bmod \ell \neq 0$. Let $K = a^{m-1}(a^m)^*$ and $L = a^{(m-1) \bmod \ell}(a^\ell)^*$ be unary languages recognized by permutation DFAs A and B of m and ℓ states, respectively. The cut automaton $A \,!\, B$ has a tail of length $m - 1$ and a loop of length m starting in the state $(m - 1, 0)$ which has in-transitions from the non-final state $(m - 2, \perp)$ and the final state $(m - 2, (m - 1) \bmod \ell)$. Let us show that the loop is minimal. If $m \leq \ell$, then the loop has just one final state, so it is minimal. Let $m > \ell$. If $m \bmod \ell = 1$, then $(m - 1) \bmod \ell = 0$, hence the two states $(m - 2, 0)$ and $(m - 1, 0)$ are final, while every other contiguous segment of final states is of length 1, so the loop is minimal again. If $2 \leq m \bmod \ell \leq \ell - 1$, then $1 \leq (m - 1) \bmod \ell \leq \ell - 2$. Then every contiguous segment of non-final states is of length $\ell - 1$, except for the segment containing $(m - 1, 0)$ which is of length $(m - 1) \bmod \ell < \ell - 1$, so the loop is minimal.

Let $m \bmod \ell = 0$ for each ℓ with $2 \leq \ell \leq n$. Then $m \bmod 2 = m \bmod 3 = 0$, so $m \geq 6$. Consider the permutation DFAs $A = (\mathbf{m}, \{a\}, \cdot_A, 0, \{m - 2, m - 1\})$ with $a\colon (0, 1, \ldots, m-1)$ and $B = (\mathbf{n}, \{a\}, \cdot_B, 0, \{0\})$ with $a\colon (0, 1)$. The DFA $A \,!\, B$ has a tail of length $m-2$ and a loop of size m starting in the state $(m-2, 0)$ which has in-transitions from the non-final state $(m-3, \perp)$ and the final state $(m-3, 0)$. In the loop, each contiguous segment of final states is of length 1, except for that consisting of $(m - 3, 0)$, $(m - 2, 0)$, and $(m - 1, 0)$. Therefore the loop is minimal, and so we have $\mathrm{sc}(L(A) \,!\, L(B)) = 2m - 2$.

Let us show that this bound is tight if $m \bmod \ell = 0$ for each ℓ with $2 \leq \ell \leq n$. Since $m \bmod n = 0$ and $m \bmod (n - 1) = 0$, we must have $m > n$. Let K and L be accepted by unary permutation DFAs A and B; these DFAs are cyclic, that is, they do not have any tail. To meet the upper bound $2m - 1$, the cut automaton $A \,!\, B$ must have $m - 1$ states of the form $(i, \perp)$. This is possible only if $A = (\mathbf{m}, \{a\}, \cdot_A, 0, \{m-1\})$ with $a\colon (0, 1, \ldots, m-1)$. Let $B = (\mathbf{n}, \{a\}, \cdot_B, 0, F)$ with $a\colon (0, 1, \ldots, \ell-1)$ for some ℓ with $1 \leq \ell \leq n$. If $\ell = 1$, then $L = \emptyset$ or $L = a^*$, and $K \,!\, L$ is either empty or equal to $a^{m-1}a^*$, so it is of state complexity 1 or m. Let $\ell \geq 2$. Then $m \bmod \ell = 0$. In the loop of the cut automaton $A \,!\, B$, we have

$$(m - 1, 0) \xrightarrow{a^\ell} (\ell - 1, 0) \xrightarrow{a^\ell} (2\ell - 1, 0) \xrightarrow{a^\ell} \cdots \xrightarrow{a^\ell} (m - \ell - 1, 0) \xrightarrow{a^\ell} (m - 1, 0).$$

Since the loop in B is of length ℓ, we have $(i, j) \xrightarrow{a^\ell} (i + \ell, j)$ for each state (i, j) in the loop; here $i + \ell$ is modulo m. The states (i, j) and $(i + \ell, j)$ have the same finality since all states with j in the second component have the same

Table 1. Closure properties and state complexity of operations on the class of permutation languages. ($\star$): there exists ℓ with $2 \le \ell \le n$ such that $m \bmod \ell \ne 0$.

Operation	Closed?	State complexity	$\lvert \Sigma \rvert$	State complexity, $\lvert \Sigma \rvert = 1$
L^c	Yes ([22])	n	1	n
$\cap, \cup, \setminus, \oplus$	Yes ([22])	mn	2	mn; $\gcd(m,n) = 1$
KL	No ([22])	$m2^n - 2^{n-1} - m + 1$	2	mn; $\gcd(m,n) = 1$
L^2	No	$n2^{n-1} - 2^{n-2}$	2	$2n - 1$
L^k	No	?		$k(n-1) + 1$
L^+	No	$(3/4)2^n - 1$	2	$(n-1)^2 + 1$
L^*	No ([22])	$(3/4)2^n$	2	$(n-1)^2 + 1$
L^R	Yes ([22])	$\binom{n}{\lceil n/2 \rceil}$	2	n
$L^{-1}K$	Yes ([22])	$\binom{m}{\lceil m/2 \rceil}$; $m \le n$	2	$\min\{m, n\}$
KL^{-1}	Yes ([22])	m; $m \le n$	1	$\min\{m, n\}$
$K!L$	No	$(m-1)n + m$	2	$2m - 1$ if ($\star$); $2m - 2$ otherwise
$K \sqcup L$	No	?		mn; $\gcd(m,n) = 1$
shift(L)	No	?		n
per(L)	No	?		n

finality as j in B. It follows that the loop can be replaced by a loop of length ℓ, so $\mathrm{sc}(K!L) \le m - 1 + \ell < 2m - 1$.   □

5   Conclusions

We examined the class of permutation languages, that is, the languages that are recognized by deterministic finite automata in which every input symbol induces a permutation on the state set. We used automata constructions and properties of permutation automata to show that the class of permutation languages is closed under Boolean operations, reversal, and left and right quotient, and it is not closed under concatenation, power, positive closure, Kleene closure, cut, shuffle, cyclic shift, and permutation.

We also studied the state complexity of operations on permutation languages. Our results are summarized in Table 1. The table also displays the size of alphabet used to describe witnesses. All our witnesses are described over a unary or binary alphabet, and the binary alphabet is always optimal except for the Boolean operations in case of $\gcd(m,n) = 1$.

We did not consider the state complexity of shuffle and cyclic shift and leave them for the future work. Although the class of regular languages is not closed under the permutation operation, we conjecture that the permutation of a permutation language is always regular. If this is the case, then the state complexity of this operation is of great interest to us as well.

References

1. Bassino, F., Giambruno, L., Nicaud, C.: Complexity of operations on cofinite languages. In: López-Ortiz, A. (ed.) LATIN 2010. LNCS, vol. 6034, pp. 222–233. Springer, Heidelberg (2010). https://doi.org/10.1007/978-3-642-12200-2_21
2. Berglund, M., Björklund, H., Drewes, F., van der Merwe, B., Watson, B.: Cuts in regular expressions. In: Béal, M.-P., Carton, O. (eds.) DLT 2013. LNCS, vol. 7907, pp. 70–81. Springer, Heidelberg (2013). https://doi.org/10.1007/978-3-642-38771-5_8
3. Brzozowski, J.A., Jirásková, G., Li, B.: Quotient complexity of ideal languages. Theor. Comput. Sci. **470**, 36–52 (2013). https://doi.org/10.1016/j.tcs.2012.10.055
4. Brzozowski, J.A., Jirásková, G., Zou, C.: Quotient complexity of closed languages. Theory Comput. Syst. **54**(2), 277–292 (2014). https://doi.org/10.1007/s00224-013-9515-7
5. Brzozowski, J.A., Liu, B.: Quotient complexity of star-free languages. Int. J. Found. Comput. Sci. **23**(6), 1261–1276 (2012). https://doi.org/10.1142/S0129054112400515
6. Câmpeanu, C., Culik, K., Salomaa, K., Yu, S.: State complexity of basic operations on finite languages. In: Boldt, O., Jürgensen, H. (eds.) WIA 1999. LNCS, vol. 2214, pp. 60–70. Springer, Heidelberg (2001). https://doi.org/10.1007/3-540-45526-4_6
7. Drewes, F., Holzer, M., Jakobi, S., van der Merwe, B.: Tight bounds for cut-operations on deterministic finite automata. Fund. Inform. **155**(1–2), 89–110 (2017). https://doi.org/10.3233/FI-2017-1577
8. Han, Y., Salomaa, K.: State complexity of basic operations on suffix-free regular languages. Theor. Comput. Sci. **410**(27–29), 2537–2548 (2009). https://doi.org/10.1016/j.tcs.2008.12.054
9. Han, Y., Salomaa, K., Wood, D.: State complexity of prefix-free regular languages. In: Leung, H., Pighizzini, G. (eds.) DCFS 2006, pp. 165–176. New Mexico State University, Las Cruces (2006)
10. Holzer, M., Hospodár, M.: The range of state complexities of languages resulting from the cut operation. In: Martín-Vide, C., Okhotin, A., Shapira, D. (eds.) LATA 2019. LNCS, vol. 11417, pp. 190–202. Springer, Cham (2019). https://doi.org/10.1007/978-3-030-13435-8_14
11. Hopcroft, J.E., Ullman, J.D.: Introduction to Automata Theory, Languages, and Computation. Addison-Wesley, Boston (1979)
12. Jirásková, G., Krajňáková, I.: NFA-to-DFA trade-off for regular operations. In: Hospodár, M., Jirásková, G., Konstantinidis, S. (eds.) DCFS 2019. LNCS, vol. 11612, pp. 184–196. Springer, Cham (2019). https://doi.org/10.1007/978-3-030-23247-4_14
13. Jirásková, G., Masopust, T.: Complexity in union-free regular languages. Int. J. Found. Comput. Sci. **22**(7), 1639–1653 (2011). https://doi.org/10.1142/S0129054111008933
14. Jirásková, G., Okhotin, A.: State complexity of cyclic shift. RAIRO Theor. Inf. Appl. **42**(2), 335–360 (2008). https://doi.org/10.1051/ita:2007038
15. Maslov, A.N.: Estimates of the number of states of finite automata. Soviet Math. Doklady **11**, 1373–1375 (1970)
16. Nicaud, C.: Average state complexity of operations on unary automata. In: Kutyłowski, M., Pacholski, L., Wierzbicki, T. (eds.) MFCS 1999. LNCS, vol. 1672, pp. 231–240. Springer, Heidelberg (1999). https://doi.org/10.1007/3-540-48340-3_21

17. Palmovský, M.: Kleene closure and state complexity. RAIRO Theor. Inform. Appl. **50**(3), 251–261 (2016). https://doi.org/10.1051/ita/2016024
18. Pighizzini, G., Shallit, J.: Unary language operations, state complexity and Jacobsthal's function. Int. J. Found. Comput. Sci. **13**(1), 145–159 (2002). https://doi.org/10.1142/S012905410200100X
19. Rabin, M.O., Scott, D.: Finite automata and their decision problems. IBM J. Res. Dev. **3**, 114–125 (1959). https://doi.org/10.1147/rd.32.0114
20. Rampersad, N.: The state complexity of L^2 and L^k. Inf. Process. Lett. **98**(6), 231–234 (2006). https://doi.org/10.1016/j.ipl.2005.06.011
21. Sipser, M.: Introduction to the Theory of Computation. Cengage Learning, Boston (2012)
22. Thierrin, G.: Permutation automata. Math. Syst. Theory **2**(1), 83–90 (1968). https://doi.org/10.1007/BF01691347
23. Yu, S., Zhuang, Q., Salomaa, K.: The state complexities of some basic operations on regular languages. Theor. Comput. Sci. **125**(2), 315–328 (1994). https://doi.org/10.1016/0304-3975(92)00011-F

Space Complexity of Stack Automata Models

Oscar H. Ibarra[1], Jozef Jirásek Jr.[2], Ian McQuillan[2]([✉]),
and Luca Prigioniero[3][iD]

[1] Department of Computer Science, University of California,
Santa Barbara, CA 93106, USA
ibarra@cs.ucsb.edu
[2] Department of Computer Science, University of Saskatchewan,
Saskatoon, SK S7N 5A9, Canada
jirasek.jozef@usask.ca, mcquillan@cs.usask.ca
[3] Dipartimento di Informatica, Università degli Studi di Milano, Milan, Italy
prigioniero@di.unimi.it

Abstract. This paper examines several measures of space complexity on variants of stack automata: non-erasing stack automata and checking stack automata. These measures capture the minimum stack size required to accept any word in a language (weak measure), the maximum stack size used in any accepting computation on any accepted word (accept measure), and the maximum stack size used in any computation (strong measure). We give a detailed characterization of the accept and strong space complexity measures for checking stack automata. Exactly one of three cases can occur: the complexity is either bounded by a constant, behaves (up to small technicalities explained in the paper) like a linear function, or it grows arbitrarily larger than the length of the input word. However, this result does not hold for non-erasing stack automata; we provide an example when the space complexity grows with the square root of the input length. Furthermore, an investigation is done regarding the best complexity of any machine accepting a given language, and on decidability of space complexity properties.

Keywords: Checking stack automata · Stack automata · Pushdown automata · Space complexity · Machine models

1 Introduction

When studying different machine models, it is common to study both time and space complexity of a machine or an algorithm. In particular, the study of complexity of Turing machines gave way to the area of computational complexity, which has been one of the most well-studied areas of theoretical computer science for the past 40 years [7]. The field of automata theory specializes in different

The research of I. McQuillan and L. Prigioniero was supported, in part, by Natural Sciences and Engineering Research Council of Canada Grant 2016-06172.

N. Jonoska and D. Savchuk (Eds.): DLT 2020, LNCS 12086, pp. 137–149, 2020.
https://doi.org/10.1007/978-3-030-48516-0_11

machine models, often with more restricted types of data stores and operations. Various models of automata differ in the languages that can be accepted by the model, in the size of the machine (e.g. the number of states), in the algorithms to decide various properties of a machine, and in the complexity of these algorithms. Some of the well-studied automata models with more restricted power than Turing machines are finite automata [5,9], pushdown automata [5,9], stack automata [4], checking stack automata [4], visibly pushdown automata [1], and many others.

For Turing machines, several different space complexity measures have been studied. Some of these complexity measures are the following [13]:

- weak measure: for an input word w, the smallest tape size required for some accepting computation on w;
- accept measure: for an input word w, the largest tape size required for any accepting computation on w;
- strong measure: for an input word w, the largest tape size required for any computation on w.

For any of these measures, the space complexity of a machine can be defined as a function of an integer n as the maximum tape size required for any input word of length n under these conditions. Finally, given a language, one can examine the space complexity of different machines accepting this language. For many of the more restricted automata models, some of these three complexity measures have not been studied as extensively as for Turing machines[1]. This paper aims to fill the gaps for several machine models.

We study the above complexity measures for machines and languages of one-way stack automata, non-erasing stack automata, and checking stack automata. One-way stack automata are, intuitively, pushdown automata with the additional ability to read letters from inside the stack; but still only push to and pop from the top of the stack. Non-erasing stack automata are stack automata without the ability to erase (pop) letters from the stack. Finally, checking stack automata are further restricted so that as soon as they read from inside of the stack, they can no longer push new letters on the stack.

It is known that checking stack languages form a proper subset of non-erasing stack languages, which form a proper subset of stack languages [4], and those in turn form a proper subset of context-sensitive languages [8]. In terms of space complexity, it is possible to study the three space complexity measures (weak, accept, and strong) as the maximum stack size required for any input of length n. It is already known that every stack language can be accepted by *some* stack automaton which operates in linear space using the weak measure [8,12]. However, this does not imply that *every* stack automaton has this property. We prove here that every checking stack automaton has this property. Further results are

---

[1] We point out that, especially in the context of Turing machines, the weak measure, corresponding to the minimal cost among all accepting computations on a given input, if any, is by far the most commonly used.

known relating one-way and two-way versions of these machines to other models, and to space complexity classes of Turing machines, e.g. [3,10,12].

For checking stack automata, we give a complete characterization of the possible accept and strong space measures. For both measures, exactly one of the following three cases must occur for every checking stack automaton:

1. The complexity is $O(1)$. Then the automaton accepts a regular language.
2. There is some word (accepted word for the accept measure) u which has computations (accepting computations, respectively) that use arbitrarily large stack space on u, and so the complexity is not $O(f(n))$ for any integer function f. The language accepted can be regular or not.
3. The complexity is $O(n)$, but it is not $o(n)$. The language accepted can be regular or not.

The third case is essentially saying that the complexity is $\Theta(n)$, except for some minor technicalities that will be discussed further in the paper. Therefore, there is a "gap" in the possible asymptotical behaviors of space complexity. No checking stack machine can have a space complexity between $\Theta(1)$ and $\Theta(n)$; or complexity above $\Theta(n)$ (as long as there is *some* function which bounds the space). The lower bound proof uses a method involving store languages of stack automata (the language of all words occurring on the stack of an accepting computation). We have not seen this technique used previously in the literature. Indeed, store languages are used in multiple proofs of this paper.

For non-erasing stack automata, there are differences with checking stack automata, as the complexity can be in $o(n)$, though not constant. We present an automaton with a weak and accept space complexity in $\Theta(\sqrt{n})$.

We also consider the following problem: Given a language (accepted by one of the stack automaton models) and one of the space complexity measures, what are the space complexities of the machines accepting it? We show that there is a checking stack language such that with the strong measure, every machine accepting it can use arbitrarily larger stack space than the input size, and therefore it is not $O(f(n))$ for any function f. Lastly, decidability questions on space complexity are addressed. It is shown that it is undecidable whether a checking stack automaton operates in constant space using the weak measure, however for both the strong and accept measures, it is decidable even for arbitrary stack automata.

2   Preliminaries

This section introduces basic notation used in this paper, and defines the three models of stack automata that we shall consider.

We assume that the reader is familiar with basics of formal language and automata theory. Please see [9] for an introduction. An *alphabet* is a finite set of *letters*. A *word* over an alphabet $\Sigma = \{a_1, \ldots, a_k\}$ is a finite sequence of letters from Σ. The set of all words over Σ is denoted by Σ^*, which includes the *empty word*, denoted by λ. A *language* L (over Σ) is any set of words $L \subseteq \Sigma^*$. The complement of L over Σ, denoted by $\overline{L}$ is equal to $\Sigma^* \setminus L$. Given a word $w \in \Sigma^*$, the *length* of w is denoted by $|w|$, and the number of occurrences of a letter a_i in w by $|w|_{a_i}$. The *Parikh image* of w is the vector $\psi(w) = (|w|_{a_1}, \ldots, |w|_{a_k})$, which is extended to a language L as $\psi(L) = \{\psi(w) \mid w \in L\}$. We do not define the concept of semilinearity formally here, but it is known that a language L is *semilinear* if and only if there is a regular language L' with $\psi(L) = \psi(L')$ [5]. Given two words $w, u \in \Sigma^*$, we say that u is a *prefix* of w if $w = uv$ for some $v \in \Sigma^*$. The *prefix closure* of a language L, $\mathrm{pref}(L)$, is the set of all prefixes of all words in L. It is known that if L is a regular language, then $\mathrm{pref}(L)$ is also regular.

2.1   Automata Models

Next, we define the three types of stack automata models discussed in this paper.

Definition 1. *A one-way nondeterministic* stack automaton *(SA for short) is a 6-tuple* $M = (Q, \Sigma, \Gamma, \delta, q_0, F)$*, where:*

- Q *is the finite set of* states.
- Σ *and* Γ *are the* input *and* stack *alphabets, respectively.*
- Γ *contains symbols* $\triangleright$ *and* $\triangleleft$*, which represent the* bottom *and* top *of the stack. We denote by* Γ_0 *the alphabet* $\Gamma \setminus \{\triangleright, \triangleleft\}$*.*
- $q_0 \in Q$ *and* $F \subseteq Q$ *are the* initial state *and the set of* final states*, respectively.*
- δ *is the nondeterministic transition function from* $Q \times (\Sigma \cup \{\lambda\}) \times \Gamma$ *into subsets of* $Q \times \{\mathtt{stay}, \mathtt{push}(x), \mathtt{pop}, -1, 0, +1 \mid x \in \Gamma_0\}$*. We use the notation* $(q, a, y) \to (p, \iota)$ *to denote that* $(p, \iota) \in \delta(q, a, y)$*.*

A *configuration* c of an SA is a triple $c = (q, w, \gamma)$, where $q \in Q$ is the current state, $w \in \Sigma^*$ is the remaining input to be read, and γ is the current stack tape. The word γ either has to be of the form $\triangleright \Gamma_0^* \downarrow \Gamma_0^* \triangleleft$, or of the form $\triangleright \Gamma_0^* \triangleleft \downarrow$. The symbol denotes the position of the stack head, which is currently scanning the symbol directly preceding it. We shall occasionally refer to the "pure" stack content, that is, the word γ without the end markers and the head symbol. We denote this word by $\hat\gamma$. The *stack size* of c is $\|c\|_\Gamma = |\hat\gamma| = |\gamma| - 3$.

We use two relations between configurations:

- The *write relation*: If $(q, a, y) \to (p, \iota)$, where $q, p \in Q$, $a \in \Sigma \cup \{\lambda\}$, $y \in \Gamma_0 \cup \{\triangleright\}$, and $\iota \in \{\mathtt{stay}, \mathtt{push}(x), \mathtt{pop}\}$; then for $u \in \Sigma^*$, $\gamma \in \Gamma^*$, with $\gamma y \in \triangleright \Gamma_0^*$:
  - $(q, au, \gamma y \downarrow \triangleleft) \vdash_w (p, u, \gamma y \downarrow \triangleleft)$ if $\iota = \mathtt{stay}$,
  - $(q, au, \gamma y \downarrow \triangleleft) \vdash_w (p, u, \gamma y x \downarrow \triangleleft)$ if $\iota = \mathtt{push}(x)$,
  - $(q, au, \gamma y \downarrow \triangleleft) \vdash_w (p, u, \gamma \downarrow \triangleleft)$ if $\iota = \mathtt{pop}$ and $y \neq \triangleright$.

Notice that the write relation is defined only if $\mathtt{stay}$, $\mathtt{push}$, and $\mathtt{pop}$ transitions are performed when the stack head is scanning the topmost symbol of the stack. If one of these operations is executed when the stack head is not on the top of the stack, the machine halts and rejects.

- The *read relation*: If $(q, a, y) \to (p, \iota)$, where $q, p \in Q$, $a \in \Sigma \cup \{\lambda\}$, $y \in \Gamma$, and $\iota \in \{-1, 0, 1\}$; then for $u \in \Sigma^*$, $\gamma_1, \gamma_2 \in \Gamma^*$, with $\gamma_1 y \gamma_2 \in \triangleright \Gamma_0^* \triangleleft$:
  - $(q, au, \gamma_1 y \downarrow \gamma_2) \vdash_r (p, u, \gamma_1 \downarrow y \gamma_2)$ if $\iota = -1$ and $y \neq \triangleright$,
  - $(q, au, \gamma_1 y \downarrow \gamma_2) \vdash_r (p, u, \gamma_1 y \downarrow \gamma_2)$ if $\iota = 0$,
  - $(q, au, \gamma_1 y \downarrow \gamma_2) \vdash_r (p, u, \gamma_1 y x \downarrow \gamma_2')$ if $\iota = +1$, $\gamma_2 = x \gamma_2'$ and $x \in \Gamma$.

The union of $\vdash_w$ and $\vdash_r$ is denoted by $\vdash$. The transitive closures of $\vdash_w$, $\vdash_r$, and $\vdash$ are denoted by $\vdash_w^+$, $\vdash_r^+$, and $\vdash^+$; and their transitive and reflexive closures by $\vdash_w^*$, $\vdash_r^*$, and $\vdash^*$, respectively.

A *partial computation* of the automaton M on an input word u is a sequence of configurations

$$\mathcal{C} : \overbrace{(p_0, u_0, \gamma_0)}^{c_0} \vdash \cdots \vdash \overbrace{(p_n, u_n, \gamma_n)}^{c_n}, \tag{1}$$

where $p_0 = q_0, u_0 = u, \gamma_0 = \triangleright \downarrow \triangleleft$. If also $u_n = \lambda$, we say that this is a *computation*; and furthermore, if also $p_n \in F$ then it is an *accepting computation*. The *stack size* of the (partial) computation $\mathcal{C}$, denoted by $\|\mathcal{C}\|_\Gamma$, is defined as $\max\{\|c_j\|_\Gamma \mid 0 \leq j \leq n\}$.

The *language accepted* by an SA M, denoted by $L(M)$, is the set of words w for which M has an accepting computation on w. The *store language* of M, $S(M)$, is the set of state and stack contents that can appear in an accepting computation: $S(M) = \{q\gamma \mid (q, u, \gamma)$ is a configuration in some accepting$\}$ computation of M. Notice that these words contain both the state and the stack head position. It is known that for every SA M, $S(M)$ is a regular language [2,11].

The accepting computation in Eq. (1) can be written uniquely as

$$c_0 \vdash_w^* d_1 \vdash_r^+ c_1 \vdash_w^+ \cdots \vdash_w^+ d_m \vdash_r^* c_m.$$

We call a sequence of transitions $c_i \vdash_w^* d_{i+1}$ a *write phase*, and a sequence of transitions $d_i \vdash_r^* c_i$ a *read phase*. By this definition, a computation always starts with a write phase and ends with a read phase. For the purpose of this paper, we can assume without loss of generality that both the first write phase and last read phase are non-empty, by altering the machine to always start by writing with a $\mathtt{stay}$ instruction, and to always read with a 0 instruction before finishing.

Furthermore, for any such SA $M = (Q, \Sigma, \Gamma, \delta, q_0, F)$ (with a non-empty initial read phase and final write phase), we can construct an SA $M' = (Q_w \cup Q_r, \Sigma, \Gamma, \delta', q_{0w}, F')$; where Q_w and Q_r are two distinct copies of the state set Q of M, with the copied states denoted by the $_w$ and $_r$ subscripts, $F' = \{q_r \mid q \in F\}$, and where δ' is a union of two transition functions:

- δ_w, which contains transitions $(q_w, a, y) \to (p_w, \iota)$ and $(q_w, a, y) \to (p_r, \iota)$; where $(q, a, y) \to (p, \iota)$ in δ, and $\iota \in \{\mathtt{stay}, \mathtt{push}(x), \mathtt{pop}\}$; and
- δ_r, which contains transitions $(q_r, a, y) \to (p_w, \iota)$ and $(q_r, a, y) \to (p_r, \iota)$; where $(q, a, y) \to (p, \iota)$ in δ, and $\iota \in \{-1, 0, 1\}$.

We call transitions in δ_w *write transitions*, and transitions in δ_r *read transitions*. Similarly, we call states in Q_w (resp. Q_r) *write states* (resp. *read states*). Observe that the language accepted by M' is the same as the one accepted by M.

Any stack machine that has states that can be partitioned into write and read ones, such that write transitions can only be applied from write states, and read transitions can only be applied from read states, is said to have *partitioned states*. In such a machine, the current state in every configuration dictates whether the next transition to be taken is a write or a read transition.

A stack automaton is called *non-erasing* (NESA) if it contains no transitions to an element of $Q \times \{\text{pop}\}$. A non-erasing stack automaton is called a *checking stack automaton* (CSA) if it has partitioned states and it contains no transitions from a read state to a write state. Every accepting computation of a checking stack automaton therefore has a single write phase followed by a single read phase.

We denote by $\mathcal{L}(\text{SA}), \mathcal{L}(\text{NESA})$, and $\mathcal{L}(\text{CSA})$ the families of languages accepted by the three types of devices.

3   Complexity Measures on Stack Automata

For an SA $M = (Q, \Sigma, \Gamma, \delta, q_0, F)$, one can consider three different space complexity measures defined similarly as for Turing machines [13]. Consider an input word $u \in \Sigma^*$ to M.

– **weak** measure:

$$\sigma_M^{\text{w}}(u) = \begin{cases} \min\left\{\|\mathcal{C}\|_\Gamma \mid \mathcal{C} \text{ an accepting computation on } u\right\}, & \text{if } u \in L(M), \\ 0, & \text{otherwise.} \end{cases}$$

– **accept** measure:

$$\sigma_M^{\text{a}}(u) = \begin{cases} \max\left\{\|\mathcal{C}\|_\Gamma \mid \mathcal{C} \text{ an accepting computation on } u\right\}, & \text{if exists } \& \\ & \qquad\qquad u \in L(M), \\ \infty & \text{if does not exist } \& \ u \in L(M), \\ 0, & \qquad\qquad u \notin L(M). \end{cases}$$

– **strong** measure:

$$\sigma_M^{\text{s}}(u) = \begin{cases} \max\left\{\|\mathcal{C}\|_\Gamma \mid \mathcal{C} \text{ is a partial computation on } u\right\} & \text{if it exists,} \\ \infty & \text{otherwise.} \end{cases}$$

Next, we are interested in studying stack sizes as a function of the length of the input. Thus, for each $z \in \{\text{w}, \text{a}, \text{s}\}$, we define the functions,

$$\sigma_M^z(n) = \max\left\{\sigma_M^z(u) \mid u \in \Sigma^* \text{ and } |u| = n\right\},$$
$$\acute{\sigma}_M^z(n) = \max\left\{\sigma_M^z(u) \mid u \in \Sigma^* \text{ and } |u| \leq n\right\}.$$

The latter essentially forces the space complexity to be a non-decreasing function. We leave off M if it is clear based on the context.

Using this notation, we can now write $\sigma^z(n) \in O(f(n))$, $o(f(n))$, $\Omega(f(n))$, etc., for some function $f(n)$ from $\mathbb{N}_0$ to $\mathbb{N}_0$, in the usual fashion.

Note that, if there is any single word u with $\sigma^z(u) = \infty$, with $z \in \{a, s\}$ (this occurs if there are infinitely many accepting computations of the word u of arbitrarily large stack sizes), then $\sigma^z(n) = \infty$ for $n = |u|$, and $\sigma^z(n)$ cannot be in $O(f(n))$ for any integer function f. If there is such a word u, then we say that M is z-unlimited, and z-limited otherwise.

Example 2. Consider the language

$$L = \{a^m b^k \mid m \leq k, m \text{ divides } m + k\}.$$

This contains for example $a^3 b^6$ because 3 divides 9. It contains ab^4 because 1 divides 5, but there are no other words accepted of length 5 since 5 is prime.

An obvious CSA M accepting L copies a^m to the stack, then verifies that m divides k by going back and forth on the checking stack while reading from the input and checking that both the stack and input reach the ends of their tape at the same time.

Here are some properties of M:

1. $\sigma^s(n), \sigma^a(n), \sigma^w(n)$ are all $O(n)$.
2. For every even n, $\sigma^a(a^{n/2} b^{n/2}) = n/2$, and thus $\sigma^a(n) \geq n/2$. Therefore, $\sigma^a(n)$ is not $o(n)$.
3. Also $\acute{\sigma}^a(n)$ is $\Omega(f(n))$ since $\acute{\sigma}^a(n)$ is non-decreasing.
4. For every prime number n, $\sigma^a(n) = 1$. Thus, $\sigma^a(n)$ is not $\Omega(n)$. Further, $\sigma^a(n)$ is not $\Omega(f(n))$ for any $f(n)$ in $\omega(1)$ (i.e. $f(n)$ is $\omega(1)$ if for any positive constant c, there exists a constant k such that $0 \leq c < f(n)$ for all $n \geq k$). So, if we use the function σ^a instead of the non-decreasing function $\acute{\sigma}^a$, then it is no longer at least linear.

4   Space Complexities of Stack Automata

In [8] it was shown that for every stack automaton M, there exists another stack automaton M' such that $L(M) = L(M')$ and $\sigma^w_{M'}(n)$ is $O(n)$. Here we show the stronger statement that for any checking checking stack automaton M, $\sigma^w_M(n)$ is in $O(n)$ (i.e. it is true for M without converting to M'). Furthermore, it is also true for the accept (and strong) measures as well as long as they are a-limited (s-limited).

The basic idea of the proof for the weak measure is the following: consider some accepting computation $\mathcal{C}$ on some string $u \in L(M)$. Consider the stack at the end of this computation. We shall look for maximal sections of the write phase whereby, from the start of pushing this section until the end of pushing this section, only λ-transitions are applied; and then in later read phases, this section of the stack is also only read on λ-transitions. Therefore, the behavior of the automaton in this section of the stack does not depend on the input string. We shall show that if this section of the stack is too large (larger than some constant z'), then we can find some smaller word that we can replace it with, without altering this behavior. This means that for every accepted string u, we can find an accepting computation in which each of these sections is at most z' letters long. Since each of these sections of the stack are surrounded by cells of the stack in which the automaton reads some input symbol, there can be at most $(|u|+2)z'$ letters on the stack in this new computation. A similar argument can be used for the accept and strong measures.

Proposition 3. *Let M be a* CSA. *The following are true:*

- $\sigma_M^{\mathrm{w}}(n)$ *is in* $O(n)$;
- *if* $\sigma_M^{\mathrm{a}}(n)$ *is* a-*limited, then* $\sigma^{\mathrm{a}}(n)$ *is* $O(n)$;
- *if* $\sigma_M^{\mathrm{s}}(n)$ *is* s-*limited, then* $\sigma^{\mathrm{s}}(n)$ *is* $O(n)$.

Whether or not the result above holds for NESA and SA generally is an open problem.

Notice that in the proposition above, it is not true that every CSA machine M has $\sigma^{\mathrm{a}}(n)$ in $O(n)$, because of the following more general fact:

Remark 4. Consider a CSA machine M accepting Σ^* that nondeterministically pushes any string onto the stack using λ-transitions, and then reads any input and accepts. Here, M is not a-limited or s-limited and $\sigma^{\mathrm{a}}(n)$ is not $O(f(n))$ for any function f.

Lower bounds on the space complexity functions can also be studied similarly to upper bounds. The next proof starts with an accepting computation using some stack word, and then finds a new accepting computation on some possibly different input word that is roughly linear in the size of the stack. It then uses the regularity of the store languages of stack automata in order to determine that for every increase in some constant c, there's at least one more input word of that length that has a stack that is linear in the size of the input. That is enough to show that the accept and strong space complexities cannot be $o(n)$, and if the non-decreasing function $\acute{\sigma}^z$ is used, then it is at least linear.

Lemma 5. *Let $z \in \{\mathrm{a},\mathrm{s}\}$. Let M be a* CSA *such that $\sigma^z(n) \notin O(1)$ and M is z-limited. The following are true:*

- *there exist c, d, e such that, for every $n \in \mathbb{N}_0$, there is some input $u \in \Sigma^*$ (with $u \in L(M)$ if $z = \mathrm{a}$) where $n \le |u| \le n + c$ and $d|u| \le \sigma^z(u) \le e|u|$,*
- *$\sigma^z(n)$ cannot be $o(n)$,*
- *$\acute{\sigma}^z(n) \in \Omega(n)$.*

Lemma 5 is the "best possible result" in that it is not always the case that $\sigma^{\mathrm{a}}(n) \in \Omega(n)$ since the space complexity can periodically go below linear infinitely often as demonstrated with Example 2. But what this lemma says is that it returns to at least linear infinitely often as well. Furthermore, there is a constant c such that it must return to at least linear for every increase of c in the length of the input. Putting together all results for CSA so far, we get the following complete characterization:

Theorem 6. *Let M be a* CSA. *For $z \in \{a, s\}$, exactly one of the following must occur.*

1. *M is z-unlimited, and so there is no f such that $\sigma^z(n) \in O(f(n))$ (and $L(M)$ can be either regular or not);*
2. *M is z-limited, $\sigma^z(n) \in O(1)$, and $L(M)$ is regular;*
3. *M is z-limited, $\sigma^z(n) \in O(n), \sigma^z \notin o(n)$, and $\acute{\sigma}^z(n) \in \Theta(n)$ (and $L(M)$ can be either regular or not).*

Proof. Consider the case $z = a$. Either M is a-limited, or not. If it is not, then $L(M)$ can be either regular or not. Moreover, both are possible, as one can take an arbitrary CSA M' (which can either be regular or not), and modify it to M'' by starting by pushing an arbitrary word over a new stack letter x on λ-transitions, then simulating M'. Thus, $L(M') = L(M'')$, and M'' is a-unlimited.

Assume that M is a-limited. And, assume that $\sigma^a(n)$ is $O(1)$. Then only a bounded amount of the stack is used, and M can therefore be simulated by an NFA, and hence $L(M)$ is regular.

Assume $\sigma^a(n)$ is not $O(1)$. Then Lemma 5 applies, and the statement follows. Also, it is possible for it to be non-regular (Example 2), or regular (by taking a DFA and simulating it with a CSA that copies the input to the stack while simulating the DFA). □

The question arises next of whether the lower bound is also true for NESA and SA. We see that this is not true.

Proposition 7. *There exists a* NESA *(and a* SA*) M that accepts a non-regular language such that $\sigma^a(n)$ and $\sigma^w(n)$ are in $\Theta(\sqrt{n})$, and $\sigma^s(n) \in O(\sqrt{n})$.*

Proof. Consider the language $L = \{a^1ba^2b \cdots a^rb \mid r \geq 1\}$, and let $L_0 = \mathrm{pref}(L)$, which is not regular. Then L_0 can be accepted as follows. Consider the input $a^{l_1}ba^{l_2}b \cdots a^{l_r}ba^l, r, l \geq 0$. M starts by reading and pushing a, then it repeats the following: It reads b and pushes one a on the stack, moves to the left end of the stack and matches the a's on the stack to the next block of a's of the input. These steps are repeated until the end of the last section, that can be (the only one) shorter than the word stored in the stack.

On an input of size n, there is one word accepted of this length and the stack γ satisfies $1/2\sqrt{n} \leq |\gamma| \leq \sqrt{n}$. Thus, $\sigma^a(n)$ and $\sigma^w(n)$ are in $\Theta(\sqrt{n})$. For the strong measure, the stack is at most $\sqrt{n}$ in size. □

It is possible to observe that there exists a NESA M that accepts a regular language such that $\sigma^a(n) \in \Theta(\sqrt{n})$ and $\sigma^s(n) \in O(\sqrt{n})$. Let us consider $L_0 \cup \{a, b\}^* = \{a, b\}^*$. A machine M' can be built that either simulates M described in the proof of Proposition 7, or it reads the input and accepts without using the stack. Thus, $\sigma_M^a(n) = \sigma_{M'}^a(n)$ and $\sigma_M^s(n) = \sigma_{M'}^s(n)$.

In conclusion, $\sqrt{n} \in o(n)$, and thus Lemma 5 cannot be generalized to NESA or SA. Therefore, unlike CSA, there is not a complete gap between $\Theta(1)$ and $\Theta(n)$, and the exact functions possible between them (besides $\sqrt{n}$) remains open.

5   Space Complexities of Languages Accepted by Stack Automata

Just as the space complexity of stack machines can be studied, it is also possible to ask the question, given a language $L \in \mathcal{L}(\mathsf{CSA})$, what are the space complexities of the machines in CSA accepting L? (and similarly for $\mathsf{NESA}, \mathsf{SA}$). It follows from Proposition 3 that for every $L \in \mathcal{L}(\mathsf{CSA})$, there exists a machine $M \in \mathsf{CSA}$ such that $\sigma^w(n)$ is $O(n)$. A similar result is also known to be true for SA generally [8,12].

For the strong measure, we will see that there are languages where all the machines that accept them are more complicated.

Example 8. Consider the language

$$L_{\mathrm{copy}} = \{u\$u\sharp v\$v \mid u, v \in \{a,b\}^*\}.$$

Certainly, there is an $M \in \mathsf{CSA}$ that accepts L as follows: on input $u'\$u''\#v'\$v''$, M guesses two words u and v in advance on λ-transitions, and pushes $u\#v$ on the stack; then M verifies that $u = u' = u''$, and $v = v' = v''$.

For the accept measure, $\sigma^a_M(n) \in O(n)$, as all computations that accept have a stack that is linear in the input. However, for the strong measure, because the machine M starts by guessing and pushing both u and v on λ-transitions, M could guess u and v that are substantially longer (arbitrarily longer) than u' and v'. This machine M is therefore s-unlimited.

The question arises as to whether *every* machine that accepts L_{copy} is s-unlimited. We will see that this is indeed the case. To show this, we first prove the following lemma, which again uses store languages

Lemma 9. *Let* $M = (Q, \Sigma, \Gamma, \delta, q_0, F)$ *be a* CSA *with partitioned states* Q_w *and* Q_r. *The language*

$$L_{w,M} = \{u \mid (q_0, uv, \rhd \lrcorner \lhd) \vdash^*_w (q, v, \gamma) \vdash^*_r (q_f, \lambda, \gamma'), q \in Q_r, q_f \in F\},$$

composed of all the input words scanned by M *during the (complete) write phase of each accepting computation, is a regular language.*

This is useful towards the following proposition.

Proposition 10. *For the language* $L_{\mathrm{copy}} \in \mathcal{L}(\mathit{CSA})$ *from Example 8, for all* M *accepting* L_{copy}, M *is s-unlimited. Thus, for each such* M *accepting* L_{copy}, *there is no function* f *such that* $\sigma^s_M(n)$ *is* $O(f(n))$.

Proof. We show that for every CSA M accepting L_{copy}, M can perform an arbitrarily long sequence of λ-transitions that write (either push or stay) where the size of the stack can grow arbitrarily without reading input letters. If there is some sequence of λ-transitions that writes (that can be reached from the initial configuration) that is bigger than the number of states of the machine multiplied by the stack alphabet, then M has a cycle in which only write λ-transitions occur. Then as long as this cycle has at least one push transition in it, the stack can grow arbitrarily. Hence, there exist infinitely many (possibly rejecting) computations during which arbitrarily many letters are pushed on the stack making the machine s-unlimited.

Let $M = (Q, \Sigma, \Gamma, \delta, q_0, F)$ be an arbitrary CSA accepting L_{copy}. Assume, by contradiction, that L_{copy} is s-limited. Consider the language $L_{w,M}$ from Lemma 9, which is a regular language. Furthermore, let $W = \text{pref}(L_{w,M}) \cap \{a, b\}^*\${a, b\}^*\sharp$, which must also be regular. Assume that W is infinite. Thus, there exist infinitely many words in W that have an accepting computation that does not enter the read phase until after $\#$. But as W is regular, and only contains words of the form $u\$u\#$, there must be some infinite subset of $\{u\$u\# \mid u \in \{a, b\}^*\}$ that is regular, a contradiction, by the pumping lemma. Thus, W must be finite.

Let u be some word such that $u\$u\# \notin W$. Thus, for all accepting computations on any word in $L_{\text{copy}} = \{u\$u\sharp v'\$v'' \mid v', v'' \in \{a, b\}^*\}$, it must enter the read phase before reaching the $\#$ symbol. Also, there must exist a constant c such that for all of these accepting computations, the stack must grow to at most $c|u|$, otherwise M could enter an infinite cycle on λ-transitions that push in the write phase, and it would be s-unlimited. Consider some word $u\$u\#v\$v \in L_{\text{copy}}$ where $|v| > |Q| \cdot c \cdot |u|$, and consider some accepting computation (where it must enter the read phase before hitting $\#$). When scanning the second v, there must be two configurations reached where M reaches the same state and stack position, and at least one letter of Σ was read between them. Hence, $u\$u\#v\$v'$ is also accepted, $v \neq v'$, a contradiction. Hence, M is s-unlimited.   □

For the accept measure, the situation is more complicated, and it is left open. However, we have the following conjecture. Consider the language

$$L = \{1^k \#v_1 \# \cdots \#v_m \mid v_i \in \{0, 1\}^*, |\{v_1, \ldots, v_m\}| \leq k\}.$$

This language can be accepted by a CSA machine that, for every 1 read, pushes a nondeterministically guessed word over $\{0, 1\}^*$ on the stack so that its contents is $u_1\# \cdots \#u_k$. Then, for each v_i on the input, it guesses some u_j on the stack and verifies that they are equal. However, this machine does not keep track of whether each u_j on the stack was matched to some v_i (and it seems to have no way of keeping track of this), and u_j could be arbitrarily long. We conjecture that every $M \in$ CSA accepting L is a-unlimited. In fact, we conjecture that this is true for every $M \in$ SA.

6   Decidability Properties Regarding Space Complexity of Stack Machines

It is an easy observation that when the space used by a checking stack automaton is constant, the device is no more powerful than a finite automaton. Nevertheless, given a checking stack automaton M, it is not possible to decide whether or not it accepts by using a constant amount of space with the weak measure. This result can be derived by adapting the argument used in [14] for proving that, when the weak measure is considered, it is not decidable whether or not a nondeterministic pushdown automaton accepts by using a constant amount of pushdown store. In that case, the authors used a technique introduced in [6], based on suitable encodings of single-tape Turing machine computations and reducing the proof of the decidability to the halting problem; this can be done here as well.

Proposition 11. *It is undecidable whether a* CSA *M accepts in space $\sigma^{\mathrm{w}}(n) \in O(1)$ or not.*

On the other hand, although it may seem counterintuitive, the same problem is decidable for the accept and strong measures, even for stack automata.

Proposition 12. *For $z \in \{\mathrm{a}, \mathrm{s}\}$, it is decidable whether an* SA *M satisfies $\sigma^z(M) \in O(1)$ or not.*

Proof. For the accept measure, first, we construct a finite automaton M' accepting the store language of M. We can then decide finiteness of $L(M')$ since it is regular, which is finite if and only if M operates in constant space.

For the strong measure, we can take M, and change it so that all states are final, then calculate the store language, and decide finiteness.   □

7   Conclusions and Future Directions

In this paper, we defined and studied the weak, accept, and strong space complexity measures for variants of stack automata. For checking stack automata with the accept or strong measures, there is "gap", and no function is possible between constant and linear, or above linear. For non-erasing stack automata, there are machines with complexity between constant and linear. Then, it is shown that for the strong measure, there is a checking stack language such that every machine accepting it is s-unlimited (there is no function bounding the strong space complexity). Lastly, it is shown that it is undecidable whether a checking stack automaton has constant space complexity with the weak measure. But, this is decidable for both the accept and strong measures even for stack automata.

Many open problems remain. It is desirable to know whether there are any gaps between constant and linear space for the weak space complexity measure for checking stacks. Also, it is open whether all stack automata have linear weak space complexity (it is known that every language has some machine that operates in linear space complexity). The exact accept and strong space complexity functions possible for non-erasing and stack automata (besides constant, square root, and linear) still need to be determined. It is also open whether there is some stack language (or non-erasing stack language) such that every machine accepting it is s-unlimited. Furthermore, for the accept measure, we conjecture that there is a CSA language whereby every machine is a-unlimited, although this is also an open problem. Answering these open questions would be of interest to the automata theory community.

References

1. Alur, R., Madhusudan, P.: Visibly pushdown languages. In: Proceedings of the Thirty-Sixth Annual ACM Symposium on Theory of Computing, pp. 202–211 (2004)
2. Bensch, S., Björklund, J., Kutrib, M.: Deterministic stack transducers. Int. J. Found. Comput. Sci. **28**(05), 583–601 (2017)
3. Engelfriet, J.: The power of two-way deterministic checking stack automata. Inf. Comput. **80**(2), 114–120 (1989)
4. Greibach, S.: Checking automata and one-way stack languages. J. Comput. Syst. Sci. **3**(2), 196–217 (1969)
5. Harrison, M.A.: Introduction to Formal Language Theory. Addison-Wesley Series in Computer Science. Addison-Wesley Pub. Co., Boston (1978)
6. Hartmanis, J.: Context-free languages and Turing machine computations. In: Mathematical Aspects of Computer Science. Proceedings of Symposia in Applied Mathematics, vol. 19, pp. 42–51. American Mathematical Society (1967)
7. Hartmanis, J.: Turing award lecture: on computational complexity and the nature of computer science. ACM Comput. Surv. **27**(1), 7–16 (1995). https://doi.org/10.1145/214037.214040
8. Hopcroft, J., Ullman, J.: Sets accepted by one-way stack automata are context sensitive. Inf. Control **13**(2), 114–133 (1968)
9. Hopcroft, J., Ullman, J.: Introduction to Automata Theory, Languages, and Computation Reading. Addison-Wesley, Boston (1979)
10. Ibarra, O.H., McQuillan, I.: Generalizations of checking stack automata: characterizations and hierarchies. In: Hoshi, M., Seki, S. (eds.) DLT 2018. LNCS, vol. 11088, pp. 416–428. Springer, Cham (2018). https://doi.org/10.1007/978-3-319-98654-8_34
11. Ibarra, O.H., McQuillan, I.: On store languages of languages acceptors. Theor. Comput. Sci. **745**, 114–132 (2018)
12. King, K., Wrathall, C.: Stack languages and $\log n$ space. J. Comput. Syst. Sci. **17**(3), 281–299 (1978)
13. Pighizzini, G.: Nondeterministic one-tape off-line Turing machines and their time complexity. J. Autom. Lang. Comb. **14**, 107–124 (2009)
14. Pighizzini, G., Prigioniero, L.: Pushdown automata and constant height: decidability and bounds. In: Hospodár, M., Jirásková, G., Konstantinidis, S. (eds.) DCFS 2019. LNCS, vol. 11612, pp. 260–271. Springer, Cham (2019). https://doi.org/10.1007/978-3-030-23247-4_20

Descriptional Complexity of Semi-simple Splicing Systems

Lila Kari[1] and Timothy Ng[1,2(✉)]

[1] School of Computer Science, University of Waterloo,
Waterloo, ON N2L 3G1, Canada
lila.kari@uwaterloo.ca, timng@uchicago.edu
[2] Department of Computer Science, University of Chicago, Chicago, IL 60637, USA

Abstract. Splicing systems are generative mechanisms introduced by Tom Head in 1987 to model the biological process of DNA recombination. The computational engine of a splicing system is the "splicing operation", a cut-and-paste binary string operation defined by a set of "splicing rules", quadruples $r = (u_1, u_2; u_3, u_4)$ where u_1, u_2, u_3, u_4 are words over an alphabet Σ. For two strings $x_1u_1u_2y_1$ and $x_2u_3u_4y_2$, applying the splicing rule r produces the string $x_1u_1u_4y_2$. In this paper we focus on a particular type of splicing systems, called (i, j) semi-simple splicing systems, $i = 1, 2$ and $j = 3, 4$, wherein all splicing rules r have the property that the two strings in positions i and j in r are singleton letters, while the other two strings are empty. The language generated by such a system consists of the set of words that are obtained starting from an initial set called "axiom set", by iteratively applying the splicing rules to strings in the axiom set as well as to intermediately produced strings. We consider semi-simple splicing systems where the axiom set is a regular language, and investigate the descriptional complexity of such systems in terms of the size of the minimal deterministic finite automata that recognize the languages they generate.

1 Introduction

Splicing systems are generative mechanisms introduced by Tom Head [7] to model the biological process of DNA recombination. A splicing system consists of an initial language called an *axiom set*, and a set of so-called *splicing rules*. The result of applying a splicing rule to a pair of operand strings is a new "recombinant" string, and the language generated by a splicing system consists of all the words that can be obtained by successively applying splicing rules to axioms and the intermediately produced words. The most natural variant of splicing systems, often referred to as *finite splicing systems*, is to consider a finite set of axioms and a finite set of rules. Several different types of splicing systems have been proposed in the literature, and Bonizzoni et al. [1] showed that the classes of languages they generate are related: the class of languages generated by finite Head splicing systems [7] is strictly contained in the class of languages generated

© Springer Nature Switzerland AG 2020
N. Jonoska and D. Savchuk (Eds.): DLT 2020, LNCS 12086, pp. 150–163, 2020.
https://doi.org/10.1007/978-3-030-48516-0_12

by finite Păun splicing systems [13], which is strictly contained in the class of languages generated by finite Pixton splicing systems [12].

In this paper we will use the Păun definition [13], which defines a splicing rule as a quadruplet of words $r = (u_1, u_2; u_3, u_4)$. This rule splices two words $x_1 u_1 u_2 y_1$ and $x_2 u_3 u_4 y_2$ as follows: The words are cut between the factors u_1, u_2, respectively u_3, u_4, and the prefix of the first word (ending in u_1) is recombined by catenation with the suffix of the second word (starting with u_4), resulting in the word $x_1 u_1 u_4 y_2$.

Culik II and Harju [3] proved that finite Head splicing systems can only generate regular languages, while [8] and [12] proved a similar result for Păun, respectively Pixton splicing systems. Gatterdam [5] gave $(aa)^*$ as an example of a regular language which cannot be generated by a finite Head splicing system, which proved that this is a strict inclusion.

As the classes of languages generated by finite splicing systems are subclasses of the family of regular languages, their descriptional complexity can be considered in terms of the finite automata that recognize them. For example, Loos et al. [10] gave a bound on the number of states required for a nondeterministic finite automaton to recognize the language generated by an equivalent Păun finite splicing system. Other descriptional complexity measures for finite splicing systems that have been investigated in the literature include the number of rules, the number of words in the initial language, the maximum length of a word in the initial axiom set, and the sum of the lengths of all words in the axiom set. Păun [13] also proposed the radius, defined to be the size of the largest u_i in a rule, as another possible measure.

In the original definition, simple splicing systems are finite splicing systems where all the words in the splicing rules are singleton letters. The descriptional complexity of simple splicing systems was considered by Mateescu et al. [11] in terms of the size of a right linear grammar that generates a simple splicing language. Semi-simple splicing systems were introduced in Goode and Pixton [6] as having a finite axiom set, and splicing rules of the form $(a, \varepsilon; b, \varepsilon)$ where a, b are singleton letters, and ε denotes the empty word.

In this paper we focus our study on some variants of semi-simple splicing systems called (i, j)-semi-simple splicing systems, $i = 1, 2$ and $j = 3, 4$, wherein all splicing rules have the property that the two strings in positions i and j are singleton letters, while the other two strings are empty. (Note that Ceterchi et al. [2] showed that all classes of languages generated by semi-simple splicing systems are pairwise incomparable[1]). In addition, in a departure from the original definition of semi-simple splicing systems [6], in this paper the axiom set is allowed to be a (potentially infinite) regular set.

More precisely, we investigate the descriptional complexity of (i, j)-semi-simple splicing systems with regular axiom sets, in terms of the size of the minimal deterministic finite automaton that recognizes the language generated by the system. The paper is organized as follows: Sect. 2 introduces definitions

---

[1] Simple splicing language classes are pairwise incomparable except for the pair (1,3) and (2,4), which are equivalent [11].

and notations, Sect. 3 defines splicing systems and outlines some basic results on simple splicing systems, Sects. 4, 5 and 6 investigate the state complexity of (2,4)-, (2,3)- respectively (1,4)-semi-simple splicing systems, and Sect. 7 summarizes our results (Table 1).

2   Preliminaries

Let Σ be a finite alphabet. We denote by Σ^* the set of all finite words over Σ, including the empty word, which we denote by ε. We denote the length of a word w by $|w| = n$. If $w = xyz$ for $x, y, z \in \Sigma^*$, we say that x is a prefix of w, y is a factor of w, and z is a suffix of w.

A deterministic finite automaton (DFA) is a tuple $A = (Q, \Sigma, \delta, q_0, F)$ where Q is a finite set of states, Σ is an alphabet, δ is a function $\delta : Q \times \Sigma \to Q$, $q_0 \in Q$ is the initial state, and $F \subseteq Q$ is a set of final states. We extend the transition function δ to a function $Q \times \Sigma^* \to Q$ in the usual way. A DFA A is complete if δ is defined for all $q \in Q$ and $a \in \Sigma$. In this paper, all DFAs are defined to be complete. We will also make use of the notation $q \xrightarrow{w} q'$ for $\delta(q, w) = q'$, where $w \in \Sigma^*$ and $q, q' \in Q$. The language recognized or accepted by A is $L(A) = \{w \in \Sigma^* \mid \delta(q_0, w) \in F\}$.

Each letter $a \in \Sigma$ defines a transformation of the state set Q. Let $\delta_a : Q \to Q$ be the transformation on Q induced by a, defined by $\delta_a(q) = \delta(q, a)$. We extend this definition to words by composing the transformations $\delta_w = \delta_{a_1} \circ \delta_{a_2} \circ \cdots \circ \delta_{a_n}$ for $w = a_1 a_2 \cdots a_n$. We denote by $\operatorname{im} \delta_a$ the image of δ_a, defined $\operatorname{im} \delta_a = \{\delta(p, a) \mid p \in Q\}$.

A state q is called *reachable* if there exists a string $w \in \Sigma^*$ such that $\delta(q_0, w) = q$. A state q is called *useful* if there exists a string $w \in \Sigma^*$ such that $\delta(q, w) \in F$. A state that is not useful is called *useless*. A complete DFA with multiple useless states can be easily transformed into an equivalent DFA with at most one useless state, which we refer to as the *sink state*.

Two states p and q of A are said to be *equivalent* or *indistinguishable* in the case that $\delta(p, w) \in F$ if and only if $\delta(q, w) \in F$ for every word $w \in \Sigma^*$. States that are not equivalent are *distinguishable*. A DFA A is minimal if each state $q \in Q$ is reachable from the initial state and no two states are equivalent. The state complexity of a regular language L is the number of states of the minimal complete DFA recognizing L [4].

A nondeterministic finite automaton (NFA) is a tuple $A = (Q, \Sigma, \delta, I, F)$ where Q is a finite set of states, Σ is an alphabet, δ is a function $\delta : Q \times \Sigma \to 2^Q$, $I \subseteq Q$ is a set of initial states, and $F \subseteq Q$ is a set of final states. The language recognized by an NFA A is $L(A) = \{w \in \Sigma^* \mid \bigcup_{q \in I} \delta(q, w) \cap F \neq \emptyset\}$. As with DFAs, transitions of A can be viewed as transformations on the state set. Let $\delta_a : Q \to 2^Q$ be the transformation on Q induced by a, defined by $\delta_a(q) = \delta(q, a)$. We define $\operatorname{im} \delta_a = \bigcup_{q \in Q} \delta_a(q)$. We make use of the notation $P \xrightarrow{w} P'$ for $P' = \bigcup_{q \in P} \delta(q, w)$, where $w \in \Sigma^*$ and $P, P' \subseteq Q$.

3   Semi-simple Splicing Systems

In this paper we will use the notation of Păun [13]. The splicing operation is defined via sets of quadruples $r = (u_1, u_2; u_3, u_4)$ with $u_1, u_2, u_3, u_4 \in \Sigma^*$ called splicing rules. For two strings $x = x_1 u_1 u_2 x_2$ and $y = y_1 u_3 u_4 y_2$, applying the rule $r = (u_1, u_2; u_3, u_4)$ produces a string $z = x_1 u_1 u_4 y_2$, which we denote by $(x, y) \vdash^r z$.

A *splicing scheme* is a pair $\sigma = (\Sigma, \mathcal{R})$ where Σ is an alphabet and $\mathcal{R}$ is a set of splicing rules. For a splicing scheme $\sigma = (\Sigma, \mathcal{R})$ and a language $L \subseteq \Sigma^*$, we denote by $\sigma(L)$ the language

$$\sigma(L) = L \cup \{z \in \Sigma^* \mid (x, y) \vdash^r z, \text{ where } x, y \in L, r \in \mathcal{R}\}.$$

Then we define $\sigma^0(L) = L$ and $\sigma^{i+1}(L) = \sigma(\sigma^i(L))$ for $i \geq 0$ and

$$\sigma^*(L) = \lim_{i \to \infty} \sigma^i(L) = \bigcup_{i \geq 0} \sigma^i(L).$$

For a splicing scheme $\sigma = (\Sigma, \mathcal{R})$ and an initial language $L \subseteq \Sigma^*$, we say the triple $H = (\Sigma, \mathcal{R}, L)$ is a *splicing system*. The language generated by H is defined by $L(H) = \sigma^*(L)$.

Goode and Pixton [6] define a restricted class of splicing systems called semi-simple splicing systems. A semi-simple splicing system is a triple $H = (\Sigma, M, I)$, where Σ is an alphabet, $M \subseteq \Sigma \times \Sigma$ is a set of markers, and I is a finite initial language over Σ. We have $(x, y) \vdash^{(a,b)} z$ if and only if $x = x_1 a x_2$, $y = y_1 b y_2$, and $z = x_1 a y_2$ for some $x_1, x_2, y_1, y_2 \in \Sigma^*$. That is, a semi-simple splicing system is a splicing system in which the set of rules is $\mathcal{M} = \{(a, \varepsilon; b, \varepsilon) \mid (a, b) \in M\}$. Since the rules are determined solely by our choice of $M \subseteq \Sigma \times \Sigma$, the set M is used in the definition of the semi-simple splicing system rather than the set of rules $\mathcal{M}$.

It is shown in [6] that the class of languages generated by semi-simple splicing systems is a subclass of the regular languages. Semi-simple splicing systems are a generalization of the class of simple splicing systems, defined by Mateescu et al. [11]. A splicing system is a simple splicing system if it is a semi-simple splicing system and all markers are of the form (a, a) for $a \in \Sigma$. It is shown in [11] that the class of languages generated by simple splicing systems is a subclass of the extended star-free languages.

Observe that the set of rules $\mathcal{M} = \{(a, \varepsilon; b, \varepsilon) \mid (a, b) \in M\}$ of a semi-simple splicing system consist of 4-tuples with symbols from Σ in positions 1 and 3 and ε in positions 2 and 4. We can call such splicing rules (1,3)-splicing rules. Then a (1,3)-splicing system is a splicing system with only (1,3)-splicing rules and ordinary semi-simple splicing systems can be considered (1,3)-semi-simple splicing systems. The state complexity of (1,3)-simple and (1,3)-semi-simple splicing systems was studied previously by the authors in [9].

We can consider variants of semi-simple splicing systems in this way by defining semi-simple (i, j)-splicing systems, for $i = 1, 2$ and $j = 3, 4$. A semi-simple (2,4)-splicing system is a splicing system (Σ, M, I) with rules $\mathcal{M} = \{(\varepsilon, a; \varepsilon, b) \mid$

$(a, b) \in M$}. A (2,3)-semi-simple splicing system is a splicing system (Σ, M, I) with rules $\mathcal{M} = \{(\varepsilon, a; b, \varepsilon) \mid (a, b) \in M\}$. A (1,4)-semi-simple splicing system is a semi-simple splicing system (Σ, M, I) with rules $\mathcal{M} = \{(a, \varepsilon; \varepsilon, b) \mid (a, b) \in M\}$.

The classes of languages generated by simple and semi-simple splicing systems and their variants have different relationships among each other. Mateescu et al. [11] show that the classes of languages generated by (1,3)-simple splicing systems (i.e. ordinary simple splicing systems) and (2,4)-simple splicing systems are equivalent, while, the classes of languages generated by (1,3)-, (1,4)-, and (2,3)-simple splicing systems are all incomparable and subregular.

The situation is different for semi-simple splicing systems. Ceterchi et al. [2] show that each of the classes of languages generated by (1,3)-, (1,4)-, (2,3)-, and (2,4)-semi-simple splicing systems are all incomparable. So unlike simple splicing systems, the (1,3)- and (2,4)- variants are *not* equivalent. They show this by showing that the language $a^+ \cup a^+ab \cup aba^+ \cup aba^+b$ is generated by the (1,3)-semi-simple splicing system $(\{a, b\}, \{(a, \varepsilon; b, \varepsilon)\}, \{abab\})$ but cannot be generated by a (2,4)-semi-simple splicing system, while the language $b^+ \cup abb^+ \cup b^+ab \cup ab^+ab$ can be generated by the (2,4)-semi-simple splicing system $(\{a, b\}, \{(\varepsilon, a; \varepsilon, b)\}, \{abab\})$ but not a (1,3)-semi-simple splicing system.

In this paper, we will relax the condition that the initial language of a semi-simple splicing system must be a finite language, and we will consider also semi-simple splicing systems with regular initial languages. By [13], it is clear that such a splicing system will also produce a regular language. In the following, we will use the convention that I denotes a finite language and L denotes an infinite language.

4   State Complexity of (2,4)-semi-simple Splicing Systems

In this section, we will consider the state complexity of (2,4)-semi-simple splicing systems. Recall that a (2,4)-semi-simple splicing system is a splicing system with rules of the form $(\varepsilon, a; \varepsilon, b)$ for $a, b \in \Sigma$. As mentioned previously, the classes of languages generated by (1,3)- and (2,4)-simple splicing systems were shown to be equivalent by Mateescu et al. [11], while the classes of languages generated by (1,3)- and (2,4)-semi-simple splicing systems were shown to be incomparable by Ceterchi et al. [2].

First, we define an NFA that recognizes the language of a given (2,4)-semi-simple splicing system. This construction is based on the construction of Head and Pixton [8] for Păun splicing rules, which is based on the construction for Pixton splicing rules by Pixton [12]. The original proof of regularity of finite splicing is due to Culik and Harju [3]. We follow the Head and Pixton construction and apply ε-transition removal on the resulting NFA to obtain an NFA for the semi-simple splicing system with the same number of states as the DFA for the initial language of the splicing system.

Proposition 1. *Let $H = (\Sigma, M, L)$ be a (2,4)-semi-simple splicing system with a regular initial language and let L be recognized by a DFA with n states. Then there exists an NFA A'_H with n states such that $L(A'_H) = L(H)$.*

The result of this construction is an NFA that "guesses" when a splicing operation occurs. Since each component of a semi-simple splicing rule is of length at most 1, the construction of the NFA need only consider the outgoing and incoming transitions of states. In the case of (2,4)-semi-simple splicing systems, for a rule (a, b), any state with an outgoing transition on a has added transitions on a to every state with an incoming transition on b.

From this NFA construction, we can obtain a DFA via subset construction. This gives an upper bound of $2^n - 1$ reachable states. This upper bound is the same for (1,3)-simple and (1,3)-semi-simple splicing systems and was shown to be tight [9]. Since (1,3)-simple splicing systems and (2,4)-simple splicing systems are equivalent, we state without proof that the same result holds for (2,4)-simple splicing systems via the same lower bound witness. Therefore, this bound is reachable for (2,4)-semi-simple splicing systems via the same lower bound witness.

Proposition 2 [9]. *For $|\Sigma| \geq 3$ and $n \geq 3$, there exists a (2,4)-simple splicing system with a regular initial language $H = (\Sigma, M, L)$ with $|M| = 1$ where L is a regular language with state complexity n such that the minimal DFA for $L(H)$ requires at least $2^n - 1$ states.*

It was also shown in [9] that if the initial language is finite, this upper bound is not reachable for (1,3)-simple and (1,3)-semi-simple splicing systems. This result holds for all variants of semi-simple splicing systems and the proof is exactly the same as in [9]. We state the result for semi-simple splicing systems for completeness.

Proposition 3 [9]. *Let $H = (\Sigma, M, I)$ be a semi-simple splicing system with a finite initial language where I is a finite language recognized by a DFA A with n states. Then a DFA recognizing $L(H)$ requires at most $2^{n-2} + 1$ states.*

This upper bound is witnessed by a (2,4)-semi-simple splicing system which requires both an alphabet and ruleset that grows exponentially with the number of states of the initial language. This is in contrast to the lower bound witness for (1,3)-semi-simple systems from [9], which requires only three letters. We also note that the initial language used for this witness is the same as that for (1,3)-simple splicing systems from [9]. From this, we observe that the choice of the visible sites for the splicing rules (i.e. (1,3) vs. (2,4)) makes a difference in the state complexity. We will see other examples of this later as we consider semi-simple splicing systems with other rule variants.

Theorem 4. *Let $H = (\Sigma, M, I)$ be a (2,4)-semi-simple splicing system with a finite initial language, where I is a finite language with state complexity n and $M \subseteq \Sigma \times \Sigma$. Then the state complexity of $L(H)$ is at most $2^{n-2} + 1$ and this bound can be reached in the worst case.*

5   State Complexity of (2,3)-semi-simple Splicing Systems

We will now consider the state complexity of (2,3)-semi-simple splicing systems. Recall that a (2,3)-semi-simple splicing system is a splicing system with rules of the form $(\varepsilon, a; b, \varepsilon)$ for $a, b \in \Sigma$. We can follow the same construction from Proposition 1 with slight modifications to account for $(2,3)$-semi-simple splicing rules to obtain an NFA for a language generated by a $(2,3)$-semi-simple splicing system with the same number of states as the DFA for the initial language of the splicing system.

Proposition 5. *Let $H = (\Sigma, M, L)$ be a $(2,3)$-semi-simple splicing system with a regular initial language and let L be recognized by a DFA with n states. Then there exists an NFA A'_H with n states such that $L(A'_H) = L(H)$.*

Note that in this NFA construction, for each $(2,3)$-semi-simple splicing rule (a, b), any state with an outgoing transition on a has additional ε-transitions to every state with an incoming transition on b. This differs from the NFA construction for $(2,4)$-semi-simple splicing systems, where the new transitions were on the symbol a. From this NFA, we then get an upper bound of $2^n - 1$ reachable states via the subset construction. However, we will show that because of the ε-transitions, this bound cannot be reached.

Proposition 6. *Let $H = (\Sigma, M, L)$ be a $(2,3)$-semi-simple splicing system with a regular initial language, where $M \subseteq \Sigma \times \Sigma$ and $L \subseteq \Sigma^*$ is recognized by a DFA with n states. Then there exists a DFA A_H such that $L(A_H) = L(H)$ and A_H has at most 2^{n-1} states.*

Proof. Let $A = (Q, \Sigma, \delta, q_0, F)$ be the DFA for L and let $B_H = (Q, \Sigma, \delta', q_0, F)$ be the NFA obtained via the construction of Proposition 5 given the $(2,3)$-semi-simple splicing system H. Let A_H be the DFA obtained by applying the subset construction to B_H. Note that the states of A_H are subsets of states of B_H.

Consider $a \in \Sigma$ with $(a, b) \in M$ and $\delta(q, a) = q'$ is defined for some $q' \in Q$. In other words, q has an outgoing transition on a. Assuming that (a, b) is non-trivial and im δ_b contains useful states, for any set $P \subseteq Q$, we must have im $\delta_b \subseteq P$ if $q \in P$. This is because for each symbol $a \in \Sigma$ for which there is a pair $(a, b) \in M$, if the NFA B_H enters a state $q \in Q$ with an outgoing transition on a, the NFA B_H also simultaneously, via ε-transitions, enters any state with an incoming transition on b. This implies that not all $2^n - 1$ non-empty subsets of Q are reachable in A_H, since the singleton set $\{q\}$ is unreachable.

Because of this construction, the number of distinct sets that contains q decreases as the size of im δ_b grows. Thus, to maximize the number of sets that can be reached, the number of states with incoming transitions on any symbol b with $(a, b) \in M$ must be minimized. Therefore, for $(a, b) \in M$, there can be only one useful state with incoming transitions on b. Let us call this state $q_b \in Q$.

We claim that to maximize the number of states, A must contain no useless states and therefore A contains no sink state. First, suppose otherwise and that A contains a sink state $q_\emptyset$. To maximize the number of states, we minimize the number of states of A with outgoing transitions, so there is only one state of A, say q', with an outgoing transition on a. We observe that $q' \neq q_b$, since otherwise, $|$ im $\delta_b| = 1$ and if the only state with an outgoing transition on a is q_b itself, then the only reachable subset that contains q_b is the singleton set $\{q_b\}$.

Now, recall that for all subsets $P \subseteq Q \setminus \{q_\emptyset\}$, the two sets P and $P \cup \{q_\emptyset\}$ are indistinguishable. Then there are at most 2^{n-2} distinguishable subsets containing q_b and at most $2^{n-3} - 1$ nonempty subsets of $Q \setminus \{q_b, q', q_\emptyset\}$. Together with the sink state, this gives a total of at most $2^{n-2} + 2^{n-3}$ states in A_H.

Now, we consider when A contains no sink state. In this case, since A must be a complete DFA, in order to satisfy the condition that $| \operatorname{im} \delta_b|$ is minimal, we must have $\delta(q, a) = q_b$ for all $q \in Q$. But this means that for any state $q \in Q$ and subset $P \subseteq Q$, if $q \in P$, then $q_b \in P$. Therefore, every reachable subset of Q must contain q_b. This gives an upper bound of 2^{n-1} states in A_H.

Since $2^{n-1} > 2^{n-2} + 2^{n-3}$ for $n \geq 3$, the DFA A_H can have at most 2^{n-1} states in the worst case. $\qquad \square$

The bound of Proposition 6 is reachable when the initial language is a regular language, even when restricted to simple splicing rules defined over an alphabet of size 3. This upper bound is met by the (2,3)-simple splicing system $H = (\Sigma, \{(c, c)\}, L(A_n))$, where $\Sigma = \{a, b, c\}$ and A_n is the DFA shown in Fig. 1. This gives us the following result.

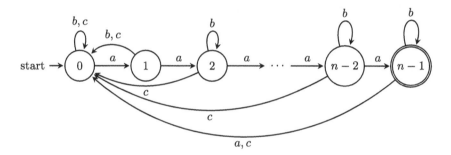

Fig. 1. The DFA A_n of Theorem 7

Theorem 7. *Let $H = (\Sigma, M, L)$ be a (2,3)-semi-simple splicing system with a regular initial language, where $L \subseteq \Sigma^*$ is a regular language with state complexity n and $M \subseteq \Sigma \times \Sigma$. Then the state complexity of $L(H)$ is at most 2^{n-1} and this bound can be reached in the worst case.*

The bound of Proposition 6 depends on whether or not the DFA for the initial language contains a sink state. Since a DFA recognizing a finite language must have a sink state, the upper bound stated in the proposition is clearly not reachable when the initial language is finite.

Proposition 8. *Let $H = (\Sigma, M, I)$ be a (2,3)-semi-simple splicing system where I is a finite language recognized by a DFA A with n states. Then a DFA recognizing $L(H)$ requires at most $2^{n-3} + 2$ states.*

Proof. Let $A = (Q, \Sigma, \delta, q_0, F)$ be the DFA for I and let A_H be the DFA obtained via the construction of Proposition 6, given the (2,3)-semi-simple splicing system

H. We will consider the number of reachable and pairwise distinguishable states of A_H.

Recall from the proof of Proposition 6 that to maximize the number of sets that can be reached in A_H, the number of states with incoming transitions on any symbol b with $(a, b) \in M$ must be minimized. Then for $(a, b) \in M$, there can be only one useful state with incoming transitions on b. Let us call this state $q_b \in Q$.

Since I is a finite language, we know that q_0, the initial state of A, is contained in exactly one reachable state in A_H. Similarly A must contain a sink state $q_\emptyset$ and for all subsets $P \subseteq Q$, we have that P and $P \cup \{q_\emptyset\}$ are indistinguishable. Finally, we observe that there must exist at least one state $q_1 \in Q$ that is directly reachable from q_0 and is not reachable by any word of length greater than 1. Therefore, in order to maximize the number of reachable subsets, we must have that $q_1 = q_b$.

Let Q^a denote the set of states for which there is an outgoing transition on the symbol a. That is, if $q \in Q^a$, we have $\delta(q, a) \leq n - 2$. Let $k_a = |Q^a|$. It is clear that $k_a \geq 1$. Now, consider a reachable subset $P \subseteq Q \setminus \{q_0, q_\emptyset\}$. We claim that if $|P| \geq 2$ and $q_b \in P$, then we must have $q \in P$ for some $q \in Q^a$.

Suppose otherwise and that $Q^a \cap P = \emptyset$. Recall that $q_b = q_1$ and the only incoming transitions to q_1 are from the initial state q_0. Then this means that $P = \{q_1\}$ and $|P| = 1$, a contradiction. Therefore, we have $Q^a \cap P \neq \emptyset$ whenever $q_b \in P$ with $|P| \geq 2$.

Now, we can count the number of reachable subsets of $Q \setminus \{q_0, q_\emptyset\}$. There are $2^{n-3-k_a}(2^{k_a} - 1)$ non-empty subsets of size greater than 1 which contain q_b and there are $2^{n-3-k_a} - 1$ non-empty subsets which do not contain q_b. Together with the initial and sink states and the set $\{q_b\}$, we have

$$2^{n-3-k_a}(2^{k_a} - 1) + 2^{n-3-k_a} - 1 + 3.$$

Thus, the DFA A_H has at most $2^{n-3} + 2$ reachable states.   □

Let $H = (\Sigma, \{(a, c)\}, L(B_n))$ be a $(2,3)$-semi-simple splicing system, where $\Sigma = \{a, b, c\}$ and B_n is a DFA for a finite language with n states. The DFA B_n is shown in Fig. 2. Then H is a $(2,3)$-semi-simple splicing system with an initial finite language that is defined over a fixed alphabet that can reach the upper bound of Proposition 8. This then gives us the following theorem.

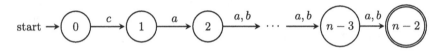

Fig. 2. The DFA B_n of Theorem 9. Transitions not shown are to the sink state $n - 1$, which is not shown.

Theorem 9. *Let $H = (\Sigma, M, I)$ be a (2,3)-semi-simple splicing system with a finite initial language, where I is a finite language with state complexity n and $M \subseteq \Sigma \times \Sigma$. Then the state complexity of $L(H)$ is at most $2^{n-3} + 2$ and this bound can be reached in the worst case.*

Unlike the situation with (2,3)-semi-simple splicing systems with regular initial languages, when we restrict (2,3)-semi-simple splicing systems with initial finite languages to allow only (2,3)-simple splicing rules, the bound of Theorem 9 is not reachable.

Proposition 10. *Let $H = (\Sigma, M, I)$ be a (2,3)-simple splicing system where I is a finite language recognized by a DFA A with n states. Then a DFA recognizing $L(H)$ requires at most $2^{n-4} + 2^{n-5} + 2$ states.*

This bound is reachable by a family of witnesses defined over an alphabet of size 7. We define the (2,3)-finite simple splicing system $H = (\Sigma, \{(c, c)\}, L(C_n))$, where $\Sigma = \{a, b, c, d, e, f, g\}$ and C_n is a DFA with n states that accepts a finite language, shown in Fig. 3. Then we have the following theorem.

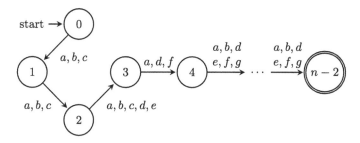

Fig. 3. The DFA C_n of Theorem 11. Transitions not shown are to the sink state $n - 1$, which is not shown.

Theorem 11. *Let $H = (\Sigma, M, I)$ be a (2,3)-simple splicing system with a finite initial language, where $I \subseteq \Sigma^*$ is a finite language with state complexity n and $M \subseteq \Sigma^* \times \Sigma^*$. Then the state complexity of $L(H)$ is at most $2^{n-4} + 2^{n-5} + 2$ and this bound can be reached in the worst case.*

6   State Complexity of (1,4)-semi-simple Splicing Systems

In this section, we consider the state complexity of (1,4)-semi-simple splicing systems. Recall that a (1,4)-semi-simple splicing system is a splicing system with rules of the form $(a, \varepsilon; \varepsilon, b)$ for $a, b \in \Sigma$. As with (2,3)-semi-simple splicing systems, we can easily modify the construction of Proposition 1 to obtain an NFA for (1,4)-semi-simple splicing systems.

Proposition 12. *Let $H = (\Sigma, M, L)$ be a (1,4)-semi-simple splicing system with a regular initial language, $M = M_1 \times M_2$ with $M_1, M_2 \subseteq \Sigma$ and let L be recognized by a DFA with n states. Then there exists an NFA A'_H with $n + m$ states such that $L(A'_H) = L(H)$, where $m = |M_1|$.*

This NFA construction differs from the constructions for (2,3)- and (2,4)-semi-simple splicing systems in that additional states are introduced for each splicing rule. For each (1,4)-semi-simple splicing rule (a, b), we add a new state p_a to which any state with an outgoing transition on a has additional transitions on a and from which there are transitions on b to every state with an incoming transition on b.

This construction immediately gives an upper bound of 2^{n+m} states necessary for an equivalent DFA via the subset construction, where m is the number of symbols on the left side of each pair of rules in M. However, we will show via the following DFA construction that the upper bound is much lower than this.

Proposition 13. *Let $H = (\Sigma, M, L)$ be a (1,4)-semi-simple splicing system with a regular initial language, where $M = M_1 \times M_2$ with $M_1, M_2 \subseteq \Sigma$ and $L \subseteq \Sigma^*$ is recognized by a DFA with n states. Then there exists a DFA A_H such that $L(A_H) = L(H)$ and A_H has at most $(2^n - 2)(|M_1| + 1) + 1$ states.*

Proof. Let $A = (Q, \Sigma, \delta, q_0, F)$ be a DFA for L. We will define the DFA $A_H = (Q', \Sigma, \delta', q_0', F')$. Then the state set of A_H is $Q' = 2^Q \times (M_1 \cup \{\varepsilon\})$, the initial state is $q_0' = \langle \{q_0\}, \varepsilon \rangle$, the set of final states is $F' = \{\langle P, a \rangle \mid P \cap F \neq \emptyset\}$, and the transition function δ' is defined

- $\delta'(\langle P, \varepsilon \rangle, a) = \langle P', \varepsilon \rangle$ if $a \notin M_1$,
- $\delta'(\langle P, \varepsilon \rangle, a) = \langle P', a \rangle$ if $a \in M_1$,
- $\delta'(\langle P, b \rangle, a) = \langle P', \varepsilon \rangle$ if $(b, a) \notin M$ and $a \notin M_1$,
- $\delta'(\langle P, b \rangle, a) = \langle P', a \rangle$ if $(b, a) \notin M$ and $a \in M_1$,
- $\delta'(\langle P, b \rangle, a) = \langle \text{ im } \delta_a, \varepsilon \rangle$ if $(b, a) \in M$ and $a \notin M_1$,
- $\delta'(\langle P, b \rangle, a) = \langle \text{ im } \delta_a, a \rangle$ if $(b, a) \in M$ and $a \in M_1$,

where $P' = \bigcup_{q \in P} \delta(q, a)$.

This construction gives an immediate upper bound of $(2^n - 1)(|M_1| + 1)$ states, however, not all of these states are distinguishable. Consider the two states $\langle Q, \varepsilon \rangle$ and $\langle Q, a \rangle$ for some $a \in M_1$. We claim that these two states are indistinguishable. This arises from the observation that $\bigcup_{q \in Q} \delta(q, a) = \text{ im } \delta_a$ for all $a \in \Sigma$. Then one of the following occurs:

- $\langle Q, \varepsilon \rangle \xrightarrow{b} \langle \text{ im } \delta_b, \varepsilon \rangle$ and $\langle Q, a \rangle \xrightarrow{b} \langle \text{ im } \delta_b, \varepsilon \rangle$ if $b \notin M_1$,
- $\langle Q, \varepsilon \rangle \xrightarrow{b} \langle \text{ im } \delta_b, b \rangle$ and $\langle Q, a \rangle \xrightarrow{b} \langle \text{ im } \delta_b, b \rangle$ if $b \in M_1$.

Note that in either case, it does not matter whether or not $(a, b) \in M$ and the two cases are distinguished solely by whether or not b is in M_1. Thus, all states $\langle Q, a \rangle$ with $a \in M_1 \cup \{\varepsilon\}$ are indistinguishable.

Thus, A_H has at most $(2^n - 2)(|M_1| + 1) + 1$ states. $\square$

When the initial language is a regular language, the upper bound is easily reached, even when we are restricted to simple splicing rules. We consider the (1,4)-simple splicing system $H = (\Sigma, \{(c, c)\}, L(D_n))$, where $\Sigma = \{a, b, c\}$ and D_n is the DFA shown in Fig. 4.

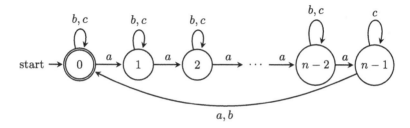

Fig. 4. The DFA D_n for Theorem 14

We note that the witness, H has $|M| = 1$ and therefore $|M_1| = 1$. We observe that we can set $|M_1|$ to be arbitrarily large by adding symbols and transitions appropriately and adding the corresponding markers to M for each new such symbol. We then have the following result.

Theorem 14. *Let $H = (\Sigma, M, L)$ be a (1,4)-semi-simple splicing system with a regular initial language, where $L \subseteq \Sigma^*$ is a regular language with state complexity n and $M = M_1 \times M_2$ with $M_1, M_2 \subseteq \Sigma$. Then the state complexity of $L(H)$ is at most $(2^n - 2)(|M_1| + 1) + 1$ and this bound can be reached in the worst case.*

We will show that this bound cannot be reached by any (1,4)-semi-simple splicing system when the initial language is finite.

Proposition 15. *Let $H = (\Sigma, M, I)$ be a (1,4)-semi-simple splicing system with a finite initial language, where $M = M_1 \times M_2$ with $M_1, M_2 \subseteq \Sigma$ and $I \subseteq \Sigma^*$ is a finite language recognized by a DFA with n states. Then there exists a DFA A_H such that $L(A_H) = L(H)$ and A_H has at most $2^{n-2} + |M_1| \cdot 2^{n-3} + 1$ states.*

Proof. Let $A = (Q, \Sigma, \delta, q_0, F)$ be a DFA for I with n states and let A_H be the DFA recognizing $L(H)$ obtained via the construction of Proposition 13. Since I is finite, the initial state of A contains no incoming transitions and A must have a sink state. Therefore, for any state $\langle S, c \rangle$, we have $S \subseteq Q \setminus \{q_0, q_\emptyset\}$ and $c \in M_1 \cup \{\varepsilon\}$, where $q_\emptyset$ is the sink state. This gives us up to $(2^{n-2}-1)(|M_1|+1)+2$ states.

We can reduce the number of reachable states further by noting that since I is finite, A must contain at least one useful state q_1 that is directly reachable only from the initial state q_0. Then there are only two ways to reach a state $\langle P, c \rangle$ in A_H with $q_1 \in P$. Either $P = \{q_1\}$ and is reached directly via a transition from $\{q_0\}$ or $|P| \geq 2$ and $P = \operatorname{im} \delta_b$ for some $(a, b) \in M$. For each $c \in M_1$, this gives a total of 2 reachable states $\langle P, c \rangle$.

Therefore, we can enumerate the reachable states of A_H as follows:

- the initial state $\langle \{q_0\}, \varepsilon \rangle$ and the sink state $\langle \{q_\emptyset\}, \varepsilon \rangle$,
- at most $2^{n-2} - 1$ states of the form $\langle P, \varepsilon \rangle$, where $P \subseteq Q \setminus \{q_0, q_\emptyset\}$,
- at most $|M_1|$ states of the form $\langle \{q_1\}, c \rangle$ with $c \in M_1$,
- at most $|M_1|$ states of the form $\langle P, c \rangle$ such that $P \subseteq Q \setminus \{q_0, q_\emptyset\}$, $|P| \geq 2$, and $q_1 \in P$ with $c \in M_1$,

– at most $|M_1|(2^{n-3} - 1)$ states of the form $\langle P, c \rangle$ such that $P \subseteq Q \setminus \{q_0, q_1, q_\emptyset\}$ with $c \in M_1$.

This gives a total of at most $2^{n-2} + |M_1| \cdot (2^{n-3} + 1) + 1$ reachable states in A_H.   □

This bound is witnessed by a (1,4)-semi-simple splicing system that is defined over an alphabet and ruleset that grows exponentially in the size of the number of states of the initial language. This is similar to the (2,4)-semi-simple case. We note also that one can arbitrarily increase the size of M by adding symbols and corresponding pairs of rules appropriately. We then get the following result.

Theorem 16. *Let $H = (\Sigma, M, I)$ be a (1,4)-semi-simple splicing system with a finite initial language, where $I \subseteq \Sigma^*$ is a finite language with state complexity n and $M = M_1 \times M_2$ with $M_1, M_2 \subseteq \Sigma$. Then the state complexity of $L(H)$ is at most $2^{n-2} + |M_1| \cdot 2^{n-3} + 1$ and this bound is reachable in the worst case.*

7   Conclusion

We have studied the state complexity of several variants of semi-simple splicing systems. Our results are summarized in Table 1 and we include the state complexity of (1,3)-semi-simple and (1,3)-simple splicing systems from [9] for comparison.

Table 1. Summary of state complexity bounds for (i, j)-simple and semi-simple splicing systems with alphabet Σ, state complexity of the axiom set n, and set of splicing rules $M = M_1 \times M_2$, with $M_1, M_2 \subseteq \Sigma$. Regular axiom sets have $|\Sigma| = 3$.

	Regular axiom set	Finite axiom set	$|\Sigma|$
(2,4)-semi.	$2^n - 1$	$2^{n-2} + 1$	$\geq 2^{n-3}$
(2,3)-semi.	2^{n-1}	$2^{n-3} + 2$	3
(1,4)-semi.	$(2^{n-2} - 2)(|M_1| + 1) + 1$	$2^{n-2} + |M_1| \cdot 2^{n-3}$	$\geq 2^{n-3}$
(1,3)-semi. [9]	$2^n - 1$	$2^{n-2} + 1$	3
(2,4)-simple	$2^n - 1$	*Same as (1,3)*	
(2,3)-simple	2^{n-1}	$2^{n-4} + 2^{n-5} + 2$	7
(1,4)-simple	$(2^{n-2} - 2)(|M_1| + 1) + 1$	?	
(1,3)-simple [9]	$2^n - 1$	$2^{n-2} + 1$	$\geq 2^{n-3}$

Observe that for all variants of semi-simple splicing systems, the state complexity bounds for splicing systems with regular initial languages are reached with simple splicing witnesses defined over a three-letter alphabet. For semi-simple splicing systems with finite initial languages, we note that the state complexity bounds for the (2,3) and (1,3) variants are reached by witnesses defined over a three-letter alphabet, while both of the (1,4) and (2,4) variants require an alphabet size that is exponential in the size of the DFA for the initial language.

We note that the witness for (2,3)-simple splicing systems with a finite initial language is defined over a fixed alphabet of size 7, while the problem remains open for (1,4)-simple splicing systems. Another problem that remains open is the state complexity of (1,4)- and (2,4)- simple and semi-simple splicing systems with finite initial languages defined over alphabets of size k for $3 < k < 2^{n-3}$. A similar question can be asked of (2,3)-simple splicing systems with a finite initial language for alphabets of size less than 7.

References

1. Bonizzoni, P., Ferretti, C., Mauri, G., Zizza, R.: Separating some splicing models. Inf. Process. Lett. **79**(6), 255–259 (2001)
2. Ceterchi, R., Martín-Vide, C., Subramanian, K.G.: On some classes of splicing languages. In: Aspects of Molecular Computing: Essays Dedicated to Tom Head, on the Occasion of His 70th Birthday, pp. 84–105 (2003)
3. Culik II, K., Harju, T.: Splicing semigroups of dominoes and DNA. Disc. Appl. Math. **31**(3), 261–277 (1991)
4. Gao, Y., Moreira, N., Reis, R., Yu, S.: A survey on operational state complexity. J. Autom. Lang. Comb. **21**(4), 251–310 (2016)
5. Gatterdam, R.: Splicing systems and regularity. Int. J. Comput. Math. **31**(1–2), 63–67 (1989)
6. Goode, E., Pixton, D.: Semi-simple splicing systems. In: Martín-Vide, C., Mitrana, V. (eds.) Where Mathematics, Computer Science, Linguistics and Biology Meet, pp. 343–352. Springer, Dordrecht (2001). https://doi.org/10.1007/978-94-015-9634-3_30
7. Head, T.: Formal language theory and DNA: an analysis of the generative capacity of specific recombinant behaviors. Bull. Math. Biol. **49**(6), 737–759 (1987)
8. Head, T., Pixton, D.: Splicing and regularity. In: Recent Advances in Formal Languages and Applications, Studies in Computational Intelligence, vol. 25, pp. 119–147. Springer (2006)
9. Kari, L., Ng, T.: State complexity of simple splicing. In: Hospodár, M., Jirásková, G., Konstantinidis, S. (eds.) DCFS 2019. LNCS, vol. 11612, pp. 197–209. Springer, Cham (2019). https://doi.org/10.1007/978-3-030-23247-4_15
10. Loos, R., Malcher, A., Wotschke, D.: Descriptional complexity of splicing systems. Int. J. Found. Comput. Sci. **19**(04), 813–826 (2008)
11. Mateescu, A., Păun, G., Rozenberg, G., Salomaa, A.: Simple splicing systems. Disc. Appl. Math. **84**(1–3), 145–163 (1998)
12. Pixton, D.: Regularity of splicing languages. Disc. Appl. Math. **69**(1–2), 101–124 (1996)
13. Păun, G.: On the splicing operation. Disc. Appl. Math. **70**(1), 57–79 (1996)

On the Degeneracy of Random Expressions Specified by Systems of Combinatorial Equations

Florent Koechlin[1], Cyril Nicaud[1(✉)], and Pablo Rotondo[2]

[1] LIGM, Univ Gustave Eiffel, CNRS, ENPC, Champs-sur-Marne, France
{florent.koechlin,cyril.nicaud}@u-pem.fr
[2] LITIS, Université de Rouen, Saint-Étienne-du-Rouvray, France
pablo.rotondo@univ-rouen.fr

Abstract. We consider general expressions, which are trees whose nodes are labeled with operators, that represent syntactic descriptions of formulas. We assume that there is an operator that has an absorbing pattern and prove that if we use this property to simplify a uniform random expression with n nodes, then the expected size of the result is bounded by a constant. In our framework, expressions are defined using a combinatorial system, which describes how they are built: one can ensure, for instance, that there are no two consecutive stars in regular expressions. This generalizes a former result where only one equation was allowed, confirming the lack of expressivity of uniform random expressions.

1 Introduction

This article is the sequel of the work started in [10], where we investigate the lack of expressivity of uniform random expressions. In our setting, we use the natural encoding of expressions as trees, which is a convenient way to manipulate them both in theory and in practice. In particular, it allows us to treat many different kinds of expressions at a general level (see Fig. 1 below): regular expressions, arithmetic expressions, boolean formulas, LTL formulas, and so on.

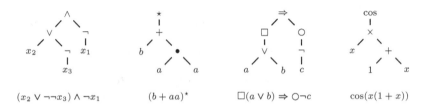

$(x_2 \lor \neg\neg x_3) \land \neg x_1 \qquad (b + aa)^\star \qquad \Box(a \lor b) \Rightarrow \bigcirc\neg c \qquad \cos(x(1 + x))$

Fig. 1. Four expression trees and their associated formulas. From left to right: a logical formula, a regular expression, an LTL formula and a function.

© Springer Nature Switzerland AG 2020
N. Jonoska and D. Savchuk (Eds.): DLT 2020, LNCS 12086, pp. 164–177, 2020.
https://doi.org/10.1007/978-3-030-48516-0_13

For this representation, some problems are solved using a simple traversal of the tree: for instance, testing whether the language of a regular expression contains the empty word, or formally differentiating a function. Sometimes however, the tree is not the best way to encode the object it represents in the computer, and we transform it into an equivalent adequate structure; in the context of formal languages, a regular expression (encoded using a tree) is typically transformed into an automaton, using one of the many known algorithms such as the Thompson construction or the Glushkov automaton.

In our setting, we assume that one wants to estimate the efficiency of an algorithm, or a tool, whose inputs are expressions. The classical theoretical framework consists in analyzing the worst case complexity, but there is often some discrepancy between this measure of efficiency and what is observed in practice. A practical approach consists in using benchmarks to test the tool on real data. But in many contexts, having access to good benchmarks is quite difficult. An alternative to these two solutions is to consider the average complexity of the algorithm, which is sometimes amenable to a mathematical analysis, and which can be studied experimentally, provided we have a random generator at hand. Going that way, we have to choose a probability distribution on size-n inputs, which can be difficult: we would like to study a "realistic" probability distribution that is also mathematically tractable. When no specific random model is available, it is classical to consider the uniform distribution, where all size-n inputs are equally likely. In many frameworks, such as sorting algorithms, studying the uniform distribution yields useful insights on the behavior of the algorithm.

Following this idea, several works have been undertaken on uniform random expressions, in various contexts. Some are done at a general level: the expected height of a uniform random expression [12] always grows in $\Theta(\sqrt{n})$, if we identify common subexpressions then the expected size of the resulting acyclic graph [7] is in $\Theta(\frac{n}{\sqrt{\log n}})$, ... There are also more specific results on the expected size of the automaton built from a uniform random regular expression, using various algorithms [4,14]. In another setting, the expected cost of the computation of the derivative of a random function was proved to be in $\Theta(n^{3/2})$, both in time and space [8]. There are also a lot of results on random boolean formulas, but the framework is a bit different (for a more detailed account on this topic, we refer the interested reader to Gardy's survey [9]).

In [10], we questioned the model of uniform random expressions. Let us illustrate the main result of [10] on an example, regular expressions on the alphabet $\{a, b\}$. The set $\mathcal{L_R}$ of regular expressions is inductively defined by

$$\mathcal{L_R} = a + b + \varepsilon + \overset{*}{\underset{\mathcal{L_R}}{|}} + \overset{\bullet}{\underset{\mathcal{L_R}\ \mathcal{L_R}}{\wedge}} + \overset{+}{\underset{\mathcal{L_R}\ \mathcal{L_R}}{\wedge}}. \tag{$\star$}$$

The formula above is an equation on trees, where the size of a tree is its number of nodes. In particular a, b and ε represent trees of size 1, reduced to a leaf, labeled accordingly. As one can see from the specification $(\star)$, leaves have labels in $\{a, b, \varepsilon\}$, unary nodes are labeled by $\star$ and binary nodes by either the concatenation $\bullet$ or the union $+$. Observe that the regular expression $\mathcal{P}$ corresponding to

$(a+b)^*$ denotes the regular language $\{a, b\}^*$ of all possible words. This language is absorbing for the union operation on regular languages. So if we start with a regular expression $\mathcal{R}$ (a tree), identify every occurrence of the pattern $\mathcal{P}$ (a subtree), then rewrite the tree (bottom-up) by using inductively the simplifications $\overset{+}{\underset{\mathcal{X}\ \mathcal{P}}{\wedge}} \to \mathcal{P}$ and $\overset{+}{\underset{\mathcal{P}\ \mathcal{X}}{\wedge}} \to \mathcal{P}$, this results in a *simplified tree* $\sigma(\mathcal{R})$ that denotes the same regular language. Of course, other simplifications could be considered, but we focus on this particular one. The theorem we proved in [10] implies that if one takes uniformly at random a regular expression of size n and applies this simplification algorithm, then the expected size of the resulting equivalent expression tends to a constant! It means that the uniform distribution on regular expressions produces a degenerated distribution on regular languages. More generally, we proved that: *For every class of expressions that admits a specification similar to Eq. ($\star$) and such that there is an absorbing pattern for some of the operations, the expected size of the simplification of a uniform random expression of size n tends to a constant as n tends to infinity.*[1] This negative result is quite general, as most examples of expressions have an absorbing pattern: for instance $x \wedge \neg x$ is always `false`, and therefore absorbing for $\wedge$.

The statement of the main theorem of [10] is general, as it can be used to discard the uniform distribution for expressions defined inductively as in Eq. ($\star$). However it is limited to that kind of simple specifications. And if we take a closer look at the regular expressions from $\mathcal{L}_{\mathcal{R}}$, we observe that nothing prevents, for instance, useless sequences of nested stars as in $(((a + bb)^*)^*)^*$. It is natural to wonder whether the result of [10] still holds when we forbid two consecutive stars in the specification. We could also use the associativity of the union to prevent different representations of the same language, as depicted in Fig. 2, or many other properties, to try to reduce the redundancy at the combinatorial level.

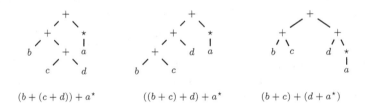

$$(b + (c + d)) + a^* \qquad\qquad ((b + c) + d) + a^* \qquad\qquad (b + c) + (d + a^*)$$

Fig. 2. Three regular expression trees whose denoted languages are equal because of the associativity of the union.

This is the question we investigate in this article: does the degeneracy phenomenon of [10] still hold for more advanced combinatorial specifications? More precisely, we now consider specifications made using a system of (inductive) combinatorial equations, instead of only one as in Eq. ($\star$). For instance, we can forbid consecutive stars using the combinatorial system:

---

[1] The idea behind our work comes from a very specific analysis of and/or formulas established in Nguyên Thê PhD's dissertation [13, Ch 4.4].

$$\begin{cases} \mathcal{L_R} = \overset{\star}{\underset{\mathcal{S}}{|}} + \mathcal{S}, \\ \mathcal{S} = a + b + \varepsilon + \overset{+}{\underset{\mathcal{L_R}\ \mathcal{L_R}}{\wedge}} + \overset{\bullet}{\underset{\mathcal{L_R}\ \mathcal{L_R}}{\wedge}}. \end{cases} \qquad (\star\star)$$

The associativity of the union (Fig. 2) can be taken into account by preventing the right child of any +-node from being also labeled by +. Clearly, systems cannot be used for forbidding intricated patterns, but they still greatly enrich the families of expressions we can deal with. Moreover that kind of systems, which has strong similarities with context-free grammars, is amenable to analytic techniques as we will see in the sequel; this was for instance used by Lee and Shallit to estimate the number of regular languages in [11].

Our contributions can be described as follows. We consider expressions defined by systems of combinatorial equations (instead of just one equation), and establish a similar degeneracy result: if there is an absorbing pattern, then the expected reduced size of a uniform random expression of size n is upper bounded by a constant as n tends to infinity.[2] Hence, even if we use the system to remove redundancy from the specification (e.g., by forbidding consecutive stars), uniform random expressions still lack expressivity. Technically, we once again rely on the framework of analytic combinatorics for our proofs. However, the generalization to systems induces two main difficulties. First, we are not dealing with the well-known *varieties of simple trees* anymore [6, VII.3], so we have to rely on much more advanced techniques of analytic combinatorics; this is sketched in Sect. 5. Second, some work is required on the specification itself, to identify suitable hypotheses for our theorem; for instance, it is easy from the specification to prevent the absorbing pattern from appearing as a subtree at all, in which case our statement does not hold anymore, since there are no simplifications taking place.

Due to the lack of space, the analytic proofs are only sketched or omitted in this extended abstract: we chose to focus on the discussion on combinatorial systems (Sect. 3) and on the presentation of our framework (Sect. 4).

2 Basic Definitions

For a given positive integer n, $[n] = \{1, \ldots, n\}$ denotes the set of the first n positive integers. If E is a finite set, $|E|$ denotes its cardinality.

A *combinatorial class* is a set $\mathcal{C}$ equipped with a *size function* $|\cdot|$ from $\mathcal{C}$ to $\mathbb{N}$ (the size of $C \in \mathcal{C}$ is $|C|$) such that for any $n \in \mathbb{N}$, the set $\mathcal{C}_n$ of size-n elements of $\mathcal{C}$ is finite. Let $C_n = |\mathcal{C}_n|$, the *generating series* $C(z)$ of $\mathcal{C}$ is the formal power series defined by

$$C(z) = \sum_{C \in \mathcal{C}} z^{|C|} = \sum_{n \geq 0} C_n z^n.$$

Generating series are tools of choice to study combinatorial objects. When their radius of convergence is not zero, they can be viewed as analytic function from

---

[2] The result holds for natural yet technical conditions on the system.

$\mathbb{C}$ to $\mathbb{C}$, and very useful theorems have been developed in the field of analytic combinatorics [6] to, for instance, easily obtain an asymptotic equivalent to C_n. We rely on that kind of techniques in Sect. 5 to prove our main theorem.

If $C(z) = \sum_{n \geq 0} C_n z^n$ is a formal power series, let $[z^n]C(z)$ denote its n-th coefficient C_n. Let ξ be a *parameter* on the combinatorial class $\mathcal{C}$, that is, a mapping from $\mathcal{C}$ to $\mathbb{N}$. Typically, ξ stands for some statistic on the objects of $\mathcal{C}$: the number of cycles in a permutation, the number of leaves in a tree, ... We define the *bivariate generating series* $C(z, u)$ associated with $\mathcal{C}$ and ξ by:

$$C(z, u) = \sum_{C \in \mathcal{C}} z^{|C|} u^{\xi(C)} = \sum_{k, n \geq 0} C_{n,k} z^n u^k,$$

where $C_{n,k}$ is the number of size-n elements C of $\mathcal{C}$ such that $\xi(C) = k$. In particular, $C(z) = C(z, 1)$. Bivariate generating series are useful to obtain information on ξ, such as its expectation or higher moments. Indeed, if $\mathbb{E}_n[\xi]$ denotes the expectation of ξ for the uniform distribution on $\mathcal{C}_n$, i.e. where all the elements of size n are equally likely, a direct computation yields:

$$\mathbb{E}_n[\xi] = \frac{[z^n] \partial_u C(z, u)\big|_{u=1}}{[z^n] C(z)}, \tag{1}$$

where $\partial_u C(z, u)\big|_{u=1}$ consists in first differentiating $C(z, u)$ with respect to u, and then setting $u = 1$. Hence, if we have an expression for $C(z, u)$ we can estimate $\mathbb{E}_n[\xi]$ if we can estimate the coefficients of the series in Eq. (1).

In the sequel, the combinatorial objects we study are trees, and we will have methods to compute the generating series directly from their specifications. Then, powerful theorems from analytic combinatorics will be used to estimate the expectation, using Eq. (1). So we delay the automatic construction and the analytic treatment to their respective sections.

3   Combinatorial Systems of Trees

3.1   Definition of Combinatorial Expressions and of Systems

In the sequel the only combinatorial objects we consider are *plane trees*. These are trees embedded in the plane, which means that the order of the children matters: the two trees ⋏ and ⋏ are different. Every node is labeled by an element in a set of symbols and the *size* of a tree is its number of nodes.

More formally, let S be a finite set, whose elements are *operator symbols*, and let a be a mapping from S to $\mathbb{N}$. The value $a(s)$ is called the *arity*[3] of the operator s. An *expression over* S is a plane tree where each node of arity i is labeled by an element $s \in S$ such that $a(s) = i$ (leaves' symbols have arity 0).

---

[3] We do not use the term *degree*, because if the tree is viewed as a graph, the degree of a node is its arity plus one (except for the root).

Example 1. In Fig. 1, the first tree is an expression over $S = \{\wedge, \vee, \neg, x_1, x_2, x_3\}$ with $a(\wedge) = a(\vee) = 2$, $a(\neg) = 1$ and $a(x_1) = a(x_2) = a(x_3) = 0$.

An *incomplete expression over* S is an expression where (possibly) some leaves are labeled with a new symbol $\square$ of arity 0. Informally, such a tree represents part of an expression, where the $\square$-nodes need to be completed by being substituted by an expression. An incomplete expression with no $\square$-leaf is called a *complete* expression, or just an expression. If T is an incomplete expression over S, its *arity* $a(T)$ is its number of $\square$-leaves. It is consistent with the definition of the arity of a symbol, by viewing a symbol s of arity $a(s)$ as an incomplete expression made of a root labeled by s with $a(s)$ $\square$-children: $\wedge$ is viewed as $\overset{\wedge}{\underset{\square\ \square}{\wedge}}$. Let $\mathcal{T}_\square(S)$ and $\mathcal{T}(S)$ be the set of incomplete and complete expressions over S.

If T is an incomplete expression over S of arity t, and $T_1, \ldots, T_t$ are expressions over S, we denote by $T[T_1, \ldots, T_t]$ the expression obtained by substituting the i-th $\square$-leaf in depth-first order by T_i, for $i \in [t]$. This notation is generalized to sets of expressions: if $\mathcal{T}_1, \ldots, \mathcal{T}_t$ are sets of expressions then $T[\mathcal{T}_1, \ldots, \mathcal{T}_t] = \{T[T_1, \ldots, T_t] : T_1 \in \mathcal{T}_1, \ldots, T_t \in \mathcal{T}_t\}$.

A *rule* of dimension $m \geq 1$ over S is an incomplete expression $T \in \mathcal{T}_\square(S)$ where each $\square$-node is replaced by an integer of $[m]$. Alternatively, a rule can be seen as a tuple $\mathcal{M} = (T, i_1, \ldots, i_t)$, where T is an incomplete expression of arity t and $i_1, \ldots, i_t$ are the values placed in its $\square$-leaves in depth-first order. The arity $a(\mathcal{M})$ of a rule $\mathcal{M}$ is the arity of its incomplete expression, and $\mathrm{IND}(\mathcal{M}) = (i_1, \ldots, i_t)$ is the tuple of integer values obtained by a depth-first traversal of $\mathcal{M}$. A *combinatorial system of trees* $\mathcal{E} = \{E_1, \ldots, E_m\}$ of dimension m over S is a system of m set equations of complete trees in $\mathcal{T}(S)$: each E_i is a non-empty finite set of rules over S, and the system in variables $\mathcal{L}_1, \ldots, \mathcal{L}_m$ is:

$$\begin{cases} \mathcal{L}_1 = \bigcup_{(T,i_1,\ldots,i_t)\in E_1} T[\mathcal{L}_{i_1}, \ldots, \mathcal{L}_{i_t}] \\ \quad\vdots \\ \mathcal{L}_m = \bigcup_{(T,i_1,\ldots,i_t)\in E_m} T[\mathcal{L}_{i_1}, \ldots, \mathcal{L}_{i_t}]. \end{cases} \tag{2}$$

Example 2. To specify the system given in Eq. (⋆⋆) using our formalism, we have $m = 2$. Its tuples representation is: $E_1 = \left\{ \left(\substack{\star\\ \shortmid}, 2\right), (\square, 2) \right\}$, and $E_2 = \left\{ \left(\substack{\bullet\\ \wedge}, 1, 1\right), \left(\substack{+\\ \wedge}, 1, 1\right), (a), (b), (\varepsilon) \right\}$, and its equivalent tree representation is $E_1 = \left\{ \substack{\star\\ \shortmid\\ 2} \right\}$, and $E_2 = \left\{ \substack{\bullet\\ \wedge\\ 1\ 1}, \substack{+\\ \wedge\\ 1\ 1}, a, b, \varepsilon \right\}$, which corresponds to Eq. (⋆⋆) with $\mathcal{L}_\mathcal{R} = \mathcal{L}_1$ and $\mathcal{S} = \mathcal{L}_2$. In practice, we prefer descriptions as in Eq. (⋆⋆), which are easier to read, but they are all equivalent.

3.2   Generating Series

If the system is not ambiguous, that is, if $\mathcal{L}_1, \ldots, \mathcal{L}_m$ is the[4] solution of the system and every tree in every $\mathcal{L}_i$ can be uniquely built from the specification, then the system can be directly translated into a system of equations on the generating series. This is a direct application of the *symbolic method* in analytic combinatorics [6, Part A] and we get the system

$$
\begin{cases}
L_1(z) = \displaystyle\sum_{(T, i_1, \ldots, i_{a(T)}) \in E_1} z^{|T| - a(T)} L_{i_1}(z) \cdots L_{i_{a(T)}}(z) \\
\quad\vdots \\
L_m(z) = \displaystyle\sum_{(T, i_1, \ldots, i_{a(T)}) \in E_m} z^{|T| - a(T)} L_{i_1}(z) \cdots L_{i_{a(T)}}(z).
\end{cases}
\tag{3}
$$

where $L_i(z)$ is the generating series of $\mathcal{L}_i$. If the system is ambiguous, the $L_i(z)$'s still have a meaning: each expression of $\mathcal{L}_i$ accounts for the number of ways it can be derived from the system. When the system is unambiguous, there is only one way to derive each expression, and $L_i(z)$ is the generating series of $\mathcal{L}_i$.

3.3   Designing Practical Combinatorial Systems of Trees

Systems of trees such as Eq. (2) are not always well-founded. Sometimes they are, but still contain unnecessary equations. It is not the topic of this article to fully characterize when a system is correct, but we nonetheless need sufficient conditions to ensure that our results hold: in this section, we just present examples to underline some bad properties that might happen. For a more detailed account on combinatorial systems, the reader is referred to [1,8,16].

Ambiguity. As mentioned above, the system can be ambiguous, in which case the combinatorial system cannot directly be translated into a system of generating series. This is a case for instance for the following system

$$
\begin{cases}
\mathcal{L}_1 = a + \overset{\star}{\underset{\mathcal{L}_1}{|}} + \overset{\star}{\underset{\mathcal{L}_2}{|}} \\
\mathcal{L}_2 = \overset{\star}{\underset{\mathcal{L}_1}{|}} + a + b + \varepsilon,
\end{cases}
$$

as the expression $\overset{\star}{\underset{a}{|}}$ can be produced in two ways for the component $\mathcal{L}_1$.

Empty Components. Some specifications produce empty $\mathcal{L}_i$'s. For instance, consider the system $\left\{ \mathcal{L}_1 = \underset{\mathcal{L}_1 \ \mathcal{L}_2}{\overset{\bullet}{\wedge}} ; \ \mathcal{L}_2 = a + b + \varepsilon + \mathcal{L}_1 \right\}$: its only solution is $\mathcal{L}_1 = \emptyset$ and $\mathcal{L}_2 = \{a, b, \varepsilon\}$.

---

[4] In all generalities, there can be several solutions to a system, but the conditions we will add prevent this from happening.

Cyclic Unit-Dependency. The *unit-dependency graph* $\mathcal{G}_\square(\mathcal{E})$ of a system $\mathcal{E}$ is the directed graph of vertex set $[m]$, with an edge $i \to j$ whenever $(\square, j) \in E_i$. Such a rule is called a *unit rule*. It means that $\mathcal{L}_i$ directly depends on $\mathcal{L}_j$. For instance $\mathcal{L}_\mathcal{R}$ directly depends on $\mathcal{S}$ in Eq. $(\star\star)$. We can work with systems having unit dependencies, provided the unit-dependency graph is acyclic. If it is not, then the equations forming a cycle are useless or badly defined for our purposes. Consider for instance the system and its unit-dependency graph depicted in Fig. 3.

$$\begin{cases} \mathcal{L}_1 = \quad \mathcal{L}_2 + \overset{\star}{\underset{\mathcal{L}_1}{\vert}} \\ \mathcal{L}_2 = a + b + \varepsilon + \mathcal{L}_1 \end{cases}$$

Fig. 3. The unit-dependency graph is not acyclic, and there are infinitely many ways to derive a from $\mathcal{L}_2$: $\mathcal{L}_2 \to a$, $\mathcal{L}_2 \to \mathcal{L}_1 \to \mathcal{L}_2 \to a \ldots$

Not Strongly Connected. The *dependency graph* $\mathcal{G}(\mathcal{E})$ of the system $\mathcal{E}$ is the directed graph of vertex set $[m]$, with an edge $i \to j$ whenever there is a rule $\mathcal{M} \in E_i$ such that $j \in \text{IND}(\mathcal{M})$: $\mathcal{L}_i$ depends on $\mathcal{L}_j$ in the specification. Some parts of the system may be unreachable from other parts, which may bring up difficulties. A sufficient condition to prevent this from happening is to ask for the dependency graph to be strongly connected; it is not necessary, but this assumption will also be useful in the proof our main theorem. See Sect. 6 for a more detailed discussion on non-strongly connected systems. In Fig. 4 is depicted a system and its associated graph.

$$\begin{cases} \mathcal{L}_1 = \quad \overset{\star}{\underset{\mathcal{L}_2}{\vert}} + \overset{\star}{\underset{\mathcal{L}_3}{\vert}} \\ \mathcal{L}_2 = a + b + \varepsilon + \overset{\bullet}{\underset{\mathcal{L}_4 \; \mathcal{L}_4}{\wedge}} \\ \mathcal{L}_3 = \quad \underset{\mathcal{L}_4 \; \mathcal{L}_1}{\overset{+}{\wedge}} + \underset{\mathcal{L}_4 \; \mathcal{L}_2}{\overset{+}{\wedge}} \\ \mathcal{L}_4 = \quad \mathcal{L}_1 + \mathcal{L}_2 + \mathcal{L}_3 \end{cases}$$

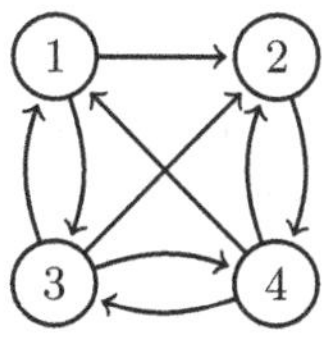

Fig. 4. A system and its associated dependency graph, which is strongly connected.

4 Settings, Working Hypothesis and Simplifications

4.1 Framework

In this section, we describe our framework: we specify the kind of systems we are going to work with, and the settings for describing syntactic simplifications.

Let $\mathcal{E}$ be a combinatorial system of trees over S of dimension m of solution $(\mathcal{L}_1, \ldots, \mathcal{L}_m)$. A set of expressions $\mathcal{L}$ over S is *defined by* $\mathcal{E}$ if there exists a non-empty subset I of $[m]$ such that $\mathcal{L} = \cup_{i \in I} \mathcal{L}_i$.

From now on we assume that we are using a system $\mathcal{E}$ of dimension m over S and that S contains an operator $\circledast$ of arity at least 2. We furthermore assume that there is a complete expression $\mathcal{P}$, such that when interpreted, every expression of root $\circledast$ having $\mathcal{P}$ as a child is equivalent to $\mathcal{P}$: the interpretation of $\mathcal{P}$ is absorbing for the operator associated with $\circledast$. The expression $\mathcal{P}$ is the *absorbing pattern* and $\circledast$ is the *absorbing operator*.

Example 3. Our main example is $\mathcal{L}$ defined by the system of Eq. $(\star\star)$ with $\mathcal{L} = \mathcal{L}_\mathcal{R}$, the regular expressions with no two consecutive stars. As regular expressions, they are interpreted as regular languages. Since the language $(a+b)^\star$ is absorbing for the union, we set the associated expression as the absorbing pattern $\mathcal{P}$ and the operator symbol $+$ as the absorbing operator.

The *simplification* of a complete expression T is the complete expression $\sigma(T)$ obtained by applying bottom-up the rewriting rule, where a is the arity of $\circledast$:

$$\begin{array}{c} \circledast \\ \diagup\,\diagdown \\ C_1 \cdots C_a \end{array} \;\rightsquigarrow\; \mathcal{P}\,,\ \text{whenever } C_i = \mathcal{P} \text{ for some } i \in \{1,\dots,a\}.$$

More formally, the simplification $\sigma(T, \mathcal{P}, \circledast)$ of T, or just $\sigma(T)$ when the context is clear, is inductively defined by: $\sigma(T) = T$ if T has size 1 and

$$\sigma((\oplus, C_1, \dots, C_d)) = \begin{cases} \mathcal{P} & \text{if } \oplus = \circledast \text{ and } \exists i, \sigma(C_i) = \mathcal{P}, \\ (\oplus, \sigma(C_1), \dots, \sigma(C_d)) & \text{otherwise.} \end{cases}$$

A complete expression T is *fully reducible* when $\sigma(T) = \mathcal{P}$.

We also need some conditions on the system $\mathcal{E}$. Some of them come from the discussion of Sect. 3.3, others are needed for the techniques from analytic combinatorics used in our proofs. A system $\mathcal{E}$ satisfies the hypothesis $(\mathbf{H})$ when:

$(\mathbf{H_1})$ The graph $\mathcal{G}(\mathcal{E})$ is strongly connected and $\mathcal{G}_\square(\mathcal{E})$ is acyclic.

$(\mathbf{H_2})$ The system is *aperiodic*: there exists N such that for all $n \geq N$, there is at least one expression of size n in every coordinate of the solution $(\mathcal{L}_1, \dots, \mathcal{L}_m)$ of the system.

$(\mathbf{H_3})$ For some j, there is a rule $T \in E_j$ of root $\circledast$, having at least two children T' and T'' such that: there is a way to produce a fully reducible expression from T' and $a(T'') \geq 1$.

$(\mathbf{H_4})$ The system is *not linear*: there is a rule of arity at least 2.

$(\mathbf{H_5})$ The system is *non-ambiguous*: each complete expression can be built in at most one way.

Conditions $(\mathbf{H_1})$ and $(\mathbf{H_5})$ were already discussed in Sect. 3.3. Condition $(\mathbf{H_4})$ prevents the system from generating only lists (trees whose internal nodes have arity 1), or more generally families that grow linearly (for instance $\mathcal{L} = \begin{smallmatrix} + \\ \diagup\,\diagdown \\ \mathcal{L}\ \ a \end{smallmatrix} + b$), which are degenerated. Without Condition $(\mathbf{H_3})$ the system could be designed in a way that prevents simplifications (in which case our result does not hold, of course). Finally, Condition $(\mathbf{H_2})$ is necessary to keep the analysis manageable (together with the strong connectivity of $\mathcal{G}(\mathcal{E})$ of Condition $(\mathbf{H_1})$).

4.2 Proper Systems and System Iteration

A combinatorial system of trees $\mathcal{E}$ is *proper* when it contains no unit rules and when the $\square$-leaves of all its rules have depth one (they are children of a root). In this section we establish the following preparatory proposition:

Proposition 1. *If $\mathcal{L}$ is defined by a system $\mathcal{E}$ that satisfies* (**H**), *then there exists a proper system $\mathcal{E}'$ that satisfies* (**H**) *such that $\mathcal{L}$ is defined by $\mathcal{E}'$.*

Proposition 1 will be important in the sequel, as proper systems are easier to deal with for the analytic analysis. One key tool to prove Proposition 1 is the notion of *system iteration*, which consists in substituting simultaneously every integer-leaf i in each rule by all the rules of E_i. For instance, if we iterate once our recurring system $\{\mathcal{L}_1 = \overset{\star}{\underset{\mathcal{L}_2}{|}} + \mathcal{L}_2;\ \mathcal{L}_2 = a + b + \varepsilon + \overset{+}{\underset{\mathcal{L}_1\,\mathcal{L}_1}{\wedge}} + \overset{\bullet}{\underset{\mathcal{L}_1\,\mathcal{L}_1}{\wedge}}\}$, we get[5]

$$
\begin{cases}
\mathcal{L}_1 = \overset{\star}{\underset{a}{|}} + \overset{\star}{\underset{b}{|}} + \overset{\star}{\underset{\varepsilon}{|}} + \overset{\star}{\underset{\mathcal{L}_1\,\mathcal{L}_1}{\overset{+}{\wedge}}} + \overset{\star}{\underset{\mathcal{L}_1\,\mathcal{L}_1}{\overset{\bullet}{\wedge}}} + a + b + \varepsilon + \overset{+}{\underset{\mathcal{L}_1\,\mathcal{L}_1}{\wedge}} + \overset{\bullet}{\underset{\mathcal{L}_1\,\mathcal{L}_1}{\wedge}} \\[2em]
\mathcal{L}_2 = a + b + \varepsilon + \overset{+}{\underset{\mathcal{L}_2\,\mathcal{L}_2}{\wedge}} + \overset{+}{\underset{\star\ \mathcal{L}_2}{\wedge}} + \overset{+}{\underset{\mathcal{L}_2\ \star}{\wedge}} + \overset{+}{\underset{\star\ \star}{\wedge}} + \overset{\bullet}{\underset{\mathcal{L}_2\,\mathcal{L}_2}{\wedge}} + \overset{\bullet}{\underset{\star\ \mathcal{L}_2}{\wedge}} + \overset{\bullet}{\underset{\mathcal{L}_2\ \star}{\wedge}} + \overset{\bullet}{\underset{\star\ \star}{\wedge}}.
\end{cases}
$$

Formally, if we iterate $\mathcal{E} = \{E_1, \ldots, E_m\}$ once, then for all $i \in [m]$ we have

$$
\mathcal{L}_i = \bigcup_{(T, i_1, \ldots, i_t) \in E_1} T\left[\bigcup_{(T_1, \mathbf{j_1}) \in E_{i_1}} T_1[\mathcal{L}_{j_{1,1}}, \ldots, \mathcal{L}_{j_{1,t_1}}], \ldots, \bigcup_{(T_t, \mathbf{j_t}) \in E_{i_t}} T_t[\mathcal{L}_{j_{t,1}}, \ldots, \mathcal{L}_{j_{t,t_t}}] \right]
$$

where $\mathbf{j_1} = (j_{1,1}, \ldots, j_{1,t_1}), \ldots, \mathbf{j_t} = (j_{t,1}, \ldots, j_{t,t_t})$.

Let $\mathcal{E}^2$ denote the system obtained after iterating $\mathcal{E}$ once; it is called the *system of order 2* (from $\mathcal{E}$). More generally $\mathcal{E}^t$ is the system of order t obtained by iterating $t-1$ times the system $\mathcal{E}$. From the definition we directly get:

Lemma 1. *If $\mathcal{L}$ is defined by a system $\mathcal{E}$, it is also defined by all its iterates $\mathcal{E}^t$. Moreover, if $\mathcal{E}$ satisfies* (**H**), *every $\mathcal{E}^t$ also satisfies* (**H**), *except that $\mathcal{G}(\mathcal{E}^t)$ may not be strongly connected.*

We can sketch the proof of Proposition 1 as follows: since $\mathcal{G}_\square(\mathcal{E})$ is acyclic, we can remove all the unit rules by iterating the system sufficiently many times. By Lemma 1, we have to be cautious, and find an order t so that $\mathcal{G}^t$ is strongly connected: a study of the cycle lengths in $\mathcal{G}(\mathcal{E})$ ensures that such a t exists. So $\mathcal{L}$ is defined by $\mathcal{E}^t$, which has no unit rules and which satisfies (**H**). To transform

---

[5] Observe that the iterated system is not strongly connected anymore. It also yields two ways of defining the set of expressions using only one equation: it is very specific to this example, no such property holds in general.

$\mathcal{E}^t$ into an equivalent proper system, we have to increase the dimension to cut the rules as needed. It is better explained on an example:

$$\mathcal{L}_1 = \overset{\star}{\underset{\mathcal{L}_3}{|}} + \overset{\bullet}{\underset{\mathcal{L}_1}{\overset{\wedge}{\star \mathcal{L}_2}}} \quad \rightarrow \quad \begin{cases} \mathcal{L}_1 = \overset{\star}{\underset{\mathcal{L}_3}{|}} + \overset{\bullet}{\underset{\mathcal{K}}{\overset{\wedge}{\mathcal{L}_2}}} \\ \mathcal{K} = \overset{\star}{\underset{\mathcal{L}_1}{|}} . \end{cases}$$

This construction can be systematized. It preserves (**H**) and introduces no unit rules, which concludes the proof sketch.

5 Main Result

Our main result establishes the degeneracy of uniform random expressions when there is an absorbing pattern, in our framework:

Theorem 1. *Let $\mathcal{E}$ be a combinatorial system of trees over S, of absorbing operator $\circledast$ and of absorbing pattern $\mathcal{P}$, that satisfies (**H**). If $\mathcal{L}$ is defined by $\mathcal{E}$ then there exists a positive constant C such that, for the uniform distribution on size-n expressions in $\mathcal{L}$, the expected size of the simplification of a random expression is smaller than C. Moreover, every moment of order t of this random variable is bounded from above by a constant C_t.*

The remainder of this section is devoted to the proof sketch of the first part of Theorem 1: the expectation of the size after simplification. The moments are handled similarly. Thanks to Proposition 1, we can assume that $\mathcal{E}$ is a proper system. By Condition (**H$_5$**), it is non-ambiguous so we can directly obtain a system of equations for the associated generating series, as explained in Sect. 3.2. From now on, for readability and succinctness, we use the vector notation (with bold characters): $\mathbf{L}(z)$ denotes the vector $(L_1(z), \ldots, L_m(z))$, and we rewrite the system of Eq. (3) in the more compact form

$$\mathbf{L}(z) = z\,\phi(z; \mathbf{L}(z)), \tag{4}$$

where $\phi = (\phi_1, \ldots, \phi_m)$ and $\phi_i(z; \mathbf{y}) = \displaystyle\sum_{(T, i_1, \ldots, i_{a(T)}) \in E_i} z^{|T|-1-a(T)} \prod_{j=1}^{a(T)} y_{i_j}$.

Under this form, and because $\mathcal{E}$ satisfies (**H**), we are in the setting of Drmota's celebrated Theorem for systems of equations (Theorem 2.33 in [5], refined in [3]), which gives the asymptotics of the coefficients of the $L_i(z)$'s. This is stated in Proposition 2 below, where $\mathsf{Jac_y}[\phi](z; \mathbf{y})$ is the Jacobian matrix of the system, which is the $m \times m$ matrix such that $\mathsf{Jac_y}[\phi](z; \mathbf{y})_{i,j} = \partial_{y_j} \phi_i(z; \mathbf{y})$.

Proposition 2. *As $\mathcal{E}$ satisfies (**H**), the solution $\mathbf{L}(z)$ of the system of equations (4) is such that all its coordinates $L_j(z)$ share the same dominant singularity $\rho \in (0, 1]$, and we have $\tau_j := L_j(\rho) < \infty$. The singularity ρ and $\tau = (\tau_j)_j$ verify the characteristic system $\{\tau = \rho\,\phi(\rho; \tau), \det(\mathrm{Id}_{m \times m} - \rho\,\mathsf{Jac_y}[\phi](\rho; \tau)) = 0\}$. Moreover, for every j, there exist two functions $g_j(z)$ and $h_j(z)$, analytic at $z = \rho$, such that locally around $z = \rho$, with $z \notin [\rho, +\infty)$,*

$$L_j(z) = g_j(z) - h_j(z)\sqrt{1 - z/\rho}, \quad \text{with } h_j(\rho) \neq 0.$$

Lastly, we have the asymptotics $[z^n]L_j(z) \sim C_j \rho^{-n}/n^{3/2}$ for some positive C_j.

The next step is to introduce the bivariate generating series associated with the size of the simplified expression $\mathbf{L}(z, u) = (L_1(z, u); \ldots, L_m(z, u))$. We rely on Eq. (1) to estimate the expectation of this statistic for uniform random expressions. Proposition 2 already gives an estimation of the denominator, so we focus on proving that for all $j \in [m]$, $[z^n]\partial_u L_j(z, u) \leq \alpha \rho^{-n} n^{-3/2}$, for some positive α.

For this purpose, let $\mathcal{R}_j$ be the set of fully reducible elements of $\mathcal{L}_j$ and let $\mathcal{G}_j = \mathcal{L}_j \setminus \mathcal{R}_j$. Let $\mathbf{R}(z)$ and $\mathbf{L}(z)$ be the vectors of the generating series $\mathcal{R}_j$ and $\mathcal{L}_j$, respectively. Let also $\mathbf{R}(z, u)$ and $\mathbf{G}(z, u)$ be the vectors of their associated bivariate generating series, where u accounts for the size of the simplified expression. Of course we have $\mathbf{R}(z, u) = u^p \mathbf{R}(z)$, where $p = |\mathcal{P}|$ is the size of the absorbing pattern. We also split the system ϕ into $\phi = \underline{\phi} + \mathbf{A} + \mathbf{B}$ where: $\underline{\phi}$ use all the rules of ϕ whose root is not $\circledast$ and $\mathbf{B}$ gathers the rules of root $\circledast$ with a constant fully reducible child; if necessary, we iterate the system to ensure that $\mathbf{B}$ is not constant as a function of $\mathbf{y}$. Using marking techniques (see [10] for a detailed presentation on expression simplification) we finally obtain:[6]

$$\mathbf{L}(z, u) = u^p \left(\mathbf{R}(z) - \mathbf{P}(z) \right) + zu \left(\underline{\phi}(zu; \mathbf{L}(z, u)) + \mathbf{A}(zu; \ \mathbf{G}(z, u)) \right), \quad (5)$$

where $\mathbf{P}(z) = (a_1 z^p, \ldots, a_m z^p)$, with $a_i = 1$ if $\mathcal{P} \in \mathcal{L}_i$ and 0 otherwise.

At this point, we can differentiate the whole equality with respect to u and set $u = 1$. But we do not have much information on $\mathbf{R}(z)$ and $\mathbf{G}(z)$, so it is not possible to conclude directly. Instead of working directly on $\mathcal{R}$ and $\mathcal{G}$, which may rise some technical difficulties, we exploit the fact that $\mathcal{G}, \mathcal{R} \subseteq \mathcal{L}$ and apply a fixed point iteration: this results in a crucial bound for $[z^n]\partial_u \mathbf{L}(z, u)\big|_{u=1}$ purely in terms of $\mathbf{L}(z)$, which is stated in the following proposition.

Proposition 3. *For some $C > 0$, the following coordinate-wise bound holds:*

$$[z^n]\left\{ \partial_u \mathbf{L}(z, u)\big|_{u=1} \right\} \leq C \cdot [z^n]\left\{ \left(\mathtt{Id}_{m \times m} - z \cdot \mathtt{Jac}_\mathbf{y}[\underline{\phi} + \mathbf{A}](z; \mathbf{L}(z)) \right)^{-1} \cdot \mathbf{L}(z) \right\}.$$

So we switch to the analysis of the right hand term in the inequality of Proposition 3. Despite its expression, it is easier to study its dominant singularities, and we do so by examining the spectrum of the matrix $J(z) = \mathtt{Jac}_\mathbf{y}[\underline{\phi} + \mathbf{A}](z; \mathbf{L}(z))$. This yields the following estimate, which concludes the whole proof:

Proposition 4. *The function $\mathbf{F}: z \mapsto (\mathtt{Id} - z \cdot J(z))^{-1} \cdot \mathbf{L}(z)$ has $\rho = \rho_\mathbf{L}$, as its dominant singularity. Further, around $z = \rho$ there exist analytic functions $\tilde{g}_j, \tilde{h}_j$ such that $F_j(z) = \tilde{g}_j(z) - \tilde{h}_j(z)\sqrt{1 - z/\rho}$ with $\tilde{h}_j(\rho) \neq 0$. Moreover, we have the asymptotics $[z^n]F_j(z) \sim D_j \rho^{-n} n^{-3/2}$, for some positive D_j.*

6   Conclusion and Discussion

To summarize our contributions in one sentence, we proved in this article that even if we use systems to specify them, uniform random expressions lack expressivity as they are drastically simplified as soon as there is an absorbing pattern.

---

[6] In fact this is the size of a less effective variation of the simplification algorithm, which is ok for our proof as we are looking for an upper bound.

This confirms and extends our previous result [10], which holds for much more simple specifications only. It questions the relevance of uniform distributions in this context, both for experiments and for theoretical analysis.

Roughly speaking, the intuition behind the surprising power of this simple simplifications is that, on the one hand the absorbing pattern appears a linear number of times, while on the other, the shape of uniform trees facilitates the pruning of huge chunks of the expression.

Mathematically speaking, Theorem 1 is not a generalization of the main result of [10]: we proved that the expectation is bounded (and not that it tends to a constant), and we only allowed finitely many rules. Obtaining that the expectation tends to a constant is doable, but technically more difficult; we do not think it is worth the effort, as our result already proves the degeneracy of the distribution. Using infinitely many rules is probably possible, under some analytic conditions, and there are other hypotheses that may be weakened: it is not difficult for instance to ask that the dependency graph has one large strongly connected component (all others having size one)[7], periodicity is also manageable, ... All of these generalizations introduce technical difficulties in the analysis, but we think that in most natural cases, unless we explicitly design the specification to prevent the simplifications from happening sufficiently often, the uniform distribution is degenerated when interpreting the expression: this phenomenon can be considered as inherent in this framework.

In our opinion, instead of generalizing the kind of specification even more, the natural continuation of this work is to investigate non-uniform distributions. The first candidate that comes in mind is what is called BST-like distributions, where the size of the children are distributed as in a binary search tree: that kind of distribution is really used to test algorithms, and it is probably mathematically tractable [15], even if it implies dealing with systems of differential equations.

Acknowledgments. The third author is funded by the Project RIN Alenor (Regional Project from French Normandy).

References

1. Aho, A.V., Ullman, J.D.: The Theory of Parsing, Translation, and Compiling. Prentice-Hall Inc., Upper Saddle River (1972)
2. Banderier, C., Drmota, M.: Formulae and asymptotics for coefficients of algebraic functions. Comb. Probab. Comput. **24**(1), 1–53 (2015)
3. Bell, J.P., Burris, S., Yeats, K.A.: Characteristic points of recursive systems. Electr. J. Comb. **17**(1) (2010)
4. Broda, S., Machiavelo, A., Moreira, N., Reis, R.: Average size of automata constructions from regular expressions. Bull. EATCS **116** (2015)
5. Drmota, M.: Random Trees: An Interplay Between Combinatorics and Probability, 1st edn. Springer, Vienna (2009). https://doi.org/10.1007/978-3-211-75357-6

---

[7] The general case with no constraint on the dependency graph can be really intricate, starting with the asymptotics that may behave differently [2].

6. Flajolet, P., Sedgewick, R.: Analytic Combinatorics. Cambridge University Press, Cambridge (2009)
7. Flajolet, P., Sipala, P., Steyaert, J.-M.: Analytic variations on the common subexpression problem. In: Paterson, M.S. (ed.) ICALP 1990. LNCS, vol. 443, pp. 220–234. Springer, Heidelberg (1990). https://doi.org/10.1007/BFb0032034
8. Flajolet, P., Steyaert, J.-M.: A complexity calculus for recursive tree algorithms. Math. Syst. Theory **19**(4), 301–331 (1987)
9. Gardy, D.: Random Boolean expressions. In: Discrete Mathematics & Theoretical Computer Science, DMTCS Proceedings Volume AF, Computational Logic and Applications (CLA 2005), pp. 1–36 (2005)
10. Koechlin, F., Nicaud, C., Rotondo, P.: Uniform random expressions lack expressivity. In: Rossmanith, P., Heggernes, P., Katoen, J.-P. (eds.) 44th International Symposium on Mathematical Foundations of Computer Science, MFCS 2019, Aachen, Germany, 26–30 August 2019. LIPIcs, vol. 138, pp. 51:1–51:14. Schloss Dagstuhl - Leibniz-Zentrum für Informatik (2019)
11. Lee, J., Shallit, J.: Enumerating regular expressions and their languages. In: Domaratzki, M., Okhotin, A., Salomaa, K., Yu, S. (eds.) CIAA 2004. LNCS, vol. 3317, pp. 2–22. Springer, Heidelberg (2005). https://doi.org/10.1007/978-3-540-30500-2_2
12. Meir, A., Moon, J.W.: On an asymptotic method in enumeration. J. Comb. Theory Ser. A **51**(1), 77–89 (1989)
13. Nguyên-Thê, M.: Distribution of valuations on trees. Theses, Ecole Polytechnique X (2004)
14. Nicaud, C.: On the average size of Glushkov's automata. In: Dediu, A.H., Ionescu, A.M., Martín-Vide, C. (eds.) LATA 2009. LNCS, vol. 5457, pp. 626–637. Springer, Heidelberg (2009). https://doi.org/10.1007/978-3-642-00982-2_53
15. Nicaud, C., Pivoteau, C., Razet, B.: Average analysis of Glushkov automata under a BST-like model. In: Lodaya, K., Mahajan, M. (eds.) IARCS Annual Conference on Foundations of Software Technology and Theoretical Computer Science, FSTTCS 2010, Chennai, India, 15–18 December 2010. LIPIcs, vol. 8, pp. 388–399. Schloss Dagstuhl - Leibniz-Zentrum fuer Informatik (2010)
16. Pivoteau, C., Salvy, B., Soria, M.: Algorithms for combinatorial structures: well-founded systems and newton iterations. J. Comb. Theory Ser. A **119**(8), 1711–1773 (2012)

Dynamics of Cellular Automata on Beta-Shifts and Direct Topological Factorizations

Johan Kopra[✉][iD]

Department of Mathematics and Statistics, University of Turku,
20014 Turku, Finland
jtjkop@utu.fi

Abstract. We consider the range of possible dynamics of cellular automata (CA) on two-sided beta-shifts S_β. We show that any reversible CA $F : S_\beta \to S_\beta$ has an almost equicontinuous direction whenever S_β is not sofic. This has the corollary that non-sofic beta-shifts are topologically direct prime, i.e. they are not conjugate to direct topological factorizations $X \times Y$ of two nontrivial subshifts X and Y. We also make some preliminary observations on direct topological factorizations of beta-shifts that are subshifts of finite type.

Keywords: Cellular automata · Beta-shifts · Sensitivity · Direct topological factorizations

1 Introduction

Let $X \subseteq A^{\mathbb{Z}}$ be a one-dimensional subshift over a finite symbol set A. A cellular automaton (CA) is a function $F : X \to X$ defined by a local rule, and it endows the space X with translation invariant dynamics given by local interactions. It is natural to ask how the structure of the underlying subshift X affects the range of possible topological dynamics that can be achieved by CA on X. Our preferred approach is via the framework of directional dynamics of Sablik [16]. This framework is concerned with the possible space-time diagrams of $x \in X$ with respect to F, in which successive iterations $F^t(x)$ are drawn on consecutive rows (see Fig. 1 for a typical space-time diagram of a configuration with respect to the CA which shifts each symbol by one position to the left). Information cannot cross the dashed gray line in the figure so we say that the slope of this line is an almost equicontinuous direction. On the other hand, a slope is called a sensitive direction if information can cross over every line having that slope.

It has been proven in Theorem 5.2.19 of [7] that every nontrivial mixing sofic subshift admits a reversible CA which is sensitive in all directions. On the other hand, Subsect. 5.4.2 of [7] presents a collection of non-sofic S-gap shifts X_S, all of them synchronizing and many with specification property, such that every

---

The work was partially supported by the Academy of Finland grant 296018.

N. Jonoska and D. Savchuk (Eds.): DLT 2020, LNCS 12086, pp. 178–191, 2020.
https://doi.org/10.1007/978-3-030-48516-0_14

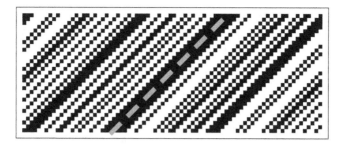

Fig. 1. A space-time diagram of the binary shift map σ. White and black squares correspond to digits 0 and 1 respectively. The dashed gray line shows an almost equicontinuous direction.

reversible CA on X_S has an almost equicontinuous direction. It would be interesting to extend the latter result to other natural classes of subshifts. The classical study of $\mathrm{Aut}(X)$, the group of reversible CA on X, is mostly not related to our line of inquiry. However, we highlight the result of [4] saying that $\mathrm{Aut}(X)/\langle\sigma\rangle$ is a periodic group if X is a transitive subshift that has subquadratic growth. This implies for such X that every $F \in \mathrm{Aut}(X)$ has an almost equicontinuous direction.

In this paper we consider two-sided beta-shifts, which form a naturally occurring class of mixing coded subshifts. We show in Theorem 3 that if S_β is a non-sofic beta-shift, then every reversible CA on S_β has an almost equicontinuous direction. As an application we use this result to show in Theorem 4 that non-sofic beta shifts are topologically direct prime, i.e. they are not conjugate to direct topological factorizations $X \times Y$ of two nontrivial subshifts X and Y, thus answering a problem suggested in the presentation [13].

The proof of Theorem 4 relies on the observation that whenever X and Y are infinite transitive subshifts, then $X \times Y$ has a very simple reversibe CA with all directions sensitive: it just shifts information into opposite directions in the X and Y components. Therefore the problem of determining whether a given subshift is topologically direct prime is closely related to the study of directional dynamics. In the last section of this paper we suggest a program of studying direct topological factorizations of sofic beta-shifts and accompany this suggestion with some preliminary remarks.

2   Preliminaries

In this section we recall some preliminaries concerning symbolic dynamics and topological dynamics in general. Good references to these topics are [8,11].

Definition 1. *If X is a compact metrizable topological space and $T : X \to X$ is a continuous map, we say that (X, T) is a (topological) dynamical system.*

When there is no risk of confusion, we may identify the dynamical system (X, T) with the underlying space or the underlying map, so we may say that X is a dynamical system or that T is a dynamical system.

Definition 2. *We write* $\psi : (X, T) \to (Y, S)$ *whenever* (X, T) *and* (Y, S) *are dynamical systems and* $\psi : X \to Y$ *is a continuous map such that* $\psi \circ T = S \circ \psi$ *(this equality is known as the* equivariance condition*). Then we say that* ψ *is a* morphism. *If* ψ *is injective, we say that* ψ *is an* embedding. *If* ψ *is surjective, we say that* ψ *is a* factor map *and that* (Y, S) *is a* factor *of* (X, T) *(via the map* ψ*). If* ψ *is bijective, we say that* ψ *is a* conjugacy *and that* (X, T) *and* (Y, S) *are* conjugate *(via* ψ*).*

A finite set A containing at least two elements (*letters*) is called an *alphabet*. In this paper the alphabet usually consists of numbers and thus for $n \in \mathbb{N}_+$ we denote $\Sigma_n = \{0, 1, \ldots, n-1\}$. The set $A^{\mathbb{Z}}$ of bi-infinite sequences (*configurations*) over A is called a *full shift*. The set $A^{\mathbb{N}}$ is the set of one-way infinite sequences over A. Formally any $x \in A^{\mathbb{Z}}$ (resp. $x \in A^{\mathbb{N}}$) is a function $\mathbb{Z} \to A$ (resp. $\mathbb{N} \to A$) and the value of x at $i \in \mathbb{Z}$ is denoted by $x[i]$. It contains finite, right-infinite and left-infinite subsequences denoted by $x[i, j] = x[i]x[i+1] \cdots x[j]$, $x[i, \infty] = x[i]x[i+1] \cdots$ and $x[-\infty, i] = \cdots x[i-1]x[i]$. Occasionally we signify the symbol at position zero in a configuration x by a dot as follows:

$$x = \cdots x[-2]x[-1].x[0]x[1]x[2] \cdots .$$

A configuration $x \in A^{\mathbb{Z}}$ or $x \in A^{\mathbb{N}}$ is *periodic* if there is a $p \in \mathbb{N}_+$ such that $x[i + p] = x[i]$ for all $i \in \mathbb{Z}$. Then we may also say that x is p-periodic or that x has period p. We say that x is *eventually periodic* if there are $p \in \mathbb{N}_+$ and $i_0 \in \mathbb{Z}$ such that $x[i + p] = x[i]$ holds for all $i \geq i_0$.

A *subword* of $x \in A^{\mathbb{Z}}$ is any finite sequence $x[i, j]$ where $i, j \in \mathbb{Z}$, and we interpret the sequence to be empty if $j < i$. Any finite sequence $w = w[1]w[2] \cdots w[n]$ (also the empty sequence, which is denoted by ϵ) where $w[i] \in A$ is a *word* over A. Unless we consider a word w as a subword of some configuration, we start indexing the symbols of w from 1 as we have done here. The concatenation of a word or a left-infinite sequence u with a word or a right-infinite sequence v is denoted by uv. A word u is a *prefix* of a word or a right-infinite sequence x if there is a word or a right-infinite sequence v such that $x = uv$. Similarly, u is a *suffix* of a word or a left-infinite sequence x if there is a word or a left-infinite sequence v such that $x = vu$. The set of all words over A is denoted by A^*, and the set of non-empty words is $A^+ = A^* \setminus \{\epsilon\}$. The set of words of length n is denoted by A^n. For a word $w \in A^*$, $|w|$ denotes its length, i.e. $|w| = n \iff w \in A^n$. For any word $w \in A^+$ we denote by $^\infty w$ and w^∞ the left- and right-infinite sequences obtained by infinite repetitions of the word w. We denote by $w^{\mathbb{Z}} \in A^{\mathbb{Z}}$ the configuration defined by $w^{\mathbb{Z}}[in, (i+1)n - 1] = w$ (where $n = |w|$) for every $i \in \mathbb{Z}$.

Any collection of words $L \subseteq A^*$ is called a *language*. For any $S \subseteq A^{\mathbb{Z}}$ the collection of words appearing as subwords of elements of S is the language of S, denoted by $L(S)$. For $n \in \mathbb{N}$ we denote $L^n(S) = L(S) \cap A^n$. For any $L \subseteq A^*$, the *Kleene star* of L is

$$L^* = \{w_1 \cdots w_n \mid n \geq 0, w_i \in L\} \subseteq A^*,$$

i.e. L^* is the set of all finite concatenations of elements of L. If $\epsilon \notin L$, define $L^+ = L^* \setminus \{\epsilon\}$ and if $\epsilon \in L$, define $L^+ = L^*$.

To consider topological dynamics on subsets of the full shifts, the sets $A^{\mathbb{Z}}$ and $A^{\mathbb{N}}$ are endowed with the product topology (with respect to the discrete topology on A). These are compact metrizable spaces. The shift map $\sigma : A^{\mathbb{Z}} \to A^{\mathbb{Z}}$ is defined by $\sigma(x)[i] = x[i+1]$ for $x \in A^{\mathbb{Z}}$, $i \in \mathbb{Z}$, and it is a homeomorphism. Also in the one-sided case we define $\sigma : A^{\mathbb{N}} \to A^{\mathbb{N}}$ by $\sigma(x)[i] = x[i+1]$. Any topologically closed *nonempty* subset $X \subseteq A^{\mathbb{Z}}$ such that $\sigma(X) = X$ is called a *subshift*. A subshift X equipped with the map σ forms a dynamical system and the elements of X can also be called *points*. Any $w \in L(X) \setminus \epsilon$ and $i \in \mathbb{Z}$ determine a *cylinder* of X

$$\mathrm{Cyl}_X(w, i) = \{x \in X \mid w \text{ occurs in } x \text{ at position } i\}.$$

Definition 3. *We say that a subshift X is* transitive *(or irreducible in the terminology of [11]) if for all words $u, v \in L(X)$ there is $w \in L(X)$ such that $uwv \in L(X)$. We say that X is* mixing *if for all $u, v \in L(X)$ there is $N \in \mathbb{N}$ such that for all $n \geq N$ there is $w \in L^n(X)$ such that $uwv \in L(X)$.*

Definition 4. *Let $X \subseteq A^{\mathbb{Z}}$ and $Y \subseteq B^{\mathbb{Z}}$ be subshifts. We say that the map $F : X \to Y$ is a* sliding block code *from X to Y (with memory m and anticipation a for integers $m \leq a$) if there exists a* local rule *$f : A^{a-m+1} \to B$ such that $F(x)[i] = f(x[i+m], \ldots, x[i], \ldots, x[i+a])$. If $X = Y$, we say that F is a* cellular automaton *(CA). If we can choose m and a so that $-m = a = r \geq 0$, we say that F is a* radius-r CA.

Note that both memory and anticipation can be either positive or negative. Note also that if F has memory m and anticipation a with the associated local rule $f : A^{a-m+1} \to A$, then F is also a radius-r CA for $r = \max\{|m|, |a|\}$, with possibly a different local rule $f' : A^{2r+1} \to A$.

The following observation of [6] characterizes sliding block codes as the class of structure preserving transformations between subshifts.

Theorem 1 (Curtis-Hedlund-Lyndon). *A map $F : X \to Y$ between subshifts X and Y is a morphism between dynamical systems (X, σ) and (Y, σ) if and only if it is a sliding block code.*

Bijective CA are called *reversible*. It is known that the inverse map of a reversible CA is also a CA. We denote by $\mathrm{End}(X)$ the monoid of CA on X and by $\mathrm{Aut}(X)$ the group of reversible CA on X (the binary operation is function composition).

The notions of almost equicontinuity and sensitivity can be defined for general topological dynamical systems. We omit the topological definitions, because for cellular automata on transitive subshifts there are combinatorial character-izations for these notions using blocking words. We present these alternative characterizations below.

Definition 5. *Let $F : X \to X$ be a radius-r CA and $w \in L(X)$. We say that w is a blocking word if there is an integer e with $|w| \geq e \geq r + 1$ and an integer $p \in [0, |w| - e]$ such that*

$$\forall x, y \in \mathrm{Cyl}_X(w, 0), \forall n \in \mathbb{N}, F^n(x)[p, p + e - 1] = F^n(y)[p, p + e - 1].$$

The following is proved in Proposition 2.1 of [16].

Proposition 1. *If X is a transitive subshift and $F : X \to X$ is a CA, then F is almost equicontinuous if and only if it has a blocking word.*

We say that a CA on a transitive subshift is *sensitive* if it is not almost equicontinuous. The notion of sensitivity is refined by Sablik's framework of directional dynamics [16].

Definition 6. *Let $F : X \to X$ be a cellular automaton and let $p, q \in \mathbb{Z}$ be coprime integers, $q > 0$. Then p/q is a sensitive direction of F if $\sigma^p \circ F^q$ is sensitive. Similarly, p/q is an almost equicontinuous direction of F if $\sigma^p \circ F^q$ is almost equicontinuous.*

As indicated in the introduction, this definition is best understood via the *space-time diagram* of $x \in X$ with respect to F, in which successive iterations $F^t(x)$ are drawn on consecutive rows (see Fig. 1 for a typical space-time diagram of a configuration with respect to the shift map). By definition $-1 = (-1)/1$ is an almost equicontinuous direction of $\sigma : A^{\mathbb{Z}} \to A^{\mathbb{Z}}$ because $\sigma^{-1} \circ \sigma = \mathrm{Id}$ is almost equicontinuous. This is directly visible in the space-time diagram of Fig. 1, because it looks like the space-time diagram of the identity map when it is followed along the dashed line. Note that the slope of the dashed line is equal to -1 with respect to the vertical axis extending downwards in the diagram.

The notions of subshifts of finite type (SFT) and sofic subshifts are well known and can be found in Chapters 2 and 3 of [11]. Any square matrix A with nonnegative entries is an adjacency matrix of a directed graph with multiple edges. The set of all bi-infinite sequences of edges forming valid paths is an *edge SFT* (associated to A), whose alphabet is the set of edges.

Some other classes of subshifts relevant to the study of beta-shifts are the following.

Definition 7. *Given a subshift X, we say that a word $w \in L(X)$ is synchroniz-ing if*

$$\forall u, v \in L(X) : uw, wv \in L(X) \implies uwv \in L(X).$$

We say that a transitive subshift X is synchronizing if $L(X)$ contains a synchro-nizing word.

Definition 8. *A language $L \subseteq A^+$ is a code if for all distinct $u, v \in L$ it holds that u is not a prefix of v. A subshift $X \subseteq A^{\mathbb{Z}}$ is a coded subshift (given by a code L) if $L(X)$ is the set of all subwords of elements of L^*.*

3   Beta-Shifts

We recall some preliminaries on beta-shifts from Blanchard's paper [2] and from Lothaire's book [12].

For $\xi \in \mathbb{R}$ we denote $\mathrm{Frac}\,(\xi) = \xi - \lfloor \xi \rfloor$, for example $\mathrm{Frac}\,(1.2) = 0.2$ and $\mathrm{Frac}\,(1) = 0$.

Definition 9. *For every real number* $\beta > 1$ *we define a dynamical system* $(\mathbb{I}, T_\beta)$, *where* $\mathbb{I} = [0,1]$ *and* $T_\beta(\xi) = \mathrm{Frac}\,(\beta\xi)$ *for every* $\xi \in \mathbb{I}$.

Definition 10. *The* β-*expansion of a number* $\xi \in \mathbb{I}$ *is the sequence* $d(\xi, \beta) \in \Sigma_{\lfloor \beta \rfloor + 1}^{\mathbb{N}}$ *where* $d(\xi, \beta)[i] = \lfloor \beta T^i(\xi) \rfloor$ *for* $i \in \mathbb{N}$.

Denote $d(1, \beta) = d(\beta)$. By this convention $d(2) = 2000\ldots$ If $d(\beta)$ ends in infinitely many zeros, i.e. $d(\beta) = d_0 \cdots d_m 0^\infty$ for $d_m \neq 0$, we say that $d(\beta)$ is finite, write $d(\beta) = d_0 \cdots d_m$, and define $d^*(\beta) = (d_0 \cdots (d_m - 1))^\infty$. Otherwise we let $d^*(\beta) = d(\beta)$. Denote by D_β the set of β-expansions of numbers from $[0, 1)$. It is the set of all infinite concatenations of words from the code

$$Y_\beta = \{d_0 d_1 \cdots d_{n-1} b \mid n \in \mathbb{N}, 0 \leq b < d_n\}$$

where $d(\beta) = d_0 d_1 d_2 \ldots$. For example, $Y_2 = \{0, 1\}$. Let S_β be the coded subshift given by the code Y_β. Since S_β is coded, it also has a natural representation by a deterministic automaton (not necessarily finite) [3,17]. These representations allow us to make pumping arguments similar to those that occur in the study of sofic shifts and regular languages.

The subshift S_β is mixing. Namely, any $u, v \in L(S_\beta)$ are subwords of $u_1 \cdots u_n$ and $v_1 \cdots v_m$ respectively for some $n, m \in \mathbb{N}_+$ and $u_i, v_i \in Y_\beta$. Because the code Y_β always contains the word 0, it follows that $u_1 \cdots u_n 0^i v_1 \cdots v_m \in L(S_\beta)$ for all $i \in \mathbb{N}$ and mixingness follows. The subshift S_β is sofic if and only if $d(\beta)$ is eventually periodic and it is an SFT if and only if $d(\beta)$ is finite.

There is a natural lexicographical ordering on $\Sigma_n^{\mathbb{N}}$ which we denote by $<$ and $\leq$. Using this we can alternatively characterize S_β as

$$S_\beta = \{x \in \Sigma_{\lfloor \beta \rfloor}^{\mathbb{Z}} \mid x[i, \infty] \leq d^*(\beta) \text{ for all } i \in \mathbb{Z}\}.$$

We call S_β a *beta-shift* (with base β). When $\beta > 1$ is an integer, the equality $S_\beta = \Sigma_\beta^{\mathbb{Z}}$ holds.

4   CA Dynamics on Beta-Shifts

In this section we study the topological dynamics of reversible CA on beta-shifts, and more precisely the possibility of them having no almost equicontinuous directions. By Theorem 5.2.19 of [7] every nontrivial mixing sofic subshift admits a reversible CA which is sensitive in all directions, and in particular this holds for mixing sofic beta-shifts. In this section we see that this result does not extend to the class of non-sofic beta-shifts.

We begin with a proposition showing that a CA on a non-sofic beta-shift has to "fix the expansion of one in the preimage" in some sense.

Proposition 2. *Let $\beta > 1$ be such that S_β is not sofic, let $F \in \mathrm{End}(S_\beta)$, let $x \in S_\beta$ be such that $x[0,\infty] = d(\beta)$ and let $y \in F^{-1}(x)$. Then there is a unique $i \in \mathbb{Z}$ such that $y[i,\infty] = d(\beta)$. Moreover, i does not depend on the choice of x or y.*

Proof. Let $r \in \mathbb{N}$ be such that F is a radius-r CA.

We first claim that i does not depend on the choice of x or y when it exists. To see this, assume to the contrary that for $j \in \{1,2\}$ there exist $x_j \in S_\beta$ with $x_j[0,\infty] = d(\beta)$, $y_j \in F^{-1}(x_j)$ and $i_j \in \mathbb{Z}$ such that $i_1 < i_2$ and $y_1[i_1,\infty] = d(\beta) = y_2[i_2,\infty]$. Then in particular for $M = \max\{i_2 - i_1, i_2\}$ it holds that $y_2[M,\infty] = y_2[M - i_2 + i_2,\infty] = y_1[M - i_2 + i_1,\infty]$ and

$$d(\beta)[M - i_2 + i_1 + r,\infty] = x_1[M - i_2 + i_1 + r,\infty] = F(y_1)[M - i_2 + i_1 + r,\infty]$$
$$= F(y_2)[M + r,\infty] = x_2[M + r,\infty] = d(\beta)[M + r,\infty].$$

Then $d(\beta)$ would be eventually periodic, contradicting the assumption that S_β is not sofic.

For the other claim, let us assume that for some x and y as in the assumption of the proposition there is no $i \in \mathbb{Z}$ such that $y[i,\infty] = d(\beta)$. We claim that the sequence $y[-r,\infty]$ can be written as an infinite concatenation of elements of Y_β. This concatenation is found inductively. By our assumption $y[-r,\infty] < d(\beta)$, so $y[-r,\infty]$ has a prefix of the form $w_1 = d_0 d_1 \cdots d_{n-1} b \in Y_\beta$ for some $n \in \mathbb{N}$, $b < d_n$. We can write $y[-r,\infty] = w_1 x_1$ for some $x_1 \in \Sigma_{\lfloor\beta\rfloor}^{\mathbb{N}}$. Because x_1 is a suffix of y, then again from our assumption it follows that $x_1 < d(\beta)$ and we can find a $w_2 \in Y_\beta$ which is a prefix of x_1. For all $i \in \mathbb{Z}$ we similarly we find $w_i \in Y_\beta$ such that $y[-r,\infty] = w_1 w_2 w_3 \ldots$.

Let r_i be such that $y[-r,r_i] = w_1 \cdots w_i$ for all $i \in \mathbb{N}$. Fix some $j, k \in \mathbb{N}$ such that $0 \le r_j < r_k$, $|r_k - r_j| \ge 2r$ and $y[r_j - r, r_j + r] = y[r_k - r, r_k + r]$. Because x is not eventually periodic, it follows that $x[r_j + 1,\infty] \ne x[r_k + 1,\infty]$.

Assume first that $x[r_j + 1,\infty] < x[r_k + 1,\infty]$. Because S_β is coded, there is a configuration $z \in S_\beta$ such that $z[-r,\infty] = w_1 \cdots w_j w_{k+1} w_{k+2} \cdots$, i.e. this suffix can be found by removing the word $w_{j+1} \cdots w_k$ from the middle of $y[-r,\infty]$. Then $F(z) \in S_\beta$ but $F(z)[0,\infty] = x[0,r_j]x[r_k + 1,\infty] > x[0,r_j]x[r_j + 1,\infty] = d(\beta)$ contradicting one of the characterizations of S_β.

Assume then alternatively that $x[r_j + 1,\infty] > x[r_k + 1,\infty]$. Because S_β is coded, there is a configuration $z \in S_\beta$ such that

$$z[-r,\infty] = w_1 \cdots w_j (w_{j+1} \cdots w_k)(w_{j+1} \cdots w_k) w_{k+1} w_{k+2} \cdots,$$

i.e. this suffix can be found by repeating the occurrence of the word $w_{j+1} \cdots w_k$ in the middle of $y[-r,\infty]$. Then $F(z) \in S_\beta$ but

$$F(z)[0,\infty] = x[0,r_j]x[r_j + 1, r_k]x[r_j + 1, r_k]x[r_k + 1,\infty]$$
$$= x[0,r_j]x[r_j + 1, r_k]x[r_j + 1,\infty] > x[0,r_j]x[r_j + 1, r_k]x[r_k + 1,\infty] = d(\beta)$$

contradicting again the characterization of S_β. □

To apply the previous proposition for a non-sofic S_β and $F \in \mathrm{End}(S_\beta)$, there must exist at least some $x, y \in S_\beta$ such that $x[0, \infty] = d(\beta)$ and $y \in F^{-1}(x)$. This happens at least when F is surjective, in which case we take the number $i \in \mathbb{Z}$ of the previous proposition and say that the intrinsic shift of F is equal to i. If the intrinsic shift of F is equal to 0, we say that F is shiftless.

In the class of non-synchronizing beta-shifts we get a very strong result on surjective CA: they are all shift maps.

Theorem 2. *If S_β is not synchronizing, then all surjective CA in $\mathrm{End}(S_\beta)$ are powers of the shift map.*

Proof. Let $F \in \mathrm{End}(S_\beta)$ be an arbitrary surjective CA and let $r \in \mathbb{N}$ be some radius of F. We may assume without loss of generality (by composing F with a suitable power of the shift if necessary) that F is shiftless. We prove that $F = \mathrm{Id}$.

Assume to the contrary that $F \neq \mathrm{Id}$, so there is $x \in S_\beta$ such that $F(x)[0] \neq x[0]$. Let $e = {}^\infty 0.d(\beta)$ and let $z \in F^{-1}(e)$ be arbitrary, so in particular $z[0, \infty] = d(\beta)$ by Proposition 2. Since S_β is not synchronizing, it follows that every word of $L(S_\beta)$ occurs in $d(\beta)$ (as explained by Kwietniak in [9], attributed to Bertrand-Mathis [1]). In particular it is possible to choose $i \geq r+1$ such that $\sigma^i(z)[-r, r] = x[-r, r]$ and $F(x)[0] = F(\sigma^i(z))[0] = \sigma^i(z)[0] = x[0]$, a contradiction.   □

Next we consider only reversible CA. They do not have to be shift maps in the class of general non-sofic beta-shifts, and in fact the group $\mathrm{Aut}(X)$ contains a copy of the free product of all finite groups whenever X is an infinite synchronizing subshift by Theorem 2.17 of [5]. Nevertheless $\mathrm{Aut}(S_\beta)$ is constrained in the sense of directional dynamics.

Theorem 3. *If S_β is not sofic and $F \in \mathrm{Aut}(S_\beta)$ is shiftless then F admits a blocking word. In particular all elements of $\mathrm{Aut}(S_\beta)$ have an almost equicontinuous direction.*

Proof. Let $r \in \mathbb{N}_+$ be a radius of both F and F^{-1}. Since $d(\beta)$ is not eventually periodic, it is easy to see (and is one formulation of the Morse-Hedlund theorem, see e.g. Theorem 7.3 of [15]) that there is a word $u \in \Sigma_{\lfloor \beta \rfloor}^{3r}$ and symbols $a < b$ such that both ua and ub are subwords of $d(\beta)$. Let $p = p'ub$ ($p, p' \in L(S_\beta)$) be some prefix of $d(\beta)$ ending in ub. We claim that p is blocking. More precisely we will show that if $x \in S_\beta$ is such that $x[0, |p| - 1] = p$ then $F^t(x)[0, |p| - 2] = p'u$ for all $t \in \mathbb{N}$.

Assume to the contrary that $t \in \mathbb{N}$ is the minimal number for which we have $F^t(x)[0, |p| - 2] \neq p'u$. We can find $w, v, v' \in L(S_\beta)$ and $c, d \in \Sigma_{\lfloor \beta \rfloor}$ ($c < d$) so that $u = wdv$, $|w| \geq 2r$ and $F^t(x)[0, |p| - 2] = p'wcv'$. Indeed, F^{-1} is shiftless because F is, and therefore the prefix $p'w$ still remains unchanged in $F^t(x)[0, \infty]$.

Now we note that x could have been chosen so that some of its suffixes is equal to 0^∞ and in particular under this choice no suffix of $F^t(x)$ is equal to $d(\beta)$. As in the proof of Proposition 2 we can represent $F^t(x)[0, \infty] = w_1 w_2 w_3 \ldots$ where $w_i \in Y_\beta$ for all $i \in \mathbb{N}$ and in fact $w_1 = p'wc$.

Now let $q = q'ua$ ($q, q' \in L(S_\beta)$) be some prefix of $d(\beta)$ ending in ua. Then also $q'wd$ is a prefix of $d(\beta)$ and thus $q'wc \in Y_\beta$. Because S_β is a coded subshift, there is a configuration $y \in S_\beta$ such that $y[0, \infty] = (q'wc)w_2w_3 \ldots$. For such y it holds that $F^{-t}(y) \in S_\beta$ but $F^{-t}(y)[0, \infty] = q'(ub)x[|p|, \infty] > d(\beta)$ contradicting the characterization of S_β. $\qquad \square$

5 Topological Direct Primeness of Beta-Shifts

We recall the terminology of Meyerovitch [14]. A *direct topological factorization* of a subshift X is a subshift $X_1 \times \cdots \times X_n$ which is conjugate to X and where each X_i is a subshift. We also say that each subshift X_i is a *direct factor* of X. The subshift X is *topologically direct prime* if it does not admit a non-trivial direct factorization, i.e. if every direct factorization contains one copy of X and the other X_i in the factorization contain just one point.

Non-sofic β-shifts turn out to be examples of topologically direct prime dynamical systems. This is an application of Theorem 3.

Theorem 4. *If S_β is a non-sofic beta-shift then it is topologically direct prime.*

Proof. Assume to the contrary that there is a topological conjugacy $\phi : S_\beta \to X \times Y$ where X and Y are non-trivial direct factors of S_β. The subshifts X and Y are mixing as factors of the mixing subshift S_β, and in particular both of them are infinite, because a mixing finite subshift can only contain one point.

Define a reversible CA $F : X \times Y \to X \times Y$ by $F(x, y) = (\sigma(x), \sigma^{-1}(y))$ for all $x \in X$, $y \in Y$. Because X and Y are infinite, it follows that F has no almost equicontinuous directions, i.e. $\sigma^r \circ F^s$ is sensitive for all coprime r and s such that $s > 0$. Then define $G = \phi^{-1} \circ F \circ \phi : S_\beta \to S_\beta$. The map G is a reversible CA on S_β and furthermore (S_β, G) and $(X \times Y, F)$ are conjugate via the map ϕ. By Theorem 3 the CA G has an almost equicontinuous direction, so we can fix coprime r and s such that $s > 0$ for which $\sigma^r \circ G^s$ is almost equicontinuous. But $\sigma^r \circ G^s$ is conjugate to $\sigma^r \circ F^s$ via the map ϕ, so $\sigma^r \circ F^s$ is also almost equicontinuous, a contradiction. $\qquad \square$

In general determining whether a given subshift is topologically direct prime or not seems to be a difficult problem. Lind gives sufficient conditions in [10] for SFTs based on their entropies: for example any mixing SFT with entropy $\log p$ for a prime number p is topologically direct prime. The paper [14] contains results on multidimensional full shifts, multidimensional 3-colored chessboard shifts and p-Dyck shifts with p a prime number.

In the class of beta-shifts the question of topological direct primeness remains open in a countable number of cases.

Problem 1. Characterize the topologically direct prime sofic beta-shifts.

Example 1. If $\beta > 1$ is an integer, then $S_\beta = \Sigma_\beta^{\mathbb{Z}}$ is topologically direct prime if and only if β is a prime number. Namely, if $\beta = nm$ for integers $n, m \geq 2$, then S_β is easily seen to be conjugate to $S_n \times S_m$ via a coordinatewise symbol permutation. The case when $\beta = p$ is a prime number follows by Lind's result because the entropy of $\Sigma_p^{\mathbb{Z}}$ is $\log p$.

In this example the existence of a direct factorization is characterized by the existence of direct factorization into beta shifts with integral base. Therefore, considering the following problem might be a good point to start with Problem 1.

Problem 2. Characterize the numbers $n, \gamma > 1$ such that n is an integer and $S_{n\gamma}$ is conjugate to $S_n \times S_\gamma$.

In Theorem 5 we consider this simpler problem in the SFT case. We start with a definition and a lemma stated in [2].

Definition 11. *Let $n > 1$ be an integer, $a \in \Sigma_n$ and $w \in \Sigma_n^*$. We say that aw is lexicographically greater than all its shifts if $aw0^\infty > \sigma^i(aw0^\infty)$ for every $i > 0$.*

Lemma 1 (Blanchard, [2]). *S_β is an SFT if and only if $\beta > 1$ is the unique positive solution of some equation $x^d = a_{d-1}x^{d-1} + \cdots + a_0$ where $d \geq 1$, $a_{d-1}, a_0 \geq 1$ and $a_i \in \mathbb{N}$ such that $a_{d-1} \cdots a_0$ is lexicographically greater than all its shifts. Then $d(\beta) = a_{d-1} \cdots a_0$.*

Proof. The polynomial equation is equivalent to $1 = a_{d-1}x^{-1} + \cdots + a_0x^{-d}$, which clearly has a unique positive solution. If β satisfies such an equation then $d(\beta) = a_{d-1} \cdots a_0$ and S_β is an SFT. On the other hand, if S_β is an SFT, then $d(\beta)$ takes the form of a word $a_{d-1} \cdots a_0$ which is lexicographically greater than all its shifts and β satisfies $1 = a_{d-1}x^{-1} + \cdots + a_0x^{-d}$. □

For the following we also recall some facts on zeta functions. The zeta function $\zeta_X(t)$ of a subshift X is a formal power series that encodes information about the number of periodic configurations in X and it is a conjugacy invariant of X (for precise definitions see Section 6.4 of [11]). Every SFT X is conjugate to an edge SFT associated to a square matrix A. Let I be an index set and let $\{\mu_i \in \mathbb{C} \setminus \{0\} \mid i \in I\}$ be the collection of non-zero eigenvalues of A with multiplicities: it is called the non-zero spectrum of X. It is known that then $\zeta_X(t) = \prod_{i \in I}(1 - \mu_i t)^{-1}$. The number of p-periodic configurations in X is equal to $\sum_{i \in I} \mu_i^p$ for $p \in \mathbb{N}_+$. If Y is also an SFT with $\zeta_Y(t) = \prod_{j \in J}(1 - \nu_j t)^{-1}$, then the zeta function of $X \times Y$ is $\zeta_X(t) \otimes \zeta_Y(t) = \prod_{i \in I, j \in J}(1 - \mu_i \nu_j t)^{-1}$ [10].

Theorem 5. *Let S_γ be an SFT with γ the unique positive solution of some equation $x^d = a_{d-1}x^{d-1} + \cdots + a_0$ where $d \geq 1$, $a_{d-1}, a_0 \geq 1$ and $a_i \geq 0$ such that $a_{d-1} \cdots a_0$ is lexicographically greater than all its shifts. If $n \geq 2$ is an integer such that also $(na_{d-1}) \cdots (n^d a_0)$ is lexicographically greater than all its shifts, then $S_{n\gamma}$ is topologically conjugate to $S_n \times S_\gamma$. The converse also holds: if $(na_{d-1}) \cdots (n^d a_0)$ is not lexicographically greater than all its shifts, then either $S_{n\gamma}$ is not an SFT or $S_{n\gamma}$ and $S_n \times S_\gamma$ have different zeta functions. In particular they are not conjugate.*

Proof. We have $d(\gamma) = a_{d-1} \cdots a_0$. The roots of $x^d = a_{d-1}x^{d-1} + \cdots + a_0$ are $\gamma_1 = \gamma, \gamma_2, \ldots, \gamma_d$. By multiplying both sides by n^d and by substituting $y = nx$ we see that the roots of $y^d = na_{d-1}y^{d-1} + \cdots + n^d a_0$ are $n\gamma_i$ and $n\gamma$ is the unique positive solution. Because multiplying $\prod_i(x - \gamma_i) = 0$ by n^d yields $\prod_i(y - n\gamma_i) = 0$, we also see that the multiplicities of the roots γ_i and $n\gamma_i$ are the same in their respective equations. If $(na_{d-1}) \cdots (n^d a_0)$ is lexicographically greater than all its shifts, then $S_{n\gamma}$ is an SFT with $d(n\gamma) = na_{d-1} \cdots n^d a_0$. As in [17], the shifts S_γ and $S_{n\gamma}$ are conjugate to the edge shifts X_C and X_B respectively given by the matrices

$$C = \begin{pmatrix} a_{d-1} & 1 & 0 & \cdots & 0 \\ a_{d-2} & 0 & 1 & \cdots & 0 \\ \vdots & \vdots & \vdots & & \vdots \\ a_0 & 0 & 0 & \cdots & 0 \end{pmatrix} \qquad B = \begin{pmatrix} na_{d-1} & 1 & 0 & \cdots & 0 \\ n^2 a_{d-2} & 0 & 1 & \cdots & 0 \\ \vdots & \vdots & \vdots & & \vdots \\ n^d a_0 & 0 & 0 & \cdots & 0 \end{pmatrix}.$$

They are also the companion matrices of the polynomials $x^d - a_{d-1}x^{d-1} - \cdots - a_0$ and $y^d - na_{d-1}y^{d-1} - \cdots - n^d a_0$, so the eigenvalues are the roots of these polynomials and the zeta functions of S_γ and $S_{n\gamma}$ are

$$\zeta_{X_C}(t) = \prod_i (1 - \gamma_i t)^{-1} \quad \text{and} \quad \zeta_{X_B}(t) = \prod_i (1 - n\gamma_i t)^{-1}.$$

In any case $\zeta_{S_n} = (1 - nt)^{-1}$, so the zeta function of $X = S_n \times S_\gamma$ is $\zeta_X(t) = \prod_i (1 - n\gamma_i t)^{-1}$, which is equal to ζ_{X_B}.

We will construct a conjugacy between $S_n \times X_C$ and X_B. We will choose the labels of the edges in X_C and X_B as in Figs. 2 and 3. The labels in the figures range according to $0 \le i_j < n$ and $0 \le k_j < a_{d-j}$ for $1 \le j \le d$.

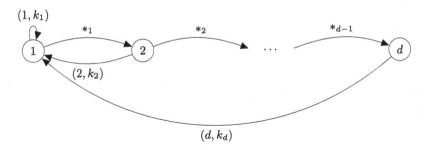

Fig. 2. The choice of labels for the graph of X_C.

The labeling has been chosen in a way that suggests the correct choice of the reversible sliding block code $\phi : S_n \times X_C \to X_B$. For any $x \in S_n \times X_C$ we make the usual identification $x = (x_1, x_2)$ where $x_1 \in S_n$, $x_2 \in X_C$ and we denote $\pi_1(x) = x_1$, $\pi_2(x) = x_2$. Then ϕ is defined by

$$\phi(x)[i] = \begin{cases} *_j & \text{when } \pi_2(x)[i] = *_j, \\ (i_1, k_1) & \text{when } \pi_2(x)[i] = (1, k_1) \text{ and } \pi_1(x)[i] = i_1, \\ (i_1, i_2, \ldots, i_j, k_j) & \text{when } \pi_2(x)[i - (j-1), i] = *_1 *_2 \cdots *_{j-1} (j, k_j) \\ & \quad \text{and } \pi_1(x)[i - (j-1), i] = i_1 i_2 \cdots i_j \end{cases}$$

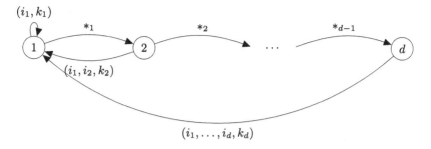

Fig. 3. The choice of labels for the graph of X_B.

The intuition here is that the sliding block code ϕ attempts to deposit all the information at $x[i]$ to $\phi(x)[i]$. This is not possible when $\pi_2(x)[x] = *_j$, so the remaining information is deposited to the nearest suitable coordinate to the right.

For the converse, assume that the word $(na_{d-1})\cdots(n^d a_0)$ is not lexicographically greater than all its shifts and that $S_{n\gamma}$ is an SFT. Then $n\gamma$ is the unique positive solution of some equation $x^e = b_{e-1}x^{e-1} + \cdots + b_0$ where $e \geq 1$, $b_{e-1}, b_0 \geq 1$ and $b_i \geq 0$ such that $b_{e-1}\cdots b_0$ is lexicographically greater than all its shifts. As above, $S_{n\gamma}$ is conjugate to an edge shift Y given by a matrix with eigenvalues $\beta_1, \beta_2, \ldots, \beta_e$ which are also the roots of the corresponding polynomial. By our assumption the polynomials $x^e - b_{e-1}x^{e-1} - \cdots - b_0$ and $y^d - na_{d-1}y^{d-1} - \cdots - n^d a_0$ are different, so they have different sets of roots (with multiplicities taken into account) and

$$\zeta_Y(t) = \prod_j (1 - \beta_i t)^{-1} \neq \prod_i (1 - n\gamma_i t)^{-1} = \zeta_X(t),$$

because $\mathbb{C}[t]$ is a unique factorization domain. $\square$

We conclude with an example concerning an SFT beta-shift $S_{\beta_1 \times \beta_2}$ where the assumption of either β_1 or β_2 being an integer is dropped.

Example 2. A beta-shift $S_{\gamma \times \gamma}$ can be topologically direct prime even if S_γ and $S_{\gamma \times \gamma}$ are SFTs (and then in particular $S_{\gamma \times \gamma}$ is not conjugate to $S_\gamma \times S_\gamma$). Denote by γ the unique positive root of $x^3 - x^2 - x - 1$. By Lemma 1 we have $d(\gamma) = 111$ and in particular S_γ is an SFT. Denote $\beta = \gamma^2$. Its minimal polynomial is $x^3 - 3x^2 - x - 1$ and by Lemma 1 $d(\beta) = 311$, so S_β is an SFT and it is conjugate to the edge SFT given by the matrix $A = \left(\begin{smallmatrix} 3 & 1 & 0 \\ 1 & 0 & 1 \\ 1 & 0 & 0 \end{smallmatrix}\right)$. It has three distinct eigenvalues $\beta_0 = \beta$, β_1 and β_2.

We claim that S_β is topologically direct prime. To see this, assume to the contrary that S_β is topologically conjugate to $X \times Y$ where X and Y are nontrivial direct factors for S_β. Since $X \times Y$ is a mixing SFT, it follows from Proposition 6 of [10] that X and Y are mixing SFTs and in particular they are infinite. The zeta functions of X and Y are of the form

$$\zeta_X(t) = \prod_i (1 - \mu_i t)^{-1} \quad \text{and} \quad \zeta_Y(t) = \prod_j (1 - \nu_j t)^{-1}$$

for some $\mu_i, \nu_j \in \mathbb{C} \setminus \{0\}$. The zeta function of S_β is

$$\zeta_{S_\beta}(t) = (1 - \beta t)^{-1}(1 - \beta_1 t)^{-1}(1 - \beta_2 t)^{-1} = \prod_{i,j}(1 - \mu_i \nu_j t)^{-1}.$$

Because $\mathbb{C}[t]$ is a unique factorization domain and because X and Y are non-trivial SFTs, we may assume without loss of generality that $\zeta_X(t) = (1 - \mu t)$ and $\zeta_Y(t) = (1 - \nu_1 t)(1 - \nu_2 t)(1 - \nu_3 t)$ for some $\mu, \nu_1, \nu_2, \nu_3 \in \mathbb{C} \setminus \{0\}$. The quantities μ and $\nu_1 + \nu_2 + \nu_3$ are the numbers of 1-periodic points of X and Y respectively and thus the number of 1-periodic points of S_β is equal to $\mu(\nu_1 + \nu_2 + \nu_3) = 3$ where μ and $\nu_1 + \nu_2 + \nu_3$ are nonnegative integers. In particular $\mu \in \{1, 3\}$.

Assume first that $\mu = 1$. Therefore X has the same zeta function as the full shift over the one letter alphabet and X has just one periodic point. As a mixing SFT X has periodic points dense so X only contains one point, contradicting the nontriviality of X.

Assume then that $\mu = 3$. Therefore X has the same zeta function as $\Sigma_3^{\mathbb{Z}}$ and X has precisely 3^n n-periodic points for all $n \in \mathbb{N}_+$. In particular the number of 2-periodic points of S_β is divisible by $3^2 = 9$. On the other hand the number of 2-periodic points of S_β is equal to $\mathrm{Tr}(A^2) = 11$, a contradiction.

References

1. Bertrand-Mathis, A.: Questions diverses relatives aux systemes codés: applications au θ-shift. Preprint
2. Blanchard, F.: β-expansions and symbolic dynamics. Theoret. Comput. Sci. **65**(2), 131–141 (1989). https://doi.org/10.1016/0304-3975(89)90038-8
3. Blanchard, F., Hansel, G.: Systèmes codés. Theoret. Comput. Sci. **44**(1), 17–49 (1986). https://doi.org/10.1016/0304-3975(86)90108-8
4. Cyr, V., Kra, B.: The automorphism group of a shift of subquadratic growth. Proc. Amer. Math. Soc. **144**(2), 613–621 (2016). https://doi.org/10.1090/proc12719
5. Fiebig, D., Fiebig, U.R.: The automorphism group of a coded system. Trans. Am. Math. Soc. **348**(8), 3173–3191 (1996)
6. Hedlund, G.A.: Endomorphisms and automorphisms of the shift dynamical system. Math. Syst. Theory **3**(4), 320–375 (1969)
7. Kopra, J.: Cellular automata with complicated dynamics. Ph.D. thesis, University of Turku (2019)
8. Kůrka, P.: Topological and Symbolic Dynamics, vol. 11. SMF (2003)
9. Kwietniak, D.: Synchronised β-shifts. MathOverflow (2018). https://mathoverflow.net/questions/297064/synchronised-beta-shifts. Accessed 20 Dec 2019
10. Lind, D.A.: The entropies of topological Markov shifts and a related class of algebraic integers. Ergod. Theory Dyn. Syst. **4**(2), 283–300 (1984). https://doi.org/10.1017/S0143385700002443

11. Lind, D., Marcus, B.: An Introduction to Symbolic Dynamics and Coding. Cambridge University Press, Cambridge (1995)

12. Lothaire, M.: Algebraic Combinatorics on Words, Encyclopedia of Mathematics and Its Applications, vol. 90. Cambridge University Press, Cambridge (2002). https://doi.org/10.1017/CBO9781107326019

13. Meyerovitch, T.: On direct product factorization of homeomorphisms. Slides (2013). http://perso.ens-lyon.fr/nathalie.aubrun/DySyCo/Meyerovitch.pdf. Accessed 19 Jan 2020

14. Meyerovitch, T.: Direct topological factorization for topological flows. Ergod. Theory Dyn. Syst. **37**(3), 837–858 (2017). https://doi.org/10.1017/etds.2015.67

15. Morse, M., Hedlund, G.A.: Symbolic dynamics. Am. J. Math. **60**(4), 815–866 (1938). https://doi.org/10.2307/2371264

16. Sablik, M.: Directional dynamics for cellular automata: a sensitivity to initial condition approach. Theoret. Comput. Sci. **400**(1–3), 1–18 (2008)

17. Thurston, W.P.: Groups, tilings and finite state automata. Lecture notes (1989). http://timo.jolivet.free.fr/docs/ThurstonLectNotes.pdf. Accessed 07 Jan 2020

Avoidability of Additive Cubes over Alphabets of Four Numbers

Florian Lietard[1,2]([⊠]) [iD] and Matthieu Rosenfeld[3]

[1] Loria, Université de Lorraine, Campus Scientifique,
54506 Vandœuvre-lès-Nancy, France
florian.lietard@univ-lorraine.fr

[2] Institut Élie Cartan de Lorraine, Université de Lorraine, Site de Nancy,
54506 Vandœuvre-lès-Nancy, France

[3] CNRS, LIS, Aix Marseille Université, Université de Toulon, Marseille, France
matthieu.rosenfeld@univ-amu.fr

Abstract. Let $\mathcal{A} \subset \mathbb{N}$ be a set of size 4 such that $\mathcal{A}$ cannot be obtained by applying the same affine function to all of the elements of $\{0, 1, 2, 3\}$. We show that there is an infinite sequence of elements of $\mathcal{A}$ that contains no three consecutive blocks of same size and same sum (additive cubes). Moreover, it is possible to replace $\mathbb{N}$ by $\mathbb{C}$ in the statement.

Keywords: Abelian/additive equivalence · Abelian/additive powers · Combinatorics on words

1 Introduction

Let $k \geq 2$ be an integer and $(G, +)$ a semigroup. An *additive kth power* is a non-empty word $w_1 \cdots w_k$ over $\mathcal{A} \subseteq G$ such for every $i \in \{2, \ldots, k\}$, $|w_i| = |w_1|$ and $\sum w_i = \sum w_1$ (where $\sum v$ denotes the sum of the letters in v seen as numbers). It is a longstanding question whether there exists an infinite word w over a finite subset of $\mathbb{N}$ that avoids additive squares (additive 2nd powers) [3, 4,6]. One motivation for studying this problem is that a positive answer to this question would imply that additive squares are avoidable over any semigroup that contains some finitely generated infinite semigroup [6] (an application of van der Waerden's theorem shows that additive powers are not avoidable over any other semigroup [4]). Cassaigne et al. [1] showed that there exists an infinite word over the finite alphabet $\{0, 1, 3, 4\} \subset \mathbb{Z}$ without additive cubes (additive 3rd powers). Rao [7] used this result to show that there exist infinite words avoiding additive cubes over any alphabet $\{0, i, j\} \subset \mathbb{N}^3$ with i and j coprime, $i < j$ and $6 \leq j \leq 9$ (and he conjectured that the second condition can be replaced by $6 \leq j$). This motivates the following more general problem:

Problem 1. Characterize the finite subsets of $\mathbb{N}$ over which additive cubes are avoidable.

The second author is supported by the ANR project CoCoGro (ANR-16-CE40-0005).

N. Jonoska and D. Savchuk (Eds.): DLT 2020, LNCS 12086, pp. 192–206, 2020.
https://doi.org/10.1007/978-3-030-48516-0_15

It seems restrictive to use $\mathbb{N}$ instead of $\mathbb{R}$ (or $\mathbb{C}$), but solving Problem 1 for alphabets of the form $\{0, a_1, \ldots, a_m\} \in \mathbb{N}$ with the a_i's being coprime completely solves the problem for any finite alphabet over $\mathbb{C}$ (if the a_i's are given in increasing order one can additionally assume a_1 be smaller than $a_m - a_{m-1}$). For the sake of completeness, we give a short proof of this fact in Sect. 2.

If Rao's conjecture were true then the only remaining 3-letter alphabets over $\mathbb{C}$ to characterize would be $\{0, 1, 2\}$, $\{0, 1, 3\}$, $\{0, 1, 4\}$ and $\{0, 2, 5\}$ (see [9, Section 2.2.2] for details). However, this conjecture is known to be true for only finitely many such alphabets (up to a trivial equivalence relation defined in Sect. 2.1). In the present paper we propose a twist on previously used ideas to show our main theorem (see Corollary 1).

Main Theorem. *Let $\mathcal{A} \subset \mathbb{C}$ be an alphabet with $|\mathcal{A}| \geq 4$. If $\mathcal{A}$ is not equivalent to $\{0, 1, 2, 3\}$ then additive cubes are avoidable over $\mathcal{A}$.*

This also implies that additive cubes are avoidable over any alphabet of complex numbers of size at least 5. Rao used the fact that additive cubes are avoidable over $\{0, 1, 3, 4\}$ to show that they are avoidable over some 3-letter alphabets [7, Section 3.2], so our result might also be of importance for tackling Problem 1 for alphabets of size 3.

The present paper is organized as follows. We first recall some notation and we define the equivalence between two alphabets. Equipped with this equivalence relation we explain why it is enough to study alphabets of integers or alphabets of the form $\{0, 1, a_2, a_3, \ldots, a_m\}$ with $m \in \mathbb{N}$ and $a_2, \ldots, a_m \in \mathbb{Q}$. Then we introduce the word $\mathbf{W}_{a,b,c,d}$, based on the construction of [1], and we show that for all but finitely (up to our equivalence relation) many values of $a, b, c,$ and d, the word $\mathbf{W}_{a,b,c,d}$ avoids additive cubes. Finally, using the literature for the remaining alphabets, we conclude that additive cubes are avoidable over all the remaining alphabets of size 4, with the sole exception of $\{0, 1, 2, 3\}$. We leave the case of $\{0, 1, 2, 3\}$ open, and comment on our calculations regarding this case in the last section.

2   Preliminaries

We use the standard notation introduced in Chapter 1 of [5]. In the rest of the present article all of our alphabets are finite sets of complex numbers. For the rest of this section, let $\mathcal{A} \subset \mathbb{C}$ be such an alphabet. We denote by ε the empty word and by $|\mathcal{A}|$ the cardinality of the alphabet $\mathcal{A}$. Given a word $w \in \mathcal{A}^*$, we denote by $|w|$ the length of w and by $|w|_\alpha$ the number of occurrences of the letter $\alpha \in \mathcal{A}$ in w. Two words u and v are *abelian equivalent*, denoted by $u \simeq_{ab} v$ if u and v are permutations of each other. They are *additively equivalent*, denoted by $u \simeq_{ad} v$, if $|u| = |v|$ and $\sum u = \sum v$, where $\sum v$ denotes the sum of the letters in v (this make sense since the letters are complex numbers). A word $uvw \in \mathcal{A}^*$ is an *abelian cube* (respectively, an *additive cube*) if $u \simeq_{ab} v \simeq_{ab} w$ (respectively, if $u \simeq_{ad} v \simeq_{ad} w$).

2.1   Alphabets in $\mathbb{N}$

For any function $h : \mathcal{A} \to \mathbb{C}$ and words w over $\mathcal{A} \subset \mathbb{C}$, the word $h(w)$ is obtained by replacing each letter of w by its image under h. We say that two alphabets $\mathcal{A}, \mathcal{A}' \subset \mathbb{C}$ of same size are *equivalent* if there is a function $h : \mathcal{A} \to \mathcal{A}'$ such that for all $u, v \in \mathcal{A}^*$,

$$u \simeq_{ad} v \iff h(u) \simeq_{ad} h(v).$$

Let us now show that for any alphabet of complex numbers, we either already know that additive cubes are avoidable or the alphabet is equivalent to an alphabet of integers. We start by giving sufficient conditions for two alphabets to be equivalent.

Lemma 1. *Let $u, v \in \mathcal{A}^*$ be two finite words, let $a \in \mathbb{C} \backslash \{0\}$, $b \in \mathbb{C}$ and $f : \mathbb{C} \to \mathbb{C}$, $x \mapsto ax + b$. Then $u \simeq_{ad} v$ if and only if $f(u) \simeq_{ad} f(v)$.*

The proof is left to the reader. Recall that two complex numbers a and b are said to be rationally independent if $k_1 a + k_2 b = 0$ for $(k_1, k_2) \in \mathbb{Z}^2$ implies $k_1 = k_2 = 0$.

Lemma 2. *Let $\mathcal{A} \subset \mathbb{C}$.*

(i) If $|\mathcal{A}| \leq 2$ then additive cubes are not avoidable over $\mathcal{A}$.

(ii) If $|\mathcal{A}| > 2$ and if there are $a, b, c \in \mathcal{A}$, such that $b - a$ and $c - a$ are rationally independent, then additive cubes are avoidable over $\mathcal{A}$.

(iii) If $|\mathcal{A}| > 2$ and if for any pairwise different $a, b, c \in \mathcal{A}$, the differences $b - a$ and $c - a$ are rationally dependent, then there exists an alphabet $\mathcal{A}' = \{0, a_1, \ldots, a_m\} \subset \mathbb{N}$ with $\gcd(a_1, \ldots, a_m) = 1$ such that $\mathcal{A}$ and $\mathcal{A}'$ are equivalent.

Proof.

(i) This statement follows from the fact that abelian cubes are not avoidable over two letters [2].

(ii) Since $b - a$ and $c - a$ are rationally independent, for any $k_1, k_2, k_3 \in \mathbb{Z}$, if $0 k_1 + (b - a) k_2 + (c - a) k_3 = 0$ then $k_2 = k_3 = 0$. Thus for any words $u, v \in \{0, b - a, c - a\}^*$, if $\sum u = \sum v$ then u has the same number of occurrences of $b - a$ (resp., $c - a$) as v; moreover, if $|u| = |v|$ then u and v also have the same number of occurrences of 0. Thus, for any word $u, v \in \{0, b - a, c - a\}^*$, if $u \simeq_{ad} v$ then $u \simeq_{ab} v$. From Lemma 1 (with $f : x \mapsto x + a$), for any $u, v \in \{a, b, c\}^*$, if $u \simeq_{ad} v$ then $u \simeq_{ab} v$. Since abelian cubes are avoidable over 3 letters [2], we deduce that additive cubes are avoidable over $\mathcal{A}$.

(iii) Let $\{b_1, \ldots, b_m\} = \mathcal{A}$. For any i, $b_i - b_1$ and $b_2 - b_1$ are rationally dependent which implies $\frac{b_i - b_1}{b_2 - b_1} \in \mathbb{Q}$. Thus there exists a $q \in \mathbb{Z}$ such that for all i, $q \frac{b_i - b_1}{b_2 - b_1} \in \mathbb{Z}$ and $\gcd \left(q \frac{b_2 - b_1}{b_2 - b_1}, q \frac{b_3 - b_1}{b_2 - b_1}, \ldots, q \frac{b_m - b_1}{b_2 - b_1} \right) = 1$. Let $s = \min_{1 \leq i \leq m} \left(q \frac{b_i - b_1}{b_2 - b_1} \right)$. Finally, we apply Lemma 1 with $f : x \mapsto q \frac{x - b_1}{b_2 - b_1} - s$ and we get that the alphabet $\{ q \frac{b_1 - b_1}{b_2 - b_1} - s, q \frac{b_2 - b_1}{b_2 - b_1} - s, q \frac{b_3 - b_1}{b_2 - b_1} - s, \ldots, q \frac{b_m - b_1}{b_2 - b_1} - s \}$ satisfies all the required conditions. This concludes the proof.

Thus solving Problem 1 for alphabets of the form $\{0, a_1, \ldots, a_m\} \subset \mathbb{N}$ with coprime a_i's completely solves the problem for any finite alphabet over $\mathbb{C}$. Notice that, in case (iii), one can add the condition that $a_1 < a_m - a_{m-1}$, (otherwise apply $f : x \mapsto a_m - x$ to this alphabet). One could also add that in the case $|\mathcal{A}| = 2$, one can avoid additive 4th powers (with an argument similar to (ii) and the fact that abelian 4th powers are avoidable over 2 letters [2]).

Remark 1. Every alphabet $\{a_0, a_1, \ldots, a_m\} \subset \mathbb{N}$ is equivalent to the alphabet $\{0, 1, f(a_2), \ldots, f(a_m)\} \subset \mathbb{Q}$, where $f : x \mapsto \frac{x - a_0}{a_1 - a_0}$. Therefore, in Sects. 3 and 4, instead of considering alphabets of four integers we consider alphabets of the form $\{0, 1, c, d\} \subset \mathbb{Q}$.

3   The Infinite Word $\mathbf{W}_{a,b,c,d}$

Let $a, b, c, d \in \mathbb{R}$ and let $\varphi_{a,b,c,d} : \{a, b, c, d\}^* \to \{a, b, c, d\}^*$ be the following morphism:

$$\varphi_{a,b,c,d}(a) = ac \quad ; \quad \varphi_{a,b,c,d}(b) = dc \quad ; \quad \varphi_{a,b,c,d}(c) = b \quad ; \quad \varphi_{a,b,c,d}(d) = ab.$$

Let $\mathbf{W}_{a,b,c,d} := \lim_{n \to +\infty} \varphi_{a,b,c,d}^n(a)$ be the infinite fixed point of $\varphi_{a,b,c,d}$. Cassaigne et al. [1] showed in 2014 that $\mathbf{W}_{0,1,3,4}$ avoids additive cubes. In particular, this implies that $\mathbf{W}_{0,1,3,4}$ avoids abelian cubes. This property does not depend on the choice of a, b, c, d, therefore we deduce the following lemma.

Lemma 3. *For any pairwise distinct a, b, c, d, the word $\mathbf{W}_{a,b,c,d}$ avoids abelian cubes.*

We define the *Parikh vector* Ψ as the map

$$\Psi : \{a, b, c, d\}^* \longrightarrow \mathbb{Z}^4$$
$$w \longmapsto {}^t\left(|w|_a \quad |w|_b \quad |w|_c \quad |w|_d\right).$$

Let $M_\varphi = \begin{pmatrix} 1 & 0 & 0 & 1 \\ 0 & 0 & 1 & 1 \\ 1 & 1 & 0 & 0 \\ 0 & 1 & 0 & 0 \end{pmatrix}$ be the adjacency matrix of $\varphi_{a,b,c,d}$ and τ be the vector

corresponding to the numerical approximation[1] $\tau \doteq \begin{pmatrix} 0.5788 - 0.5749i \\ -0.3219 + 0.2183i \\ -0.0690 + 0.6165i \\ -0.1662 - 0.6810i \end{pmatrix}$, which is

---

[1] We stress the fact that this is not an issue to use numerical approximation. Indeed, all our computations are numerically stable (additions, multiplications and no divisions by numbers close to zero) and if we start with sufficiently accurate approximations, we get sufficiently accurate approximations at the end (see footnote 2 for the only case where it matters that a coefficient is exactly 0). Moreover, there is an algebraic extension of $\mathbb{Q}$ of degree 24 that contains all the eigenvalues of the matrices (according to mathematica) and thus we could use the original proof of [1, Theorem 8] to get an exact value for C and only use exact computation in our article. However, one might think that this is convenient to use the fact that these roots can be expressed with radicals, but maintaining exact expressions involving radicals is much more inefficient and would lead to even more unreadable computations.

related to the eigenvalue $0.4074 + 0.4766i$ of M_φ and precisely defined in Sect. 2.1 of [1]. For the sake of conciseness, the definition is omitted here. We recall the following result from [1].

Theorem 1 ([1, **Theorem 3.1**]). *There exists a positive real constant C such that for any two factors of $\mathbf{W}_{a,b,c,d}$ (not necessarily adjacent) u and v*

$$|\tau \cdot (\Psi(u) - \Psi(v))| < C,$$

where $2.175816 < C < 2.175817$.

Let us summarize the main idea behind Theorem 1. The asymptotic behavior of the Parikh vectors of factors is closely related to the asymptotic behavior of the iterations of the matrix M_φ (since $\Psi(\varphi(u)) = M_\varphi(\Psi(u))$). Moreover, the eigenvalue corresponding to this eigenvector is of norm less than 1 and thus the associated subspace is contracting. We deduce that $\tau(\Psi(u))$ is bounded for any factor u. Theorem 1 provides good bounds in this particular case. Equipped with Lemma 3 and Theorem 1 we deduce the following one.

Lemma 4. *For any pairwise distinct $a, b, c, d \in \mathbb{R}$, let $M_{a,b,c,d} = \left(\begin{smallmatrix} 1 & 1 & 1 & 1 \\ a & b & c & d \end{smallmatrix} \right)$. Suppose that $\mathbf{W}_{a,b,c,d}$ contains an additive cube, then there exists a vector $x \in \ker(M_{a,b,c,d}) \cap \mathbb{Z}^4 \setminus \{0\}$ such that $|\tau \cdot x| < C$, where C is given in Theorem 1.*

Proof. Let uvw be an additive cube factor of $\mathbf{W}_{a,b,c,d}$. By Lemma 3, uvw cannot be an abelian cube. Thus either $\Psi(u) \neq \Psi(v)$ or $\Psi(v) \neq \Psi(w)$. Without loss of generality, $\Psi(u) \neq \Psi(v)$. In this case, let $x = \Psi(u) - \Psi(v) \neq 0$. Since x is the difference of two Parikh vectors we get $x \in \mathbb{Z}^4$. Since uvw is an additive cube, $|u| = |v|$ and $|u|_a a + |u|_b b + |u|_c c + |u|_d d = |v|_a a + |v|_b b + |v|_c c + |v|_d d$. This implies that $M_{a,b,c,d}(\Psi(u) - \Psi(v)) = 0$ which can be rewritten as $x \in \ker(M_{a,b,c,d})$. Therefore, $x \in \ker(M_{a,b,c,d}) \cap \mathbb{Z}^4 \setminus \{0\}$. By assumption u and v are two factors of $\mathbf{W}_{a,b,c,d}$ and by Theorem 1 we get $|\tau \cdot x| < C$, which concludes the proof.

This Lemma contains the main idea of the present work. If we want to know for which choices of a, b, c and d, the word $\mathbf{W}_{a,b,c,d}$ avoids additive cubes, it is sufficient to study the behavior of the lattice $\ker(M_{a,b,c,d}) \cap \mathbb{Z}^4 \setminus \{0\}$.

4   The Case of $\mathbf{W}_{0,1,c,d}$

Let us first study the lattice $\ker(M_{0,1,c,d}) \cap \mathbb{Z}^4 \setminus \{0\}$ for $c, d \in \mathbb{R}$. We show that in many cases additive cubes are avoidable over $\{0, 1, c, d\}$.

Theorem 2. *Let $c, d \in \mathbb{R}$. Suppose $d > c > 1$, $c \notin \{5/4, 4/3, 3/2, 2\}$ and $d \notin \{6 - 4c, 5 - 3c, 4 - 2c, 3 - c, 2c - 3, 2c - 2, 2c - 1, 3c - 3, 2\}$. Then $\mathbf{W}_{0,1,c,d}$ avoids additive cubes.*

Proof. From Lemma 4, it is sufficient to show that under the assumptions on c and d we get $|\tau \cdot x| \geq C$ for any $x \in \ker(M_{0,1,c,d}) \cap \mathbb{Z}^4 \setminus \{0\}$. Let us first express this set of vectors in a more convenient way. It is straightforward to check that if $\alpha = (c-1, -c, 1, 0)$ and $\beta = (d-1, -d, 0, 1)$, then $\{\alpha, \beta\}$ is a basis of $\ker(M_{0,1,c,d})$. For any reals m and n, if $m\alpha + n\beta$ is an integral vector then $m \in \mathbb{N}$ (resp., $n \in \mathbb{N}$) because otherwise its third (resp., its fourth) coordinate is not an integer and $mc + nd \in \mathbb{Z}$, otherwise the first and second coordinates are not integers. We deduce that

$$\ker(M_{0,1,c,d}) \cap \mathbb{Z}^4 = \{m\alpha + n\beta \mid m, n \in \mathbb{Z}, mc + nd \in \mathbb{Z}\}.$$

Thus, we only need to show that, under the assumptions, for any $m, n \in \mathbb{Z}$ with $mc + nd \in \mathbb{Z}$ and $(m, n) \neq (0, 0)$, we get

$$|\tau \cdot (m\alpha + n\beta)| \geq C. \tag{1}$$

Let us show that (1) holds if $n = 0$. In this case, $m \neq 0$, $|\tau \cdot m\alpha| = |m||\tau \cdot \alpha|$ and $mc \in \mathbb{Z}$. Numerical computation gives $f_0(c) := |\tau \cdot \alpha| \doteq \sqrt{1.83908 + c(-3.05698 + 1.44043c)}$. The minimum of f_0 is reached at $c \doteq \frac{3.05698}{2 \times 1.44043} \doteq 1.06114$. Thus for any numbers $x, y \in \mathbb{R}$ with $x < y$ and $1.06114 < y$ the minimum of f_0 over the interval $[x, y]$ is given by $f_0(\max(1.06114, x))$. We distinguish several cases depending on the value of c.

- If $c > 2.85$ a straightforward computation gives $|\tau \cdot \alpha| > C$ and $|\tau \cdot m\alpha| > C$.
- If $c \in [1.9, 2.9] \setminus \{2\}$, a computation gives $|\tau \cdot \alpha| > \frac{C}{2}$. Moreover, in this case $m \in \mathbb{Z}$ and $mc \in \mathbb{Z}$ imply $|m| \geq 2$ (since $c \notin \mathbb{Z}$) and $|\tau \cdot m\alpha| > C$.
- If $c \in [1.55, 1.95]$, a computation gives $|\tau \cdot \alpha| > \frac{C}{3}$. Moreover, in this case $m \in \mathbb{Z}$ and $mc \in \mathbb{Z}$ imply $|m| \geq 3$ (since $2c \notin \mathbb{Z}$) and we get $|\tau \cdot m\alpha| > C$.
- If $c \in [1.3, 1.65] \setminus \{4/3, 3/2\}$, a computation gives $|\tau \cdot \alpha| > \frac{C}{4}$. Moreover, in this case $m \in \mathbb{Z}$ and $mc \in \mathbb{Z}$ imply $|m| \geq 4$ (since $3c, 2c \notin \mathbb{Z}$) and we get $|\tau \cdot m\alpha| > C$.
- If $c \in]1, 1.35] \setminus \{5/4, 4/3\}$, a computation gives $|\tau \cdot \alpha| > \frac{C}{5}$. Moreover, in this case $m \in \mathbb{Z}$ and $mc \in \mathbb{Z}$ imply $|m| \geq 5$ (since $4c, 3c, 2c \notin \mathbb{Z}$) and we get $|\tau \cdot m\alpha| > C$.

Let us show that (1) is true if $|n| \geq 4$ and $m \in \mathbb{Z}$. We have

$$|m\tau \cdot \alpha + n\tau \cdot \beta| = |n||\tau \cdot \alpha| \left| \frac{m}{n} + \frac{\tau \cdot \beta}{\tau \cdot \alpha} \right| \geq |n||\tau \cdot \alpha| \left| \mathrm{Im}\left(\frac{m}{n} + \frac{\tau \cdot \beta}{\tau \cdot \alpha} \right) \right| \geq k|n|, \tag{2}$$

where $k = |\tau \cdot \alpha| \left| \mathrm{Im}\left(\frac{\tau \cdot \beta}{\tau \cdot \alpha} \right) \right|$. Numerical computations give:

$$k^2 \doteq \frac{1}{c^2 - 2.12228\,c + 1.27676} \Big(0.217137\,d^2 + 0.533079\,dc + 0.327181c^2$$
$$+ 0.217127\,d - 0.911556c + 0.634921 \Big),$$

$$k^2 - \left(\frac{C}{4} \right)^2 \doteq \frac{1}{c^2 - 2.12228\,c + 1.27676} \Big(0.257151 + 0.0312991c^2$$
$$+ c(-0.283614 + 0.533079d) + (-0.742604 + 0.217137d)d \Big).$$

The denominator $c^2 - 2.12228\,c + 1.27676$ is positive for any real c. Thus the sign of $k^2 - \left(\frac{C}{4}\right)^2$ is the same as the sign of the numerator. For a given d, the minimum of the numerator is reached for $c \doteq 0.00443843 - 0.00834245d < 0$ (since $d > 1$). Thus the numerator is an increasing function of c for $c > 0$ and in particular for fixed d and $1 \le c < d$ the minimum is reached at $c = 1$ and is given by $0.00483619 + (-0.209525 + 0.217137d)d$ which is positive since $d > 1$. We conclude that $k > \frac{C}{4}$. We use Eq. (2) to get that if $|n| \ge 4$, then $|m\tau \cdot \alpha + n\tau \cdot \beta| > C$.

It remains to deal with the cases $|n| \in \{1, 2, 3\}$. It is enough in (1) to consider the cases $n \in \{1, 2, 3\}$. We treat each case in a similar way. Let us start with the case $n = 1$. We get numerically

$$
\begin{aligned}
P_{c,d,1}(m) &:= |\tau \cdot (m\alpha + \beta)|^2 - C^2 \\
&\doteq -4.16782 + 0.712407m - 1.17373cm + 1.83908m^2 - 3.05698cm^2 \\
&\quad + 1.44043c^2m^2 + (-1.17373 - 3.05698m + 2.88085cm)d + 1.44043d^2.
\end{aligned}
$$

$P_{c,d,1}(m)$ is a quadratic polynomial in d. Computing the discriminant yields $P_{c,d,1}(m) > 0$, for all $c \in \mathbb{R}$ if and only if $\Delta_c(d) \doteq 25.3914 + 3.07144m - 1.25108m^2 < 0$. This is a quadratic inequality[2] in m and solving it yields

$$
m \notin [-3.44178, 5.89681] \implies |\tau \cdot (m\alpha + \beta)| > C.
$$

Thus we only need to check that for every $m \in \{5, 4, 3, 2, 1, 0, -1, -2, -3\}$ such that $mc + d \in \mathbb{Z}$, $P_{c,d,1}(m) > 0$. Let us detail the cases $m = -3$ and $m = 4$. Numerically, we get $P_{c,d,1}(-3) \doteq 10.2467 + 12.9638c^2 + c(-23.9917 - 8.64256d) + d(7.99723 + 1.44043d)$. This is a quadratic polynomial in d and we deduce[3] that

$$
P_{c,d,1}(-3) > 0 \iff d \in \,]-\infty, 3c - 3.54573[\,\cup\,]3c - 2.00625, \infty[.
$$

Thus, in particular, since by hypothesis $d \ne 3c - 3$ then either $P_{c,d,1}(-3) > 0$ or $d \in [3c - 3.54573, 3c - 2.00625]$ and then $-3c + d \notin \mathbb{Z}$. The condition $P_{c,d,1}(4) > 0$ is equivalent to $d \in \,]6.1107 - 4c, \infty[$. Since $d > c > 1$ and $d \ne 6 - 4c$ then either $P_{c,d,1}(4) > 0$ or $d + 4c \notin \mathbb{Z}$. The other cases are similar. We give, in Table 1, for each of them the condition on the reals and the assumptions that allow us to conclude.

---

[2] We remark that it is no numerical coincidence that c does not appear in the expression of $\Delta_c(d)$. It follows from $P_{c,d,1}(m) = (x + ym + z(d + cm))^2 + (x' + y'm + z'(d + cm))^2 - C^2$ with $x, y, z \in \mathbb{R}$.

[3] As for the previous note, it is no numerical coincidence that there is no complicated square root involving c since c does not appear in the discriminant.

Table 1. Study of $P_{c,d,1}(m)$ for $m \in \{5, 4, 3, 2, 1, 0, -1, -2, -3\}$.

(A): an equivalent condition for d	A sufficient condition to get (A)
$P_{c,d,1}(5) > 0 \Leftrightarrow d \in]6.78141 - 5c, \infty[$	$d > c > 1$
$P_{c,d,1}(4) > 0 \Leftrightarrow d \in]6.1107 - 4c, \infty[$	$d > c > 1$ and $d \neq 6 - 4c$
$P_{c,d,1}(3) > 0 \Leftrightarrow d \in]5.26804 - 3c, \infty[$	$d > c > 1$ and $d \neq 5 - 3c$
$P_{c,d,1}(2) > 0 \Leftrightarrow d \in]4.31762 - 2c, \infty[$	$d > c > 1$ and $d \neq 4 - 2c$
$P_{c,d,1}(1) > 0 \Leftrightarrow d \in]3.27931 - c, \infty[$	$d > c > 1$ and $d \neq 3 - c$
$P_{c,d,1}(0) > 0 \Leftrightarrow d \notin [-1.34171, 2.15655]$	$d > c > 1$ and $d \neq 2$
$P_{c,d,1}(-1) > 0 \Leftrightarrow d \in]c + 0.939592, \infty[$	$d > c$
$P_{c,d,1}(-2) > 0$ $\Leftrightarrow d \notin [2c - 3.02493, 2c - 0.404774]$	$d \notin \{2c - 3, 2c - 2, 2c - 1\}$
$P_{c,d,1}(-3) > 0$ $\Leftrightarrow d \notin [3c - 3.54573, 3c - 2.00625]$	$d \neq 3c - 3$

The next case is $n = 2$ and we treat it in a similar fashion. We get numerically

$$P_{c,d,2}(m) := |\tau \cdot (m\alpha + 2\beta)|^2 - C^2$$
$$\doteq -2.46898 + 1.42481m - 2.34745cm + 1.83908m^2 - 3.05698cm^2$$
$$+ 1.44043c^2m^2 + (-4.6949 - 6.11397m + 5.76171cm)d + 5.76171d^2.$$

Computing the discriminant yields $P_{c,d,2}(m) > 0$, for all $d \in \mathbb{R}$ if and only if $\Delta_c(d) := 78.9442 + (24.5715 - 5.00433m)m < 0$. This is a quadratic inequality in m and solving it yields $m \notin [-2.21427, 7.12433] \implies |\tau \cdot (m\alpha + \beta)| > C$. Thus we only need to check that, under the assumptions, for every $m \in \{7, 6, 5, 4, 3, 2, 1, 0, -1, -2\}$ such that $mc + 2d \in \mathbb{Z}$, $P_{c,d,2}(m) > 0$. Each case is similar to the cases with $n = 1$. We give, in Table 2, for each of them the condition on the reals and the assumptions that allow us to conclude. The only remaining case is $n = 3$ and we treat it in a similar fashion. We get numerically

$$P_{c,d,3}(m) := |\tau \cdot (m\alpha + 2\beta)|^2 - C^2$$
$$\doteq 0.362434 + 2.13722m - 3.52118bm + 1.83908m^2 - 3.05698cm^2$$
$$+ 1.44043c^2m^2 + (-10.5635 - 9.17095m + 8.64256cm)d + 12.9638d^2.$$

Computing the discriminant yields $P_{c,d,3}(m) > 0$, for all $d \in \mathbb{R}$ if and only if $\Delta_c(d) := 92.7941 + 82.929m - 11.2597m^2 < 0$. This is a quadratic inequality in m and solving it yields $m \notin [-0.986756, 8.35184] \implies |\tau \cdot (m\alpha + \beta)| > C$. Thus we only need to check that, under the assumptions, for every $m \in \{8, 7, 6, 5, 4, 3, 2, 1, 0\}$ such that $mc + 3d \in \mathbb{Z}$, $P_{c,d,3}(m) > 0$. Each case is similar to the cases $n = 1, 2$. We give, in Table 3, for each of them the condition on the reals and the assumptions that allow us to conclude. This finishes the proof of Theorem 2.

Table 2. Theorem 2: Study of $P_{c,d,2}(m)$ for $m \in \{7,6,5,4,3,2,1,0,-1,-2\}$.

(B_1): an equivalent condition for d	A sufficient condition to get (B_1)
$P_{c,d,2}(7) > 0 \Leftrightarrow d \notin]3.91363 - 3.5c, 4.32919 - 3.5c[$	$d > c > 1$
$P_{c,d,2}(6) > 0 \Leftrightarrow d \notin]3.00088 - 3c, 4.1808 - 3c[$	$d > c > 1$
$P_{c,d,2}(5) > 0 \Leftrightarrow d \notin]2.30029 - 2.5c, 3.82024 - 2.5c[$	$d > c > 1$
$P_{c,d,2}(4) > 0 \Leftrightarrow d \notin]1.67431 - 2c, 3.38509 - 2c[$	$d > c > 1$
$P_{c,d,2}(3) > 0 \Leftrightarrow d \notin]1.09888 - 1.5c, 2.89938 - 1.5c[$	$d > c > 1$
$P_{c,d,2}(2) > 0 \Leftrightarrow d \notin]0.566427 - c, 2.3707 - c[$	$d > c > 1$
$P_{c,d,2}(1) > 0 \Leftrightarrow d \notin]0.0766769 - 0.5c, 1.79931 - 0.5c[$	$d > c > 1$
$P_{c,d,2}(0) > 0 \Leftrightarrow d \notin] - 0.363621, 1.17847[$	$d > c > 1$
$P_{c,d,2}(-1) > 0 \Leftrightarrow d \notin [-0.732884 + 0.5c, 0.486592 + 0.5c]$	$d > c$
$P_{c,d,2}(-2) > 0 \Leftrightarrow d \notin [-0.925154 + c, -0.382276 + c]$	$d > c$

In fact, using a symmetry argument we can improve the previous result.

Theorem 3. *For any* $(c,d) \in \mathbb{R}^2 \backslash \mathcal{F}$ *additive cubes are avoidable over* $\{0,1,c,d\}$ *where*

$$\mathcal{F} = \left\{ \left(\frac{10}{9}, \frac{14}{9}\right), \left(\frac{9}{8}, \frac{3}{2}\right), \left(\frac{9}{8}, \frac{13}{8}\right), \left(\frac{8}{7}, \frac{10}{7}\right), \left(\frac{8}{7}, \frac{11}{7}\right), \left(\frac{8}{7}, \frac{12}{7}\right), \left(\frac{7}{6}, \frac{11}{6}\right), \right.$$
$$\left(\frac{7}{6}, \frac{3}{2}\right), \left(\frac{7}{6}, \frac{5}{3}\right), \left(\frac{6}{5}, \frac{8}{5}\right), \left(\frac{6}{5}, \frac{9}{5}\right), \left(\frac{6}{5}, 2\right), \left(\frac{5}{4}, \frac{7}{4}\right), \left(\frac{5}{4}, 2\right), \left(\frac{5}{4}, \frac{9}{4}\right),$$
$$\left(\frac{5}{4}, \frac{5}{2}\right), \left(\frac{5}{4}, \frac{11}{4}\right), \left(\frac{5}{4}, 3\right), \left(\frac{5}{4}, \frac{13}{4}\right), \left(\frac{5}{4}, \frac{7}{2}\right), \left(\frac{4}{3}, 2\right), \left(\frac{4}{3}, \frac{7}{3}\right), \left(\frac{4}{3}, \frac{8}{3}\right),$$
$$\left(\frac{4}{3}, 3\right), \left(\frac{4}{3}, \frac{10}{3}\right), \left(\frac{4}{3}, \frac{11}{3}\right), \left(\frac{4}{3}, 4\right), \left(\frac{3}{2}, \frac{5}{2}\right), \left(\frac{3}{2}, 3\right), \left(\frac{3}{2}, \frac{7}{2}\right), \left(\frac{3}{2}, 4\right),$$
$$\left.\left(\frac{3}{2}, \frac{9}{2}\right), \left(\frac{3}{2}, 5\right), (4,5) \right\}$$
$$\cup \left(\{(2,t), (t, 2t-2), (t, 2t-1), (t, 3t-3) : t \in \mathbb{R}\} \cap \{(c,d) : d > c > 1\} \right).$$

Proof. Let $\mathcal{X}$ be the following set of pairs of parametric equations:

$$\mathcal{X} = \{(5/4, t), (4/3, t), (3/2, t), (2, t), (t, 6-4t), (t, 5-3t), (t, 4-2t),$$
$$(t, 3-t), (t, 2t-3), (t, 2t-2), (t, 2t-1), (t, 3t-3), (t, 2) : t \in \mathbb{R}\}.$$

For any pair $e = (x(t), y(t))$ of parametric equations, we denote by $\mathcal{C}(e)$ the associated parametric curve (that is the set of points defined by $\{(x(t), y(t)) : t \in \mathbb{R}\}$). By the Theorem 2 for any $c, d \in \mathbb{R}$ with $c > d > 1$ and $(c,d) \notin \bigcup_{e \in \mathcal{X}} \mathcal{C}(e)$ additive cubes are avoidable over $\{0,1,c,d\}$. Moreover, for any $c, d \in \mathbb{R}$ with $d > c > 1$, the alphabet $\{0,1,c,d\}$ is equivalent to the alphabet $\{0, 1, \frac{d-1}{d-c}, \frac{d}{d-c}\}$ (via the affine map $x \mapsto \frac{d-x}{d-c}$). Let $f : \mathbb{R}^2 \to \mathbb{R}^2$, $(x,y) \mapsto \left(\frac{y-1}{y-x}, \frac{y}{y-x}\right)$. We

Table 3. Theorem 2: Study of $P_{c,d,3}(m)$ for $m \in \{8, 7, 6, 5, 4, 3, 2, 1, 0\}$.

(B_2): an equivalent condition for d	A sufficient condition to get (B_2)
$P_{c,d,3}(8) > 0 \Leftrightarrow d \notin]3.00699 - 2.66667c, 3.46726 - 2.66667c[$	$d > c > 1$
$P_{c,d,3}(7) > 0 \Leftrightarrow d \notin]2.45816 - 2.33333c, 3.30867 - 2.33333c[$	$d > c > 1$
$P_{c,d,3}(6) > 0 \Leftrightarrow d \notin]2.00508 - 2c, 3.05432 - 2c[$	$d > c > 1$
$P_{c,d,3}(5) > 0$ $\Leftrightarrow d \notin]1.59624 - 1.66667c, 2.75573 - 1.66667c[$	$d > c > 1$
$P_{c,d,3}(4) > 0$ $\Leftrightarrow d \notin]1.21937 - 1.33333c, 2.42518 - 1.33333c[$	$d > c > 1$
$P_{c,d,3}(3) > 0 \Leftrightarrow d \notin]0.870753 - c, 2.06637 - c[$	$d > c > 1$
$P_{c,d,3}(2) > 0$ $\Leftrightarrow d \notin]0.551146 - 0.666667c, 1.67855 - 0.666667c[$	$d > c > 1$
$P_{c,d,3}(1) > 0$ $\Leftrightarrow d \notin]0.266517 - 0.333333c, 1.25575 - 0.333333c[$	$d > c > 1$
$P_{c,d,3}(0) > 0 \Leftrightarrow d \notin]0.0358908, 0.778955[$	$d > c > 1$

deduce that for any $c, d \in \mathbb{R}$ with $d > c > 1$ and $(c, d) \notin \bigcup_{e \in \mathcal{X}} \mathcal{C}(f \circ e)$ additive cubes are avoidable over $\{0, 1, c, d\}$. Let

$$\mathcal{F} = \left(\bigcup_{e \in \mathcal{X}} \mathcal{C}(f \circ e) \right) \cap \left(\bigcup_{e \in \mathcal{X}} \mathcal{C}(e) \right) \cap \{(c, d) : d > c > 1\}. \tag{3}$$

Then, for any $c, d \in \mathbb{R}$ with $d > c > 1$ and $(c, d) \notin \mathcal{F}$ additive cubes are avoidable over $\{0, 1, c, d\}$. Let us now compute $\mathcal{F}$. First, one computes

$$\mathcal{C}(\{f \circ e : e \in \mathcal{X}\}) = \mathcal{C}\Big(\Big\{(t, 6t - 4), (t, 5t - 3), (t, 4t - 2), (t, 3t - 1), (t, \tfrac{3}{2}t - 1),$$
$$(t, 2(t - 1)), (2, t), (t, 3(t - 1)), (t, 2t), (t, 5t - 4),$$
$$(t, 4t - 3), (t, 3t - 2), (t, 2t - 1)\Big\}\Big).$$

We get the set from Theorem 3 by simply computing the intersection of the two sets in (3) (this is done by solving the 169 equations).

5   The Case of $\mathbf{W}_{1,0,c,d}$

We show the next result by using a similar procedure as the one in the proof of Theorem 2 in Sect. 4.

Theorem 4. Let $c, d \in \mathbb{R}$. Suppose we have $d > c > 1$, $d \notin \{2, c + 1, c + 2, 2c + 2, 2c + 1, 2c, 3c, 3c + 1, 1 + \tfrac{c}{2}, \tfrac{1}{2} + c\}$. Then $\mathbf{W}_{1,0,c,d}$ avoids additive cubes.

Proof. Following the proof of Theorem 2, we only need to show that, under the assumptions, for any $m, n \in \mathbb{Z}$ with $mc + nd \in \mathbb{Z}$, we have $|\tau \cdot (m\alpha + n\beta)| > C$, where $\alpha = (-c, c-1, 1, 0)$ and $\beta = (-d, d-1, 0, 1)$.

Let us first show that this is the case if $n = 0$. Two subcases occur:

- If $c > 1.71$ a computation gives $|\tau \cdot \alpha| > C$ and $|\tau \cdot m\alpha| > C$.
- If $c \in]1, 2[$, a computation gives $|\tau \cdot \alpha| > \frac{C}{2}$. Moreover, in this case $m \in \mathbb{Z}$ and $mc \in \mathbb{Z}$ imply $|m| \geq 2$ (since $c \notin \mathbb{Z}$) and we get $|\tau \cdot m\alpha| \geq C$.

Let us now show that $|\tau \cdot (m\alpha + n\beta)| > C$ if $|n| \geq 4$ and $m \in \mathbb{Z}$. The same computation as (2) gives:

$$|m\tau \cdot \alpha + n\tau \cdot \beta| \geq k|n|, \quad \text{where} \quad k = |\tau \cdot \alpha| \left| \operatorname{Im} \left(\frac{\tau \cdot \beta}{\tau \cdot \alpha} \right) \right|. \tag{4}$$

The same approach as in proof of Theorem 2 can be used to verify that $k^2 - (\frac{C}{4})^2 > 0$ for any $d > c > 1$. This gives with inequality (4) that if $|n| \geq 4$, then $|m\tau \cdot \alpha + n\tau \cdot \beta| > C$. It remains to treat the cases $|n| \in \{1, 2, 3\}$ but it is enough to consider the cases $n \in \{1, 2, 3\}$, as previously. We start with the case $n = 1$. Once again $P_{c,d,1}(m) := |\tau \cdot (m\alpha + \beta)|^2 - C^2$ is a quadratic polynomial in d. Computing its discriminant yields $P_{c,d,1}(m) > 0$, for all $c \in \mathbb{R}$ if and only if $\Delta_c(d) := 25.3914 + 3.07144m - 1.25108m^2 < 0$. This is a quadratic inequality in m and solving it yields $m \notin [-3.44178, 5.89681] \implies |\tau \cdot (m\alpha + \beta)| > C$ (the conditions on m happen to be exactly the same as in Sect. 4). Thus we only need to check that for every $m \in \{5, 4, 3, 2, 1, 0, -1, -2, -3\}$ such that $mc + d \in \mathbb{Z}$, we have $P_{c,d,1}(m) > 0$.

All the cases are similar to what we did in the previous proof. We give, in Table 4, for each of them the condition on the reals and the assumptions that allow us to conclude. The next case is $n = 2$ and we treat it in a similar fashion.

We verify that the only interesting cases are $m \notin [-2.21427, 7.12433]$. Thus we only need to check that for every $m \in \{7, 6, 5, 4, 3, 2, 1, 0, -1, -2\}$ such that $mc + 2d \in \mathbb{Z}$, we get $P_{c,d,2}(m) > 0$. Each case is similar to the cases with $n = 1$. We omit here, because of the lack of space, the table that give for each of them the condition on the reals and the assumptions that allow us to conclude. This table can be found at https://members.loria.fr/FLietard/tables-of-values/.

The only remaining case is $n = 3$. We once again compute the discriminant of $P_{c,d,3}(m)$ seen as a polynomial in d. We deduce that $m \notin [-0.986756, 8.35184] \implies |\tau \cdot (m\alpha + \beta)| > C$.

Summing up, we only need to check that for every $m \in \{8, 7, 6, 5, 4, 3, 2, 1, 0\}$ such that $mc + 3d \in \mathbb{Z}$, $P_{c,d,3}(m) > 0$. We prove this statement by solving each of the corresponding 9 equations and this concludes the proof of Theorem 4.

We could improve this result with the same approach as the one we used in the proof of Theorem 3, but we already have a strong enough result for our purpose.

Table 4. Theorem 4: Study of $P_{c,d,1}(m)$ for $m \in \{5, 4, 3, 2, 1, 0, -1, -2, -3\}$.

(C_1): an equivalent condition for d	A sufficient condition to get (C_1)
$P_{c,d,1}(5) > 0 \Leftrightarrow d \notin] - 0.781405 - 5c, 1.35518 - 5c[$	$d > c > 1$
$P_{c,d,1}(4) > 0 \Leftrightarrow d \notin] - 1.1107 - 4c, 1.80675 - 4c[$	$d > c > 1$
$P_{c,d,1}(3) > 0 \Leftrightarrow d \notin] - 1.26804 - 3c, 2.08636 - 3c[$	$d > c > 1$
$P_{c,d,1}(2) > 0 \Leftrightarrow d \notin] - 1.31762 - 2c, 2.25822 - 2c[$	$d > c > 1$
$P_{c,d,1}(1) > 0 \Leftrightarrow d \notin] - 1.27931 - c, d > 2.34218 - c[$	$d > c > 1$
$P_{c,d,1}(0) > 0 \Leftrightarrow d \notin [-1.15655, 2.34171]$	$d > c > 1$ and $d \neq 2$
$P_{c,d,1}(-1) > 0 \Leftrightarrow d \notin] - 0.939592 + c, 2.24702 + c[$	$d > c$ and $d \notin \{c+1, c+2\}$
$P_{c,d,1}(-2) > 0 \Leftrightarrow d \notin] - 0.595226 + 2c, 2.02493 + 2c[$	$d \notin \{2c + 2, 2c + 1, 2c\}$
$P_{c,d,1}(-3) > 0 \Leftrightarrow d \notin]0.00625218 + 3c, 1.54573 + 3c)[$	$d \notin \{3c, 3c + 1\}$

6   The Remaining Alphabets

We use Theorem 3 and Theorem 4 to get the following result:

Theorem 5. *Let*

$$\mathcal{F} = \left\{ \left(\frac{10}{9}, \frac{14}{9}\right), \left(\frac{9}{8}, \frac{13}{8}\right), \left(\frac{8}{7}, \frac{11}{7}\right), \left(\frac{7}{6}, \frac{5}{3}\right), \left(\frac{6}{5}, \frac{8}{5}\right), \left(\frac{6}{5}, 2\right), \left(\frac{5}{4}, \frac{7}{4}\right), \right.$$
$$\left(\frac{5}{4}, 2\right), \left(\frac{5}{4}, \frac{9}{4}\right), \left(\frac{5}{4}, \frac{5}{2}\right), \left(\frac{5}{4}, \frac{13}{4}\right), \left(\frac{5}{4}, \frac{7}{2}\right), \left(\frac{4}{3}, 2\right), \left(\frac{4}{3}, \frac{7}{3}\right),$$
$$\left(\frac{4}{3}, \frac{8}{3}\right), \left(\frac{4}{3}, \frac{10}{3}\right), \left(\frac{4}{3}, \frac{11}{3}\right), \left(\frac{3}{2}, \frac{5}{2}\right), \left(\frac{3}{2}, 3\right), \left(\frac{3}{2}, \frac{7}{2}\right), (4, 5), \left(\frac{4}{3}, \frac{5}{3}\right),$$
$$\left(\frac{3}{2}, 2\right), \left(\frac{8}{5}, \frac{9}{5}\right), \left(\frac{5}{3}, 2\right), \left(\frac{7}{4}, \frac{9}{4}\right), \left(2, \frac{5}{2}\right), (2, 3), (2, 4), (2, 5), \left(\frac{5}{2}, 3\right),$$
$$\left. \left(\frac{5}{2}, \frac{9}{2}\right), (3, 4), (3, 5), (3, 6), (4, 6), (4, 9) \right\}.$$

and $(c, d) \in \mathbb{R}^2 \setminus \mathcal{F}$. *Then additive cubes are avoidable over* $\{0, 1, c, d\}$.

Proof. This set is obtained by taking the intersection of the sets of forbidden pairs from Theorem 3 and Theorem 4.

In order to study the remaining alphabets (those of the form $\{0, 1, c, d\}$ with $c, d \in \mathcal{F}$) let us recall the following results from the literature.

Theorem 6 ([7, Section 3.2]). *Additive cubes are avoidable over the following alphabets:* $\{0, 1, 5\}, \{0, 1, 6\}, \{0, 1, 7\}, \{0, 2, 7\}, \{0, 3, 7\}, \{0, 1, 8\}, \{0, 3, 8\}, \{0, 1, 9\}, \{0, 2, 9\}, \{0, 4, 9\}.$

Theorem 7 ([8, **Theorem 9**]). *Additive cubes are avoidable over the following alphabets:* $\{0, 2, 3, 6\}, \{0, 1, 2, 4\}, \{0, 2, 3, 5\}$.

We use the fact that all but one of the remaining alphabets contain an alphabet equivalent to an alphabet from Theorem 6 or Theorem 7. Our main result after this reduction is the following:

Theorem 8. *For any rational numbers* c *and* d *with* $c < d$ *and* $(c, d) \neq (2, 3)$ *additive cubes are avoidable over* $\{0, 1, c, d\}$.

Proof. $\{0, 1, \frac{10}{9}, \frac{14}{9}\}$ contains an alphabet equivalent to $\{0, 1, 5\}$ (apply $x \mapsto 9x - 9$ to $\{1, \frac{10}{9}, \frac{14}{9}\}$). We deduce from Theorem 6 that additive cubes are avoidable over both alphabets. We proceed in a same way for the other alphabets and we provide for each of them the alphabet from Theorem 6 or from Theorem 7 in Table 5. This concludes the proof.

Table 5. Each remaining alphabet, with exception of $\{0, 1, 2, 3\}$, contains an alphabet equivalent to an alphabet from Theorems 6 or 7.

$(\frac{10}{9}, \frac{14}{9}), (\frac{5}{4}, \frac{7}{2}), (\frac{4}{3}, \frac{5}{3}), (4, 5), (2, \frac{5}{2}), (2, 5), (3, 5)$	$\{0, 1, 5\}$
$(\frac{6}{5}, \frac{8}{5}), (\frac{6}{5}, 2), (\frac{5}{4}, \frac{5}{2}), (\frac{5}{3}, 2), (\frac{5}{2}, 3), (3, 6), (4, 6)$	$\{0, 1, 6\}$
$(\frac{7}{6}, \frac{5}{3}), (\frac{4}{3}, \frac{10}{3})$	$\{0, 1, 7\}$
$(\frac{3}{2}, \frac{7}{2})$	$\{0, 2, 7\}$
$(\frac{5}{4}, \frac{7}{4}), (\frac{4}{3}, \frac{7}{3})$	$\{0, 3, 7\}$
$(\frac{8}{7}, \frac{11}{7}), (\frac{4}{3}, \frac{11}{3}), (\frac{3}{2}, \frac{5}{2})$	$\{0, 1, 8\}$
$(\frac{5}{4}, 2), (\frac{4}{3}, \frac{8}{3})$	$\{0, 3, 8\}$
$(\frac{9}{8}, \frac{13}{8}), (\frac{5}{4}, \frac{13}{4}), (\frac{8}{5}, \frac{9}{5})$	$\{0, 1, 9\}$
$(\frac{5}{2}, \frac{9}{2})$	$\{0, 2, 9\}$
$(\frac{5}{4}, \frac{9}{4}), (\frac{7}{4}, \frac{9}{4}), (4, 9)$	$\{0, 4, 9\}$
$(\frac{4}{3}, 2), (\frac{3}{2}, 3)$	$\{0, 2, 3, 6\}$
$(\frac{3}{2}, 2), (2, 4)$	$\{0, 1, 2, 4\}$
$(3, 4)$	$\{0, 1, 3, 4\}$

We reformulate this result in terms of Problem 1.

Corollary 1. *Let* $\mathcal{A} \subset \mathbb{C}$ *be an alphabet with* $|\mathcal{A}| \geq 4$. *If* $\mathcal{A}$ *is not equivalent to* $\{0, 1, 2, 3\}$ *then additive cubes are avoidable over* $\mathcal{A}$. *In particular, if* $|\mathcal{A}| \geq 5$, *then additive cubes are avoidable over* $\mathcal{A}$.

Note that we have shown that for all but finitely many integral alphabets of size 4 (up to the equivalence relation given in Sect. 2.1) the word $\mathbf{W}_{a,b,c,d}$ can be used to avoid additive cubes. This is probably not the only fixed point of a morphism with this property. Indeed, as long as the adjacency matrix of a morphism has at most two eigenvalues of norm at least 1, we can deduce an inequality similar to that of Theorem 1 (see Proposition 7 in [8] for details). If the word also avoids abelian cubes, we can show an inequality similar to this in Lemma 4. The conditions of this Lemma should be strong enough to study the lattice in a similar way to what we did.

Let us conclude by restating two remaining related open questions. First it is natural to ask whether additive cubes are avoidable over the only remaining alphabet (see Problem 7 in [8]).

Question 1. Are additive cubes avoidable over $\{0, 1, 2, 3\}$?

On the one hand, Rao [7] claims that he got a word of length 1.4×10^5 over the alphabet $\{0, 1, 2, 3\}$ without additive cubes. Damien Jamet, the first author and Thomas Stoll constructed over this alphabet several words of length greater than 10^7 without additive cubes (see https://members.loria.fr/FLietard/un-mot-sur-0123/ for such a word). Therefore, it seems to be reasonable to believe that there exists an infinite word without additive cubes over $\{0, 1, 2, 3\}$. On the other hand, for every alphabet $\{a, b, c, d\}$ different from $\{0, 1, 2, 3\}$ it is possible to provide a short morphism with the same eigenvalues as those of $\varphi_{a,b,c,d}$ or $\varphi_{a,b,c,d}^2$ with an infinite fixed point avoiding additive cubes. An exhaustive research shows, however, that every morphism over $\{0, 1, 2, 3\}$ with images of size at most 7 fails to provide an infinite fixed point without additive cubes. We do not dare to conjecture whether or not a morphism providing such an infinite word exists.

It seems that additive cubes are avoidable over most alphabets of size 3. Our result might stimulate research to treat the following question.

Question 2. Can we characterize the sets of integers of size 3 over which additive cubes are avoidable?

In fact, with the exception of $\{0, 1, 2, 3\}$, the alphabets of size three are the only remaining case of Problem 1 due to Lemma 2 and Theorem 8.

References

1. Cassaigne, J., Currie, J.D., Schaeffer, L., Shallit, J.: Avoiding three consecutive blocks of the same size and same sum. J. ACM **61**(2), 1–17 (2014). https://doi.org/10.1145/2590775
2. Dekking, F.: Strongly non-repetitive sequences and progression-free sets. J. Comb. Theory Ser. A **27**(2), 181–185 (1979). https://doi.org/10.1016/0097-3165(79)90044-X
3. Halbeisen, L., Hungerbühler, N.: An application of Van der Waerden's theorem in additive number theory. INTEGERS: Electron. J. Comb. Number Theory, **0**(A07) (2000)
4. Justin, J.: Généralisation du théorème de van der Waerden sur les semi-groupes répétitifs. J. Comb. Theory Ser. A **12**(3), 357–367 (1972). https://doi.org/10.1016/0097-3165(72)90101-X
5. Lothaire, M.: Algebraic Combinatorics on Words, Encyclopedia of Mathematics and its Applications, vol. 90. Cambridge University Press, Cambridge (2002)

6. Pirillo, G., Varricchio, S.: On uniformly repetitive semigroups. Semigroup Forum **49**(1), 125–129 (1994). https://doi.org/10.1007/BF02573477
7. Rao, M.: On some generalizations of abelian power avoidability. Theor. Comput. Sci. **601**, 39–46 (2015). https://doi.org/10.1016/j.tcs.2015.07.026
8. Rao, M., Rosenfeld, M.: Avoiding two consecutive blocks of same size and same sum over $\mathbb{Z}^2$. SIAM J. Discrete Math. **32**(4), 2381–2397 (2018). https://doi.org/10.1137/17M1149377
9. Rosenfeld, M.: Avoidability of Abelian Repetitions in Words. Ph.D. thesis, École Normale Supérieure de Lyon (2017)

Equivalence of Linear Tree Transducers with Output in the Free Group

Raphaela Löbel$^{(\boxtimes)}$, Michael Luttenberger, and Helmut Seidl

TU München, Munich, Germany
{loebel,luttenbe,seidl}@in.tum.de

Abstract. We show that equivalence of deterministic linear tree transducers can be decided in polynomial time when their outputs are interpreted over the free group. Due to the cancellation properties offered by the free group, the required constructions are not only more general, but also simpler than the corresponding constructions for proving equivalence of deterministic linear tree-to-word transducers.

Keywords: Linear tree transducer · Free group · Equivalence problem · Polynomial time

1 Introduction

In 2009, Staworko and Niehren observed that equivalence for *sequential* tree-to-word transducers [13] can be reduced to the morphism equivalence problem for context-free languages. Since the latter problem is decidable in polynomial time [10], they thus proved that equivalence of sequential tree-to-word transducers is decidable in polynomial time. This decision procedure was later accompanied by a canonical normal form which can be applied to learning [3,4]. Sequentiality of transducers means that subtrees must always be processed from left to right. This restriction was lifted by Boiret who provided a canonical normal form for unrestricted linear tree-to-word transducers [1]. Construction of that normal form, however, may require exponential time implying that the corresponding decision procedure requires exponential time as well. In order to improve on that, Palenta and Boiret provided a polynomial time procedure which just normalizes the *order* in which an unrestricted linear tree-to-word transducer processes the subtrees of its input [2]. They proved that after that normalization, equivalent transducers are necessarily *same-ordered*. As a consequence, equivalence of linear tree-to-word transducers can thus also be reduced to the morphism equivalence problem for context-free languages and thus can be decided in polynomial time. Independently of that, Seidl, Maneth and Kemper showed by algebraic means, that equivalence of general (possibly non-linear) tree-to-word transducers is decidable [11]. Their techniques are also applicable if the outputs of transducers are not just in a free monoid of words, but also if outputs are in a free group. The latter means that output words are considered as equivalent not just when they are literally equal, but also when they

© Springer Nature Switzerland AG 2020
N. Jonoska and D. Savchuk (Eds.): DLT 2020, LNCS 12086, pp. 207–221, 2020.
https://doi.org/10.1007/978-3-030-48516-0_16

become equal after cancellation of matching positive and negative occurrences of letters. For the special case of *linear* tree transducers with outputs either in a free monoid or a free group, Seidl et al. provided a *randomized* polynomial time procedure for in-equivalence. The question remained open whether for outputs in a free group, randomization can be omitted. Here, we answer this question to the affirmative. In fact, we follow the approach of [2] to normalize the order in which tree transducers produce their outputs. For that normalization, we heavily rely on commutation laws as provided for the free group. Due to these laws, the construction as well as the arguments for its correctness, are not only more general but also much cleaner than in the case of outputs in a free monoid only. The observation that reasoning over the free group may simplify arguments has also been made, e.g., by Tomita and Seino and later by Senizergues when dealing with the equivalence problem for deterministic pushdown transducers [12,14]. As morphism equivalence on context-free languages is decidable in polynomial time—even if the morphism outputs are in a free group [10], we obtain a polynomial time algorithm for equivalence of tree transducers with output in the free group.

Missing proofs can be found in the extended version of this paper [5].

2   Preliminaries

We use Σ to denote a finite *ranked* alphabet, while A is used for an *unranked* alphabet. $\mathcal{T}_\Sigma$ denotes the set of all trees (or terms) over Σ. The *depth* $\mathsf{depth}(t)$ of a tree $t \in \mathcal{T}_\Sigma$ equals 0, if $t = f()$ for some $f \in \Sigma$ of rank 0, and otherwise, $\mathsf{depth}(t) = 1 + \max\{\mathsf{depth}(t_i) \mid i = 1, \ldots, m\}$ for $t = f(t_1, \ldots, t_m)$. We denote by $\mathcal{F}_\mathsf{A}$ the representation of the free group generated by A where the carrier is the set of *reduced* words instead of the usual quotient construction: For each $a \in \mathsf{A}$, we introduce its *inverse* a^-. The set of elements of $\mathcal{F}_\mathsf{A}$ then consists of all words over the alphabet $\{a, a^- \mid a \in \mathsf{A}\}$ which do not contain $a\,a^-$ or a^-a as factors. These words are also called *reduced*. In particular, $\mathsf{A}^* \subseteq \mathcal{F}_\mathsf{A}$. The group operation "$\cdot$" of $\mathcal{F}_\mathsf{A}$ is concatenation, followed by *reduction*, i.e., repeated cancellation of subwords $a\,a^-$ or a^-a. Thus, $a\,b\,c^- \cdot c\,b^-\,a =_{\mathcal{F}_\mathsf{A}} a\,a$. The *neutral element* w.r.t. this operation is the empty word ε, while the inverse w^- of some element $w \in \mathcal{F}_\mathsf{A}$ is obtained by reverting the order of the letters in w while replacing each letter a with a^- and each a^- with a. Thus, e.g., $(a\,b\,c^-)^- = c\,b^-\,a^-$.

In light of the inverse operation $(.)^-$, we have that $v \cdot w =_{\mathcal{F}_\mathsf{A}} v'w'$ where $v = v'u$ (as words) for a maximal suffix u so that u^- is a prefix of w with $w = u^-w'$. For an element $w \in \mathcal{F}_\mathsf{A}$, $\langle w \rangle = \{w^l \mid l \in \mathbb{Z}\}$ denotes the cyclic subgroup of $\mathcal{F}_\mathsf{A}$ generated from w. As usual, we use the convention that $w^0 = \varepsilon$, and $w^{-l} = (w^-)^l$ for $l > 0$. An element $p \in \mathcal{F}_\mathsf{A}$ different from ε, is called *primitive* if $w^l =_{\mathcal{F}_\mathsf{A}} p$ for some $w \in \mathcal{F}_\mathsf{A}$ and $l \in \mathbb{Z}$ implies that $w =_{\mathcal{F}_\mathsf{A}} p$ or $w =_{\mathcal{F}_\mathsf{A}} p^-$, i.e., p and p^- are the only (trivial) roots of p. Thus, primitive elements generate *maximal* cyclic subgroups of $\mathcal{F}_\mathsf{A}$. We state two crucial technical lemmas.

Lemma 1. *Assume that* $y^m =_{\mathcal{F}_\mathsf{A}} \beta \cdot y^n \cdot \beta^-$ *with* $y \in \mathcal{F}_\mathsf{A}$ *primitive. Then* $m = n$, *and* $\beta =_{\mathcal{F}_\mathsf{A}} y^k$ *for some* $k \in \mathbb{Z}$.

Proof. Since $\beta \cdot y^n \cdot \beta^- =_{\mathcal{F}_A} (\beta \cdot y \cdot \beta^-)^n$, we find by [8, Proposition 2.17] a primitive element c such that y and $\beta \cdot y \cdot \beta^-$ are powers of c. As y is primitive, c can be chosen as y. Accordingly,

$$y^j =_{\mathcal{F}_A} \beta \cdot y \cdot \beta^- \tag{1}$$

holds for some j. If β is a power of y, then $\beta \cdot y \cdot \beta^- =_{\mathcal{F}_A} y$, and the assertion of the lemma holds. Likewise if $j = 1$, then β and y commute. Since y is primitive, then β necessarily must be a power of y.

For a contradiction, therefore now assume that β is not a power of y and $j \neq 1$. W.l.o.g., we can assume that $j > 1$. First, assume now that y is *cyclically reduced*, i.e., the first and last letters, a and b, respectively, of y are not mutually inverse. Then for each $n > 0$, y^n is obtained from y by n-concatenation of y as a word (no reduction taking place). Likewise, either the last letter of β is different a^- or the first letter of β^- is different from b^- because these two letters are mutually inverse. Assume that the former is the case. Then $\beta \cdot y$ is obtained by concatenation of β and y as words (no reduction taking place). By (1), $\beta \cdot y^n =_{\mathcal{F}_A} y^{j \cdot n} \cdot \beta$. for every $n \geq 1$. Let $m > 0$ denote the length of β as a word. Since β can cancel only a suffix of $y^{j \cdot n}$ of length at most m, it follows, that the word βy must a prefix of the word y^{m+1}. Since β is not a power of y, the word y can be factored into $y = y'c$ for some non-empty suffix c such that $\beta = y^{j'} y'$, implying that $yc = cy$ holds. As a consequence, $y = c^l$ for some $l > 1$—in contradiction to the irreducibility of y.

If on the other hand, the first letter of β^- is not the inverse of the last letter of y, then $y \cdot \beta^-$ is obtained as the concatenation of y and β^- as words. As a consequence, $y\beta^-$ is a suffix of y^{m+1}, and we arrive at a contradiction.

We conclude that the statement of the lemma holds whenever y is cyclically reduced. Now assume that y is not yet cyclically reduced. Then we can find a maximal suffix r of y (considered as a word) such that $y = r^- sr$ holds and s is cyclically reduced. Then s is also necessarily primitive. (If $s =_{\mathcal{F}_A} c^n$, then $y =_{\mathcal{F}_A} (r^- cr)^n$). Then assertion (1) can be equivalently formulated as

$$s^j =_{\mathcal{F}_A} (r \cdot \beta \cdot r^-) \cdot y \cdot (r \cdot \beta \cdot r^-)^-$$

We conclude that $r \cdot \beta \cdot r^- =_{\mathcal{F}_A} s^l$ for some $l \in \mathbb{Z}$. But then $\beta =_{\mathcal{F}_A} (r^- \cdot s \cdot r)^l =_{\mathcal{F}_A} y^l$, and the claim of the lemma follows.

Lemma 2. *Assume that x_1, x_2 and y_1, y_2 are distinct elements in $\mathcal{F}_A$ and that*

$$x_i \cdot \alpha \cdot y_j \cdot \beta =_{\mathcal{F}_A} \gamma \cdot y_j' \cdot \alpha' \cdot x_i' \cdot \beta' \tag{2}$$

holds for $i = 1, 2$ and $j = 1, 2$. Then there is some primitive element p and exponents $r, s \in \mathbb{Z}$ such that $x_1 \cdot \alpha =_{\mathcal{F}_A} x_2 \cdot \alpha \cdot p^r$ and $y_1 =_{\mathcal{F}_A} p^s \cdot y_2$.

Proof. By the assumption (2),

$$\gamma =_{\mathcal{F}_A} (x_1 \cdot \alpha \cdot y_j \cdot \beta) \cdot (y_j' \cdot \alpha' \cdot x_1' \cdot \beta')^-$$
$$=_{\mathcal{F}_A} (x_2 \cdot \alpha \cdot y_j \cdot \beta) \cdot (y_j' \cdot \alpha' \cdot x_2' \cdot \beta')^-$$

for all $j = 1, 2$. Thus,

$$x_1 \cdot \alpha \cdot y_j \cdot \beta\beta'^- x_1'^- \alpha'^- y_j'^- =_{\mathcal{F}_A} x_2 \cdot \alpha \cdot y_j \cdot \beta \cdot \beta'^- \cdot x_2'^- \cdot \alpha'^- \cdot y_j'^- \quad \text{implying}$$
$$y_j^- \cdot \alpha^- \cdot x_2^- \cdot x_1 \cdot \alpha \cdot y_j =_{\mathcal{F}_A} \beta \cdot \beta'^- \cdot x_2'^- \cdot x_1' \cdot \beta' \cdot \beta^-$$

for $j = 1, 2$. Hence,

$$y_1^- \cdot \alpha^- \cdot x_2^- \cdot x_1 \cdot \alpha \cdot y_1 =_{\mathcal{F}_A} y_2^- \cdot \alpha^- \cdot x_2^- \cdot x_1 \cdot \alpha \cdot y_2 \quad \text{implying}$$
$$(x_2 \cdot \alpha)^- x_1 \cdot \alpha =_{\mathcal{F}_A} (y_1 \cdot y_2^-) \cdot ((x_2 \cdot \alpha)^- \cdot x_1 \cdot \alpha) \cdot (y_1 \cdot y_2^-)^-$$

Since x_1 is different from x_2, also $x_1 \cdot \alpha$ is different from $x_2 \cdot \alpha$. Let p denote a primitive root of $(x_2 \cdot \alpha)^- \cdot x_1 \cdot \alpha$. Then by Lemma 1,

$$x_1 \cdot \alpha =_{\mathcal{F}_A} x_2 \cdot \alpha \cdot p^r$$
$$y_1 =_{\mathcal{F}_A} p^s \cdot y_2$$

for suitable exponents $r, s \in \mathbb{Z}$.

As the elements of $\mathcal{F}_A$ are words, they can be represented by *straight-line* programs (SLPs). An SLP is a context-free grammar where each non-terminal occurs as the left-hand side of exactly one rule. We briefly recall basic complexity results for operations on elements of $\mathcal{F}_A$ when represented as SLPs [7].

Lemma 3. *Let U, V be SLPs representing words $w_1, w_2 \in \{a, a^- \mid a \in A\}$, respectively. Then the following computations/decision problems can be realized in polynomial time*

- *compute an SLP for w_1^-;*
- *compute the primitive root of w_1 if $w_1 \neq \varepsilon$;*
- *compute an SLP for $w =_{\mathcal{F}_A} w_1$ with w reduced;*
- *decide whether $w_1 =_{\mathcal{F}_A} w_2$;*
- *decide whether it exists $g \in \mathcal{F}_A$, such that $w_1 \in g \cdot \langle w_2 \rangle$ and compute an SLP for such g.*

In the following, we introduce *deterministic linear tree transducers* which produce outputs in the free group $\mathcal{F}_A$. For convenience, we follow the approach in [11] where only *total* deterministic transducers are considered—but equivalence is relative w.r.t. some *top-down deterministic* domain automaton B. A top-down deterministic automaton (DTA) B is a tuple $(H, \Sigma, \delta_B, h_0)$ where H is a finite set of states, Σ is a finite ranked alphabet, $\delta_B : H \times \Sigma \to H^*$ is a partial function where $\delta_B(h, f) \in H^k$ if the the rank of f equals k, and h_0 is the start state of B. For every $h \in H$, we define the set $\mathrm{dom}(h) \subseteq \mathcal{T}_\Sigma$ by $f(t_1, \ldots, t_m) \in \mathrm{dom}(h)$ iff $\delta_B(h, f) = h_1 \ldots h_m$ and $t_i \in \mathrm{dom}(h_i)$ for all $i = 1, \ldots, k$. B is called *reduced* if $\mathrm{dom}(h) \neq \emptyset$ for all $h \in H$. The *language* $\mathcal{L}(B)$ accepted by B is the set $\mathrm{dom}(h_0)$. We remark that for every DTA B with $\mathcal{L}(B) \neq \emptyset$, a reduced DTA B' can be constructed in polynomial time with $\mathcal{L}(B) = \mathcal{L}(B')$. Therefore, we subsequently assume w.l.o.g. that each DTA B is reduced.

A (total deterministic) *linear tree transducer* with output in $\mathcal{F}_A$ (LT$_A$ for short) is a tuple $M = (\Sigma, A, Q, S, R)$ where Σ is the ranked alphabet for the input trees, A is the finite (unranked) output alphabet, Q is the set of *states*, S is the *axiom* of the form u_0 or $u_0 \cdot q_0(x_0) \cdot u_1$ with $u_0, u_1 \in \mathcal{F}_A$ and $q_0 \in Q$, and R is the set of *rules* which contains for each state $q \in Q$ and each input symbol $f \in \Sigma$, one rule of the form

$$q(f(x_1,\ldots,x_m)) \rightarrow u_0 \cdot q_1(x_{\sigma(1)}) \cdot \ldots \cdot u_{n-1} \cdot q_n(x_{\sigma(n)}) \cdot u_n \qquad (3)$$

Here, m is the rank of f, $n \leq m$, $u_0,\ldots,u_n \in \mathcal{F}_A$ and σ is an injective mapping from $\{1,\ldots,n\}$ to $\{1,\ldots,m\}$. The *semantics* of a state q is the function $\llbracket q \rrbracket : \mathcal{T}_\Sigma \rightarrow \mathcal{F}_A$ defined by

$$\llbracket q \rrbracket(f(t_1,\ldots,t_m)) =_{\mathcal{F}_A} u_0 \cdot \llbracket q_1 \rrbracket(t_{\sigma(1)}) \cdot \ldots \cdot u_{n-1} \cdot \llbracket q_n \rrbracket(t_{\sigma(n)}) \cdot u_n$$

if there is a rule of the form (3) in R. Then the *translation* of M is the function $\llbracket M \rrbracket : \mathcal{T}_\Sigma \rightarrow \mathcal{F}_A$ defined by $\llbracket M \rrbracket(t) =_{\mathcal{F}_A} u_0$ if the axiom of M equals u_0, and $\llbracket M \rrbracket(t) =_{\mathcal{F}_A} u_0 \cdot \llbracket q \rrbracket(t) \cdot u_1$ if the axiom of M is given by $u_0 \cdot q(x_0) \cdot u_1$.

Example 1. Let $A = \{a, b\}$. As a running example we consider the LT_A M with input alphabet $\Sigma = \{f^2, g^1, k^0\}$ where the superscripts indicate the rank of the input symbols. M has axiom $q_0(x_0)$ and the following rules

$$\begin{array}{lll}
q_0(f(x_1,x_2)) \rightarrow q_1(x_2)bq_2(x_1) & q_0(g(x_1)) \rightarrow q_0(x_1) & q_0(k) \rightarrow \varepsilon \\
q_1(f(x_1,x_2)) \rightarrow q_0(x_1)q_0(x_2) & q_1(g(x_1)) \rightarrow abq_1(x_1) & q_1(k) \rightarrow a \\
q_2(f(x_1,x_2)) \rightarrow q_0(x_1)q_0(x_2) & q_2(g(x_1)) \rightarrow abq_2(x_1) & q_2(k) \rightarrow ab
\end{array}$$

Two LT_As M, M' are *equivalent* relative to the DTA B iff their translations coincide on all input trees accepted by B, i.e., $\llbracket M \rrbracket(t) =_{\mathcal{F}_A} \llbracket M' \rrbracket(t)$ for all $t \in \mathcal{L}(B)$.

To relate the computations of the LT_A M and the domain automaton B, we introduce the following notion. A mapping $\iota : Q \rightarrow H$ from the set of states of M to the set of states of B is called *compatible* if either the set of states of M is empty (and thus the axiom of M consists of an element of $\mathcal{F}_A$ only), or the following holds:

1. $\iota(q_0) = h_0$;
2. If $\iota(q) = h$, $\delta_B(h, f) = h_1 \ldots h_m$, and there is a rule in M of the form (3) then $\iota(q_i) = h_{\sigma(i)}$ for all $i = 1,\ldots,n$;
3. If $\iota(q) = h$ and $\delta_B(h, f)$ is undefined for some $f \in \Sigma$ of rank $m \geq 0$, then M has the rule $q(f(x_1,\ldots,x_m)) \rightarrow \bot$ for some dedicated symbol $\bot$ which does not belong to A.

Lemma 4. *For an LT_A M and a DTA $B = (H, \Sigma, \delta_B, h_0)$, an LT_A M' with a set of states Q' together with a mapping $\iota : Q' \rightarrow H$ can be constructed in polynomial time such that the following holds:*

1. *M and M' are equivalent relative to B;*
2. *ι is compatible.*

Example 2. Let $\mathsf{LT_A}$ M be defined as in Example 1. Consider DTA B with start state h_0 and the transition function $\delta_B = \{(h_0, f) \mapsto h_1 h_1, (h_1, g) \mapsto h_1, (h_1, h) \to \varepsilon\}$. According to Lemma 4, $\mathsf{LT_A}$ M' for M then is defined as follows. M' has axiom $\langle q_0, h_0 \rangle (x_0)$ and the rules

$$\langle q_0, h_0 \rangle (f(x_1, x_2)) \to \langle q_1, h_1 \rangle (x_2) \, b \, \langle q_2, h_1 \rangle (x_1)$$
$$\langle q_1, h_1 \rangle (g(x_1)) \to ab \, \langle q_1, h_1 \rangle (x_1) \qquad \langle q_1, h_1 \rangle (k) \to a$$
$$\langle q_2, h_1 \rangle (g(x_1)) \to ab \, \langle q_2, h_1 \rangle (x_1) \qquad \langle q_2, h_1 \rangle (k) \to ab$$

where the rules with left-hand sides $\langle q_0, h_0 \rangle (g(x_1))$, $\langle q_0, h_0 \rangle (h)$, $\langle q_1, h_1 \rangle$ $(f(x_1, x_2))$, $\langle q_2, h_1 \rangle (f(x_1, x_2))$, all have right-hand-sides $\bot$. The compatible map ι is then given by $\iota = \{\langle q_0, h_0 \rangle \mapsto h_0, \langle q_1, h_1 \rangle \mapsto h_1, \langle q_2, h_1 \rangle \mapsto h_1\}$. For convenience, we again denote the pairs $\langle q_0, h_0 \rangle, \langle q_1, h_1 \rangle, \langle q_2, h_1 \rangle$ with q_0, q_1, q_2, respectively.

Subsequently, we w.l.o.g. assume that each $\mathsf{LT_A}$ M with corresponding DTA B for its domain, comes with a compatible map ι. Moreover, we define for each state q of M, the set $\mathcal{L}(q) = \{[\![q]\!](t) \mid t \in \mathsf{dom}(\iota(q))\}$ of all outputs produced by state q (on inputs in $\mathsf{dom}(\iota(q))$), and $\mathcal{L}^{(i)}(q) = \{[\![q]\!](t) \mid t \in \mathsf{dom}(\iota(q)), \mathsf{depth}(t) < i\}$ for $i \geq 0$.

Beyond the availability of a compatible map, we also require that all states of M are *non-trivial* (relative to B). Here, a state q of M is called *trivial* if $\mathcal{L}(q)$ contains a single element only. Otherwise, it is called *non-trivial*. This property will be established in Theorem 1.

3   Deciding Equivalence

In the first step, we show that equivalence relative to the DTA B of *same-ordered* $\mathsf{LT_A}$s is decidable. For a DTA B, consider the $\mathsf{LT_A}$s M and M' with compatible mappings ι and ι', respectively. M and M' are *same-ordered* relative to B if they process their input trees in the same order. We define set of pairs $\langle q, q' \rangle$ of *co-reachable* states of M and M'. Let $u_0 \cdot q_0(x_1) \cdot u_1$ and $u_0' \cdot q_0'(x_1) \cdot u_1'$ be the axioms of M and M', respectively, where $\iota(q_0) = \iota'(q_0')$ is the start state of B. Then the pair $\langle q_0, q_0' \rangle$ is co-reachable. Let $\langle q, q' \rangle$ be a pair of co-reachable states. Then $\iota(q) = \iota'(q')$ should hold. For $f \in \Sigma$, assume that $\delta_B(\iota(q), f)$ is defined. Let

$$
\begin{aligned}
q(f(x_1, \ldots, x_m)) &\to u_0 q_1(x_{\sigma(1)}) u_1 \ldots u_{n-1} q_n(x_{\sigma(n)}) u_n \\
q'(f(x_1, \ldots, x_m)) &\to u_0' q_1'(x_{\sigma'(1)}) u_1' \ldots u_{n'-1}' q_{n'}'(x_{\sigma'(n')}) u_{n'}'
\end{aligned}
\tag{4}
$$

be the rules of q, q' for f, respectively. Then $\langle q_j, q_{j'}' \rangle$ is co-reachable whenever $\sigma(j) = \sigma'(j')$ holds. In particular, we then have $\iota(q_j) = \iota'(q_{j'}')$.

The pair $\langle q, q' \rangle$ of co-reachable states is called same-ordered, if for each corresponding pair of rules (4), $n = n'$ and $\sigma = \sigma'$. Finally, M and M' are *same-ordered* if for every co-reachable pair $\langle q, q' \rangle$ of states of M, M', and every $f \in \Sigma$, each pair of rules (4) is same-ordered whenever $\delta_B(\iota(q), f)$ is defined.

Given that the $\mathsf{LT_A}$s M and M' are same-ordered relative to B, we can represent the set of pairs of runs of M and M' on input trees by means of a single context-free grammar G. The set of nonterminals of G consists of a distinct start nonterminal S together with all co-reachable pairs $\langle q, q' \rangle$ of states q, q' of M, M', respectively. The set of terminal symbols T of G is given by $\{a, a^-, \bar{a}, \bar{a}^- \mid a \in \mathsf{A}\}$ for fresh distinct symbols $\bar{a}, \bar{a}^-, a \in \mathsf{A}$. Let $\langle q, q' \rangle$ be a co-reachable pair of states of M, M', and $f \in \Sigma$ such that $\delta_B(\iota(q), f)$ is defined. For each corresponding pair of rules (4), G receives the rule

$$\langle q, q' \rangle \to u_0 \bar{u}'_0 \langle q_1, q'_1 \rangle u_1 \bar{u}'_1 \ldots u_{n-1} \bar{u}'_{n-1} \langle q_n, q'_n \rangle u_n \bar{u}'_n$$

where $\bar{u}'_i$ is obtained from u'_i by replacing each output symbol $a \in \mathsf{A}$ with its barred copy $\bar{a}$ as well as each inverse a^- with its barred copy $\bar{a}^-$. For the axioms $u_0 q(x_1) u_1$ and $u'_0 q'(x_1) u'_1$ of M, M', respectively, we introduce the rule $S \to u_0 \bar{u}'_0 \langle q, q' \rangle u_1 \bar{u}'_1$ where again $\bar{u}'_i$ are the barred copies of u'_i. We define morphisms $f, g : T^* \to \mathcal{F}_\mathsf{A}$ by

$$\begin{array}{llll} f(a) = a & f(a^-) = a^- & f(\bar{a}) = f(\bar{a}^-) = \varepsilon \\ g(\bar{a}) = a & g(\bar{a}^-) = a^- & g(a) = g(a^-) = \varepsilon \end{array}$$

for $a \in \mathsf{A}$. Then M and M' are equivalent relative to B iff $g(w) =_{\mathcal{F}_\mathsf{A}} f(w)$ for all $w \in \mathcal{L}(G)$. Combining Plandowski's polynomial construction of a test set for a context-free language to check morphism equivalence over finitely generated free groups [10, Theorem 6], with Lohrey's polynomial algorithm for checking equivalence of SLPs over the free group [6], we deduce that the equivalence of the morphisms f and g on all words generated by the context-free grammar G, is decidable in polynomial time. Consequently, we obtain:

Corollary 1. *Equivalence of same-ordered* $\mathsf{LT_A}$*s relative to a* DTA B *is decidable in polynomial time.*   □

Next, we observe that for every $\mathsf{LT_A}$ M with compatible map ι and non-trivial states only, a *canonical* ordering can be established. We call M *ordered* (relative to B) if for all rules of the form (3), with $\mathcal{L}(q_i) \cdot u_i \cdot \ldots \cdot u_{j-1} \cdot \mathcal{L}(q_j) \subseteq v \cdot \langle p \rangle$, $p \in \mathcal{F}_\mathsf{A}$ the ordering $\sigma(i) < \ldots < \sigma(j)$ holds. Here we have naturally extended the operation "$\cdot$" to sets of elements.

We show that two ordered $\mathsf{LT_A}$s, when they are equivalent, are necessarily *same-ordered*. The proof of this claim is split in two parts. First, we prove that the *set* of indices of subtrees processed by equivalent co-reachable states are identical and second, that the order is the same.

Lemma 5. *Let* M, M' *be* $\mathsf{LT_A}$*s with compatible maps* ι *and* ι', *respectively, and non-trivial states only so that* M *and* M' *are equivalent relative to the* DTA B. *Let* $\langle q, q' \rangle$ *be a pair of co-reachable states of* M *and* M'. *Assume that* $\delta_B(\iota(q), f)$ *is defined for some* $f \in \Sigma$ *and consider the corresponding pair of rules* (4). *Then the following holds:*

1. $\{\sigma(1), \ldots, \sigma(n)\} = \{\sigma'(1), \ldots, \sigma'(n')\}$;
2. $\sigma = \sigma'$.

Proof. Since $\langle q, q' \rangle$ is a co-reachable pair of states, there are elements $\alpha, \alpha', \beta, \beta' \in \mathcal{F}_A$ such that

$$\alpha \cdot [\![q]\!](t) \cdot \beta =_{\mathcal{F}_A} \alpha' \cdot [\![q']\!](t) \cdot \beta'$$

holds for all $t \in \mathrm{dom}(\iota(q))$. Consider the first statement. Assume for a contradiction that $q_k(x_j)$ occurs on the right-hand side of the rule for q but x_j does not occur on the right-hand side of the rule for q'. Then, there are input trees $t = f(t_1, \ldots, t_m)$ and $t' = f(t'_1, \ldots, t'_m)$, both in $\mathrm{dom}(\iota(q))$, such that $[\![q_k]\!](t_j) \neq_{\mathcal{F}_A} [\![q_k]\!](t'_j)$ and $t_i = t'_i$ for all $i \neq j$. Moreover, there are $\mu_1, \mu_2 \in \mathcal{F}_A$ s.t.

$$\alpha \cdot [\![q]\!](t) \cdot \beta =_{\mathcal{F}_A} \alpha \cdot \mu_1 \cdot [\![q_k]\!](t_j) \cdot \mu_2 \cdot \beta \neq_{\mathcal{F}_A} \alpha \cdot \mu_1 \cdot [\![q_k]\!](t'_j) \cdot \mu_2 \cdot \beta =_{\mathcal{F}_A} \alpha \cdot [\![q]\!](t') \cdot \beta$$

But then,

$$\alpha \cdot [\![q]\!](t) \cdot \beta =_{\mathcal{F}_A} \alpha' \cdot [\![q']\!](t) \cdot \beta' =_{\mathcal{F}_A} \alpha' \cdot [\![q']\!](t') \cdot \beta' =_{\mathcal{F}_A} \alpha \cdot [\![q]\!](t') \cdot \beta$$

—a contradiction. By an analogous argument for some x_j only occurring in the right-hand side of the rule for q' the first statement follows.

Assume for contradiction that the mappings σ and σ' in the corresponding rules (4) differ. Let k denote the minimal index so that $\sigma(k) \neq \sigma'(k)$. W.l.o.g., we assume that $\sigma'(k) < \sigma(k)$. By the first statement, $n = n'$ and $\{\sigma(1), \ldots, \sigma(n)\} = \{\sigma'(1), \ldots, \sigma'(n)\}$. Then there are $\ell, \ell' > k$ such that

$$\sigma'(k) = \sigma(\ell) < \sigma(k) = \sigma'(\ell')$$

Let $t = f(t_1, \ldots, t_n) \in \mathrm{dom}(\iota(q))$ be an input tree. For that we obtain

$$\mu_0 := u_0 \cdot [\![q_1]\!](t_{\sigma(1)}) \cdot \ldots \cdot u_{k-1} \qquad \mu'_0 := u'_0 \cdot [\![q'_1]\!](t_{\sigma'(1)}) \cdot \ldots \cdot u'_{k-1}$$
$$\mu_1 := u_k \cdot [\![q_{k+1}]\!](t_{\sigma(k+1)}) \cdot \ldots \cdot u_{\ell-1} \qquad \mu'_1 := u'_k \cdot [\![q'_{k+1}]\!](t_{\sigma'(k+1)}) \cdot \ldots \cdot u'_{\ell'-1}$$
$$\mu_2 := u_\ell \cdot [\![q_\ell]\!](t_{\sigma(\ell)}) \cdot \ldots \cdot u_n \qquad \mu'_2 := u'_{\ell'} \cdot [\![q'_{\ell'}]\!](t_{\sigma'(\ell')}) \cdot \ldots \cdot u'_n$$

Then for all input trees $t' \in \mathrm{dom}(\iota(q_k))$, $t'' \in \mathcal{L}(\mathrm{dom}(\iota(q'_k)))$,

$$\alpha \cdot \mu_0 \cdot [\![q_k]\!](t') \cdot \mu_1 \cdot [\![q_\ell]\!](t'') \cdot \mu_2 \cdot \beta =_{\mathcal{F}_A} \alpha' \cdot \mu'_0 \cdot [\![q'_k]\!](t'') \cdot \mu'_1 \cdot [\![q'_{\ell'}]\!](t') \cdot \mu'_2 \cdot \beta'$$

Let $\gamma' = \mu_0^- \alpha^- \alpha' \mu'_0$. Then

$$[\![q_k]\!](t') \cdot \mu_1 \cdot [\![q_\ell]\!](t'') \cdot \mu_2 \cdot \beta =_{\mathcal{F}_A} \gamma' \cdot [\![q'_k]\!](t'') \cdot \mu'_1 \cdot [\![q'_{\ell'}]\!](t') \cdot \mu'_2 \cdot \beta'$$

By Lemma 2, we obtain that for all $w_1, w_2 \in \mathcal{L}(q_k)$ and $v_1, v_2 \in \mathcal{L}(q_\ell)$, $w_2^- \cdot w_1 \in \mu_1 \cdot \langle p \rangle \cdot \mu_1^-$ and $v_1 \cdot v_2^- \in \langle p \rangle$ for some primitive p.

If $\ell = k + 1$, i.e., there is no further state between $q_k(x_{\sigma(k)})$ and $q_\ell(x_{\sigma(\ell)})$, then $\mu_1 =_{\mathcal{F}_A} u_k$, $\mathcal{L}(q_k) \subseteq w \cdot u_k \cdot \langle p \rangle \cdot u_k^-$ and $\mathcal{L}(q_\ell) \subseteq \langle p \rangle \cdot v$ for some fixed $w \in \mathcal{L}(q_k)$ and $v \in \mathcal{L}(q_\ell)$. As $\sigma(k) > \sigma'(k) = \sigma(\ell)$, this contradicts M being ordered.

For the case that there is at least one occurrence of a state between $q_k(x_{\sigma(k)})$ and $q_\ell(x_{\sigma(\ell)})$, we show that for all $\alpha_1, \alpha_2 \in u_k \cdot \mathcal{L}(q_{k+1}) \cdot \ldots \cdot u_{\ell-1} =_{\mathcal{F}_A}: \hat{L}$,

$\alpha_1^- \alpha_2 \in \langle p \rangle$ holds. We fix $w_1, w_2 \in \mathcal{L}(q_k)$ and $v_1, v_2 \in \mathcal{L}(q_\ell)$ with $w_1 \neq w_2$ and $v_1 \neq v_2$. For every $\alpha \in \hat{L}$, we find by Lemma 2, primitive p_α and exponent $r_\alpha \in \mathbb{Z}$ such that $v_1 \cdot v_2^- =_{\mathcal{F}_A} p_\alpha^{r_\alpha}$ holds. Since p_α is primitive, this means that $p_\alpha =_{\mathcal{F}_A} p$ or $p_\alpha =_{\mathcal{F}_A} p^-$. Furthermore, there must be some exponent r'_α such that $w_1^- \cdot w_2 =_{\mathcal{F}_A} \alpha \cdot p^{r_\alpha} \cdot \alpha^-$. For $\alpha_1, \alpha_2 \in \hat{L}$, we therefore have that

$$p^{r'_{\alpha_1}} =_{\mathcal{F}_A} (\alpha_1^- \cdot \alpha_2) \cdot p^{r'_{\alpha_2}} \cdot (\alpha_1^- \cdot \alpha_2)^-$$

Therefore by Lemma 1, $\alpha_1^- \cdot \alpha_2 \in \langle p \rangle$. Let us fix some $w_k \in \mathcal{L}(q_k)$, $\alpha \in \hat{L} =_{\mathcal{F}_A} u_k \cdot \mathcal{L}(q_{k+1}) \cdot \ldots \cdot u_{\ell-1}$, and $w_l \in \mathcal{L}(q_l)$. Then $\mathcal{L}(q_k) \subseteq w_k \cdot \alpha \cdot \langle p \rangle \cdot \alpha^-$, $\hat{L} \subseteq \alpha \cdot \langle p \rangle$ and $\mathcal{L}(q_l) \subseteq \langle p \rangle \cdot w_l$. Therefore,

$$\mathcal{L}(q_k) \cdot u_k \cdot \ldots \cdot \mathcal{L}(q_\ell) \subseteq w_k \cdot \alpha \cdot \langle p \rangle \cdot \alpha^- \cdot \alpha \cdot \langle p \rangle \cdot \langle p \rangle \cdot w_l =_{\mathcal{F}_A} w_k \cdot \alpha \cdot \langle p \rangle \cdot w_l$$

As $\sigma(k) > \sigma'(k) = \sigma(\ell)$, this again contradicts M being ordered.

It remains to show that every LT_A can be ordered in polynomial time. For that, we rely on the following characterization.

Lemma 6. *Assume that $L_1, \ldots, L_n$ are neither empty nor singleton subsets of $\mathcal{F}_A$ and $u_1, \ldots, u_{n-1} \in \mathcal{F}_A$. Then there are $v_1, \ldots, v_n \in \mathcal{F}_A$ such that*

$$L_1 \cdot u_1 \cdot \ldots \cdot L_{n-1} \cdot u_{n-1} \cdot L_n \subseteq v \cdot \langle p \rangle \tag{5}$$

holds if and only if for $i = 1, \ldots, n$, $L_i \subseteq v_i \cdot \langle p_i \rangle$ with

$$\begin{aligned} p_n &=_{\mathcal{F}_A} p \\ p_i &=_{\mathcal{F}_A} (u_i \cdot v_{i+1}) \cdot p_{i+1} \cdot (u_i \cdot v_{i+1})^- \quad \text{for } i < n \end{aligned}$$

and

$$v^- \cdot v_1 \cdot u_1 \cdot \ldots \cdot v_{n-1} \cdot u_{n-1} \cdot v_n \in \langle p \rangle \tag{6}$$

Proof. Let $s_1 = \varepsilon$. For $i = 2, \ldots, n$ we fix some word $s_i \in L_1 \cdot u_1 \cdot L_2 \cdot \ldots \cdot L_{i-1} \cdot u_{i-1}$. Likewise, let $t_n = \varepsilon$ and for $i = 1, \ldots, n-1$ fix some word $t_i \in u_i \cdot L_{i+1} \cdot \ldots \cdot L_n$, and define $v_i =_{\mathcal{F}_A} s_i^- \cdot v \cdot t_i^-$.

First assume that the inclusion (5) holds. Let $p'_i =_{\mathcal{F}_A} t_i \cdot p \cdot t_i^-$. Then for all i, $s_i \cdot L_i \cdot t_i \subseteq v \cdot \langle p \rangle$, and therefore

$$L_i \subseteq s_i^- \cdot v \cdot \langle p \rangle \cdot t_i^- =_{\mathcal{F}_A} s_i^- \cdot v \cdot t_i^- \cdot t_i \cdot \langle p \rangle \cdot t_i^- =_{\mathcal{F}_A} v_i \langle p'_i \rangle$$

We claim that $p'_i = p_i$ for all $i = 1, \ldots, n$. We proceed by induction on $n - i$. As $t_n = \varepsilon$, we have that $p'_n = p = p_n$. For $i < n$, we can rewrite $t_i =_{\mathcal{F}_A} u_i \cdot w_{i+1} \cdot t_{i+1}$ where $w_{i+1} \in L_{i+1}$ and thus is of the form $v_{i+1} \cdot p_{i+1}^{k_{i+1}}$ for some exponent k_{i+1}.

$$\begin{aligned} p'_i &=_{\mathcal{F}_A} t_i \cdot p \cdot t_i^- \\ &=_{\mathcal{F}_A} u_i \cdot w_{i+1} \cdot t_{i+1} \cdot p \cdot t_{i+1}^- \cdot w_{i+1}^- \cdot u_i^- \\ &=_{\mathcal{F}_A} u_i \cdot w_{i+1} \cdot p_{i+1} \cdot w_{i+1}^- \cdot u_i^- \qquad \text{by I.H.} \\ &=_{\mathcal{F}_A} u_i \cdot v_{i+1} \cdot p_{i+1} \cdot v_{i+1}^- \cdot u_i^- \\ &=_{\mathcal{F}_A} p_i \end{aligned}$$

It remains to prove the inclusion (6). Since $w_i \in L_i$, we have by (5) that $v^- w_1 \cdot u_1 \cdot \ldots w_n \cdot u_n \in \langle p \rangle$ holds. Now we calculate:

$$
\begin{aligned}
v^- \cdot w_1 \cdot u_1 \cdot \ldots u_{n-1} \cdot w_n &=_{\mathcal{F}_A} v^- \cdot v_1 \cdot p_1^{k_1} \cdot u_1 \cdot \ldots \cdot u_{n-1} \cdot v_n \cdot p_n^{k_n} \\
&=_{\mathcal{F}_A} v^- \cdot v_1 \cdot u_1 \cdot v_2 \cdot p_2^{k_1+k_2} \cdot u_2 \cdot \ldots \cdot u_{n-1} \cdot v_n \cdot p_n^{k_n} \\
&\ldots \\
&=_{\mathcal{F}_A} v^- \cdot v_1 \cdot u_1 \cdot \ldots v_{n-1} \cdot u_{n-1} \cdot v_n \cdot p_n^{k}
\end{aligned}
$$

where $k = k_1 + \ldots + k_n$. Since $p_n = p$, the claim follows.

The other direction of the claim of the lemma follows directly:

$$
\begin{aligned}
L_1 u_1 \ldots L_{n-1} u_{n-1} L_n &\subseteq v_1 \cdot \langle p_1 \rangle \cdot u_1 \cdot \ldots \cdot v_{n-1} \cdot \langle p_{n-1} \rangle \cdot u_{n-1} \cdot v_n \cdot \langle p_n \rangle \\
&=_{\mathcal{F}_A} v_1 \cdot u_1 \cdot v_2 \cdot \langle p_2 \rangle \cdot \langle p_2 \rangle \cdot u_2 \cdot \ldots \cdot v_{n-1} \cdot \langle p_{n-1} \rangle \cdot u_{n-1} \cdot v_n \cdot \langle p_n \rangle \\
&=_{\mathcal{F}_A} v_1 \cdot u_1 \cdot v_2 \cdot \langle p_2 \rangle \cdot u_2 \cdot \ldots \cdot v_{n-1} \cdot \langle p_{n-1} \rangle \cdot u_{n-1} \cdot v_n \cdot \langle p_n \rangle \\
&\ldots \\
&=_{\mathcal{F}_A} v_1 \cdot u_1 \cdot v_2 \cdot \ldots \cdot u_{n-1} \cdot v_n \cdot \langle p_n \rangle \\
&=_{\mathcal{F}_A} v_1 \cdot u_1 \cdot v_2 \cdot \ldots \cdot u_{n-1} \cdot v_n \cdot \langle p \rangle \\
&\subseteq v \cdot \langle p \rangle
\end{aligned}
$$

where the last inclusion follows from (6).

Let us call a non-empty, non-singleton language $L \subseteq \mathcal{F}_A$ *periodic*, if $L \subseteq v \cdot \langle p \rangle$ for some $v, p \in \mathcal{F}_A$. Lemma 6 then implies that if a concatenation of languages and elements from $\mathcal{F}_A$ is periodic, then so must be all non-singleton component languages. In fact, the languages in the composition can then be arbitrarily permuted.

Corollary 2. *Assume for non-empty, nonsingleton languages $L_1, \ldots, L_n \subseteq \mathcal{F}_A$ and $u_1, \ldots, u_{n-1} \in \mathcal{F}_A$ that property (5) holds. Then for every permutation π, there are elements $u_{\pi,0}, \ldots, u_{\pi,n} \in \mathcal{F}_A$ such that*

$$
L_1 \cdot u_1 \cdot \ldots \cdot L_{n-1} \cdot u_{n-1} \cdot L_n = u_{\pi,0} \cdot L_{\pi(1)} \cdot u_{\pi,1} \cdot \ldots \cdot u_{\pi_n-1} \cdot L_{\pi(n)} \cdot u_{\pi,n}
$$

Example 3. We reconsider LT$_A$ M' and DTA B from Example 2. We observe that $\mathcal{L}(q_1) \subseteq a \cdot \langle ba \rangle$, $\mathcal{L}(q_2) \subseteq \langle ab \rangle$, and thus $\mathcal{L}(q_0) = \mathcal{L}(q_1) \cdot b \cdot \mathcal{L}(q_2) \subseteq \langle ab \rangle$. Accordingly, the rule for state q_0 and input symbol f is not ordered. Following the notation of Corollary 2, we find $v_1 = a$, $u_1 = b$ and $v_2 = \varepsilon$, and the rule for q_0 and f can be reordered to

$$
q_0(f(x_1, x_2)) \rightarrow ab \cdot q_2(x_1) \cdot b^- a^- \cdot q_1(x_2) b
$$

This example shows major improvements compared to the construction in [2]. Since we have inverses at hand, only *local* changes must be applied to the subsequence $q_1(x_2) \cdot b \cdot q_2(x_1)$. In contrast to the construction in [2], neither auxiliary states nor further changes to the rules of q_1 and q_2 are required.

By Corollary 2, the order of occurrences of terms $q_k(x_{\sigma(k)})$ can be permuted in every sub-sequence $q_i(x_{\sigma(i)}) \cdot u_i \cdot \ldots \cdot u_{j-1} q_j(x_{\sigma(j)})$ where $\mathcal{L}(q_i) \cdot u_i \cdot \ldots \cdot u_{j-1} \cdot \mathcal{L}(q_j) \in u \cdot \langle p \rangle$ is periodic, to satisfy the requirements of an ordered LT$_A$. A sufficient condition for that is, according to Lemma 6, that $\mathcal{L}(q_k)$ is periodic for each q_k occurring in that sub-sequence. Therefore we will determine the subset of *all* states q where $\mathcal{L}(q)$ is periodic, and if so elements v_q, p_q such that $\mathcal{L}(q) \subseteq v_q \cdot \langle p_q \rangle$. In order to do so we compute an *abstraction* of the sets $\mathcal{L}(q)$ by means of a complete lattice which both reports constant values and also captures periodicity.

Let $\mathcal{D} = 2^{\mathcal{F}_A}$ denote the complete lattice of subsets of the free group $\mathcal{F}_A$. We define a *projection* $\alpha : \mathcal{D} \to \mathcal{D}$ by $\alpha(\emptyset) = \emptyset$, $\alpha(\{g\}) = \{g\}$, and for languages L with at least two elements,

$$\alpha(L) = \begin{cases} g\langle p \rangle & \text{if } L \subseteq g\langle p \rangle \text{ and } p \text{ is primitive} \\ \mathcal{F}_A & \text{otherwise} \end{cases}$$

The projection α is a *closure* operator, i.e., is a monotonic function with $L \subseteq \alpha(L)$, and $\alpha(\alpha(L)) = \alpha(L)$. The image of α can be considered as an *abstract* complete lattice $\mathcal{D}^\sharp$, partially ordered by subset inclusion. Thereby, the abstraction α commutes with least upper bounds as well as with the group operation. For that, we define *abstract* versions $\sqcup, \star : (\mathcal{D}^\sharp)^2 \to \mathcal{D}^\sharp$ of set union and the group operation by

$$A_1 \sqcup A_2 = \alpha(A_1 \cup A_2) \qquad A_1 \star A_2 = \alpha(A_1 \cdot A_2)$$

In fact, "$\sqcup$" is the least upper bound operation for $\mathcal{D}^\sharp$. The two abstract operators can also be more explicitly defined by:

$$\begin{aligned}
\emptyset \sqcup L &= L \sqcup \emptyset &= L \\
\mathcal{F}_A \sqcup L &= L \sqcup \mathcal{F}_A &= \mathcal{F}_A
\end{aligned}$$

$$\{g_1\} \sqcup \{g_2\} = \begin{cases} \{g_1\} & \text{if } g_1 = g_2 \\ g_1 \cdot \langle p \rangle & \text{if } g_1 \neq g_2, p \text{ primitive root of } g_1^- \cdot g_2 \end{cases}$$

$$\{g_1\} \sqcup g_2 \cdot \langle p \rangle = g_2 \cdot \langle p \rangle \sqcup \{g_1\} = \begin{cases} g_2 \cdot \langle p \rangle & \text{if } g_1 \in g_2 \cdot \langle p \rangle \\ \mathcal{F}_A & \text{otherwise} \end{cases}$$

$$g_1 \cdot \langle p_1 \rangle \sqcup g_2 \cdot \langle p_2 \rangle = \begin{cases} g_1 \cdot \langle p_1 \rangle & \text{if } p_2 \in \langle p_1 \rangle \text{ and } g_2^- \cdot g_1 \in \langle p_1 \rangle \\ \mathcal{F}_A & \text{otherwise} \end{cases}$$

$$\begin{aligned}
\emptyset \star L &= L \star \emptyset &= \emptyset \\
\mathcal{F}_A \star L &= L \star \mathcal{F}_A = \mathcal{F}_A & \text{for } L \neq \emptyset \\
\{g_1\} \star \{g_2\} &= \{g_1 \cdot g_2\} \\
\{g_1\} \star g_2 \cdot \langle p \rangle &= (g_1 \cdot g_2) \cdot \langle p \rangle \\
g_1 \cdot \langle p \rangle \star \{g_2\} &= (g_1 \cdot g_2) \cdot \langle g_2^- \cdot p \cdot g_2 \rangle
\end{aligned}$$

$$g_1 \cdot \langle p_1 \rangle \star g_2 \cdot \langle p_2 \rangle = \begin{cases} (g_1 \cdot g_2) \cdot \langle p_2 \rangle & \text{if } g_2^- \cdot p_1 \cdot g_2 \in \langle p_2 \rangle \\ \mathcal{F}_A & \text{otherwise} \end{cases}$$

Lemma 7. *For all subsets $L_1, L_2 \subseteq \mathcal{F}_A$, $\alpha(L_1 \cup L_2) = \alpha(L_1) \sqcup \alpha(L_2)$ and $\alpha(L_1 \cdot L_2) = \alpha(L_1) \star \alpha(L_2)$.*

We conclude that α in fact represents a *precise* abstract interpretation in the sense of [9]. Accordingly, we obtain:

Lemma 8. *For every* LT$_A$ *M and* DTA *B with compatible map ι, the sets $\alpha(\mathcal{L}(q))$, q state of M, can be computed in polynomial time.*

Proof. We introduce one unknown X_q for every state q of M, and one constraint for each rule of M of the form (3) where $\delta(\iota(q), f)$ is defined in B. This constraint is given by:

$$X_q \sqsupseteq u_0 \star X_{q_1} \star \ldots \star u_{n-1} \star X_{q_n} \star u_n \tag{7}$$

As the right-hand sides of the constraints (7) all represent monotonic functions, the given system of constraints has a *least* solution. In order to obtain this solution, we consider for each state q of M, the sequence $X_q^{(i)}, i \geq 0$ of values in $\mathcal{D}^\sharp$ where $X_q^{(0)} = \emptyset$, and for $i > 0$, we set $X_q^{(i)}$ as the least upper bound of the values obtained from the constraints with left-hand side X_q of the form (7) by replacing the unknowns X_{q_j} on the right-hand side with the values $X_{q_j}^{(i-1)}$. By induction on $i \geq 0$, we verify that for all states q of M,

$$X_q^{(i)} = \alpha(\mathcal{L}^{(i)}(q))$$

holds. Note that the induction step thereby, relies on Lemma 7.

As each strictly increasing chain of elements in $\mathcal{D}^\sharp$ consists of at most four elements, we have that the least solution of the constraint system is attained after at most $3 \cdot N$ iterations, if N is the number of states of M, i.e., for each state q of M, $X_q^{(3N)} = X_q^{(i)}$ for all $i \geq 3N$. The elements of $\mathcal{D}^\sharp$ can be represented by SLPs where the operations $\star$ and $\sqcup$ run in polynomial time, cf. Lemma 3. Since each iteration requires only a polynomial number of operations $\star$ and $\sqcup$, the statement of the lemma follows.

We now exploit the information provided by the $\alpha(\mathcal{L}(q))$ to remove trivial states as well as order subsequences of right-hand sides which are periodic.

Theorem 1. *Let B be a* DTA *such that $\mathcal{L}(B) \neq \emptyset$. For every* LT$_A$ *M with compatible map ι, an* LT$_A$ *M' with compatible map ι' can be constructed in polynomial time such that*

1. *M and M' are equivalent relative to B;*
2. *M' has no trivial states;*
3. *M' is ordered.*

Proof. By Lemma 8, we can, in polynomial time, determine for every state q of M, the value $\alpha(\mathcal{L}(q))$. We use this information to remove from M all trivial states. W.l.o.g., assume that the axiom of M is given by $u_0 \cdot q_0(x_0) \cdot u_1$. If the state q_0 occurring in the axiom of M is trivial with $\mathcal{L}(q_0) = \{v\}$, then M_1 has no states or rules, but the axiom $u_0 \cdot v \cdot u_1$.

Therefore now assume that q_0 is non-trivial. We then construct an LT$_A$ M_1 whose set of states Q_1 consists of all *non-trivial* states q of M where the compatible map ι_1 of M_1 is obtained from ι by restriction to Q_1. Since $\mathcal{L}(M) \neq \emptyset$, the state of M occurring in the axiom is non-trivial. Accordingly, the axiom of M is also used as axiom for M_1. Consider a non-trivial state q of M and $f \in \Sigma$. If $\delta(\iota(q), f)$ is not defined M_1 has the rule $q(f(x_1, \ldots, x_m) \to \bot$. Assume that $\delta(\iota(q), f)$ is defined and M has a rule of the form (3). Then M_1 has the rule

$$q(f(x_1, \ldots, x_m)) \to u_0 \cdot g_1 \cdot \ldots \cdot u_{n-1} \cdot g_n \cdot u_n$$

where for $i = 1, \ldots, n$, g_i equals $q_i(x_{\sigma(i)})$ if q_i is non-trivial, and equals the single word in $\mathcal{L}(q_i)$ otherwise. Obviously, M and M_1 are equivalent relative to B where M_1 now has no trivial states, while for every non-trivial state q, the semantics of q in M and M_1 are the same relative to B. Our goal now is to equivalently rewrite the right-hand side of each rule of M_1 so that the result is ordered. For each state q of the LT_A we determine whether there are $v, p \in \mathsf{B}^*$ such that $\mathcal{L}(q) \subseteq v\langle p\rangle$, cf. Lemma 8. So consider a rule of M_1 of the form (3). By means of the values $\alpha(\mathcal{L}(q_i))$, $i = 1, \ldots, n$, together with the abstract operation "$\star$", we can determine maximal intervals $[i, j]$ such that $\mathcal{L}(q_i) \cdot u_i \cdot \ldots \cdot u_{j-1} \cdot \mathcal{L}(q_j)$ is periodic, i.e., $\alpha(\mathcal{L}(q_i)) \star u_i \cdot \ldots \star u_{j-1} \star \alpha(\mathcal{L}(q_j)) \subseteq v \cdot \langle p\rangle$ for some $v, p \in \mathcal{F}_\mathsf{A}$. We remark that these maximal intervals are necessarily disjoint. By Corollary 2, for every permutation $\pi : [i, j] \to [i, j]$, elements $u', u'_i, \ldots, u'_j, u'' \in \mathcal{F}_\mathsf{A}$ can be found so that $q_i(x_{\sigma(i)}) \cdot u_i \cdot \ldots \cdot u_{j-1} \cdot q_j(x_{\sigma(j)})$ is equivalent to $u' \cdot q_{\pi(i)}(x_{\sigma(\pi(i))}) \cdot u'_i \cdot \ldots \cdot u'_{j-1} \cdot q_{\pi(j)}(x_{\sigma(\pi(j))}) \cdot u''$.

In particular, this is true for the permutation π with $\sigma(\pi(i)) < \ldots < \sigma(\pi(j))$. Assuming that all group elements are represented as SLPs, the overall construction runs in polynomial time.

In summary, we arrive at the main theorem of this paper.

Theorem 2. *The equivalence of $\mathsf{LT}_\mathsf{A}s$ relative to some DTA B can be decided in polynomial time.*

Proof. Assume we are given $\mathsf{LT}_\mathsf{A}s$ M, M' with compatible maps (relative to B). By Theorem 1, we may w.l.o.g. assume that M and M' both have no trivial states and are ordered. It can be checked in polynomial time whether or not M and M' are same-ordered. If they are not, then by Lemma 5, they cannot be equivalent relative to B. Therefore now assume that M and M' are same-ordered. Then their equivalence relative to B is decidable in polynomial time by Corollary 1. Altogether we thus obtain a polynomial decision procedure for equivalence of $\mathsf{LT}_\mathsf{A}s$ relative to some DTA B.

4 Conclusion

We have shown that equivalence of $\mathsf{LT}_\mathsf{A}s$ relative to a given DTA B can be decided in polynomial time. For that, we considered *total* transducers only, but defined the domain of allowed input trees separately by means of the DTA. This does not impose any restriction of generality, since any (possibly partial) linear deterministic top-down tree transducer can be translated in polynomial time to a corresponding *total* LT$_A$ together with a corresponding DTA (see, e.g., [11]). The required constructions for LT$_A$s which we have presented here, turn out to be more general than the constructions provided in [2] since they apply to transducers which may not only output symbols $a \in A$, but also their inverses a^-. At the same time, they are *simpler* and easier to be proven correct due to the combinatorial and algebraic properties provided by the free group.

Acknowledgements. We also like to thank the anonymous reviewers for their detailed comments and valuable advice.

References

1. Boiret, A.: Normal form on linear tree-to-word transducers. In: Dediu, A.-H., Janoušek, J., Martín-Vide, C., Truthe, B. (eds.) LATA 2016. LNCS, vol. 9618, pp. 439–451. Springer, Cham (2016). https://doi.org/10.1007/978-3-319-30000-9_34

2. Boiret, A., Palenta, R.: Deciding equivalence of linear tree-to-word transducers in polynomial time. In: Brlek, S., Reutenauer, C. (eds.) DLT 2016. LNCS, vol. 9840, pp. 355–367. Springer, Heidelberg (2016). https://doi.org/10.1007/978-3-662-53132-7_29

3. Laurence, G., Lemay, A., Niehren, J., Staworko, S., Tommasi, M.: Normalization of sequential top-down tree-to-word transducers. In: Dediu, A.-H., Inenaga, S., Martín-Vide, C. (eds.) LATA 2011. LNCS, vol. 6638, pp. 354–365. Springer, Heidelberg (2011). https://doi.org/10.1007/978-3-642-21254-3_28

4. Laurence, G., Lemay, A., Niehren, J., Staworko, S., Tommasi, M.: Learning sequential tree-to-word transducers. In: Dediu, A.-H., Martín-Vide, C., Sierra-Rodríguez, J.-L., Truthe, B. (eds.) LATA 2014. LNCS, vol. 8370, pp. 490–502. Springer, Cham (2014). https://doi.org/10.1007/978-3-319-04921-2_40

5. Löbel, R., Luttenberger, M., Seidl, H.: Equivalence of linear tree transducers with output in the free group. CoRR abs/2001.03480 (2020). http://arxiv.org/abs/2001.03480

6. Lohrey, M.: Word problems on compressed words. In: Díaz, J., Karhumäki, J., Lepistö, A., Sannella, D. (eds.) ICALP 2004. LNCS, vol. 3142, pp. 906–918. Springer, Heidelberg (2004). https://doi.org/10.1007/978-3-540-27836-8_76

7. Lohrey, M.: The Compressed Word Problem for Groups. SM. Springer, New York (2014). https://doi.org/10.1007/978-1-4939-0748-9

8. Lyndon, R.C., Schupp, P.E.: Combinatorial Group Theory. CM, vol. 89. Springer, Heidelberg (2001). https://doi.org/10.1007/978-3-642-61896-3

9. Müller-Olm, M.: Variations on Constants. LNCS, vol. 3800. Springer, Heidelberg (2006). https://doi.org/10.1007/11871743

10. Plandowski, W.: Testing equivalence of morphisms on context-free languages. In: van Leeuwen, J. (ed.) ESA 1994. LNCS, vol. 855, pp. 460–470. Springer, Heidelberg (1994). https://doi.org/10.1007/BFb0049431

11. Seidl, H., Maneth, S., Kemper, G.: Equivalence of deterministic top-down tree-to-string transducers is decidable. J. ACM **65**(4), 21:1–21:30 (2018). https://doi.org/10.1145/3182653

12. Sénizergues, G.: T(A) = T(B)? In: Wiedermann, J., van Emde Boas, P., Nielsen, M. (eds.) ICALP 1999. LNCS, vol. 1644, pp. 665–675. Springer, Heidelberg (1999). https://doi.org/10.1007/3-540-48523-6_63

13. Staworko, S., Laurence, G., Lemay, A., Niehren, J.: Equivalence of deterministic nested word to word transducers. In: Kutyłowski, M., Charatonik, W., Gębala, M. (eds.) FCT 2009. LNCS, vol. 5699, pp. 310–322. Springer, Heidelberg (2009). https://doi.org/10.1007/978-3-642-03409-1_28
14. Tomita, E., Seino, K.: A direct branching algorithm for checking the equivalence of two deterministic pushdown transducers, one of which is real-time strict. Theory Comput. Sci. **64**(1), 39–53 (1989). https://doi.org/10.1016/0304-3975(89)90096-0

On the Balancedness of Tree-to-Word Transducers

Raphaela Löbel$^{(\boxtimes)}$, Michael Luttenberger, and Helmut Seidl

TU München, Munich, Germany
{loebel,luttenbe,seidl}@in.tum.de

Abstract. A language over an alphabet $B = A \cup \overline{A}$ of opening (A) and closing ($\overline{A}$) brackets, is balanced if it is a subset of the Dyck language $\mathbb{D}_B$ over B, and it is well-formed if all words are prefixes of words in $\mathbb{D}_B$. We show that well-formedness of a context-free language is decidable in polynomial time, and that the longest common reduced suffix can be computed in polynomial time. With this at a hand we decide for the class 2-TW of non-linear tree transducers with output alphabet B^* whether or not the output language is balanced.

Keywords: Balancedness of tree-to-word transducer · Equivalence · Longest common suffix/prefix of a CFG

1 Introduction

Structured text requires that pairs of opening and closing brackets are properly nested. This applies to text representing program code as well as to XML or HTML documents. Subsequently, we call properly nested words over an alphabet B of opening and closing brackets *balanced*. Balanced words, i.e. structured text, need not necessarily be constructed in a structured way. Therefore, it is a non-trivial problem whether the set of words produced by some kind of text processor, consists of balanced words only. For the case of a single pair of brackets and context-free languages, decidability of this problem has been settled by Knuth [3] where a polynomial time algorithm is presented by Minamide and Tozawa [9]. Recently, these results were generalized to the output languages of monadic second-order logic (MSO) definable tree-to-word transductions [8]. The case when the alphabet B consists of *multiple* pairs of brackets, though, seems to be more intricate. Still, balancedness for context-free languages was shown to be decidable by Berstel and Boasson [1] where a polynomial time algorithm again has been provided by Tozawa and Minamide [13]. Whether or not these results for B can be generalized to MSO definable transductions as e.g. done by finite copying macro tree transducers with regular look-ahead, remains as an open problem. Reynier and Talbot [10] considered visibly pushdown transducers and showed decidability of this class with well-nested output in polynomial time.

Here, we provide a first step to answering this question. We consider deterministic tree-to-word transducers which process their input at most twice by

© Springer Nature Switzerland AG 2020
N. Jonoska and D. Savchuk (Eds.): DLT 2020, LNCS 12086, pp. 222–236, 2020.
https://doi.org/10.1007/978-3-030-48516-0_17

calling in their axioms at most two *linear* transductions of the input. Let 2-TW denote the class of these transductions. Note that the output languages of *linear* deterministic tree-to-word transducers is context-free, which does not need to be the case for 2-TW transducers. 2-TW forms a subclass of MSO definable transductions which allows to specify transductions such as *prepending* an XML document with the list of its section headings, or *appending* such a document with the list of figure titles. For 2-TW transducers we show that balancedness is decidable—and this in polynomial time. In order to obtain this result, we first generalize the notion of balancedness to the notion of *well-formedness* of a language, which means that each word is a *prefix* of a balanced word. Then we show that well-formedness for context-free languages is decidable in polynomial time. A central ingredient is the computation of the *longest common suffix* of a context-free language L over B *after reduction* i.e. after canceling all pairs of matching brackets. While the proof shares many ideas with the computation of the longest common prefix of a context-free language [7] we could not directly make use of the results of [7] s.t. the results of this paper fully subsume the results of [7]. Now assume that we have verified that the output language of the first linear transduction called in the axiom of the 2-TW transducer and the *inverted* output language of the second linear transformation both are well-formed. Then balancedness of the 2-TW transducer in question, effectively reduces to the *equivalence* of two deterministic linear tree-to-word transducers—modulo the reduction of opening followed by corresponding closing brackets. Due to the well-formedness we can use the equivalence of linear tree-to-word transducers over the *free group* which can be decided in polynomial time [5].

This paper is organized as follows. After introducing basic concepts in Sect. 2, Sect. 3 shows how balancedness for 2-TW transducers can be reduced to equivalence over the free group and well-formedness of LT_Δs. Section 4 considers the problem of deciding well-formedness of context-free languages in general.

Missing proofs can be found in the extended version of this paper [4].

2   Preliminaries

As usual, $\mathbb{N}$ ($\mathbb{N}_0$) denotes the natural numbers (including 0). The power set of a set S is denoted by 2^S. Σ denotes some generic (nonempty) alphabet, Σ^* and Σ^ω denote the set of all finite words and the set of all infinite words, respectively. Then $\Sigma^\infty = \Sigma^* \cup \Sigma^\omega$ is the set of all countable words. Note, that the transducers considered here output finite words only; however, for the operations needed to analyze the output infinite words are very helpful. We denote the empty word by ε. For a finite word $w = w_0 \ldots w_l$, its reverse w^R is defined by $w^R = w_l \ldots w_1 w_0$; as usual, set $L^R := \{w^R \mid w \in L\}$ for $L \subseteq \Sigma^*$. A is used to denote an alphabet of *opening brackets* with $\overline{A} = \{\overline{a} \mid a \in A\}$ the derived alphabet of *closing brackets*, and $B := A \cup \overline{A}$ the resulting alphabet of *opening and closing brackets*.

Longest Common Prefix and Suffix. Let Σ be an alphabet. We first define the *longest common prefix* of a language, and then reduce the definition of the *longest* *common suffix* to it by means of the reverse. We write $\stackrel{p}{\sqsubseteq}$ to denote the prefix relation on Σ^∞, i.e. we have $u \stackrel{p}{\sqsubseteq} w$ if either (i) $u, w \in \Sigma^*$ and there exists $v \in \Sigma^*$ s.t. $w = uv$, or (ii) $u \in \Sigma^*$ and $w \in \Sigma^\omega$ and there exists $v \in \Sigma^\omega$ s.t. $w = uv$, or (iii) $u, w \in \Sigma^\omega$ and $u = w$. We extend Σ^∞ by a greatest element $\top \notin \Sigma^\infty$ w.r.t. $\stackrel{p}{\sqsubseteq}$ s.t. $u \stackrel{p}{\sqsubseteq} \top$ for all $u \in \Sigma^\infty_\top := \Sigma^\infty \cup \{\top\}$. Then every set $L \subseteq \Sigma^\infty_\top$ has an infimum w.r.t. $\stackrel{p}{\sqsubseteq}$ which is called the *longest common prefix* of L, abbreviated by $\mathsf{lcp}(L)$. Further, define $\varepsilon^\omega := \top$, $\top^R := \top$, and $\top w := \top =: w\top$ for all $w \in \Sigma^\infty_\top$.

In Sect. 4 we will need to study the *longest common suffix* (lcs) of a language L. For $L \subseteq \Sigma^*$, we can simply set $\mathsf{lcs}(L) := \mathsf{lcp}(L^R)^R$, but also certain infinite words are very useful for describing how the lcs changes when concatenating two languages (see e.g. Example 2). Recall that for $u, w \in \Sigma^*$ and $w \neq \varepsilon$ the ω-regular expression uw^ω denotes the unique infinite word $uwww \ldots$ in $\bigcap_{k \in \mathbb{N}_0} uw^k \Sigma^\omega$; such a word is also called *ultimately periodic*. For the lcs we will use the expression $w^\frown u$ to denote the *ultimately left-periodic* word $\ldots wwwu$ that ends on the suffix u with infinitely many copies of w left of u; these words are used to abbreviate the fact that we can generate a word $w^k u$ for unbounded $k \in \mathbb{N}_0$. As we reduce the lcs to the lcp by means of the reverse, we define the reverse of $w^\frown u$, denoted by $(w^\frown u)^R$, by means of $(w^\frown u)^R := u^R(w^R)^\omega$.

Definition 1. *Let Σ^{ulp} denote the set of all expressions of the form $w^\frown u$ with $u \in \Sigma^*$ and $w \in \Sigma^+$. Σ^{ulp} is called the set of* ultimately left-periodic words. *Define the reverse of an expression $w^\frown u \in \Sigma^{ulp}$ by means of $(w^\frown u)^R := u^R(w^R)^\omega$. Accordingly, set $(uw^\omega)^R := (w^R)^\frown u^R$ for $u \in \Sigma^*$, $w \in \Sigma^+$.*

The suffix order *on $\Sigma^* \cup \Sigma^{ulp} \cup \{\top\}$ is defined by $u \stackrel{s}{\sqsubseteq} v :\Leftrightarrow u^R \stackrel{p}{\sqsubseteq} v^R$. The longest common suffix (lcs) of a language $L \subseteq \Sigma^* \cup \Sigma^{ulp}$ is $\mathsf{lcs}(L) := \mathsf{lcp}(L^R)^R$.*

For instance, we have $\mathsf{lcs}((bba)^\frown, (ba)^\frown a) = a$, and $\mathsf{lcs}((ab)^\frown, (ba)^\frown b) = (ab)^\frown$.

As usual, we write $u \stackrel{s}{\sqsubset} v$ if $u \stackrel{s}{\sqsubseteq} v$, but $u \neq v$. As the lcp is the infimum w.r.t. $\stackrel{p}{\sqsubseteq}$, we also have for $x, y, z \in \{\top\} \cup \Sigma^* \cup \Sigma^{ulp}$ and $L, L' \subseteq \{\top\} \cup \Sigma^* \cup \Sigma^{ulp}$ that (i) $\mathsf{lcs}(x, y) = \mathsf{lcs}(y, x)$, (ii) $\mathsf{lcs}(x, \mathsf{lcs}(y, z)) = \mathsf{lcs}(x, y, z)$, (iii) $\mathsf{lcs}(L) \stackrel{s}{\sqsubseteq} \mathsf{lcs}(L')$ for $L \supseteq L'$, and (iv) $\mathsf{lcs}(Lx) = \mathsf{lcs}(L)x$ for $x \in \{\top\} \cup \Sigma^*$. In Lemma 8 in the appendix of the extended version [4] we derive further equalities for lcs that allow to simplify its computation. In particular, the following two equalities (for $x, y \in \Sigma^*$) are very useful:

$$\mathsf{lcs}(x, xy) = \mathsf{lcs}(x, y^\frown) = \mathsf{lcs}(x, xy^k) \quad \text{for every } k \geq 1$$

$$\mathsf{lcs}(x^\frown, y^\frown) = \begin{cases} (xy)^\frown & \text{if } xy = yx \\ \mathsf{lcs}(xy, x^\frown) = \mathsf{lcs}(xy, yx^k) & \text{if } xy \neq yx, \text{ for every } k \geq 1 \end{cases}$$

For instance, we have $\mathsf{lcs}((ab)^\frown, (bab)^\frown) = bab = \mathsf{lcs}(abbab, (ab)^\frown)$. Note also that by definition we have $\varepsilon^\frown = \top$ s.t. $\mathsf{lcs}(x^\frown, \varepsilon^\frown) = (x\varepsilon)^\frown$. We will use the following observation frequently:

Lemma 1. *Let $L \subseteq \Sigma^*$ be nonempty. Then for any $x \in L$ we have $\mathsf{lcs}(L) = \mathsf{lcs}(\mathsf{lcs}(x, z) \mid z \in L)$; in particular, there is some witness $y \in L$ (w.r.t. x) s.t. $\mathsf{lcs}(L) = \mathsf{lcs}(x, y)$.*

Involutive Monoid. We briefly recall the basic definitions and properties of the finitely generated involutive monoid, but refer the reader for details and a formal treatment to e.g. [11]. Let A be a finite alphabet (of opening brackets/letters). From A we derive the alphabet $\overline{\mathsf{A}} := \{\overline{a} \mid a \in \mathsf{A}\}$ (of closing brackets/letters) where we assume that $\mathsf{A} \cap \overline{\mathsf{A}} = \emptyset$. Set $\mathsf{B} := \mathsf{A} \cup \overline{\mathsf{A}}$. We use roman letters $p, q, \dots$ to denote words over A, while Greek letters $\alpha, \beta, \gamma, \dots$ will denote words over B.

We extend $\overline{\cdot}$ to an involution on B^* by means of $\overline{\varepsilon} := \varepsilon$, $\overline{\overline{a}} := a$ for all $a \in \mathsf{A}$, and $\overline{\alpha\beta} := \overline{\beta}\,\overline{\alpha}$ for all other $\alpha, \beta \in \mathsf{B}^*$. Let $\xrightarrow{\rho}$ be the binary relation on B^* defined by $\alpha a \overline{a} \beta \xrightarrow{\rho} \alpha\beta$ for any $\alpha, \beta \in \mathsf{B}^*$ and $a \in \mathsf{A}$, i.e. $\xrightarrow{\rho}$ cancels nondeterministically one pair of matching opening and closing brackets. A word $\alpha \in \mathsf{B}^*$ is *reduced* if it does not contain any infix of the form $a\overline{a}$ for any $a \in \mathsf{A}$, i.e. α is reduced if and only if it has no direct successor w.r.t. $\xrightarrow{\rho}$. For every $\alpha \in \mathsf{B}^*$ canceling all matching brackets in any arbitrary order always results in the same unique reduced word which we denote by $\rho(\alpha)$; we write $\alpha \stackrel{\rho}{=} \beta$ if $\rho(\alpha) = \rho(\beta)$. Then $\mathsf{B}^* / \stackrel{\rho}{=}$ is the free involutive monoid generated by A, and $\rho(\alpha)$ is the shortest word in the $\stackrel{\rho}{=}$-equivalence class of α. For $L \subseteq \mathsf{B}^*$ we set $\rho(L) := \{\rho(w) \mid w \in L\}$.

Well-Formed Languages and Context-Free Grammars. We are specifically interested in context-free grammars (CFG) G over the alphabet B. We write $\to_G$ for the rewrite rules of G. We assume that G is reduced to the productive nonterminals that are reachable from its axiom S. For simplicity, we assume for the proofs and constructions that the rules of G are of the form

$$X \to_G YZ \qquad X \to_G Y \qquad X \to_G \overline{u}\,v$$

for nonterminals X, Y, Z and $u, v \in \mathsf{A}^*$. We write $L_X := \{\alpha \in \mathsf{B}^* \mid X \to_G^* \alpha\}$ for the language generated by the nonterminal X. Specifically for the axiom S of G we set $L := L_S$. The height of a derivation tree w.r.t. G is measured in the maximal number of nonterminals occurring along a path from the root to any leaf, i.e. in our case any derivation tree has height at least 1. We write $L_X^{\leq h}$ for the subset of L_X of words that possess a derivation tree of height at most h s.t.:

$$L_X^{\leq 1} = \{\overline{u}\,v \mid X \to_G \overline{u}\,v\} \quad L_X^{\leq h+2} = L_X^{\leq h+1} \cup \bigcup_{X \to_G YZ} L_Y^{\leq h+1} L_Z^{\leq h+1} \cup \bigcup_{X \to_G Y} L_Y^{\leq h+1}$$

We will also write $L_X^{<h}$ for $L_X^{\leq h-1}$ and $L_X^{=h}$ for $L_X^{\leq h} \setminus L_X^{<h}$. The *prefix closure* of $L \subseteq \mathsf{B}^*$ is denoted by $\mathsf{Prf}(L) := \{\alpha' \mid \alpha'\alpha'' \in L\}$.

Definition 2. *Let $\alpha \in \mathsf{B}^*$ and $L \subseteq \mathsf{B}^*$.*

1. *Let $\Delta(\alpha) := |\alpha|_\mathsf{A} - |\alpha|_{\overline{\mathsf{A}}}$ be the difference of opening brackets to closing brackets. α is nonnegative if $\forall \alpha' \stackrel{p}{\sqsubseteq} \alpha \colon \Delta(\alpha') \geq 0$. $L \subseteq \mathsf{B}^*$ is nonnegative if every $\alpha \in L$ is nonnegative.*

2. *A context-free grammar G with $L(G) \subseteq B^*$ is* nonnegative *if $L(G)$ is nonnegative. For a nonterminal X of G let $d_X := \sup(\{-\Delta(\alpha') \mid \alpha'\alpha'' \in L_X\})$.*

3. *A word α is* well-formed (short: wwf) *resp.* well-formed (short: wf) *if $\rho(\alpha) \in \overline{A}^*A^*$ resp. if $\rho(\alpha) \in A^*$. A context-free grammar G is* wf *if $L(G)$ is wf. $L \subseteq B^*$ is* wwf *resp.* wf *if every word of L is wwf resp. wf.*

4. *A context-free grammar G is* bounded well-formed (bwf) *if it is wwf and for every nonterminal X there is a (shortest) word $r_X \in A^*$ with $|r_X| = d_X$ s.t. $r_X L_X$ is wf.*

Note that $d_X \geq 0$ as we can always choose $\alpha' = \varepsilon$ in the definition of d_X.

As already mentioned in the abstract and the introduction, we have that L is wf *iff* $\mathsf{Prf}(L)$ is wf *iff* L is a subset of the prefix closure of the Dyck language generated by $S \to \varepsilon$, $S \to SS$, $S \to aS\overline{a}$ (for $a \in A$). We state some further direct consequences of above definition: (i) L is nonnegative *iff* the image of L under the homomorphism that collapses A to a singleton is wf. Hence, if L is wf, then L is nonnegative. Δ is an ω-continuous homomorphism from the language semiring generated by B to the tropical semiring $\langle \mathbb{Z} \cup \{-\infty\}, \min, + \rangle$. Thus it is decidable in polynomial time if G is nonnegative using the Bellman-Ford algorithm [2]. (ii) If L is not wf, then there exists some $\alpha \in \mathsf{Prf}(L) \setminus \{\varepsilon\}$ s.t. $\Delta(\alpha) < 0$ or $\alpha \stackrel{\rho}{=} ua\overline{b}$ for $u \in A^*$ and $a, b \in A$ (with $a \neq b$). (iii) If L_X is wwf, then $d_X = \sup\{|y| \mid \gamma \in L_X, \rho(\gamma) = \overline{y}\,z\}$.

In particular, because of context-freeness, it follows that, if G is wf, then for every nonterminal X there is $r_X \in A^*$ s.t. (i) $\overline{r_X} \in \rho(\mathsf{Prf}(L_X))$, (ii) $|r_X| = d_X$ and (iii) $r_X L_X$ is wf. Hence:

Lemma 2. *A context-free grammar G is wf iff G is bwf with $r_S = \varepsilon$ for S the axiom of G.*

The words r_X mentioned in the definition of bounded well-formedness can be computed in polynomial time using the Bellman-Ford algorithm similar to [13]; more precisely, a *straight-line program* (SLP) (see e.g. [6] for more details on SLPs), i.e. a context-free grammar generating exactly one derivation tree and thus word, can be extracted from G for each r_X.

Lemma 3. *Let $L = L(G)$ be wf. Let X be some nonterminal of G. Let $r_X \in A^*$ be the shortest word s.t. $r_X L_X$ is wf. We can compute an SLP for r_X from G in polynomial time.*

Tree-to-Word Transducers. We define a *linear* tree-to-word transducer (LT$_B$) $M = (\Sigma, B, Q, S, R)$ where Σ is a finite ranked input alphabet, B is the finite (unranked) output alphabet, Q is a finite set of states, the axiom S is of the form u_0 or $u_0 q(x_1) u_1$ with $u_0, u_1 \in B^*$ and R is a set of rules of the form $q(f(x_1, \ldots, x_m)) \to u_0 q_1(x_{\sigma(1)}) u_1 \ldots q_n(s_{\sigma(n)}) u_n$ with $q, q_i \in Q$, $f \in \Sigma$, $u_i \in B^*$, $n \leq m$ and σ an injective mapping from $\{1, \ldots, n\}$ to $\{1, \ldots, m\}$. Since non-deterministic choices of linear transducers can be encoded into the input symbols, we may, w.l.o.g., consider *deterministic* transducers only. For simplicity, we moreover assume the transducers to be *total*. This restriction can be lifted by additionally taking a top-down deterministic tree automaton for the domain into account. The constructions introduced in Sect. 3 would then have to be applied w.r.t. such a domain tree automaton. As we consider total deterministic transducers there is exactly one rule for each pair $q \in Q$ and $f \in \Sigma$.

A 2-copy tree-to-word transducer (2-TW) is a tuple $N = (\Sigma, \mathsf{B}, Q, S, R)$ that is defined in the same way as an $\mathsf{LT_B}$ but the axiom S is of the form u_0 or $u_0 q_1(x_1) u_1 q_2(x_1) u_2$, with $u_i \in \mathsf{B}^*$.

$\mathcal{T}_\Sigma$ denotes the set of all trees/terms over Σ. We define the semantics $[\![q]\!]$: $\mathcal{T}_\Sigma \to \mathsf{B}^*$ of a state q with rule $q(f(t_1, \ldots, t_m)) \to u_0 q_1(t_{\sigma(1)}) u_1 \ldots q_n(t_{\sigma(n)}) u_n$ inductively by

$$[\![q]\!](f(t_1, \ldots, t_m)) = \rho(u_0 [\![q_1]\!](t_{\sigma(1)}) u_1 \ldots [\![q_n]\!](t_{\sigma(n)}) u_n)$$

The semantics $[\![M]\!]$ of an $\mathsf{LT_B}$ M with axiom u_0 is given by $\rho(u_0)$; if the axiom is of the form $u_0 q(x_1) u_1$ it is defined by $\rho(u_0 [\![q]\!](t) u_1)$ for all $t \in \mathcal{T}_\Sigma$; while the semantics $[\![N]\!]$ of a 2-TW N with axiom u_0 is again given by $\rho(u_0)$ and for axiom $u_0 q_1(x_1) u_1 q_2(x_1) u_2$ it is defined by $\rho(u_0 [\![q_1]\!](t) u_1 [\![q_2]\!](t) u_2)$ for all $t \in \mathcal{T}_\Sigma$. For a state q we define the output language $\mathcal{L}(q) = \{[\![q]\!](t) \mid t \in \mathcal{T}_\Sigma\}$; For a 2-TW M we let $\mathcal{L}(M) = \{[\![M]\!](t) \mid t \in \mathcal{T}_\Sigma\}$. Note that the output language of an $\mathsf{LT_B}$ is context-free and a corresponding context-free grammar for this language can directly read from the rules of the transducer.

Additionally, we may assume w.l.o.g. that all states q of an $\mathsf{LT_B}$ are *non-singleton*, i.e., $\mathcal{L}(q)$ contains at least two words. We call a 2-TW M *balanced* if $\mathcal{L}(M) = \{\varepsilon\}$. We say an $\mathsf{LT_B}$ M is *well-formed* if $\mathcal{L}(M) \subseteq \mathsf{A}^*$. Balanced and well-formed states are defined analogously. We use $\overline{q}$ to denote the inverse transduction of q which is obtained from a copy of the transitions reachable from q by involution of the right-hand side of each rule. As a consequence, $[\![\overline{q}]\!](t) = \overline{[\![q]\!](t)}$ for all $t \in \mathcal{T}_\Sigma$, and thus, $\mathcal{L}(\overline{q}) = \overline{\mathcal{L}(q)}$. We say that two states q, q' are *equivalent* iff for all $t \in \mathcal{T}_\Sigma$, $[\![q]\!](t) = [\![q']\!](t)$. Accordingly, two 2-TWs M, M' are equivalent iff for all $t \in \mathcal{T}_\Sigma$, $[\![M]\!](t) = [\![M']\!](t)$.

3   Balancedness of 2-TWs

Let M denote a 2-TW. W.l.o.g., we assume that the axiom of M is of the form $q_1(x_1) q_2(x_1)$ for two states q_1, q_2. If this is not yet the case, an equivalent 2-TW with this property can be constructed in polynomial time. We reduce balancedness of M to decision problems for *linear* tree-to-word transducers alone.

Proposition 1. *The 2-TW M is balanced iff the following two properties hold:*

- *Both $\mathcal{L}(q_1)$ and $\overline{\mathcal{L}(q_2)}$ are well-formed;*
- *q_1 and $\overline{q_2}$ are equivalent.*

Proof. Assume first that M with axiom $q_1(x_1) q_2(x_1)$ is balanced, i.e., $\mathcal{L}(M) = \varepsilon$. Then for all w', w'' with $w = w'w'' \in \mathcal{L}(M)$, $\rho(w') = u \in \mathsf{A}^*$ and $\rho(w'') = \overline{u}$. Thus, both $\mathcal{L}(q_1)$ and $\overline{\mathcal{L}(q_2)}$ consist of well-formed words only. Assume for a contradiction that q_1 and $\overline{q_2}$ are not equivalent. Then there is some $t \in \mathcal{T}_\Sigma$ such that $[\![q_1]\!](t) \not\stackrel{\rho}{=} [\![\overline{q_2}]\!](t)$. Let $[\![q_1]\!](t) = u \in A^*$ and $[\![\overline{q_2}]\!](t) = \overline{[\![q_2]\!](t)} = v$ with $v \in A^*$ and $u \neq v$. Then $\rho([\![q_1]\!](t)[\![q_2]\!](t)) = \rho(u\overline{v}) \neq \varepsilon$ as $u \neq v$, $u, v \in A^*$. Since M is balanced, this is not possible.

Now, assume that $\mathcal{L}(q_1)$ and $\overline{\mathcal{L}(q_2)}$ are well-formed, i.e., for all $t \in \mathcal{T}_\Sigma$, $[\![q_1]\!](t) \in A^*$ and $[\![q_2]\!](t) \in \overline{A}^*$. Additionally assume that q_1 and $\overline{q_2}$ are equivalent, i.e., for all $t \in \mathcal{T}_\Sigma$, $[\![q_1]\!](t) = [\![\overline{q_2}]\!](t) = \overline{[\![q_2]\!](t)}$. Therefore for all $t \in \mathcal{T}_\Sigma$, $[\![q_2]\!](t) = \overline{[\![q_1]\!](t)}$ and hence,

$$\rho([\![q_1]\!](t)[\![q_2]\!](t)) = \rho([\![q_1]\!](t)\overline{[\![q_1]\!](t)}) = \varepsilon$$

Therefore, the 2-TW M must be balanced.   □

The output languages of states q_1 and $\overline{q_2}$ are generated by means of context-free grammars of polynomial size.

Example 1. Consider $\mathsf{LT_B}$ M with input alphabet $\Sigma = \{f^{(2)}, g^{(0)}\}$ (the superscript denotes the rank), output alphabet $B = \{a, \overline{a}\}$, axiom $q_3(x_1)$ and rules

$$\begin{aligned}
q_3(f(x_1, x_2)) &\rightarrow aq_2(x_1)q_2(x_2)\overline{a} & q_2(g) &\rightarrow \varepsilon \\
q_2(f(x_1, x_2)) &\rightarrow aq_1(x_1)q_1(x_2)\overline{a} & q_2(g) &\rightarrow \varepsilon \\
q_1(f(x_1, x_2)) &\rightarrow q_3(x_1)q_3(x_2) & q_1(g) &\rightarrow aa
\end{aligned}$$

We obtain a CFG producing exactly the output language of M by nondeterministically guessing the input symbol, i.e. the state q_i becomes the nonterminal W_i. The axiom of this CFG is then W_3, and as rules we obtain

$$W_3 \rightarrow aW_2W_2\overline{a} \mid \varepsilon \quad W_2 \rightarrow aW_1W_1\overline{a} \mid \varepsilon \quad W_1 \rightarrow W_3W_3 \mid aa$$

Note that the rules of M and the associated CFG use a form of iterated squaring, i.e. $W_3 \rightarrow^2 W_3^4$, that allows to encode potentially exponentially large outputs within the rules (see also Example 4). In general, words thus have to be stored in compressed form as SLPs [6].

Therefore, Theorem 2 of Sect. 4 implies that well-formedness of $q_1, \overline{q_2}$ can be decided in polynomial time. Accordingly, it remains to consider the equivalence problem for well-formed $\mathsf{LT_Bs}$. Since the two transducers in question are well-formed, they are equivalent as $\mathsf{LT_Bs}$ iff they are equivalent when their outputs are considered over the free group $\mathcal{F}_A$. In the free group $\mathcal{F}_A$, we additionally have that $\overline{a}\, a \stackrel{\rho}{=} \varepsilon$—which does not hold in our rewriting system. If sets $\mathcal{L}(q_1), \mathcal{L}(\overline{q_2})$ of outputs for q_1 and $\overline{q_2}$, however, are well-formed, it follows for all $u \in \mathcal{L}(q_1), v \in \mathcal{L}(\overline{q_2})$ that $\rho(u\overline{v}) = \rho(\rho(u)\rho(\overline{v}))$ cannot contain $\overline{a}\, a$. Therefore, $\rho(u\overline{v}) = \varepsilon$ iff $u\overline{v}$ is equivalent to ε over the free group $\mathcal{F}_A$. In [5, Theorem 2], we have proven that equivalence of $\mathsf{LT_Bs}$ where the output is interpreted over the free group, is decidable in polynomial time. Thus, we obtain our main theorem.

Theorem 1. *Balancedness of 2-TWs is decidable in polynomial time.*

4 Deciding Well-Formedness of Context-Free Grammars

As described in the preceding sections, given a 2-TW we split it into the two underlying LT$_{BS}$s that process a copy of the input tree. We then check that each of these two LT$_{BS}$s are equivalent w.r.t. the free group. As sketched in Example 1 we obtain a context-free grammar for the output language of each of these LT$_{BS}$s. It then remains to check that both context-free grammars are well-formed. In order to prove that we can decide in polynomial time whether a context-free grammar is well-formed (short: wf), we proceed as follows:

First, we introduce in Definition 3 the *maximal suffix extension* of a language $L \subseteq \Sigma^*$ w.r.t. the lcs (denoted by $\mathsf{lcsx}(L)$), i.e. the longest word $u \in \Sigma^\infty$ s.t. $\mathsf{lcs}(uL) = u\,\mathsf{lcs}(L)$. We then show that the relation $L \approx_{\mathsf{lcs}} L' :\Leftrightarrow \mathsf{lcs}(L) = \mathsf{lcs}(L') \wedge \mathsf{lcsx}(L) = \mathsf{lcsx}(L')$ is an equivalence relation on Σ^* that respects both union and concatenation of languages (see Lemma 5). It then follows that for every language $L \subseteq \Sigma^*$ there is some subset $\mathsf{T}_{\mathsf{lcs}}(L) \subseteq L$ of size at most 3 with $L \approx_{\mathsf{lcs}} \mathsf{T}_{\mathsf{lcs}}(L)$.

We then use $\mathsf{T}_{\mathsf{lcs}}$ to compute a finite $\approx_{\mathsf{lcs}}$-equivalent representation $\mathsf{T}_X^{\leq h}$ of the *reduced* language generated by each nonterminal X of the given context-free grammar inductively for increasing derivation height h. In particular, we show that we only have to compute up to derivation height $4N + 1$ (with N the number of nonterminals) in order to decide whether G is wf: In Lemma 7 we show that, if G is wf, then we have to have $\mathsf{T}_X^{\leq 4N+1} \approx_{\mathsf{lcs}} \mathsf{T}_X^{\leq 4N}$ for all nonterminals X of G. The complementary result is then shown in Lemma 6, i.e. if G is not wf, then we either cannot compute up to $\mathsf{T}_X^{\leq 4N+1}$ as we discover some word that is not wf, or we have $\mathsf{T}_X^{\leq 4N} \not\approx_{\mathsf{lcs}} \mathsf{T}_X^{\leq 4N+1}$ for at least one nonterminal X.

Maximal Suffix Extension and lcs-Equivalence. We first show that we can compute the longest common suffix of the union $L \cup L'$ and the concatenation LL' of two languages $L, L' \subseteq \Sigma^*$ if we know both $\mathsf{lcs}(L)$ and $\mathsf{lcs}(L')$, and in addition, the longest word $\mathsf{lcsx}(L)$ resp. $\mathsf{lcsx}(L')$ by which we can extend $\mathsf{lcs}(L)$ resp. $\mathsf{lcs}(L')$ when concatenating another language from left. In contrast to the computation of the lcp presented in [7], we have to take the maximal extension lcsx explicitly into account. In this paragraph we do not consider the involution, thus let Σ denote an arbitrary alphabet.

Definition 3. *For $L \subseteq \Sigma^*$ with $R = \mathsf{lcs}(L)$ the maximal suffix extension (lcsx) of L is defined by $\mathsf{lcsx}(L) := \mathsf{lcs}(z^m \mid zR \in L)$.*

Recall that by definition $\mathsf{lcsx}(\emptyset) = \mathsf{lcs}(\emptyset) = \top$ and $\mathsf{lcsx}(\{R\}) = \mathsf{lcs}(\varepsilon^m) = \top$. The following example motivates the definition of lcsx:

Example 2. Consider the language $L = \{R, xR, yR\}$ with $\mathsf{lcs}(L) = R$ and $\mathsf{lcsx}(L) = \mathsf{lcs}(x^m, y^m)$. Assume we prepend some word $u \in \Sigma^*$ to L resulting in the language $uL = \{uR, uxR, uyR\}$, see the following picture for an illustration (dotted boxes represent copies of $z \in \{x, y\}$ stemming from the usual line of argumentation that, if z is a suffix of $u = u'z$, then $uzR = u'zzR$, and thus eventually covering all of u by z^m):

x	x	x	x	R
u			x	R
		u		R
u			y	R
y		y		R
y		y	y	R

As motivated by the picture, $\mathsf{lcs}(uL)$ is given by $\mathsf{lcs}(u, x^m, y^m)R$. Using the concept of ultimately left-periodic words, we may also formalize this as follows:

$$
\begin{aligned}
\mathsf{lcs}(u\{xR, yR, R\}) &= \mathsf{lcs}(u, ux, uy)R \\
&= \mathsf{lcs}(\mathsf{lcs}(u, ux), \mathsf{lcs}(u, uy))R \quad (\text{as } \mathsf{lcs}(u, ux) = \mathsf{lcs}(u, x^m)) \\
&= \mathsf{lcs}(\mathsf{lcs}(u, x^m), \mathsf{lcs}(u, y^m))R \\
&= \mathsf{lcs}(u, \mathsf{lcs}(x^m, y^m))R \qquad\qquad = \mathsf{lcs}(u, \mathsf{lcsx}(L))\mathsf{lcs}(L)
\end{aligned}
$$

In particular, if $xy = yx$, we can extend lcs by any finite suffix of $\mathsf{lcsx}(L) = (xy)^m$ (note that, if $x = \varepsilon = y$, then $\mathsf{lcsx}(L) = \varepsilon^m = \top$ is defined to be the greatest element w.r.t. $\sqsubseteq^s$); but if $xy \neq yx$, we can extend it at most to $\mathsf{lcsx}(L) = \mathsf{lcs}(x^m, y^m) = \mathsf{lcs}(xy, yx) \sqsubset^s xy$. Essentially, only three cases can arise as illustrated by the following three examples:

First, consider $L_1 = \{ab, cb\}$ with $\mathsf{lcs}(L_1) = b$. Obviously, for every word $u \in \Sigma^*$ we have that $\mathsf{lcs}(uL_1) = \mathsf{lcs}(L_1)$ and so we should have $\mathsf{lcsx}(L_1) = \varepsilon$. Instantiating the definition we obtain indeed $\mathsf{lcsx}(L_1) = \mathsf{lcs}(a^m, b^m) = \mathsf{lcs}(\varepsilon) = \varepsilon$.

As another example consider $L_2 = \{a, baa\}$ with $\mathsf{lcs}(L_2) = a$. Here, we obtain $\mathsf{lcsx}(L_2) = \mathsf{lcs}(\varepsilon^m, (ba)^m) = \mathsf{lcs}(\top, (ba)^m) = (ba)^m$, i.e. the suffix of L_2 can be extended by any finite suffix of $(ba)^m = \ldots bababa$.

Finally, consider $L_3 = \{b, ba^n b, aba^n b\}$ with $\mathsf{lcs}(L_3) = b$ for some fixed $n \in \mathbb{N}$. As mentioned in Sect. 2, we have $\mathsf{lcs}(x^m, y^m) = \mathsf{lcs}(xy, yx)$ for $xy \neq yx$. We thus obtain in this case $\mathsf{lcs}((ba^n)^m, (aba^n)^m) = \mathsf{lcs}(ba^n\, aba^n, aba^n\, ba^n) = a^n ba^n$. The classic result by Fine and Wilf states that, if $xy \neq yx$, then $|\mathsf{lcs}(x^m, y^m)| < |x| + |y| - \gcd(|x|, |y|)$. Thus $x = ba^n$ and $y = aba^n$ constitute an extremal case where the lcs is only finitely extendable.

If $\mathsf{lcs}(L)$ is not contained in L, then $\mathsf{lcs}(L)$ has to be a strict suffix of every shortest word in L, and thus immediately $\mathsf{lcsx}(L) = \varepsilon$. As in the case of the lcs, also $\mathsf{lcsx}(L)$ is already defined by two words in L:

Lemma 4. *Let $L \subseteq \Sigma^*$ with $|L| \geq 2$ and $R := \mathsf{lcs}(L)$. Fix any $xR \in L \setminus \{R\}$. Then there is some $yR \in L \setminus \{R\}$ s.t. $\mathsf{lcsx}(L) = \mathsf{lcs}(x^m, y^m) = \mathsf{lcs}(x^m, y^m, z^m)$ for all $zR \in L$. If $xy = yx$, then $R \in L$.*

We show that we can compute the lcs and the extension lcsx of the union resp. the concatenation of two languages solely from their lcs and lcsx. To this end, we define the lcs-*summary* of a language as:

Definition 4. *For $L \subseteq \Sigma^*$ set $\pi_{lcs}(L) := (\mathsf{lcs}(L), \mathsf{lcsx}(L))$. The equivalence relation $\approx_{lcs}$ on 2^{Σ^*} is defined by: $L \approx_{lcs} L'$ iff $\pi_{lcs}(L) = \pi_{lcs}(L')$.*

Lemma 5. *Let $L, L' \subseteq \Sigma^*$ with $\pi_{lcs}(L) = (R, E)$ and $\pi_{lcs}(L') = (R', E')$. If $L = \emptyset$ or $L' = \emptyset$, then $\pi_{lcs}(L \cup L') = (\mathsf{lcs}(R, R'), \mathsf{lcs}(E, E'))$, and $\pi_{lcs}(LL') = (\top, \top)$. Assume thus $L \neq \emptyset \neq L'$ which implies $R \neq \top \neq R'$. Then:*

- $\text{lcs}(L \cup L') = \text{lcs}(R, R')$ and $\text{lcs}(LL') = \text{lcs}(R, E')R'$.
- If $\text{lcs}(R, R') \notin \{R, R'\}$, then $\text{lcsx}(L \cup L') = \varepsilon$; else w.l.o.g. $R' = \delta R$ and $\text{lcsx}(L \cup L') = \text{lcs}(E, \text{lcs}(E', E'\delta)\delta)$.
- If $\text{lcs}(R, E') \stackrel{s}{\sqsubseteq} R$, then $\text{lcsx}(LL') = \varepsilon$; else $E' = \delta R$ and $\text{lcsx}(LL') = \text{lcs}(E, \delta)$.

Example 3. Lemma 5 can be illustrated as follows:

	lcsx(L)				lcs(L)
(L ∪ L')	δ	δ	δ	δ	lcs(L')
		lcsx(L')			lcs(L')

	lcsx(L)		lcs(L)	lcs(L')
(LL')		δ	lcs(L)	lcs(L')
	lcsx(L')			lcs(L')

For instance, consider $L = \{a, baa\}$ and $L' = \{aa, baaa\}$ s.t. $\pi_{\text{lcs}}(L) = (a, (ba)^m)$ and $\pi_{\text{lcs}}(L') = (aa, (ba)^m)$. Applying Lemma 5, we obtain for the union $\text{lcs}(L \cup L') = \text{lcs}(a, aa) = a$ and $\text{lcsx}(L \cup L') = \text{lcs}((ba)^m, \text{lcs}((ba)^m, (ba)^m a)a) = a$. In case of the concatenation, Lemma 5 yields $\text{lcs}(LL') = \text{lcs}(a, (ba)^m)aa = aaa$ and $\text{lcsx}(LL') = \text{lcs}((ba)^m, (ab)^m) = \varepsilon$.

As both the lcs and the lcsx are determined by already two words (cf. Lemmas 1 and 4), it follows that every $L \subseteq \Sigma^*$ is $\approx_{\text{lcs}}$-equivalent to some sublanguage $\mathsf{T}_{\text{lcs}}(L) \subseteq L$ consisting of at most three words where the words xR, yR can be chosen arbitrarily up to the stated constraints (with $R = \text{lcs}(L)$):

$$\mathsf{T}_{\text{lcs}}(L) := \begin{cases} L & \text{if } |L| \le 2 \\ \{R, xR, yR\} & \text{if } \{R, xR, yR\} \subseteq L \wedge \text{lcsx}(L) = \text{lcs}(x^m, y^m) \\ \{xR, yR\} & \text{if } R = \text{lcs}(xR, yR) \wedge R \notin L \wedge \{xR, yR\} \subseteq L \end{cases}$$

Deciding Well-Formedness. For the following, we assume that G is a context-free grammar over $\mathsf{B} = \mathsf{A} \cup \overline{\mathsf{A}}$ with nonterminals $\mathfrak{X}$. Set $N := |\mathfrak{X}|$. We further assume that G is nonnegative, and that we have computed for every nonterminal X of G a word $r_X \in \mathsf{A}^*$ (represented as an SLP) s.t. $|r_X| = d_X$ and $\overline{r_X} \in \text{Prf}(\rho(L_X))$.[1] In order to decide whether G is wf we compute the languages $\rho(r_X L_X^{\le h})$ modulo $\approx_{\text{lcs}}$ for increasing derivation height h using fixed-point iteration. Assume inductively that (i) $r_X L_X^{\le h}$ is wf and (ii) that we have computed $\mathsf{T}_X^{\le h} := \mathsf{T}_{\text{lcs}}(\rho(r_X L_X^{\le h})) \approx_{\text{lcs}} \rho(r_X L_X^{\le h})$ for all $X \in \mathfrak{X}$ up to height h. Then we can compute $\mathsf{T}_{\text{lcs}}(\rho(r_X L_X^{\le h+1}))$ for each nonterminal as follows:

$$\rho(r_X L_X^{\le h+1})$$
$$= \rho(r_X L_X^{\le h}) \cup \bigcup_{X \to_G Y} \rho(r_X \overline{r_Y} \ r_Y L_Y^{\le h}) \cup \bigcup_{X \to_G YZ} \rho(r_X \overline{r_Y} \ r_Y L_Y^{\le h} \ \overline{r_Z} \ r_Z L_Z^{\le h})$$
$$\approx_{\text{lcs}} \mathsf{T}_X^{\le h} \cup \bigcup_{X \to_G Y} \rho(r_X \overline{r_Y} \ \mathsf{T}_Y^{\le h}) \cup \bigcup_{X \to_G YZ} \rho(r_X \overline{r_Y} \ \mathsf{T}_Y^{\le h} \ \overline{r_Z} \ \mathsf{T}_Z^{\le h})$$
$$\approx_{\text{lcs}} \mathsf{T}_{\text{lcs}}\Big(\rho\Big(\mathsf{T}_X^{\le h} \cup \bigcup_{X \to_G Y} r_X \overline{r_Y} \ \mathsf{T}_Y^{\le h} \cup \bigcup_{X \to_G YZ} r_X \overline{r_Y} \ \mathsf{T}_Y^{\le h} \ \overline{r_Z} \ \mathsf{T}_Z^{\le h}\Big)\Big) =: \mathsf{T}_X^{\le h+1}$$

---

[1] $\overline{r_X}$ is (after reduction) a longest word of closing brackets in $\rho(L_X)$ (if G is wf, then r_X is unique). An SLP encoding r_X can be computed in polynomial time while checking that G is nonnegative; see Definition 2 and the subsequent explanations, and the proof of Lemma 3 in the appendix of the extended version [4]. All required operations on words run in time polynomial in the size of the SLPs representing the words, see e.g. [6].

Note that, if all constants $r_X \overline{r_Y}$ and all $\mathsf{T}_{\overline{X}}^{\leq h}$ are wf, but G is not wf, then the computation has to fail while computing $r_X \overline{r_Y} \, \mathsf{T}_{\overline{Y}}^{\leq h} \overline{r_Z}$; see the following example.

Example 4. Consider the nonnegative context-free grammar G given by the rules (with the parameter $n \in \mathbb{N}$ fixed)

$$S \to Uc \quad U \to AV \mid W_n \quad V \to U\overline{B} \quad W_i \to W_{i-1}W_{i-1} \quad (2 \leq i \leq n)$$
$$A \to a \qquad B \to b \qquad \overline{B} \to \overline{b} \qquad W_1 \to BB$$

with axiom S. Except for $\overline{B}$ all nonterminals generate nonnegative languages. Note that the nonterminals W_n to W_1 form an SLP that encodes the word b^{2^n} by means of iterated squaring which only becomes productive at height $h = n + 1$. For $h \geq n + 3$ we have:

$$L_S^{\leq h} = \{a^k b^{2^n} \overline{b}^k c \mid k \leq \lfloor \tfrac{h-(n+3)}{2} \rfloor\}$$
$$L_U^{\leq h} = \{a^k b^{2^n} \overline{b}^k \mid k \leq \lfloor \tfrac{h-(n+2)}{2} \rfloor\} \qquad L_{W_i}^{\leq h} = \{b^{2^i}\} \quad L_B^{\leq h} = \{b\}$$
$$L_V^{\leq h} = \{a^k b^{2^n} \overline{b}^{k+1} \mid k \leq \lfloor \tfrac{h-(n+3)}{2} \rfloor\} \quad L_A^{\leq h} = \{a\} \quad L_{\overline{B}}^{\leq h} = \{\overline{b}\}$$

Here the words r_X used to cancel the longest prefix of closing brackets (after reduction) are $r_S = r_U = r_V = r_W = r_A = r_B = \varepsilon$ and $r_{\overline{B}} = b$. Note that $r_X L_{\overline{X}}^{\leq h}$ is wf for all nonterminals X up to $h \leq h_0 = 2^{n+1} + (n+2)$ s.t. $\mathsf{T}_{\mathsf{lcs}}(\rho(r_S L_{\overline{S}}^{\leq h})) \approx_{\mathsf{lcs}} \mathsf{T}_{\overline{S}}^{\leq h} = \{b^{2^n} c, a^k b^{2^n - k(h)} c\}$ for $k(h) = \lfloor (h - (n+3))/2 \rfloor$ and $n + 3 \leq h \leq h_0$; in particular, the lcs of $\mathsf{T}_{\overline{S}}^{\leq h}$ has already converged to c at $h = n+3$, only its maximal extension lcsx changes for $n + 3 \leq h \leq h_0$. We discover the first counterexample $a^{2^n} \overline{b}$ that G is not wf while computing $\mathsf{T}_{\overline{V}}^{\leq h_0 + 1} = \mathsf{T}_{\mathsf{lcs}}(\rho(\mathsf{T}_{\overline{U}}^{\leq h_0} \overline{b}))$.

As illustrated in Example 4, if G is not wf, then the minimal derivation height $h_0 + 1$ at which we discover a counterexample might be exponential in the size of the grammar. The following lemma states that up to this derivation height h_0 the representations $\mathsf{T}_{\overline{X}}^{\leq h}$ cannot have converged (modulo $\approx_{\mathsf{lcs}}$).

Lemma 6. *If $L = L(G)$ is not wf, then there is some least h_0 s.t. $r_X L_{\overline{Y}}^{\leq h_0} \overline{r_Z}$ is not wf with $X \to_G YZ$. For $h \leq h_0$, all $r_X L_{\overline{X}}^{\leq h}$ are wf s.t. $\mathsf{T}_{\overline{X}}^{\leq h} \approx_{\mathsf{lcs}} \rho(r_X L_{\overline{X}}^{\leq h})$. If $h_0 \geq 4N + 1$, then at least for one nonterminal X we have $\mathsf{T}_{\overline{X}}^{\leq 4N+1} \not\approx_{\mathsf{lcs}} \mathsf{T}_{\overline{X}}^{\leq 4N}$.*

The following Lemma 7 states the complementary result, i.e. if G is wf then the representations $\mathsf{T}_{\overline{X}}^{\leq h}$ have converged at the latest for $h = 4N$ modulo $\approx_{\mathsf{lcs}}$. The basic idea underlying the proof of Lemma 7 is similar to [7]: we show that from every derivation tree of height at least $4N + 1$ we can construct a derivation tree of height at most $4N$ such that both trees carry the same information w.r.t. the lcs (after reduction). In contrast to [7] we need not only to show that $\mathsf{T}_{\overline{X}}^{\leq 4N}$ has the same lcs as $\rho(r_X L_{\overline{X}}^{\leq 4N})$, but that $\mathsf{T}_{\overline{X}}^{\leq 4N}$ has converged modulo $\approx_{\mathsf{lcs}}$ if G is wf; to this end, we need to explicitly consider lcsx, and re-prove stronger versions of the results regarding the combinatorics on words which take the involution into account (see A.6 in the appendix of the extended version [4]).

Lemma 7. *Let G be a context-free grammar with N nonterminals and $L(G)$ be wf. For every nonterminal X let $r_X \in A^*$ s.t. $|r_X| = d_X$ and $r_X L_X$ wf. Then $\rho(r_X L_X) \approx_{lcs} \rho(r_X L_X^{\leq 4N})$, and $T_X^{\leq 4N} \approx_{lcs} T_X^{\leq 4N+1}$ for every nonterminal X.*

The following example sketches the main idea underlying the proof of Lemma 7.

Example 5. The central combinatorial observation[2] is that for any well-formed language $\mathcal{L} \subseteq B^*$ of the form

$$\mathcal{L} = (\alpha, \beta)[(\mu_1, \nu_1) + (\mu_2, \nu_2)]^* \gamma := \{\alpha \mu_{i_1} \dots \mu_{i_l} \gamma \nu_{i_l} \dots \nu_{i_1} \beta \mid i_1 \dots i_l \in \{1,2\}^*\}$$

we have that its *longest common suffix after reduction* $\mathsf{lcs}^\rho(\mathcal{L}) := \mathsf{lcs}(\rho(\mathcal{L}))$ is determined by the reduced longest common suffix of $\alpha \gamma \beta$ and either $(\alpha, \beta)(\mu_i, \nu_i)\gamma = \alpha \mu_i \gamma \nu_i \beta$ or $(\alpha, \beta)(\mu_i, \nu_i)(\mu_j, \nu_j)\gamma = \alpha \mu_i \mu_j \gamma \nu_j \nu_i \beta$ for some $i \in \{1,2\}$ but *arbitrary* $j \in \{1,2\}$ in the latter case.[3]

Assume now we are given a context-free grammar G with N variables. Further assume that $L := L(G)$ is well-formed. W.l.o.g. G is in Chomsky normal form and reduced to the productive nonterminals reachable from the axiom of G. Let $L^{\leq 4N}$ denote the sublanguage of words generated by G with a derivation tree of height at most $4N$. Pick a shortest (before reduction) word $\kappa_0 \in L := L(G)$. Then there is some $\kappa_1 \in L$ with $R := \mathsf{lcs}^\rho(L) = \mathsf{lcs}^\rho(\kappa_0, \kappa_1)$; we will call any such word a *witness (w.r.t. κ_0)* in the following. If $R \in L$, then $\kappa_0 \stackrel{\rho}{=} R$, and any word in L is a witness. In particular, there is a witness in $L^{\leq 4N}$. So assume $R \notin L$. Then we may factorize (in a unique way) $\kappa_0 = \kappa_0'' a \kappa_0'$ and $\kappa_1 = \kappa_1'' b \kappa_1'$ such that $\rho(\kappa_0') = R = \rho(\kappa_1')$ where $a, b \in A$ with $a \neq b$. Then $\rho(\kappa_0) = z_0' a R$ and $\rho(\kappa_1) = z_1' b R$. Further assume that $\kappa_1 \notin L^{\leq 4N}$, otherwise we are done. Fix any derivation tree t of κ_1, and fix within t the *main* path from the root of t to the last letter b of the suffix $b\kappa_1'$ of $\rho(\kappa_1)$ (the dotted path in Fig. 1). We may assume that any path starting at a node on this main path and then immediately turning left towards a letter within the prefix κ_1'' consists of at most N nonterminals: if any nonterminal occurs twice the induced pumping tree can be pruned without changing the suffix $b\kappa_1'$; as the resulting tree is still a valid derivation tree w.r.t. G, we obtain another witness w.r.t. κ_0. Thus consider any path (including the main path) in t that leads from its root to a letter within the suffix $b\kappa_1'$. If every such path consists of at most $3N$ nonterminals, then every path in t consists of at most $4N$ nonterminals so that $\kappa_1 \in L^{\leq 4N}$ follows. Hence, assume there is at least one such path consisting of $3N + 1$ nonterminals. Then there is some nonterminal X occurring at least four times on this path. Fix four occurrences of X and factorize κ_1 accordingly

$$\kappa_1 = u s_1 s_2 s_3 w \tau_3 \tau_2 \tau_1 v =: (u, v)(s_1, \tau_1)(s_2, \tau_2)(s_3, \tau_3)w$$

---

[2] This observation strengthens the combinatorial results in [7] and also allows to greatly simplify the original proof of convergence given there.

[3] To clarify notation, we set $(\alpha, \beta)(\mu, \nu) := (\alpha\mu, \nu\beta)$ and $(\alpha, \beta)\gamma := \alpha\gamma\beta$, i.e. the pair (α, β) is treated as a word with a "hole" into which the pair or word on the right-hand side is substituted.

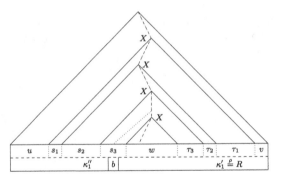

Fig. 1. Factorization of a witness $\kappa_1 = (u,v)(s_1,t_1)(s_2,t_2)(s_3,t_3)w = \kappa_1''b\kappa_1'$ w.r.t. a nonterminal X occurring at least four times a long the dashed path in a derivation tree of κ_1 leading to a letter within the suffix $b\kappa_1'$. The dotted path depicts the main path leading to the lcs^ρ-delimiting occurrence of the letter b.

In the proof of Lemma 7 we show that we may assume—as L is well-formed—that $u,v,w,s_1,s_2,s_3 \in \mathsf{A}^*$ with only $\tau_1,\tau_2,\tau_3 \in \mathsf{B}^*$. From this factorization we obtain the sublanguage $L' := (u,v)[(s_1,\tau_1) + (s_2,\tau_2) + (s_3,\tau_3)]^*w$. Our goal is to show that $(u,v)w$ or $(u,v)(s_i,\tau_i)w$ or $(u,v)(s_i,\tau_i)(s_j,\tau_j)w$ (for $i \neq j$) is a witness w.r.t. κ_0: note that each of these words result from pruning at least one pumping tree from t which inductively leads to a procedure to reduce t to a derivation tree of height at most $4N$ that still yields a witness for $R = \mathsf{lcs}^\rho(L)$ w.r.t. κ_0. Assume thus specifically that neither $(u,v)w = uwv$ nor $(u,v)(s_3,\tau_3)w = us_3w\tau_3v$ nor $(u,v)(s_i,\tau_i)(s_3,\tau_3)w = us_is_3w\tau_3\tau_iv$ for $i \in \{1,2\}$ is a witness w.r.t. κ_0, i.e. each of these words end on aR after reduction. Apply now the result mentioned at the beginning of this example to the language $L'' := (u,v)[(s_1,\tau_1) + (s_2,\tau_2)]^*(s_3,\tau_3)w$: by our assumptions $\kappa_1 \in L''$ is a witness w.r.t. $us_3w\tau_3v \in L''$ so that both $\mathsf{lcs}^\rho(L'') = \mathsf{lcs}^\rho(L)$ and also $(u,v)(s_1,\tau_1)^2(s_3,\tau_3)w$ is a witness w.r.t. $us_3w\tau_3v$ as we may choose $j = 1$. Thus also $\mathsf{lcs}^\rho(L) = \mathsf{lcs}^\rho(L''')$ for $L''' := (u,v)[(s_1,\tau_1) + (s_3,\tau_3)]^*w$ as $L''' \subseteq L$ and both $(u,v)w \in L'''$ and $(u,v)(s_1,\tau_1)(s_1,\tau_1)(s_3,\tau_3)w \in L'''$. Applying the same argument now to L''', but *choosing* $j \neq i$ it follows that $(u,v)(s_1,\tau_1)w$ or $(u,v)(s_3,\tau_3)(s_1,\tau_1)w$ has to be a witness w.r.t. κ_0.

The sketched argument can be adapted so that it also allows to conclude the maximal extension after reduction $\mathsf{lcsx}(\rho(L))$ has to have converged at derivation height $4N$ the latest, if L is well-formed. For details, see the proof of Lemma 7 in the appendix of the extended version [4].

As $|\mathsf{T}_{\overline{X}}^{\leq h}| \leq 3$, a straight-forward induction also shows that every word in $\mathsf{T}_{\overline{X}}^{\leq h}$ can be represented by an SLP that we can compute in time polynomial in G for $h \leq 4N + 1$; together with the preceding Lemmas 7 and 6 we thus obtain the main result of this section:

Theorem 2. *Given a context-free grammar G over B we can decide in time polynomial in the size of G whether G is wf.*

5   Conclusion

We have shown that well-formedness for context-free languages is decidable in polynomial time. This allowed us to decide in polynomial time whether or not a 2-TW is balanced. The presented techniques, however, are particularly tailored for 2-TWs. It is unclear how a generalization to transducers processing three or more copies of the input would look like. Thus, the question remains whether balancedness is decidable for general MSO definable transductions. It is also open whether even the single bracket case can be generalized beyond MSO definable transduction, e.g. to output languages of top-down tree-to-word transducers [12].

Acknowledgements. We also like to thank the anonymous reviewers for their detailed comments and valuable advice.

References

1. Berstel, J., Boasson, L.: Formal properties of XML grammars and languages. Acta Inf. **38**(9), 649–671 (2002). https://doi.org/10.1007/s00236-002-0085-4
2. Esparza, J., Kiefer, S., Luttenberger, M.: Derivation tree analysis for accelerated fixed-point computation. Theor. Comput. Sci. **412**(28), 3226–3241 (2011). https://doi.org/10.1016/j.tcs.2011.03.020
3. Knuth, D.E.: A characterization of parenthesis languages. Inf. Control **11**(3), 269–289 (1967)
4. Löbel, R., Luttenberger, M., Seidl, H.: On the balancedness of tree-to-word transducers. CoRR abs/1911.13054 (2019). http://arxiv.org/abs/1911.13054
5. Löbel, R., Luttenberger, M., Seidl, H.: Equivalence of linear tree transducers with output in the free group. CoRR abs/2001.03480 (2020). http://arxiv.org/abs/2001.03480
6. Lohrey, M.: Algorithmics on SLP-compressed strings: a survey. Groups Compl. Cryptol. **4**(2) (2012). https://doi.org/10.1515/gcc-2012-0016
7. Luttenberger, M., Palenta, R., Seidl, H.: Computing the longest common prefix of a context-free language in polynomial time. In: Niedermeier, R., Vallée, B. (eds.) STACS 2018. LIPIcs, vol. 96, pp. 48:1–48:13. Schloss Dagstuhl - Leibniz-Zentrum fuer Informatik (2018). https://doi.org/10.4230/LIPIcs.STACS.2018.48
8. Maneth, S., Seidl, H.: Balancedness of MSO transductions in polynomial time. Inf. Process. Lett. **133**, 26–32 (2018). https://doi.org/10.1016/j.ipl.2018.01.002
9. Minamide, Y., Tozawa, A.: XML validation for context-free grammars. In: Kobayashi, N. (ed.) APLAS 2006. LNCS, vol. 4279, pp. 357–373. Springer, Heidelberg (2006). https://doi.org/10.1007/11924661_22
10. Reynier, P., Talbot, J.: Visibly pushdown transducers with well-nested outputs. Int. J. Found. Comput. Sci. **27**(2), 235–258 (2016). https://doi.org/10.1142/S0129054116400086
11. Sakarovitch, J.: Elements of Automata Theory. Cambridge University Press, Cambridge (2009)

12. Seidl, H., Maneth, S., Kemper, G.: Equivalence of deterministic top-down tree-to-string transducers is decidable. J. ACM **65**(4), 21:1–21:30 (2018). https://doi.org/10.1145/3182653
13. Tozawa, A., Minamide, Y.: Complexity results on balanced context-free languages. In: Seidl, H. (ed.) FoSSaCS 2007. LNCS, vol. 4423, pp. 346–360. Springer, Heidelberg (2007). https://doi.org/10.1007/978-3-540-71389-0_25

On Tree Substitution Grammars

Andreas Maletti and Kevin Stier[(✉)]

Faculty of Mathematics and Computer Science, Universität Leipzig,
PO Box 100 920, 04009 Leipzig, Germany
{maletti,stier}@informatik.uni-leipzig.de

Abstract. Tree substitution grammars are formal models that are used extensively in natural language processing. It is demonstrated that their expressive power is located strictly between the local tree grammars and the regular tree grammars. A decision procedure for the problem of determining whether a tree substitution grammar generates a local tree language is provided. Unfortunately, the class of tree substitution languages is neither closed under union, nor intersection, nor complements. Indeed unions of tree substitution languages even generate an infinite hierarchy. However, all finite and all co-finite tree languages are tree substitution languages.

1 Introduction

Trees are a fundamental data structure in computer science and are used in many application areas like natural language processing [12], database theory [1], and compiler construction [17]. All the mentioned applications as well as others [6,7] require effective representations of sets of trees, also called tree languages. These requirements triggered detailed investigations of various classes of tree languages since the 1960s and by now there exists an abundance of models [5].

The most robust of those classes of tree languages are the regular tree languages [6,7], which are generated by finite-state tree automata, which are a natural extension of the finite-state string automata that generate the regular string languages [18]. Most standard problems are decidable for the regular tree languages and they generally enjoy the same nice algorithmic properties as the regular string languages. The main feature of those automata are their finitely many states, which enable most of the positive properties. However, these states are not exhibited directly in the trees generated. In application areas like natural language processing, in which representations of tree languages have to be inferred from finite sets of trees, practitioners often resorted to simpler models, in which the representation can more readily be induced from the sample.

Tree substitution grammars were originally introduced as a special case of tree-adjoining grammars [9,11], in which no adjunction is allowed. This restriction proved useful in the lexicalization of context-free grammars [10]. However, tree substitution grammar soon became popular in the parsing community [15]

K. Stier—Supported by DFG Research Training Group 1763 (QuantLA).

N. Jonoska and D. Savchuk (Eds.): DLT 2020, LNCS 12086, pp. 237–250, 2020.
https://doi.org/10.1007/978-3-030-48516-0_18

under the approach called data-oriented parsing [3] and were the formal model of many state-of-the-art parsers [16]. Similarly, synchronous tree substitution grammars, which are the same as the syntax-directed translation schemes of [2], are used in many statistical machine translation models [4,8,13,14]. Despite the multitude of applications, a fundamental study of their expressive power is missing. Rather they are attributed properties like "extended domain of locality", which provides some intuition, but has no formal definition.

A tree substitution grammar G is essentially a finite set F of tree fragments together with a set R of permissible root labels. Those tree fragments can be arbitrarily tall or large, which distinguishes tree substitution grammars from local tree grammars [6,7]. In addition, the fragments can contain leaves that are labeled by internal symbols. Leaves with such labels are called open and can be expanded further by fragments of F that have the same symbol as root label. Indeed G generates trees from a permissible root label of R by successively expanding open leaves with fragments of F until no open leaves remain. The set of all trees derivable in this manner is called the tree language generated by G. The tree languages that can be generated by some tree substitution grammar are called the tree substitution languages.

In this contribution we start a fundamental study of the expressive power of tree substitution grammars. We show that tree substitution grammars are strictly more expressive than local tree grammars [6,7], but strictly less expressive than finite-state tree automata (see Corollary 10). This, in particular, yields that most standard decision problems are also decidable for tree substitution languages because they are regular. In addition, it is decidable to determine whether a given tree substitution language is local (see Theorem 8). The decidability status of the related question whether a given regular tree language is a tree substitution language remains open. It is interesting to note that all finite and co-finite tree languages are tree substitution languages (see Theorem 6), which makes them much more useful for the approximation of finite samples of trees than the local tree languages, which do not contain all finite tree languages.

We also investigate the closure properties of the tree substitution languages. Unfortunately, they are neither closed under union (see Theorem 9), nor under intersection (see Theorem 13), nor under complement (see Theorem 14). In fact, unions of tree substitution languages even form a strict hierarchy (see Theorem 11), so unions of k tree substitution languages are strictly less expressive than unions of $k + 1$ tree substitution languages. A similar hierarchy is significantly more difficult to prove for intersections and remains an open problem because intersections break the "extended domain of locality" (as shown in the proof of Theorem 13) and can manage a non-explicit information transport over unbounded distances in the trees. Indeed the trivial union construction, which just takes the union of the fragments of the individual tree substitution grammars $G_1, \ldots, G_n$, does yield a tree substitution grammar G that can generate each tree that can be generated by some G_i. However, G might overgeneralize in the sense that it may also generate trees that cannot be generated by any $G_1, \ldots, G_n$. This property is utilized in grammar induction to generalize beyond the seen data. Overall, the expressive power of tree substitution grammars is interesting and offers new challenging problems because they are used extensively in real-world applications despite their brittle expressive power. It is exactly this absence of good closure properties, which requires separate arguments for each individual problem and thus makes several problems challenging as outlined in the open problems section.

2   Preliminaries

We denote the set of nonnegative integers (including 0) by $\mathbb{N}$. For every $k \in \mathbb{N}$, we use the subset $[k] = \{i \in \mathbb{N} \mid 1 \leq i \leq k\}$. An alphabet A is simply a finite set and $A^* = \bigcup_{k \in \mathbb{N}} A^k$ is the set of all finite words over A, where $A^k = A \times \cdots \times A$ containing k factors A and $A^0 = \{\varepsilon\}$, of which ε is called the empty word. The length $|w|$ of a word $w = a_1 \cdots a_k \in A^*$ with $a_1, \ldots, a_k \in A$ is $|w| = k$; i.e. the number of symbols making up w. Given words $v, w \in A^*$, their concatenation is written $v.w$ or simply vw. We write $v \preceq w$ provided that there exists $u \in A^*$ such that $vu = w$. The relation $\preceq$ is actually a partial order, called the prefix order.

Let S be a set and $R \subseteq S \times S$ be a relation. The identity on S is the relation $\mathrm{id}_S = \{(s, s) \mid s \in S\}$. Given another relation $R' \subseteq S \times S$, the composition $R \,;\, R'$ is given by $R \,;\, R' = \{(s_1, s_3) \mid \exists s_2 \in S: (s_1, s_2) \in R, (s_2, s_3) \in R'\}$. The relation R is reflexive if $\mathrm{id}_S \subseteq R$, and it is transitive if $R \,;\, R \subseteq R$. The reflexive, transitive closure of R is $R^* = \bigcup_{k \in \mathbb{N}} R^k$ and the transitive closure of R is $R^+ = \bigcup_{k \geq 1} R^k$, where $R^0 = \mathrm{id}_S$ and $R^k = R \,;\, \cdots \,;\, R$ containing k times the relation R.

A ranked alphabet (Σ, rk) is a pair consisting of an alphabet Σ and a mapping $\mathrm{rk}\colon \Sigma \to \mathbb{N}$ that assigns a rank to each symbol of Σ. We usually denote a ranked alphabet (Σ, rk) by just Σ alone when the ranks are clear. We also write $\sigma^{(k)}$ to indicate that $\mathrm{rk}(\sigma) = k$. Moreover, for every $k \in \mathbb{N}$, we let $\Sigma_k = \{\sigma \in \Sigma \mid \mathrm{rk}(\sigma) = k\}$. Given a ranked alphabet Σ and a set Z, the set $T_\Sigma(Z)$ of Σ" trees indexed by Z is the smallest set T such that $Z \subseteq T$ and $\sigma(t_1, \ldots, t_k) \in T$ for every $k \in \mathbb{N}$, $\sigma \in \Sigma_k$, and $t_1, \ldots, t_k \in T$. We abbreviate $T_\Sigma(\emptyset)$ simply to T_Σ, and any subset $L \subseteq T_\Sigma$ is called tree language. It is co-finite if $T_\Sigma \setminus L$ is finite.

Next, we recall some common notions and notations for trees. In the following, let $t \in T_\Sigma(Z)$ be a tree for a ranked alphabet Σ and a set Z. The set $\mathrm{pos}(t)$ of positions of t is inductively defined by $\mathrm{pos}(z) = \{\varepsilon\}$ for all $z \in Z$, and $\mathrm{pos}(\sigma(t_1, \ldots, t_k)) = \{\varepsilon\} \cup \{i.p \mid i \in [k], p \in \mathrm{pos}(t_i)\}$ for every $k \in \mathbb{N}$, $\sigma \in \Sigma_k$, and $t_1, \ldots, t_k \in T_\Sigma(Z)$. The height of t is defined by $\mathrm{ht}(t) = \max_{p \in \mathrm{pos}(t)} |p|$, and the size of t is defined by $|t| = |\mathrm{pos}(t)|$. A leaf is a position $p \in \mathrm{pos}(t)$ such that $p.1 \notin \mathrm{pos}(t)$. We denote the subset of leaves of $\mathrm{pos}(t)$ by $\mathrm{leaf}(t)$. Given a position $p \in \mathrm{pos}(t)$, the label $t(p)$ of t at p and the subtree $t|_p$ of t at p are defined by $z(\varepsilon) = z|_\varepsilon = z$ for all $z \in Z$, and

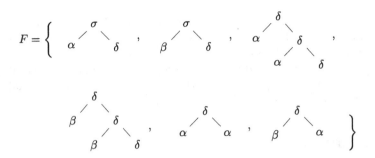

Fig. 1. Fragments of the TSG of Example 2.

$$\big(\sigma(t_1,\ldots,t_k)\big)(p) = \begin{cases} \sigma & \text{if } p = \varepsilon \\ t_i(p') & \text{if } p = i.p' \text{ with } i \in \mathbb{N} \text{ and } p' \in \text{pos}(t_i) \end{cases}$$

$$\sigma(t_1,\ldots,t_k)|_p = \begin{cases} \sigma(t_1,\ldots,t_k) & \text{if } p = \varepsilon \\ t_i|_{p'} & \text{if } p = i.p' \text{ with } i \in \mathbb{N} \text{ and } p' \in \text{pos}(t_i) \end{cases}$$

for all $k \in \mathbb{N}$, $\sigma \in \Sigma_k$, and $t_1,\ldots,t_k \in T_\Sigma(Z)$. Finally, the replacement $t[u]_p$ of the leaf $p \in \text{leaf}(t)$ by another tree $u \in T_\Sigma(Z)$ is given by $\alpha[u]_\varepsilon = u$ for every $\alpha \in Z \cup \Sigma_0$, and $\sigma(t_1,\ldots,t_k)[u]_{i.p'} = \sigma(t_1,\ldots,t_{i-1},t_i[u]_{p'},t_{i+1},\ldots,t_k)$ for every $k \in \mathbb{N}$, $i \in [k]$, $\sigma \in \Sigma_k$, $t_1,\ldots,t_k \in T_\Sigma(Z)$, and $p' \in \text{pos}(t_i)$.

We reserve the use of the special symbol $\square$. A tree $t \in T_\Sigma(Z \cup \{\square\})$ is a context, if there exists exactly one $p \in \text{pos}(t)$ with $t(p) = \square$; i.e., there is exactly one occurrence of $\square$ in t. The set of all such contexts is denoted by $C_\Sigma(Z)$. Given a context $c \in C_\Sigma(Z)$ and a tree $t \in T_\Sigma(Z \cup \{\square\})$, the substitution $c[t]$ of t into c yields the tree $c[t]_p$, where p is the unique position $p \in \text{pos}(c)$ with $c(p) = \square$. Note that given $c, c' \in C_\Sigma(Z)$, also $c[c'] \in C_\Sigma(Z)$. Similarly, we write $c^k[t]$ for $c[c[\cdots c[t] \cdots]]$ containing the context c a total of k times.

Finally, let us recall regular tree grammars (RTGs) [6,7]. An RTG is a tuple $G = (Q, \Sigma, Q_0, P)$, where Q is a finite set of states such that $Q \cap \Sigma = \emptyset$, Σ is a ranked alphabet of input symbols, $Q_0 \subseteq Q$ is a set of initial states, and $P \subseteq Q \times T_\Sigma(Q)$ is a finite set of productions. We also write productions (q, t) as $q \to t$. The derivation relation for $\xi, \zeta \in T_\Sigma(Q)$ is defined for every $\xi, \zeta \in T_\Sigma(Q)$ by $\xi \Rightarrow_G \zeta$ if and only if there exists a production $q \to t \in P$ and a context $c \in C_\Sigma(Q)$ such that $\xi = c[q]$ and $\zeta = c[t]$. The tree language generated by G is $L(G) = \bigcup_{q \in Q_0}\{t \in T_\Sigma \mid q \Rightarrow_G^+ t\}$. A tree language L is regular if there exists an RTG G such that $L(G) = L$. The class of regular tree languages is denoted by RTL. We note that RTL coincides with the class of tree languages generated by tree automata [6,7].

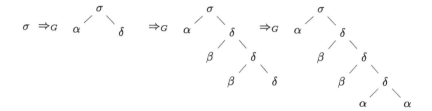

Fig. 2. Example derivation steps using the TSG G of Example 2.

3   Tree Substitution Grammars

Let us start with the formal definition of tree substitution grammars (TSGs) taken essentially from the natural language processing community [10,11]. TSGs have been applied to various tasks including parsing [16] and machine translation [19]. Consequently, the definitions of TSGs vary, but our definition captures the essence of the notion, while still being convenient to work with.

Definition 1. *A* tree substitution grammar *(TSG) is a tuple $G = (\Sigma, R, F)$, in which Σ is a ranked alphabet of input symbols, $R \subseteq \Sigma$ is a set of* root labels, *and $F \subseteq T_\Sigma(\Sigma) \setminus \Sigma$ is a finite set of* fragments. *The TSG G is a* local tree grammar *(LTG) if $\mathrm{ht}(f) \leq 1$ for all $f \in F$.*

Example 2. Consider the ranked alphabet $\Sigma = \{\sigma^{(2)}, \delta^{(2)}, \alpha^{(0)}, \beta^{(0)}\}$ and the TSG $G = (\Sigma, \{\sigma\}, F)$ with the fragments displayed in Fig. 1. Clearly, this TSG is not an LTG due to the third and fourth fragment.

Next we present the derivation semantics for a TSG $G = (\Sigma, R, F)$. Essentially we start the derivation process with a tree consisting solely of a root label of R and then iteratively replace a leaf by a fragment of F with the same root label. This process can be repeated until no replacements are possible anymore. If the such obtained tree t contains only leaves that are labeled by nullary symbols, then t is part of the tree language generated by G.

Definition 3. *Let $G = (\Sigma, R, F)$ be a TSG. For any two trees $\xi, \zeta \in T_\Sigma(\Sigma)$, we write $\xi \Rightarrow_G \zeta$ if there exists a fragment $f \in F$ and a context $c \in C_\Sigma(\Sigma)$ such that $\xi = c[f(\varepsilon)]$ and $\zeta = c[f]$. The TSG G generates the tree language $L(G) = \{t \in T_\Sigma \mid \exists \sigma \in R \colon \sigma \Rightarrow_G^* t\}$.*

Example 4. Let $\Sigma = \{\sigma^{(2)}, \gamma^{(1)}, \alpha^{(0)}\}$ and consider the TSG $G = (\Sigma, \{\sigma\}, F)$ with the fragments displayed in Fig. 3. The derivation presented in Fig. 3 illustrates that a derived tree can contain several leaves that still need to be independently replaced. More precisely, both occurrences of γ in the tree $\sigma(\gamma, \gamma)$ are independently replaced in the displayed derivation.

$$F = \left\{ \quad \overset{\sigma}{\underset{\gamma \quad \diagdown \quad \gamma}{\diagup}} \quad , \quad \overset{\gamma}{\underset{\gamma}{|}} \quad , \quad \overset{\gamma}{\underset{\alpha}{|}} \quad \right\}$$

$$\sigma \Rightarrow_G \quad \overset{\sigma}{\underset{\gamma \quad \diagdown \quad \underset{|}{\underset{\gamma}{\gamma}}}{\diagup}} \quad \Rightarrow_G \quad \overset{\sigma}{\underset{\gamma \quad \diagdown \quad \underset{|}{\underset{\alpha}{\gamma}}}{\diagup}} \quad \Rightarrow_G \quad \overset{\sigma}{\underset{\underset{|}{\underset{\gamma}{\gamma}} \quad \diagdown \quad \gamma}{\diagup}} \quad \Rightarrow_G \quad \overset{\sigma}{\underset{\underset{|}{\underset{\alpha}{\gamma}} \quad \diagdown \quad \underset{|}{\underset{\underset{|}{\alpha}}{\gamma}}}{\diagup}}$$

Fig. 3. Fragments of the TSG G of Example 4 and example derivation steps.

Example 5. Consider the TSG G from Example 2. A few derivation steps are displayed in Fig. 2. Let $c_\alpha = \delta(\alpha, \delta(\alpha, \square))$ and $c_\beta = \delta(\beta, \delta(\beta, \square))$. Overall, this TSG generates the tree language

$$\{\sigma(x, c_1[\cdots c_n[\delta(y, \alpha)] \cdots]) \mid x, y \in \{\alpha, \beta\}, \; n \in \mathbb{N}, \; \forall i \in [n] : c_i \in \{c_\alpha, c_\beta\}\}.$$

Two TSGs G and G' are *equivalent* if $L(G) = L(G')$. A tree language L is a *tree substitution language* if there exists a TSG G such that $L = L(G)$, and it is *local* [6,7] if there exists a local tree grammar G such that $L = L(G)$. The classes of all tree substitution languages and all local tree languages are denoted by TSL and LTL, respectively.

4   Expressive Power

In this section, we investigate the expressive power of tree substitution grammars and start with some simple tree languages that are contained in TSL. To this end, let FIN and co-FIN be the classes of all finite and all co-finite tree languages, respectively.

Theorem 6. FIN $\cup$ co-FIN $\subseteq$ TSL.

Proof. Every finite tree language $L \subseteq T_\Sigma$ is trivially a tree substitution language via the TSG (Σ, R, L) with $R = \{t(\varepsilon) \mid t \in L\}$.

Now, let $L \subseteq T_\Sigma$ be a co-finite tree language and $T_\Sigma \setminus L = \{t_1, \ldots, t_k\}$ be the finitely many trees outside L. Moreover, let $n > \max_{i \in [k]} \mathrm{ht}(t_i)$ be larger than the height of the tallest tree from $\{t_1, \ldots, t_k\}$. We construct the TSG (Σ, R, F) with

- $R = \{t(\varepsilon) \mid t \in L\}$ and
- $F = \{t \in L \mid \mathrm{ht}(t) \leq 2n\} \cup \{t \in T_\Sigma(\Sigma) \mid n \leq \mathrm{ht}(t) \leq 2n\}$.

Clearly, F is finite. Now we prove $L(G) = L$. For $L(G) \subseteq L$ it is sufficient to show that $t_i \notin L(G)$ for every $i \in [k]$. Obviously, the fragments of F are either in L or have height at least n, which proves $L(G) \subseteq L$. We prove the converse

$L \subseteq L(G)$ by contradiction, so suppose that there exists $t \in L$ with $t \notin L(G)$. Then there also exists a smallest $t' \in L$ with $t' \notin L(G)$. Since all trees $t' \in L$ with $\mathrm{ht}(t') \leq 2n$ can be generated directly using a single fragment from F, we must have $\mathrm{ht}(t') > 2n$. Let

$$P = \{p \in \mathrm{pos}(t') \mid |p| \leq n, \exists p' \in \mathrm{pos}(t') \colon p \preceq p', |p'| > 2n\}$$

be the short positions that are prefixes to long positions, and let $C = \max_{\preceq} P$ be the maximal (with respect to $\preceq$) elements of P. We construct the unique tree $f \in T_\Sigma(\Sigma)$ with positions

$$\mathrm{pos}(f) = \{p \in \mathrm{pos}(t') \mid |p| \leq 2n\} \setminus \{p \in \mathrm{pos}(t') \mid \exists c \in C \colon c \prec p\}$$

and labels $f(p) = t'(p)$ for all $p \in \mathrm{pos}(f)$. In other words, we obtain f by cutting all paths in t' that have length more than $2n$ at length n. Obviously, $f \in F$. In addition, we observe that $\mathrm{ht}(t'|_p) > n$ for all $p \in C$. For every $p \in C$, we thus obtain $t'|_p \in L$ and $t'|_p \in L(G)$ since $|t'|_p| < |t'|$ and t' is the smallest counterexample. However, this yields that $t'(\varepsilon) \Rightarrow_G f$ as well as $f(p) \Rightarrow_G^* t'|_p$ for all $p \in C$. Altogether $t'(\varepsilon) \Rightarrow_G^* t'$, which proves that $t' \in L(G)$ contradicting the assumption. $\qquad\square$

Next we relate the class of tree substitution languages to the well-known classes of local and regular tree languages, respectively. Unsurprisingly, they are situated strictly between them, but the second strictness will be established later (see Corollary 10).

Theorem 7. LTL $\subsetneq$ TSL $\subseteq$ RTL.

Proof. The first inclusion holds by definition. For the latter, let $G = (\Sigma, R, F)$ be a TSG and $S \notin \Sigma$ a new symbol. We construct an RTG $G' = (\overline{\Sigma} \cup \{S\}, \Sigma, S, P)$ such that $L(G') = L(G)$. To this end, we use copies $\overline{\Sigma} = \{\overline{\sigma} \mid \sigma \in \Sigma\}$ of the input symbols of Σ as states. The productions are given by $P = P_S \cup P'$ with

$$P_S = \{S \to \mathrm{rel}(\sigma) \mid \sigma \in R\}$$
$$P' = \{\overline{f(\varepsilon)} \to \mathrm{rel}(f) \mid f \in F\},$$

where $\mathrm{rel} \colon T_\Sigma(\Sigma) \to T_\Sigma(\overline{\Sigma})$ is inductively defined by

$$\mathrm{rel}(\sigma) = \begin{cases} \sigma & \text{if } \sigma \in \Sigma_0 \\ \overline{\sigma} & \text{otherwise} \end{cases}$$

for every $\sigma \in \Sigma$ and $\mathrm{rel}(\sigma(t_1, \ldots, t_k)) = \sigma(\mathrm{rel}(t_1), \ldots, \mathrm{rel}(t_k))$ for all $k \in \mathbb{N} \setminus \{0\}$, $\sigma \in \Sigma_k$, and $t_1, \ldots, t_k \in T_\Sigma(\Sigma)$. Clearly any derivation $\xi_0 \Rightarrow_G \xi_1 \Rightarrow_G \cdots \Rightarrow_G \xi_n$ of G yields a corresponding derivation $\mathrm{rel}(\xi_0) \Rightarrow_{G'} \mathrm{rel}(\xi_1) \Rightarrow_{G'} \cdots \Rightarrow_{G'} \mathrm{rel}(\xi_n)$ of G'. Together with $\mathrm{rel}(t) = t$ for all $t \in T_\Sigma$ and the new initial states, we obtain $L(G) \subseteq L(G')$. The converse is proved similarly.

The first inclusion is strict because FIN $\subseteq$ TSL by Theorem 6, but it is well-known [6,7] that FIN $\nsubseteq$ LTL. $\qquad\square$

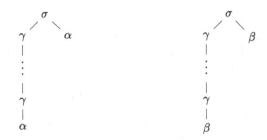

Fig. 4. The tree languages $L(G_1)$ and $L(G_2)$ used in the proof of Theorem 9.

The inclusion TSL $\subseteq$ RTL immediately yields that most interesting problems are decidable for tree substitution languages. For example, the emptiness, finiteness, inclusion, and equivalence problems are all decidable because they are decidable for regular tree languages [6,7]. We proceed with a subclass definability problem: Is it decidable whether an effectively presented tree substitution language is local? Whenever we speak about an effectively presented tree substitution language L, we assume that we are actually given a tree substitution grammar G such that $L(G) = L$. Let $G = (\Sigma, R, F)$ be a TSG. A fragment $f \in F$ is *useless* if G and $(\Sigma, R, F \setminus \{f\})$ are equivalent. The TSG G is *reduced* if no fragment $f \in F$ is useless. Clearly, for every TSG we can construct an equivalent reduced TSG.

Theorem 8. *For every effectively presented $L \in$ TSL, it is decidable whether $L \in$ LTL.*

Proof. Let $G = (\Sigma, R, F)$ be a reduced tree substitution grammar such that $L(G) = L$. We construct the local tree grammar $G' = (\Sigma, R, F')$ with

$$F' = \{f(p)(f(p.1), \ldots, f(p.k)) \mid f \in F, \ k \in \mathbb{N}, \ p \in \mathrm{pos}(f) \setminus \mathrm{leaf}(f), \ f(p) \in \Sigma_k\}.$$

Obviously, $L = L(G) \subseteq L(G')$ and all fragments of F' are essential for this property. Consequently, L is local if and only if $L(G') \subseteq L$. Since both $L(G')$ and L are regular by Theorem 7 and inclusion is decidable for regular tree languages [6,7], we obtain the desired statement. □

5   Closure Properties

In this section, we investigate the closure properties of the class of tree substitution languages. More specifically, we investigate the Boolean operations and the hierarchy for union. Unfortunately, the results are all negative, but they and, in particular, their proofs shed additional light on the expressive power of tree substitution languages. Let us start with union.

Theorem 9. TSL *is <u>not</u> closed under union.*

Proof. Consider the ranked alphabet $\Sigma = \{\sigma^{(2)}, \gamma^{(1)}, \alpha^{(0)}, \beta^{(0)}\}$ and the LTGs

$$G_1 = (\Sigma, \{\sigma\}, \{\sigma(\gamma, \alpha), \gamma(\gamma), \gamma(\alpha)\})$$
$$G_2 = (\Sigma, \{\sigma\}, \{\sigma(\gamma, \beta), \gamma(\gamma), \gamma(\beta)\}),$$

which generate the local tree languages (see Fig. 4)

$$L(G_1) = \{\sigma(c^n[\alpha], \alpha) \mid n \in \mathbb{N}\} \qquad \text{and} \qquad L(G_2) = \{\sigma(c^n[\beta], \beta) \mid n \in \mathbb{N}\}$$

with $c = \gamma(\square)$. Now suppose that their union $L = L(G_1) \cup L(G_2)$ is a tree substitution language; i.e., $L \in$ TSL. Hence there exists a TSG $G = (\Sigma, R, F)$ such that $L(G) = L$. Let $n \in \mathbb{N}$ be such that $n > \max_{f \in F} \text{ht}(f)$. Since $t = \sigma(c^n[\alpha], \alpha) \in L$, there must exist a derivation $\sigma \Rightarrow_G^* t$ and $\sigma \in R$. Since $\text{ht}(t) > n$ at least two derivation steps are required, so $\sigma \Rightarrow_G \sigma(c^k[\gamma], \alpha) \Rightarrow_G^+ t$ for some $0 \leq k < n$, which yields the subderivation $\gamma \Rightarrow_G^+ c^{n-k}[\alpha]$. In the same manner we consider the tree $t' = \sigma(c^n[\beta], \beta) \in L$, for which the derivation $\sigma \Rightarrow_G \sigma(c^\ell[\gamma], \beta) \Rightarrow_G^+ t'$ for some $0 \leq \ell < n$ and the subderivation $\gamma \Rightarrow_G^+ c^{n-\ell}[\beta]$ must exist. However, exchanging the subderivations yields the derivation

$$\sigma \Rightarrow_G \sigma(c^k[\gamma], \alpha) \Rightarrow_G^+ \sigma(c^k[c^{n-\ell}[\beta]], \alpha),$$

which shows $\sigma(c^{n-\ell+k}[\beta], \alpha) \in L(G) = L$ contradicting $L = L(G_1) \cup L(G_2)$. $\square$

Since the class of regular tree languages is closed under union [6,7], we obtain the following corollary from Theorems 7 and 9.

Corollary 10. LTL $\subsetneq$ TSL $\subsetneq$ RTL.

We demonstrated that the union of two tree substitution languages need not be a tree substitution language. Next, we ask ourselves whether additional unions increase the expressive power even further. For every $k \in \mathbb{N}$ let

$$\cup_k\text{-TSL} = \{L_1 \cup \cdots \cup L_k \mid L_1, \ldots, L_k \in \text{TSL}\}$$

be the class of those tree languages that can be presented as unions of k tree substitution languages. Since $\emptyset \in$ TSL (see Theorem 6), we obtain $\cup_0$-TSL $= \emptyset$, $\cup_1$-TSL $=$ TSL, and $\cup_k$-TSL $\subseteq \cup_{k+1}$-TSL for every $k \in \mathbb{N}$. Next, we show that the mentioned inclusion is actually strict, so that we obtain an infinite hierarchy.

Theorem 11. $\cup_k$-TSL $\subsetneq \cup_{k+1}$-TSL *for all* $k \in \mathbb{N}$.

Proof. The statement is clear for $k = 0$, so let $k \geq 1$. Consider the ranked alphabet $\Sigma = \{\sigma^{(2)}, \delta^{(2)}, \alpha^{(0)}\}$ and the TSG $G_i = (\Sigma, \{\sigma\}, F_i)$ for every $i \in [k+1]$, where

$$F_i = \{\sigma(\delta, s_i), s_i, \delta(\delta, \alpha), \delta(s_i, \alpha)\}$$

and $s_i = c_r^i[\alpha]$ with $c_r = \delta(\alpha, \square)$. Clearly, $L(G_i) = \{\sigma(c_\ell^n[s_i], s_i) \mid n \in \mathbb{N}\}$ with $c_\ell = \delta(\square, \alpha)$. The tree substitution language $L(G_i)$ and the tree s_i are illustrated in Fig. 5.

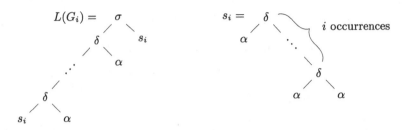

Fig. 5. Illustration of the tree substitution languages used in the proof of Theorem 11.

Obviously, $L = L(G_1) \cup \cdots \cup L(G_{k+1}) \in \cup_{k+1}$-TSL and those individual tree languages are infinite and pairwise disjoint. For the sake of a contradiction, assume that $L \in \cup_k$-TSL; i.e. there exist $L'_1, \ldots, L'_k \in$ TSL such that $L = L'_1 \cup \cdots \cup L'_k$. The pigeonhole principle establishes that there exist $i \in [k]$ and $m, n \in [k+1]$ with $m \neq n$ such that $L_m \cap L'_i$ and $L_n \cap L'_i$ are infinite. Let $G = (\Sigma, R, F)$ be a TSG such that $L(G) = L'_i$. Let $z > \max_{f \in F} \mathrm{ht}(f)$. Since $L_m \cap L(G)$ is infinite, there exists $x > z$ such that $\sigma(c^x_\ell[s_m], s_m) \in L(G)$. Similarly, there exists $y > z$ such that $\sigma(c^y_\ell[s_n], s_n) \in L(G)$ because $L_n \cap L(G)$ is infinite. Inspecting the derivations for those trees there exist $x', y' \in \mathbb{N}$ such that

$$\sigma \Rightarrow_G \sigma(c^{x'}_\ell[\delta], s_m) \Rightarrow^*_G \sigma(c^x_\ell[s_m], s_m) \qquad \text{with subderivation} \qquad \delta \Rightarrow^+_G c^{x-x'}_\ell[s_m]$$

$$\sigma \Rightarrow_G \sigma(c^{y'}_\ell[\delta], s_n) \Rightarrow^*_G \sigma(c^y_\ell[s_n], s_n) \qquad \text{with subderivation} \qquad \delta \Rightarrow^+_G c^{y-y'}_\ell[s_n]$$

Exchanging the subderivations we obtain

$$\sigma \Rightarrow_G \sigma(c^{x'}_\ell[\delta], s_m) \Rightarrow^*_G \sigma(c^{x'+y-y'}_\ell[s_n], s_m)$$

and thus $\sigma(c^{x'+y-y'}_\ell[s_n], s_m) \in L(G) \subseteq L$, which is a contradiction because $m \neq n$. □

Corollary 12 (of Theorem 11).

$$\cup_0\text{-TSL} \subsetneq \cup_1\text{-TSL} \subsetneq \cup_2\text{-TSL} \subsetneq \cup_3\text{-TSL} \subsetneq \cup_4\text{-TSL} \subsetneq \cdots$$

Let us move on to intersection. Unfortunately, TSL is not closed under intersection, but intersections of TSL become quite powerful. In particular, they allow information to be transported over unbounded distances, which can be observed from the proof.

Theorem 13. TSL is _not_ closed under intersection.

Proof. Recall the ranked alphabet $\Sigma = \{\sigma^{(2)}, \delta^{(2)}, \alpha^{(0)}, \beta^{(0)}\}$ and the TSG G of Example 2 as well as the contexts $c_\alpha = \delta(\alpha, \delta(\alpha, \square))$ and $c_\beta = \delta(\beta, \delta(\beta, \square))$ from Example 5. Additionally, let $G' = (\Sigma, \{\sigma\}, F')$ with F' displayed in Fig. 6. The generated tree substitution languages $L(G)$ and $L(G')$ are

$$\{\sigma(x, c_1[\cdots c_n[\delta(y, \alpha)] \cdots]) \mid x, y \in \{\alpha, \beta\}, n \in \mathbb{N}, \forall i \in [n]: c_i \in \{c_\alpha, c_\beta\}\}$$

$$\{\sigma(x, \delta(x, c_1[\cdots c_n[\alpha] \cdots])) \mid x \in \{\alpha, \beta\}, n \in \mathbb{N}, \forall i \in [n]: c_i \in \{c_\alpha, c_\beta\}\}$$

$$F' = \{ \quad \langle\text{image}\rangle \quad , \quad \langle\text{image}\rangle \quad , \quad \langle\text{image}\rangle \quad ,$$

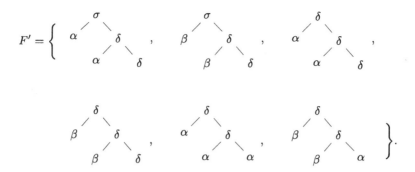

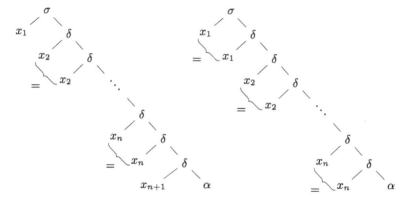

Fig. 6. Fragments of the TSG G' used in the proof of Theorem 13.

Fig. 7. Tree substitution languages $L(G)$ and $L(G')$ used in the proof of Theorem 13.

respectively, which are also illustrated in Fig. 7. Their intersection

$$L(G) \cap L(G') = \{\sigma(\alpha, \delta(\alpha, c_\alpha[\cdots c_\alpha[\alpha]\cdots])) \mid n \in \mathbb{N}\} \cup$$
$$\{\sigma(\beta, \delta(\beta, c_\beta[\cdots c_\beta[\alpha]\cdots])) \mid n \in \mathbb{N}\}$$

contains only trees, in which all left children along the spine carry the same label. This tree language is not a tree substitution language, which can be proved using the subderivation exchange technique used in the proof of Theorem 9.   □

Note how the intersection achieves a global synchronization in the proof of Theorem 13. This power makes the investigation of the intersection hierarchy difficult. We leave the strictness of the intersection hierarchy as an open problem and conclude by considering the complement.

Theorem 14. TSL *is not closed under complements.*

Proof. Consider the ranked alphabet $\Sigma = \{\gamma^{(1)}, A^{(1)}, B^{(1)}, \alpha^{(0)}, \beta^{(0)}\}$ and the LTG $G = (\Sigma, \{\gamma\}, F)$ with fragments

$$F = \{\gamma(A), A(A), A(\alpha)\} \cup \{\gamma(B), B(B), B(\beta)\}.$$

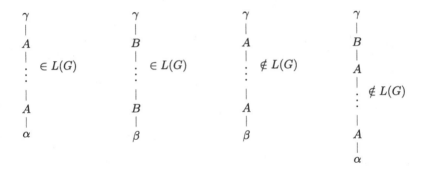

Fig. 8. Trees used in the proof of Theorem 14.

The generated tree language is illustrated in Fig. 8. Now suppose that its complement $L = T_\Sigma(\Sigma) \setminus L(G)$ is a tree substitution language; i.e., $L \in TSL$. Hence there exists a TSG $G' = (\Sigma, R, F)$ such that $L = L(G')$. Let $n \in \mathbb{N}$ be such that $n > \max_{f \in F} \mathrm{ht}(f)$. Since $t = \gamma(A^n(\beta)) \in L$ (see Fig. 8) there must exist a derivation $\gamma \Rightarrow_G^* t$ and $\gamma \in R$. Since $\mathrm{ht}(t) > n$ at least two derivation steps are required, so $\gamma \Rightarrow_G \gamma(A^k) \Rightarrow_G^+ t$ for some $0 \le k < n$, which yields the subderivation $A \Rightarrow_G^+ A^{n-k}(\beta)$. Similarly, we consider the tree $t' = \gamma(B(A^n(\alpha))) \in L$ (see Fig. 8), for which the derivation $\gamma \Rightarrow_G^+ \gamma(B(A^\ell)) \Rightarrow_G^+ t'$ for some $0 \le \ell < n$ and the subderivation $A \Rightarrow_G^+ A^{n-\ell}(\alpha)$ must exist. However, exchanging the subderivations yields the derivation

$$\gamma \Rightarrow_G \gamma(A^k) \Rightarrow_G^+ \gamma(A^k(A^{n-\ell}(\alpha)),$$

which shows $\gamma(A^k(A^{n-\ell}(\alpha)) \in L(G') = L$ contradicting $L = T_\Sigma(\Sigma) \setminus L(G)$. $\quad\square$

6   Open Problems

We showed that it is decidable whether a given tree substitution language is local. It remains open if we can also decide whether a given regular tree language is a tree substitution language. Progress on this problem will probably provide additional fine-grained insight into the expressive power of tree substitution grammars in comparison to the regular tree grammars.

Another open problem concerns the intersection hierarchy. We showed that unions of tree substitution languages can progressively express more and more tree languages. A similar hierarchy also exists for intersections of tree substitution languages and we showed that the intersection of two tree substitution languages is not necessarily a tree substitution languages. However, it remains open whether there is an infinite intersection hierarchy or whether it collapses at some level.

Acknowledgements. The authors gratefully acknowledge the financial support of the Research Training Group 1763 (QuantLA: Quantitative Logics and Automata), which is funded by the German Research Foundation (DFG). In addition, the authors would like to thank the anonymous reviewers for the careful reading of the manuscript and their valuable feedback.

References

1. Abiteboul, S., Hull, R., Vianu, V.: Foundations of Databases. Addison Wesley, Boston (1994)
2. Aho, A.V., Ullman, J.D.: The Theory of Parsing, Translation, and Compiling. Prentice Hall, Upper Saddle River (1972)
3. Bod, R.: The data-oriented parsing approach: theory and application. In: Fulcher, J., Jain, L.C. (eds.) Computational Intelligence: A Compendium. SCI, vol. 115, pp. 307–348. Springer, Heidelberg (2008). https://doi.org/10.1007/978-3-540-78293-3_7
4. Chiang, D., Knight, K.: An introduction to synchronous grammars (2006). Tutorial at 44th ACL. https://www3.nd.edu/~dchiang/papers/synchtut.pdf
5. Comon, H., et al.: Tree automata techniques and applications (2007). http://tata.gforge.inria.fr
6. Gécseg, F., Steinby, M.: Tree languages. In: Rozenberg, G., Salomaa, A. (eds.) Handbook of Formal Languages, vol. 3, pp. 1–68. Springer, Heidelberg (1997). https://doi.org/10.1007/978-3-642-59126-6_1
7. Gécseg, F., Steinby, M.: Tree Automata. arXiv, 2nd edn. (2015). https://arxiv.org/abs/1509.06233
8. Howcroft, D.M., Klakow, D., Demberg, V.: Toward Bayesian synchronous tree substitution grammars for sentence planning. In: Proceedings of the 11th NLG, pp. 391–396. ACL (2018)
9. Joshi, A.K., Levy, L.S., Takahashi, M.: Tree adjunct grammars. J. Comput. Syst. Sci. **10**(1), 136–163 (1975)
10. Joshi, A.K., Schabes, Y.: Tree-adjoining grammars and lexicalized grammars. Technical report, MS-CIS-91-22, University of Pennsylvania (1991)
11. Joshi, A.K., Schabes, Y.: Tree-adjoining grammars. In: Rozenberg, G., Salomaa, A. (eds.) Handbook of Formal Languages, vol. 3, pp. 69–123. Springer, Heidelberg (1997). https://doi.org/10.1007/978-3-642-59126-6_2
12. Jurafsky, D., Martin, J.H.: Speech and Language Processing, 2nd edn. Prentice Hall, Upper Saddle River (2008)
13. Maletti, A.: Why synchronous tree substitution grammars? In: Proceedings of the 2010 NAACL, pp. 876–884. ACL (2010)
14. Maletti, A.: An alternative to synchronous tree substitution grammars. J. Nat. Lang. Eng. **17**(2), 221–242 (2011)
15. Sangati, F., Keller, F.: Incremental tree substitution grammar for parsing and sentence prediction. Trans. ACL **1**, 111–124 (2013)
16. Shindo, H., Miyao, Y., Fujino, A., Nagata, M.: Bayesian symbol-refined tree substitution grammars for syntactic parsing. In: Proceedings of the 50th ACL, pp. 440–448. ACL (2012)
17. Wilhelm, R., Seidl, H., Hack, S.: Compiler Design. Springer, Heidelberg (2013). https://doi.org/10.1007/978-3-642-17540-4

18. Yu, S.: Regular languages. In: Rozenberg, G., Salomaa, A. (eds.) Handbook of Formal Languages, vol. 1, pp. 41–110. Springer, Heidelberg (1997). https://doi.org/10.1007/978-3-642-59136-5_2
19. Zhang, J., Zhai, F., Zong, C.: Syntax-based translation with bilingually lexicalized synchronous tree substitution grammars. IEEE Trans. Audio Speech Lang. Proc. **21**(8), 1586–1597 (2013)

Sublinear-Time Language Recognition and Decision by One-Dimensional Cellular Automata

Augusto Modanese[(⊠)]

Karlsruhe Institute of Technology (KIT), Karlsruhe, Germany
modanese@kit.edu

Abstract. After an apparent hiatus of roughly 30 years, we revisit a seemingly neglected subject in the theory of (one-dimensional) cellular automata: sublinear-time computation. The model considered is that of ACAs, which are language acceptors whose acceptance condition depends on the states of all cells in the automaton. We prove a time hierarchy theorem for sublinear-time ACA classes, analyze their intersection with the regular languages, and, finally, establish strict inclusions in the parallel computation classes SC and (uniform) AC. As an addendum, we introduce and investigate the concept of a decider ACA (DACA) as a candidate for a decider counterpart to (acceptor) ACAs. We show the class of languages decidable in constant time by DACAs equals the locally testable languages, and we also determine $\Omega(\sqrt{n})$ as the (tight) time complexity threshold for DACAs up to which no advantage compared to constant time is possible.

1 Introduction

While there have been several works on linear- and real-time language recognition by cellular automata over the years (see, e.g., [14,24] for an overview), interest in the sublinear-time case has been scanty at best. We can only speculate this has been due to a certain obstinacy concerning what is now the established acceptance condition for cellular automata, namely that the first cell determines the automaton's response, despite alternatives being long known [18]. Under this condition, only a constant-size prefix can ever influence the automaton's decision, which effectively dooms sublinear time to be but a trivial case just as it is for (classical) Turing machines, for example. Nevertheless, at least in the realm of Turing machines, this shortcoming was readily circumvented by adding a random access mechanism to the model, thus sparking rich theories on parallel computation [5,20], probabilistically checkable proofs [23], and property testing [8,19].

In the case of cellular automata, the adaptation needed is an alternate (and by all means novel) acceptance condition, covered in Sect. 2. Interestingly, in the

---

Some proofs have been omitted due to page constraints. These may be found in the full version of the paper [17].

© Springer Nature Switzerland AG 2020
N. Jonoska and D. Savchuk (Eds.): DLT 2020, LNCS 12086, pp. 251–265, 2020.
https://doi.org/10.1007/978-3-030-48516-0_19

resulting model, called ACA, parallelism and local behavior seem to be more marked features, taking priority over cell communication and synchronization algorithms (which are the dominant themes in the linear- and real-time constructions). As mentioned above, the body of theory on sublinear-time ACAs is very small and, to the best of our knowledge, resumes itself to [10,13,21]. Ibarra et al. [10] show sublinear-time ACAs are capable of recognizing non-regular languages and also determine a threshold (namely $\Omega(\log n)$) up to which no advantage compared to constant time is possible. Meanwhile, Kim and McCloskey [13] and Sommerhalder and Westrhenen [21] analyze the constant-time case subject to different acceptance conditions and characterize it based on the locally testable languages, a subclass of the regular languages.

Indeed, as covered in Sect. 3, the defining property of the locally testable languages, that is, that words which locally appear to be the same are equivalent with respect to membership in the language at hand, effectively translates into an inherent property of acceptance by sublinear-time ACAs. In Sect. 4, we prove a time hierarchy theorem for sublinear-time ACAs as well as further relate the language classes they define to the regular languages and the parallel computation classes SC and (uniform) AC. In the same section, we also obtain an improvement on a result of [10]. Finally, in Sect. 5 we consider a plausible model of ACAs as language deciders, that is, machines which must not only accept words in the target language but also explicitly reject those which do not. Section 6 concludes.

2   Definitions

We assume the reader is familiar with the theory of formal languages and cellular automata as well as with computational complexity theory (see, e.g., standard references [1,6]). This section reviews basic concepts and introduces ACAs.

$\mathbb{Z}$ denotes the set of integers, $\mathbb{N}_+$ that of (strictly) positive integers, and $\mathbb{N}_0 = \mathbb{N}_+ \cup \{0\}$. B^A is the set of functions $f \colon A \to B$. For a word $w \in \Sigma^*$ over an alphabet Σ, $w(i)$ is the i-th symbol of w (starting with the 0-th symbol), and $|w|_x$ is the number of occurrences of $x \in \Sigma$ in w. For $k \in \mathbb{N}_0$, $p_k(w)$, $s_k(w)$, and $I_k(w)$ are the prefix, suffix and set of infixes of length k of w, respectively, where $p_{k'}(w) = s_{k'}(w) = w$ and $I_{k'}(w) = \{w\}$ for $k' \geq |w|$. $\Sigma^{\leq k}$ is the set of words $w \in \Sigma^*$ for which $|w| \leq k$. Unless otherwise noted, n is the input length.

2.1   (Strictly) Locally Testable Languages

The class REG of regular languages is defined in terms of (deterministic) automata with finite memory and which read their input in a single direction (i.e., from left to right), one symbol at a time; once all symbols have been read, the machine outputs a single bit representing its decision. In contrast, a *scanner* is a memoryless machine which reads a span of $k \in \mathbb{N}_+$ symbols at a time of an input provided with start and end markers (so it can handle prefixes and suffixes separately); the scanner validates every such substring it reads using the same predicate, and it accepts if and only if all these validations are successful. The languages accepted by these machines are the strictly locally testable languages.[1]

Definition 1 (strictly locally testable). *Let Σ be an alphabet. A language $L \subseteq \Sigma^*$ is* strictly locally testable *if there is some $k \in \mathbb{N}_+$ and sets $\pi, \sigma \subseteq \Sigma^{\leq k}$ and $\mu \subseteq \Sigma^k$ such that, for every word $w \in \Sigma^*$, $w \in L$ if and only if $p_k(w) \in \pi$, $I_k(w) \subseteq \mu$, and $s_k(w) \in \sigma$.* SLT *is the class of strictly locally testable languages.*

A more general notion of locality is provided by the locally testable languages. Intuitively, L is locally testable if a word w being in L or not is entirely dependent on a property of the substrings of w of some constant length $k \in \mathbb{N}_+$ (that depends only on L, not on w). Thus, if any two words have the same set of substrings of length k, then they are equivalent with respect to being in L:

Definition 2 (locally testable). *Let Σ be an alphabet. A language $L \subseteq \Sigma^*$ is* locally testable *if there is some $k \in \mathbb{N}_+$ such that, for every $w_1, w_2 \in \Sigma^*$ with $p_k(w_1) = p_k(w_2)$, $I_k(w_1) = I_k(w_2)$, and $s_k(w_1) = s_k(w_2)$ we have that $w_1 \in L$ if and only if $w_2 \in L$.* LT *denotes the class of locally testable languages.*

LT is the *Boolean closure* of SLT, that is, its closure under union, intersection, and complement [16]. In particular, SLT $\subsetneq$ LT (i.e., the inclusion is proper [15]).

2.2   Cellular Automata

In this paper, we are strictly interested in one-dimensional cellular automata with the standard neighborhood. For $r \in \mathbb{N}_0$, let $N_r(z) = \{z' \in \mathbb{Z} \mid |z - z'| \leq r\}$ denote the *extended neighborhood of radius r* of the cell $z \in \mathbb{Z}$.

Definition 3 (cellular automaton). *A cellular automaton (CA) C is a triple (Q, δ, Σ) where Q is a finite, non-empty set of states, $\delta \colon Q^3 \to Q$ is the* local *transition function, and $\Sigma \subseteq Q$ is the input alphabet. An element of Q^3 (resp., $Q^{\mathbb{Z}}$) is called a* local *(resp.,* global*) configuration of C. δ induces the* global *transition function $\Delta \colon Q^{\mathbb{Z}} \to Q^{\mathbb{Z}}$ on the configuration space $Q^{\mathbb{Z}}$ by $\Delta(c)(z) = \delta(c(z-1), c(z), c(z+1))$, where $z \in \mathbb{Z}$ is a cell and $c \in Q^{\mathbb{Z}}$.*

Our interest in CAs is as machines which receive an input and process it until a final state is reached. The input is provided from left to right, with one cell for each input symbol. The surrounding cells are inactive and remain so for the entirety of the computation (i.e., the CA is bounded). It is customary for CAs to have a distinguished cell, usually cell zero, which communicates the machine's output. As mentioned in the introduction, this convention is inadequate for computation in sublinear time; instead, we require the finality condition to depend on the entire (global) configuration (modulo inactive cells):

---

[1] The term "(locally) testable in the strict sense" ((L)TSS) is also common [13,15,16].

q	0	1	0	1	0	1	q
q	a	a	a	a	a	a	q

✓

q	0	0	1	0	1	0	q
q	0	0	a	a	a	0	q

Fig. 1. Computation of an ACA which recognizes $L = \{01\}^+$. The input words are $010101 \in L$ and $001010 \notin L$, respectively.

Definition 4 (CA computation). *There is a distinguished state $q \in Q \setminus \Sigma$, called the* inactive *state, which, for every $z_1, z_2, z_3 \in Q$, satisfies $\delta(z_1, z_2, z_3) = q$ if and only if $z_2 = q$. A cell not in state q is said to be* active. *For an input $w \in \Sigma^*$, the* initial configuration $c_0 = c_0(w) \in Q^{\mathbb{Z}}$ *of C for w is $c_0(i) = w(i)$ for $i \in \{0, \ldots, |w| - 1\}$ and $c_0(i) = q$ otherwise. For $F \subseteq Q \setminus \{q\}$, a configuration $c \in Q^{\mathbb{Z}}$ is F-final (for w) if there is a (minimal) $\tau \in \mathbb{N}_0$ such that $c = \Delta^{\tau}(c_0)$ and c contains only states in $F \cup \{q\}$. In this context, the sequence $c_0, \ldots, \Delta^{\tau}(c_0) = c$ is the* trace *of w, and τ is the* time complexity *of C (with respect to F and w).*

Because we effectively consider only bounded CAs, the computation of w involves exactly $|w|$ active cells. The surrounding inactive cells are needed only as markers for the start and end of w. As a side effect, the initial configuration $c_0 = c_0(\varepsilon)$ for the empty word ε is stationary (i.e., $\Delta(c_0) = c_0$) regardless of the choice of δ. Since this is the case only for ε, we disregard it for the rest of the paper, that is, we assume it is not contained in any of the languages considered.

Finally, we relate final configurations and computation results. We adopt an acceptance condition as in [18,21] and obtain a so-called ACA; here, the "A" of "ACA" refers to the property that all (active) cells are relevant for acceptance.

Definition 5 (ACA). *An ACA is a CA C with a non-empty subset $A \subseteq Q \setminus \{q\}$ of* accept *states. For $w \in \Sigma^+$, if C reaches an A-final configuration, we say C* accepts *w. $L(C)$ denotes the set of words accepted by C. For $t \colon \mathbb{N}_+ \to \mathbb{N}_0$, we write $\mathsf{ACA}(t)$ for the class of languages accepted by an ACA with time complexity bounded by t, that is, for which the time complexity of accepting w is $\leq t(|w|)$.*

$\mathsf{ACA}(t_1) \subseteq \mathsf{ACA}(t_2)$ is immediate for functions $t_1, t_2 \colon \mathbb{N}_+ \to \mathbb{N}_0$ with $t_1(n) \leq t_2(n)$ for every $n \in \mathbb{N}_+$. Because Definition 5 allows multiple accept states, it is possible for each (non-accepting) state z to have a corresponding accept state z_A. In the rest of this paper, when we say a cell becomes (or marks itself as) accepting (without explicitly mentioning its state), we intend to say it changes from such a state z to z_A.

Figure 1 illustrates the computation of an ACA with input alphabet $\Sigma = \{0, 1\}$ and which accepts $\{01\}^+$ with time complexity equal to one (step). The local transition function is such that $\delta(0, 1, 0) = \delta(1, 0, 1) = \delta(q, 0, 1) = \delta(0, 1, q) = a$, a being the (only) accept state, and $\delta(z_1, z_2, z_3) = z_2$ for $z_2 \neq a$ and arbitrary z_1 and z_3.

3   First Observations

This section recalls results on sublinear-time ACA computation (i.e., $\mathsf{ACA}(t)$ where $t \in o(n)$) from [10,13,21] and provides some additional remarks. We start with the constant-time case (i.e., $\mathsf{ACA}(O(1))$). Here, the connection between scanners and ACAs is apparent: If an ACA accepts an input w in time $\tau = \tau(w)$, then w can be verified by a scanner with an input span of $2\tau + 1$ symbols and using the predicate induced by the local transition function of the ACA (i.e., the predicate is true if and only if the symbols read correspond to $N_\tau(z)$ for some cell z in the initial configuration and z is accepting after τ steps).

Constant-time ACA computation has been studied in [13,21]. Although in [13] we find a characterization based on a hierarchy over SLT, the acceptance condition there differs slightly from that in Definition 5; in particular, the automata there run for a number of steps which is fixed for each automaton, and the outcome is evaluated (only) in the final step. In contrast, in [21] we find the following, where $\mathsf{SLT}_\vee$ denotes the closure of SLT under union:

Theorem 6 ([21]). $\mathsf{ACA}(O(1)) = \mathsf{SLT}_\vee$.

Thus, $\mathsf{ACA}(O(1))$ is closed under union. In fact, more generally:

Proposition 7. *For any* $t\colon \mathbb{N}_+ \to \mathbb{N}_+$, $\mathsf{ACA}(O(t))$ *is closed under union.*

$\mathsf{ACA}(O(1))$ is closed under intersection [21]. It is an open question whether $\mathsf{ACA}(O(t))$ is also closed under intersection for every $t \in o(n)$.

Moving beyond constant time, in [10] we find the following:

Theorem 8 ([10]). *For* $t \in o(\log n)$, $\mathsf{ACA}(t) \subseteq \mathsf{REG}$.

In [10] we find an example for a non-regular language in $\mathsf{ACA}(O(\log n))$ which is essentially a variation of the language

$$\mathsf{BIN} = \{\mathrm{bin}_k(0)\#\mathrm{bin}_k(1)\# \cdots \#\mathrm{bin}_k(2^k - 1) \mid k \in \mathbb{N}_+\}$$

where $\mathrm{bin}_k(m)$ is the k-digit binary representation of $m \in \{0, \ldots, 2^k - 1\}$.

To illustrate the ideas involved, we present an example related to BIN (though it results in a different time complexity) and which is also useful in later discussions in Sect. 5. Let $w_k(i) = 0^i 10^{k-i-1}$ and consider the language

$$\mathsf{IDMAT} = \{w_k(0)\#w_k(1)\# \cdots \#w_k(k - 1) \mid k \in \mathbb{N}_+\}$$

of all identity matrices in line-for-line representations, where the lines are separated by $\#$ symbols.[2]

We now describe an ACA for IDMAT; the construction closely follows the aforementioned one for BIN found in [10] (and the difference in complexity is only due to the different number and size of blocks in the words of IDMAT and BIN). Denote each group of cells initially containing a (maximally long) $\{0,1\}^+$ substring of $w \in \mathsf{IDMAT}$ by a *block*. Each block of size b propagates its contents to the neighboring blocks (in separate registers); using a textbook CA technique, this requires exactly $2b$ steps. Once the strings align, a block initially

---

[2] Alternatively, one can also think of IDMAT as a (natural) problem on graphs presented in the adjacency matrix representation.

containing $w_k(i)$ verifies it has received $w_k(i-1)$ and $w_k(i+1)$ from its left and right neighbor blocks (if either exists), respectively. The cells of a block and its delimiters become accepting if and only if the comparisons are successful and there is a single # between the block and its neighbors. This process takes linear time in b; since any $w \in$ IDMAT has $O(\sqrt{|w|})$ many blocks, each with $b \in O(\sqrt{|w|})$ cells, it follows that IDMAT $\in$ ACA($O(\sqrt{n})$).

To show the above construction is time-optimal, we use the following observation, which is also central in proving several other results in this paper:

Lemma 9. *Let C be an ACA, and let w be an input which C accepts in (exactly) $\tau = \tau(w)$ steps. Then, for every input w' such that $p_{2\tau}(w) = p_{2\tau}(w')$, $I_{2\tau+1}(w') \subseteq I_{2\tau+1}(w)$, and $s_{2\tau}(w) = s_{2\tau}(w')$, C accepts w' in at most τ steps.*

The lemma is intended to be used with $\tau < \frac{|w|}{2}$ since otherwise $w = w'$. It can be used, for instance, to show that SOMEONE $= \{w \in \{0,1\}^+ \mid |w|_1 \geq 1\}$ is not in ACA(t) for any $t \in o(n)$ (e.g., set $w = 0^k 10^k$ and $w' = 0^{2k+1}$ for large $k \in \mathbb{N}_+$). It follows REG $\not\subseteq$ ACA(t) for $t \in o(n)$.

Since the complement of SOMEONE (respective to $\{0,1\}^+$) is $\{0\}^+$ and $\{0\}^+ \in$ ACA($O(1)$) (e.g., simply set 0 as the ACA's accepting state), ACA(t) is not closed under complement for any $t \in o(n)$. Also, SOMEONE is a regular language and BIN $\in$ ACA($O(\log n)$) is not, so we have:

Proposition 10. *For $t \in \Omega(\log n) \cap o(n)$, ACA($t$) and REG are incomparable.*

If the inclusion of infixes in Lemma 9 is strengthened to an equality, one may apply it in both directions and obtain the following stronger statement:

Lemma 11. *Let C be an ACA with time complexity bounded by $t\colon \mathbb{N}_+ \to \mathbb{N}_0$ (i.e., C accepts any input of length n in at most $t(n)$ steps). Then, for any two inputs w and w' with $p_{2\mu}(w) = p_{2\mu}(w')$, $I_{2\mu+1}(w) = I_{2\mu+1}(w')$, and $s_{2\mu}(w) = s_{2\mu}(w')$ where $\mu = \max\{t(|w|), t(|w'|)\}$, we have that $w \in L(C)$ if and only if $w' \in L(C)$.*

Finally, we can show our ACA for IDMAT is time-optimal:

Proposition 12. *For any $t \in o(\sqrt{n})$, IDMAT $\notin$ ACA(t).*

4 Main Results

In this section, we present various results regarding ACA(t) where $t \in o(n)$. First, we obtain a time hierarchy theorem, that is, under plausible conditions, ACA(t') $\subsetneq$ ACA(t) for $t' \in o(t)$. Next, we show ACA(t) $\cap$ REG is (strictly) contained in LT and also present an improvement to Theorem 8. Finally, we study inclusion relations between ACA(t) and the SC and (uniform) AC hierarchies. Save for the material covered so far, all three subsections stand out independently from one another.

4.1 Time Hierarchy

For functions $f, t\colon \mathbb{N}_+ \to \mathbb{N}_0$, we say f is *time-constructible by CAs in* $t(n)$ *time* if there is a CA C which, on input 1^n, reaches a configuration containing the value $f(n)$ (binary-encoded) in at most $t(n)$ steps.[3] Note that, since CAs can simulate (one-tape) Turing machines in real-time, any function constructible by Turing machines (in the corresponding sense) is also constructible by CAs.

Theorem 13. *Let* $f \in \omega(n)$ *with* $f(n) \le 2^n$, $g(n) = 2^{n-\lfloor \log f(n) \rfloor}$, *and let* f *and* g *be time-constructible (by CAs) in* $f(n)$ *time. Furthermore, let* $t\colon \mathbb{N}_+ \to \mathbb{N}_0$ *be such that* $3f(k) \le t(f(k)g(k)) \le cf(k)$ *for some constant* $c \ge 3$ *and all but finitely many* $k \in \mathbb{N}_+$. *Then, for every* $t' \in o(t)$, $\mathsf{ACA}(t') \subsetneq \mathsf{ACA}(t)$.

Given $a > 1$, this can be used, for instance, with any time-constructible $f \in \Theta(n^a)$ (resp., $f \in \Theta(2^{n/a})$, in which case $a = 1$ is also possible) and $t \in \Theta((\log n)^a)$ (resp., $t \in \Theta(n^{1/a})$). The proof idea is to construct a language L similar to BIN (see Sect. 3) in which every $w \in L$ has length exponential in the size of its blocks while the distance between any two blocks is $\Theta(t(|w|))$. Due to Lemma 9, the latter implies L is not recognizable in $o(t(|w|))$ time.

Proof. For simplicity, let $f(n) > n$. Consider $L = \{w_k \mid k \in \mathbb{N}_+\}$ where

$$w_k = \mathrm{bin}_k(0)\#^{f(k)-k}\mathrm{bin}_k(1)\#^{f(k)-k} \cdots \mathrm{bin}_k(g(k)-1)\#^{f(k)-k}$$

and note $|w_k| = f(k)g(k)$. Because $t(|w_k|) \in O(f(k))$ and $f(k) \in \omega(k)$, given any $t' \in o(t)$, setting $w = w_k$, $w' = 0^k\#^{|w_k|-k}$, and $\tau = t'(|w_k|)$ and applying Lemma 9 for sufficiently large k yields $L \notin \mathsf{ACA}(t')$.

By assumption it suffices to show $w = w_k \in L$ is accepted by an ACA C in at most $3f(k) \le t(|w|)$ steps for sufficiently large $k \in \mathbb{N}_+$. The cells of C perform two procedures P_1 and P_2 simultaneously: P_1 is as in the ACA for BIN (see Sect. 3) and ensures that the blocks of w have the same length, that the respective binary encodings are valid, and that the last value is correct (i.e., equal to $g(k) - 1$). In P_2, each block computes $f(k)$ as a function of its block length k. Subsequently, the value $f(k)$ is decreased using a real-time counter (see, e.g., [12] for a construction). Every time the counter is decremented, a signal starts from the block's leftmost cell and is propagated to the right. This allows every group of cells of the form bs with $b \in \{0,1\}^+$ and $s \in \{\#\}^+$ to assert there are precisely $f(k)$ symbols in total (i.e., $|bs| = f(k)$). A cell is accepting if and only if it is accepting both in P_1 and P_2. The proof is complete by noticing either procedure takes a maximum of $3f(k)$ steps (again, for sufficiently large k). □

4.2 Intersection with the Regular Languages

In light of Proposition 10, we now consider the intersection $\mathsf{ACA}(t) \cap \mathsf{REG}$ for $t \in o(n)$ (in the same spirit as a conjecture by Straubing [22]). For this section,

---

[3] Just as is the case for Turing machines, there is not a single definition for time-constructibility by CAs (see, e.g., [12] for an alternative). Here, we opt for a plausible variant which has the benefit of simplifying the ensuing line of argument.

we assume the reader is familiar with the theory of syntactic semigroups (see, e.g., [7] for an in-depth treatment).

Given a language L, let $\mathrm{SS}(L)$ denote the syntactic semigroup of L. It is well-known that $\mathrm{SS}(L)$ is finite if and only if L is regular. A semigroup S is a *semilattice* if $x^2 = x$ and $xy = yx$ for every $x, y \in S$. Additionally, S is *locally semilattice* if eSe is a semilattice for every *idempotent* $e \in S$, that is, $e^2 = e$. We use the following characterization of locally testable languages:

Theorem 14 ([3,15]). $L \in \mathsf{LT}$ *if and only if* $\mathrm{SS}(L)$ *is finite and locally semilattice.*

In conjunction with Lemma 9, this yields the following, where the strict inclusion is due to SOMEONE $\notin \mathsf{ACA}(t)$ (since SOMEONE $\in \mathsf{LT}$; see Sect. 3):

Theorem 15. *For every* $t \in o(n)$, $\mathsf{ACA}(t) \cap \mathsf{REG} \subsetneq \mathsf{LT}$.

Proof. Let $L \in \mathsf{ACA}(t)$ be a language over the alphabet Σ and, in addition, let $L \in \mathsf{REG}$, that is, $S = \mathrm{SS}(L)$ is finite. By Theorem 14, it suffices to show S is locally semilattice. To that end, let $e \in S$ be idempotent, and let $x, y \in S$.

To show $(exe)(eye) = (eye)(exe)$, let $a, b \in \Sigma^*$ and consider the words $u = a(exe)(eye)b$ and $v = a(eye)(exe)b$. For $m \in \mathbb{N}_+$, let $u'_m = a(e^m x e^m)(e^m y e^m)b$, and let $r \in \mathbb{N}_+$ be such that $r > \max\{|x|, |y|, |a|, |b|\}$ and also $t(|u'_{2r+1}|) < \frac{1}{16|e|}|u'_{2r+1}| < r$. Since e is idempotent, $u' = u'_{2r+1}$ and u belong to the same class in S, that is, $u' \in L$ if and only if $u \in L$; the same is true for $v' = a(e^{2r+1} y e^{2r+1})(e^{2r+1} x e^{2r+1})b$ and v. Furthermore, $p_{2r}(u') = p_{2r}(v')$, $I_{2r+1}(u') = I_{2r+1}(v')$, and $s_{2r}(u') = s_{2r}(v')$ hold. Since $L \in \mathsf{ACA}(t)$, Lemma 11 applies.

The proof of $(exe)(exe) = exe$ is analogous. Simply consider the words $a(e^m x e^m)b$ and $a(e^m x e^m)(e^m x e^m)b$ for sufficiently large $m \in \mathbb{N}_+$ and use, again, Lemma 11 and the fact that e is idempotent. $\qquad\square$

Using Theorems 8 and 15, we have $\mathsf{ACA}(t) \subsetneq \mathsf{LT}$ for $t \in o(\log n)$. We can improve this bound to $\mathsf{ACA}(O(1)) = \mathsf{SLT}_\vee$, which is a proper subset of LT:

Theorem 16. *For every* $t \in o(\log n)$, $\mathsf{ACA}(t) = \mathsf{ACA}(O(1))$.

Proof. We prove every ACA C with time complexity at most $t \in o(\log n)$ actually has $O(1)$ time complexity. Let Q be the state set of C and assume $|Q| \geq 2$, and let $n_0 \in \mathbb{N}_+$ be such that $t(n) < \frac{\log n}{9 \log |Q|}$ for $n \geq n_0$. Letting $k(n) = 2t(n) + 1$ and assuming $t(n) \geq 1$, we then have $|Q|^{3k(n)} \leq |Q|^{9t(n)} < n$ $(\star)$. We shall use this to prove that, for any word $w \in L$ of length $|w| \geq n_0$, there is a word $w' \in L$ of length $|w'| \leq n_0$ as well as $r < n_0$ such that $p_r(w) = p_r(w')$, $I_{r+1}(w) = I_{r+1}(w')$, and $s_r(w) = s_r(w')$. By Lemma 9, C must have $t(|w'|)$ time complexity on w and, since the set of all such w' is finite, it follows that C has $O(1)$ time complexity.

Now let w be as above and let C accept w in (exactly) $\tau = \tau(w) \leq t(|w|)$ steps. We prove the claim by induction on $|w|$. The base case $|w| = n_0$ is trivial, so let $n > n_0$ and assume the claim holds for every word in L of length strictly less than n. Consider the De Bruijn graph G over the words in $|Q|^\kappa$ where $\kappa = 2\tau + 1$. Then, from the infixes of w of length κ (in order of appearance in w) one obtains a path P in G by starting at the leftmost infix and visiting every subsequent one, up to the rightmost one. Let G' be the induced subgraph of G containing exactly the nodes visited by P, and notice P visits every node in G' at least once. It is not hard to show that, for every such P and G', there is a path P' in G' with the same starting and ending points as P and that visits every node of G' at least once while having length at most $m^2 \leq |Q|^{2\kappa}$, where m is the number of nodes in G'.[4] To this P' corresponds a word w' of length $|w'| \leq \kappa + |Q|^{2\kappa} < |Q|^{3\kappa}$ for which, by construction of P' and G', $p_{\kappa-1}(w') = p_{\kappa-1}(w)$, $I_\kappa(w') = I_\kappa(w)$, and $s_{\kappa-1}(w') = s_{\kappa-1}(w')$. Since $\kappa \leq k(|w|)$, using $(\star)$ we have $|w'| < |w|$, and then either $|w'| \leq n_0$ and $\kappa < n_0$ (since otherwise $w = w'$, which contradicts $|w'| < |w|$), or we may apply the induction hypothesis; in either case, the claim follows. $\square$

4.3   Relation to Parallel Complexity Classes

In this section, we relate $\mathsf{ACA}(t)$ to other classes which characterize parallel computation, namely the SC and (uniform) AC hierarchies. In this context, SC^k is the class of problems decidable by Turing machines in $O((\log n)^k)$ space and polynomial time, whereas AC^k is that decidable by Boolean circuits with polynomial size, $O((\log n)^k)$ depth, and gates with unbounded fan-in. SC (resp., AC) is the union of all SC^k (resp., AC^k) for $k \in \mathbb{N}_0$. Here, we consider only uniform versions of AC; when relevant, we state the respective uniformity condition. Although $\mathsf{SC}^1 = \mathsf{L} \subseteq \mathsf{AC}^1$ is known, it is unclear whether any other containment holds between SC and AC.

One should not expect to include SC or AC in $\mathsf{ACA}(t)$ for any $t \in o(n)$. Conceptually speaking, whereas the models of SC and AC are capable of random access to their input, ACAs are inherently local (as evinced by Lemmas 9 and 11). Explicit counterexamples may be found among the unary languages: For any fixed $m \in \mathbb{N}_+$ and $w_1, w_2 \in \{1\}^+$ with $|w_1|, |w_2| \geq m$, trivially $p_{m-1}(w_1) = p_{m-1}(w_2)$, $I_m(w_1) = I_m(w_2)$, and $s_{m-1}(w_1) = s_{m-1}(w_2)$ hold. Hence, by Lemma 9, if an ACA C accepts $w \in \{1\}^+$ in $t \in o(n)$ time and $|w|$ is large (e.g., $|w| > 4t(|w|)$), then C accepts any $w' \in \{1\}^+$ with $|w'| \geq |w|$. Thus, extending a result from [21]:

Proposition 17. *If $t \in o(n)$ and $L \in \mathsf{ACA}(t)$ is a unary language (i.e., $L \subseteq \Sigma^+$ and $|\Sigma| = 1$), then L is either finite or co-finite.*

In light of the above, the rest of this section is concerned with the converse type of inclusion (i.e., of $\mathsf{ACA}(t)$ in the SC or AC hierarchies). For $f, s, t : \mathbb{N}_+ \to \mathbb{N}_0$ with $f(n) \leq s(n)$, we say f is *constructible (by a Turing machine) in $s(n)$ space and $t(n)$ time* if there is a Turing machine T which, on input 1^n, outputs

---

[4] Number the nodes of G' from 1 to m according to the order in which they are first visited by P. Then, there is a path in G' from i to $i+1$ for every $i \in \{1, \ldots, m-1\}$, and a shortest such path has length at most m. Piecing these paths together along with a last (shortest) path from m to the ending point of P, we obtain a path of length at most m^2 with the purported property.

$f(n)$ in binary using at most $s(n)$ space and $t(n)$ time. Also, recall a Turing machine can simulate τ steps of a CA with m (active) cells in $O(m)$ space and $O(\tau m)$ time.

Proposition 18. *Let C be an ACA with time complexity bounded by $t \in o(n)$, $t(n) \geq \log n$, and let t be constructible in $t(n)$ space and $\mathsf{poly}(n)$ time. Then, there is a Turing machine which decides $L(C)$ in $O(t(n))$ space and $\mathsf{poly}(n)$ time.*

Thus, for polylogarithmic t (where the strict inclusion is due to Proposition 17):

Corollary 19. *For $k \in \mathbb{N}_+$, $\mathsf{ACA}(O((\log n)^k)) \subsetneq \mathsf{SC}^k$.*

Moving on to the AC classes, we employ some notions from descriptive complexity theory (see, e.g., [11] for an introduction). Let $\mathsf{FO_L}[t]$ be the class of languages describable by first-order formulas with numeric relations in L (i.e., logarithmic space) and quantifier block iterations bounded by $t \colon \mathbb{N}_+ \to \mathbb{N}_0$.

Theorem 20. *Let $t \colon \mathbb{N}_+ \to \mathbb{N}_0$ with $t(n) \geq \log n$ be constructible in logarithmic space (and arbitrary time). For any ACA C whose time complexity is bounded by t, $L(C) \in \mathsf{FO_L}[O(\frac{t}{\log n})]$.*

Since $\mathsf{FO_L}[O((\log n)^k)]$ equals L-uniform AC^k [11], by Proposition 17 we have:

Corollary 21. *For $k \in \mathbb{N}_+$, $\mathsf{ACA}(O((\log n)^k)) \subsetneq \mathsf{L}$-uniform AC^{k-1}.*

Because $\mathsf{SC}^1 \not\subseteq \mathsf{AC}^0$ (regardless of non-uniformity) [9], this is an improvement on Corollary 19 at least for $k = 1$. Nevertheless, note the usual uniformity condition for AC^0 is not L- but the more restrictive $\mathsf{DLOGTIME}$-uniformity [25], and there is good evidence that these two versions of AC^0 are distinct [4]. Using methods from [2], Corollary 21 may be rephrased for AC^0 in terms of $\mathsf{TIME}(\mathsf{polylog}(n))$- or even $\mathsf{TIME}((\log n)^2)$-uniformity, but the $\mathsf{DLOGTIME}$-uniformity case remains unclear.

5   Decider ACA

So far, we have considered ACAs strictly as language acceptors. As such, their time complexity for inputs not in the target language (i.e., those which are not accepted) is entirely disregarded. In this section, we investigate ACAs as *deciders*, that is, as machines which must also (explicitly) reject invalid inputs. We analyze the case in which these decider ACAs must reject under the same condition as acceptance (i.e., all cells are simultaneously in a final rejecting state):

Definition 22 (DACA). *A decider ACA (DACA) is an ACA C which, in addition to its set A of accept states, has a non-empty subset $R \subseteq Q \setminus \{q\}$ of reject states that is disjoint from A (i.e., $A \cap R = \varnothing$). Every input $w \in \Sigma^+$ of C must lead to an A- or an R-final configuration (or both). C accepts w if it leads*

q	0	0	0	0	0	0	q
q	r	r	r	r	r	r	q

✗

q	0	0	1	0	1	0	q
q	r	r	a	r	a	r	q
q	a	a	a	a	a	a	q

✓

Fig. 2. Computation of a DACA C which decides **SOMEONE**. The inputs words are $000000 \in L(C)$ and $001010 \notin L(C)$, respectively.

to an A-final configuration c_A and none of the configurations prior to c_A are R-final. Similarly, C rejects w if it leads to an R-final configuration c_R and none of the configurations prior to c_R are A-final. The time complexity of C (with respect to w) is the number of steps elapsed until C reaches an R- or A-final configuration (for the first time). DACA(t) *is the DACA analogue of* ACA(t)*.*

In contrast to Definition 5, here we must be careful so that the accept and reject results do not overlap (i.e., a word cannot be both accepted and rejected). We opt for interpreting the first (chronologically speaking) of the final configurations as the machine's response. Since the outcome of the computation is then irrelevant regardless of any subsequent configurations (whether they are final or not), this is equivalent to requiring, for instance, that the DACA must halt once a final configuration is reached.

One peculiar consequence of Definition 22 is the relation between languages which can be recognized by acceptor ACAs and DACAs (i.e., the classes ACA(t) and DACA(t)). As it turns out, the situation is quite different from what is usually expected of restricting an acceptor model to a decider one, that is, that deciders yield a (possibly strictly) more restricted class of machines. In fact, one can show DACA(t) $\not\subseteq$ ACA(t) holds for $t \in o(n)$ since **SOMEONE** $\notin$ ACA($O(1)$) (see discussion after Lemma 9); nevertheless, **SOMEONE** $\in$ DACA($O(1)$). For example, the local transition function δ of the DACA can be chosen as $\delta(z_1, 0, z_2) = r$ and $\delta(z_1, z, z_2) = a$ for $z \in \{1, a, r\}$, where z_1 and z_2 are arbitrary states, and a and r are the (only) accept and reject states, respectively; see Fig. 2. Choosing the same δ for an (acceptor) ACA does *not* yield an ACA for **SOMEONE** since then all words of the form 0^+ are accepted in the second step (as they are not rejected in the first one). We stress this rather counterintuitive phenomenon occurs only in the case of sublinear time (as ACA(t) = CA(t) = DACA(t) for $t \in \Omega(n)$).

Similar to (acceptor) ACAs (Lemma 9), sublinear-time DACAs operate locally:

Lemma 23. *Let C be a DACA and let $w \in \{0, 1\}^+$ be a word which C decides in exactly $\tau = \tau(w)$ steps. Then, for every word $w' \in \{0, 1\}^+$ with $p_{2\tau}(w) = p_{2\tau}(w')$, $I_{2\tau+1}(w') = I_{2\tau+1}(w)$, and $s_{2\tau}(w) = s_{2\tau}(w')$, C decides w' in $\leq \tau$ steps, and $w \in L(C)$ holds if and only if $w' \in L(C)$.*

One might be tempted to relax the requirements above to $I_{2\tau+1}(w') \subseteq I_{2\tau+1}(w)$ (as in Lemma 9). We stress, however, the equality $I_{2\tau+1}(w) = I_{2\tau+1}(w')$ is crucial; otherwise, it might be the case that C takes strictly less than τ steps to decide w' and, hence, $w \in L(C)$ may not be equivalent to $w' \in L(C)$.

We note that, in addition to Lemmas 9 and 11, the results from Sect. 4 are extendable to decider ACAs; a more systematic treatment is left as a topic for future work. The remainder of this section is concerned with characterizing $\mathsf{DACA}(O(1))$ computation (as a parallel to Theorem 6) as well as establishing the time threshold for DACAs to decide languages other than those in $\mathsf{DACA}(O(1))$ (as Theorem 16 and the result $\mathsf{BIN} \in \mathsf{ACA}(O(\log n))$ do for acceptor ACAs).

5.1   The Constant-Time Case

First notice that, for any DACA C, swapping the accept and reject states yields a DACA with the same time complexity and which decides the complement of $L(C)$. Hence, in contrast to ACAs (see discussion following Lemma 9):

Proposition 24. *For any $t \colon \mathbb{N}_+ \to \mathbb{N}_+$, $\mathsf{DACA}(t)$ is closed under complement.*

Using this, we can prove the following, which characterizes constant-time DACA computation as a parallel to Theorem 6:

Theorem 25. $\mathsf{DACA}(O(1)) = \mathsf{LT}$.

Hence, we obtain the rather surprising inclusion $\mathsf{ACA}(O(1)) \subsetneq \mathsf{DACA}(O(1))$, that is, for constant time, DACAs constitute a *strictly more* powerful model than their acceptor counterparts.

5.2   Beyond Constant Time

Theorem 16 establishes a logarithmic time threshold for (acceptor) ACAs to recognize languages not in $\mathsf{ACA}(O(1))$. We now turn to obtaining a similar result for DACAs. As it turns out, in this case the bound is considerably larger:

Theorem 26. *For any $t \in o(\sqrt{n})$, $\mathsf{DACA}(t) = \mathsf{DACA}(O(1))$.*

One immediate implication is that $\mathsf{DACA}(t)$ and $\mathsf{ACA}(t)$ are *incomparable* for $t \in o(\sqrt{n}) \cap \omega(1)$ (since, e.g., $\mathsf{BIN} \in \mathsf{ACA}(\log n)$; see Sect. 3). The proof idea is that any DACA whose time complexity is not constant admits an infinite sequence of words with increasing time complexity; however, the time complexity of each such word can be traced back to a critical set of cells which prevent the automaton from either accepting or rejecting. By contracting the words while keeping the extended neighborhoods of these cells intact, we obtain a new infinite sequence of words which the DACA necessarily takes $\Omega(\sqrt{n})$ time to decide:

Proof. Let C be a DACA with time complexity bounded by t and assume $t \notin O(1)$; we show $t \in \Omega(\sqrt{n})$. Since $t \notin O(1)$, for every $i \in \mathbb{N}_0$ there is a w_i such that C takes strictly more than i steps to decide w_i. In particular, when C receives w_i as input, there are cells x_j^i and y_j^i for $j \in \{0, \dots, i\}$ such that x_j^i (resp., y_j^i) is not accepting (resp., rejecting) in step j. Let J_i be the set of all $z \in \{0, \dots, |w_i| - 1\}$ for which $\min\{|z - x_j^i|, |z - y_j^i|\} \le j$, that is, $z \in N_j(x_j^i) \cup N_j(y_j^i)$ for some j. Consider the restriction w_i' of w_i to the symbols having index in J_i, that is,

$w_i'(k) = w_i(j_k)$ for $J_i = \{j_0, \ldots, j_{m-1}\}$ and $j_0 < \cdots < j_{m-1}$, and notice w_i' has the same property as w_i (i.e., C takes strictly more than i steps to decide w_i). Since $|w_i'| = |J_i| \leq 2(i+1)^2$, C has $\Omega(\sqrt{n})$ time complexity on the (infinite) set $\{w_i' \mid i \in \mathbb{N}_0\}$. □

Using IDMAT (see Sect. 3), we show the bound in Theorem 26 is optimal:

Proposition 27. IDMAT $\in$ DACA$(O(\sqrt{n}))$.

We have IDMAT $\in$ ACA$(O(\sqrt{n}))$ (see Sect. 3); the non-trivial part is ensuring the DACA also rejects every $w \notin$ IDMAT in $O(\sqrt{|w|})$ time. In particular, in such strings the $\#$ delimiters may be an arbitrary number of cells apart or even absent altogether; hence, naively comparing every pair of blocks is not an option. Rather, we check the existence of a particular set of substrings of increasing length and which must present if the input is in IDMAT. Every $O(1)$ steps the existence of a different substring is verified; the result is that the input length must be at least quadratic in the length of the last substring tested (and the input is timely rejected if it does not contain any one of the required substrings).

6 Conclusion and Open Problems

Following the definition of ACAs in Sect. 2, Sect. 3 reviewed existing results on ACA(t) for sublinear t (i.e., $t \in o(n)$); we also observed that sublinear-time ACAs operate in an inherently local manner (Lemmas 9 and 11). In Sect. 4, we proved a time hierarchy theorem (Theorem 13), narrowed down the languages in ACA$(t) \cap$ REG (Theorem 15), improved Theorem 8 to ACA$(o(\log n)) =$ ACA$(O(1))$ (Theorem 16), and, finally, obtained (strict) inclusions in the parallel computation classes SC and AC (Corollaries 19 and 21, respectively). The existence of a hierarchy theorem for ACAs is of interest because obtaining an equivalent result for NC and AC is an open problem in computational complexity theory. Also of note is that the proof of Theorem 13 does not rely on diagonalization (the prevalent technique for most computational models) but, rather, on a quintessential property of sublinear-time ACA computation (i.e., locality as in the sense of Lemma 9).

In Sect. 5, we considered a plausible definition of ACAs as language deciders as opposed to simply acceptors, obtaining DACAs. The respective constant-time class is LT (Theorem 25), which surprisingly is a (strict) superset of ACA$(O(1)) =$ SLT$_\vee$. Meanwhile, $\Omega(\sqrt{n})$ is the time complexity threshold for deciding languages other than those in LT (Theorem 26 and Proposition 27).

As for future work, the primary concern is extending the results of Sect. 4 to DACAs. DACA$(O(1)) =$ LT is closed under union and intersection and we saw that DACA(t) is closed under complement for any $t \in o(n)$; a further question would be whether DACA(t) is also closed under union and intersection. Finally, we have ACA$(O(1)) \subsetneq$ DACA$(O(1))$, ACA$(O(n)) =$ CA$(O(n)) =$ DACA$(O(n))$, and that ACA(t) and DACA(t) are incomparable for $t \in o(\sqrt{n}) \cap \omega(1)$; it remains open what the relation between the two classes is for $t \in \Omega(\sqrt{n}) \cap o(n)$.

Acknowledgments. I would like to thank Thomas Worsch for the fruitful discussions and feedback during the development of this work. I would also like to thank the DLT 2020 reviewers for their valuable comments and suggestions and, in particular, one of the reviewers for pointing out a proof idea for Theorem 16, which was listed as an open problem in a preliminary version of the paper.

References

1. Arora, S., Barak, B.: Computational Complexity - A Modern Approach. Cambridge University Press, Cambridge (2009)
2. Mix Barrington, D.A.: Extensions of an idea of McNaughton. Math. Syst. Theory **23**(3), 147–164 (1990). https://doi.org/10.1007/BF02090772
3. Brzozowski, J.A., Simon, I.: Characterizations of locally testable events. Discrete Math. **4**(3), 243–271 (1973). https://doi.org/10.1016/S0012-365X(73)80005-6
4. Caussinus, H., et al.: Nondeterministic NC^1 computation. J. Comput. Syst. Sci. **57**(2), 200–212 (1998). https://doi.org/10.1006/jcss.1998.1588
5. Cook, S.A.: A taxonomy of problems with fast parallel algorithms. Inf. Control **64**(1–3), 2–21 (1985). https://doi.org/10.1016/S0019-9958(85)80041-3
6. Delorme, M., Mazoyer, J. (eds.): Cellular Automata. A Parallel Model. Mathematics and Its Application, vol. 460. Springer, Netherlands (1999). https://doi.org/10.1007/978-94-015-9153-9
7. Eilenberg, S.: Automata, Languages, and Machines. Pure and Applied Mathematics, vol. B. Academic Press, New York (1976)
8. Fischer, E.: The art of uninformed decisions. In: Bulletin of the EATCS 75, p. 97 (2001)
9. Furst, M.L., et al.: Parity, circuits, and the polynomial-time hierarchy. Math. Syst. Theory **17**(1), 13–27 (1984). https://doi.org/10.1007/BF01744431
10. Ibarra, O.H., et al.: Fast parallel language recognition by cellular automata. Theor. Comput. Sci. **41**, 231–246 (1985). https://doi.org/10.1016/0304-3975(85)90073-8
11. Immerman, N.: Descriptive Complexity. Texts in Computer Science. Springer, New York (1999). https://doi.org/10.1007/978-1-4612-0539-5
12. Iwamoto, C., et al.: Constructible functions in cellular automata and their applications to hierarchy results. Theor. Comput. Sci. **270**(1–2), 797–809 (2002). https://doi.org/10.1016/S0304-3975(01)00112-8
13. Kim, S., McCloskey, R., Sam Kim and Robert McCloskey: A characterization of constant-time cellular automata computation. Phys. D **45**(1–3), 404–419 (1990). https://doi.org/10.1016/0167-2789(90)90198-X
14. Kutrib, M.: Cellular automata and language theory. In: Meyers, R. (ed.) Encyclopedia of Complexity and Systems Science, pp. 800–823. Springer, New York (2009). https://doi.org/10.1007/978-0-387-30440-3
15. McNaughton, R.: Algebraic decision procedures for local testability. Math. Syst. Theory **8**(1), 60–76 (1974). https://doi.org/10.1007/BF01761708
16. McNaughton, R., Papert, S.: Counter-Free Automata. The MIT Press, Cambridge, MA (1971)
17. Modanese, A.: Sublinear-Time Language Recognition and Decision by One-Dimensional Cellular Automata. CoRR abs/1909.05828 (2019). arXiv: 1909.05828
18. Rosenfeld, A.: Picture Languages: Formal Models for Picture Recognition. Academic Press, New York (1979)

19. Rubinfeld, R., Shapira, A., Ronitt Rubinfeld and Asaf Shapira: Sublinear time algorithms. SIAM J. Discrete Math. **25**(4), 1562–1588 (2011). https://doi.org/10.1137/100791075

20. Ruzzo, W.L.: On uniform circuit complexity. J. Comput. Syst. Sci. **22**(3), 365–383 (1981). https://doi.org/10.1016/0022-0000(81)90038-6

21. Sommerhalder, R., van Westrhenen, S.C.: Parallel language recognition in constant time by cellular automata. Acta Inf. **19**, 397–407 (1983). https://doi.org/10.1007/BF00290736

22. Straubing, H.: Finite Automata, Formal Logic, and Circuit Complexity. Progress in Theoretical Computer Science. Birkhäuser, Boston, MA (1994). https://doi.org/10.1007/978-1-4612-0289-9

23. Sudan, M.: Probabilistically checkable proofs. Commun. ACM **52**(3), 76–84 (2009). https://doi.org/10.1145/1467247.1467267

24. Terrier, V.: Language recognition by cellular automata. In: Rozenberg, G., Back, T., Kok, J.N. (eds.) Handbook of Natural Computing, pp. 123–158. Springer, Heidelberg (2012). https://doi.org/10.1007/978-3-540-92910-9_4

25. Vollmer, H.: Introduction to Circuit Complexity - A Uniform Approach. Springer, Heidelberg (1999). https://doi.org/10.1007/978-3-662-03927-4

Complexity of Searching for 2 by 2 Submatrices in Boolean Matrices

Daniel Průša[1] and Michael Wehar[2]([✉])

[1] Czech Technical University, Prague, Czech Republic
daniel.prusa@fel.cvut.cz
[2] Swarthmore College, Swarthmore, PA, USA
mwehar1@swarthmore.edu

Abstract. We study the problem of finding a given 2×2 matrix as a submatrix of a given Boolean matrix. Three variants are considered: search for a matching submatrix of any area, of minimum area, or of maximum area. The problem relates to 2D pattern matching, and to fields such as data mining, where the search for submatrices plays an important role. Besides these connections, the problem itself is very natural and its investigation helps to demonstrate differences between search tasks in one-dimensional and multidimensional topologies.

Our results reveal that the problem variants are of different complexities. First, we show that given an $m \times n$ Boolean matrix, the any variant can be solved in $\widetilde{O}(mn)$ time for any given 2×2 matrix, but requires various strategies for different 2×2 matrices. This contrasts with the complexity of the task over matrices with entries from the set $\{0, 1, 2\}$, where the problem is Triangle Finding-hard and hence no algorithm with similar running time is known for it. Then, we show that the minimization variant in the case of Boolean matrices can also be solved in $\widetilde{O}(mn)$ time. Finally, in contrast, we prove Triangle Finding-hardness for the maximization variant and show that there is a rectangular matrix multiplication-based algorithm solving it in $O\left(mn(\min\{m, n\})^{0.5302}\right)$ time.

Keywords: Boolean matrix · Submatrices · Two-dimensional pattern matching · Local picture language · Triangle Finding-hard problem · Fast matrix multiplication

1 Introduction

We study the complexity of Four Corner Problems. A Four Corner Problem is concerned with finding a given 2×2 matrix as a submatrix of a given Boolean matrix. By submatrix, we mean a matrix that is formed by restricting the original matrix to a subset of its rows and columns. For a matrix $\mathbf{B} = \left(\begin{smallmatrix} a & b \\ c & d \end{smallmatrix}\right)$, where $a, b, c, d \in \{0, 1\}$, we consider three kinds of Four Corner Problems:

1. In the input Boolean matrix $\mathbf{M}$, search for any 2×2 submatrix of $\mathbf{M}$ that matches $\mathbf{B}$ (we abbreviate this task as $\mathsf{ANY}\left[\begin{smallmatrix} a & b \\ c & d \end{smallmatrix}\right]$)

© Springer Nature Switzerland AG 2020
N. Jonoska and D. Savchuk (Eds.): DLT 2020, LNCS 12086, pp. 266–279, 2020.
https://doi.org/10.1007/978-3-030-48516-0_20

2. Search for a submatrix that matches **B** and encloses the minimum area of **M** (abbreviated as MIN $\left[\begin{smallmatrix} a & b \\ c & d \end{smallmatrix}\right]$)
3. Search for a submatrix that matches **B** and encloses the maximum area of **M** (abbreviated as MAX $\left[\begin{smallmatrix} a & b \\ c & d \end{smallmatrix}\right]$).

In general, the problem of finding a specific submatrix in a larger matrix is of importance in several computer science disciplines. For example Boolean matrices, and their associated submatrices of 1's, play a central role in *data mining* problems such as *frequent itemset mining* [13]. Moreover, finding a submatrix of 1's in the adjacency matrix of a graph G corresponds to finding a biclique of G [13]. As the maximum edge biclique problem is NP-complete [10], the complexity of searching for a $k \times k$ submatrix is expected to grow as k grows. In this paper, we deal with the simplest case when $k = 2$. An example of its use is as follows. Given m respondents answering n yes/no questions in a questionnaire, are there two respondents who answered yes on two of the same questions?

The above tasks ANY $\left[\begin{smallmatrix} a & b \\ c & d \end{smallmatrix}\right]$, MIN $\left[\begin{smallmatrix} a & b \\ c & d \end{smallmatrix}\right]$ and MAX $\left[\begin{smallmatrix} a & b \\ c & d \end{smallmatrix}\right]$ can also be viewed as two-dimensional pattern matching: we search for any/min/max rectangular block of a matrix that matches a given template. In only one dimension, similar pattern matching problems can be described using *regular languages* [2]. In this case, all the any/min/max tasks are solvable by a finite-state automaton-based algorithm in time linear in the input length [8]. In two dimensions, these problems are easily definable via the notion of *local picture languages* [5]. This is a formalism defining sets of two-dimensional arrays (so called *pictures*) for which the membership problem can be determined by looking at a window of size 2×2. These picture languages are a straightforward generalization of the well known *local (string) languages* [12], which form a proper subset of the family of regular languages.

We introduced in [8] a general algorithm solving two-dimensional pattern matching against local picture languages in time $O(mn \min\{m, n\})$ for $m \times n$ input matrices. Further, for a specific local picture language, we investigated the pattern matching problem which is precisely ANY $\left[\begin{smallmatrix} 1 & 1 \\ 1 & 1 \end{smallmatrix}\right]$ and showed it to be solvable in linear time in the input matrix area. Here our goal is to propose more efficient algorithms for a specialized subclass of local picture language pattern matching problems over Boolean matrices called Four Corner Problems. In particular, we show that the problem ANY $\left[\begin{smallmatrix} a & b \\ c & d \end{smallmatrix}\right]$ is solvable in $\widetilde{O}(mn)$ time for any $a, b, c, d \in \{0, 1\}$ (Theorem 1). This result is surprising because it was proven in [8] that searching for a submatrix matching $\left(\begin{smallmatrix} 1 & 0 \\ 1 & 1 \end{smallmatrix}\right)$ in an $n \times n$ matrix over $\{0, 1, 2\}$ is Triangle Finding-hard. In other words, the proof introduced a *fine-grained reduction* [15] from Triangle Finding to the search problem for $\left(\begin{smallmatrix} 1 & 0 \\ 1 & 1 \end{smallmatrix}\right)$ over $\{0, 1, 2\}$ suggesting that Four Corner Problems are harder over larger alphabets.

The Triangle Finding problem is to decide whether a given undirected graph $G = (V, E)$ is triangle-free or not. It is a classic algorithmic problem which can be reduced to Boolean Matrix Multiplication (see [6]) and solved in time $O(n^\omega)$, where $n = |V|$ and $\omega < 2.373$ denotes the matrix multiplication exponent [14]. However, it is currently unknown whether Triangle Finding can be solved in time $\widetilde{O}(n^2)$. Note that conditional lower bounds based on Triangle Finding are known for several problems (see, e.g., [1,7,9,11]).

We further investigate the minimization and maximization variants of the search problem over Boolean matrices. For the min variant, we improve on Theorem 1 by showing that the problem $\mathsf{MIN}\left[\begin{smallmatrix} a & b \\ c & d \end{smallmatrix}\right]$ is solvable in $\widetilde{O}(mn)$ time for any $a,b,c,d \in \{0,1\}$ (Theorem 4). For the max variant, we prove that $\mathsf{MAX}\left[\begin{smallmatrix} a & b \\ c & d \end{smallmatrix}\right]$ is Triangle Finding-hard for any $a,b,c,d \in \{0,1\}$ (Theorem 5). Also, we present an algorithm that solves $\mathsf{MAX}\left[\begin{smallmatrix} a & b \\ c & d \end{smallmatrix}\right]$ in $O\left(mn(\min\{m,n\})^{0.5302}\right)$ time (Theorem 6). This algorithm is based on computing a *minimum witness* for Boolean matrix multiplication [4]. However, it is likely impractical because it uses a fast rectangular matrix multiplication algorithm that involves a large constant factor.

The paper is structured as follows. Section 2 establishes some required notions. Then, Sects. 3, 4 and 5 gradually present results for the problems $\mathsf{ANY}\left[\begin{smallmatrix} a & b \\ c & d \end{smallmatrix}\right]$, $\mathsf{MIN}\left[\begin{smallmatrix} a & b \\ c & d \end{smallmatrix}\right]$ and $\mathsf{MAX}\left[\begin{smallmatrix} a & b \\ c & d \end{smallmatrix}\right]$.

2   Preliminaries

$\mathbb{N} = \{0,1,2,\ldots\}$ is the set of natural numbers and $\mathbb{N}^+ = \mathbb{N} \setminus \{0\}$ is the set of positive integers. For functions $f,g : \mathbb{N} \times \mathbb{N} \to \mathbb{N}$, we write $f(m,n) = \widetilde{O}(g(m,n))$ if and only if there are numbers $p,q \in \mathbb{N}$ such that $f(m,n) = O\left(g(m,n)\log^p(m)\log^q(n)\right)$.

Let $\mathbf{M}$ be an $m \times n$ Boolean matrix. We write $\overline{\mathbf{M}}$ to denote the matrix obtained from $\mathbf{M}$ by negating its entries (i.e., we have $\overline{\mathbf{M}}_{i,j} = 1 - \mathbf{M}_{i,j}$ for every entry). We consider that rows and columns of $\mathbf{M}$ are indexed from 1 to m and n, respectively. A $k \times \ell$ (rectangular) *block* of $\mathbf{M}$ at a position (r,c) is denoted as $B = \mathbf{M}[r,c;k,\ell]$, where $1 \leq k \leq m$, $1 \leq \ell \leq n$, $1 \leq r \leq m-k+1$, $1 \leq c \leq n-\ell+1$. Its entries coincide with the entries of the submatrix obtained from $\mathbf{M}$ by deleting rows $1,\ldots,r-1$ and $r+k,\ldots,m$, and columns $1,\ldots,c-1$ and $c+\ell,\ldots,n$. We use $B_{i,j}$ to refer to the entry in the i-th row and j-th column of B. We have $B_{i,j} = \mathbf{M}_{r+i-1,c+j-1}$. We define the *area* of B as $\mathrm{a}(B) = k\ell$, and the 2×2 *corners submatrix* of B as

$$\varkappa(B) = \begin{pmatrix} B_{1,1} & B_{1,\ell} \\ B_{k,1} & B_{k,\ell} \end{pmatrix}.$$

The set of all blocks of $\mathbf{M}$ is denoted by $\mathcal{B}_{\mathbf{M}}$.

For $a,b,c,d \in \{0,1\}$, we define the following search problems (also known as Four Corner Problems) for an input Boolean matrix $\mathbf{M}$.

- $\mathsf{ANY}\left[\begin{smallmatrix} a & b \\ c & d \end{smallmatrix}\right]$: find $B \in \mathcal{B}_{\mathbf{M}}$ such that $\varkappa(B) = \left(\begin{smallmatrix} a & b \\ c & d \end{smallmatrix}\right)$,
- $\mathsf{MIN}\left[\begin{smallmatrix} a & b \\ c & d \end{smallmatrix}\right]$: find $B \in \arg\min_{B \in \mathcal{B}_{\mathbf{M}}}\{\mathrm{a}(B) \mid \varkappa(B) = \left(\begin{smallmatrix} a & b \\ c & d \end{smallmatrix}\right)\}$,
- $\mathsf{MAX}\left[\begin{smallmatrix} a & b \\ c & d \end{smallmatrix}\right]$: find $B \in \arg\max_{B \in \mathcal{B}_{\mathbf{M}}}\{\mathrm{a}(B) \mid \varkappa(B) = \left(\begin{smallmatrix} a & b \\ c & d \end{smallmatrix}\right)\}$.

3   Searching for Any Matching Submatrix

This section presents algorithms for $\mathsf{ANY}\left[\begin{smallmatrix} a & b \\ c & d \end{smallmatrix}\right]$ that run in nearly linear time in the input matrix area, for every $a,b,c,$ and d. In some cases an efficient algorithm is achieved by using properties of the minimum matching submatrix, so these algorithms also solve the corresponding MIN $\left[\begin{smallmatrix} a & b \\ c & d \end{smallmatrix}\right]$ problem (see Lemmas 2 and 3).

Out of all ANY $\left[\begin{smallmatrix} a & b \\ c & d \end{smallmatrix}\right]$ problems, ANY $\left[\begin{smallmatrix} 1 & 1 \\ 1 & 1 \end{smallmatrix}\right]$ and ANY $\left[\begin{smallmatrix} 0 & 0 \\ 0 & 0 \end{smallmatrix}\right]$ are easiest to solve. It has already been shown in [8] that ANY $\left[\begin{smallmatrix} 1 & 1 \\ 1 & 1 \end{smallmatrix}\right]$ reduces to finding a four-cycle in a bipartite graph. Here we give a more straightforward algorithm.

Lemma 1. ANY $\left[\begin{smallmatrix} 1 & 1 \\ 1 & 1 \end{smallmatrix}\right]$ *is solvable in time* $O(mn)$ *for* m *by* n *Boolean matrices.*

Proof. Let an $m \times n$ Boolean matrix $\mathbf{M}$ be given. Without loss of generality, suppose that $m \geq n$. The algorithm is as follows. We create a set S of pairs of column indexes. Initially, the set is empty. The matrix is traversed row by row. For each row i, we find the set C_i of all column indexes j such that $\mathbf{M}_{i,j} = 1$. Then, for every pair $\{c_1, c_2\} \in \binom{C_i}{2}$, we check whether $\{c_1, c_2\}$ is in S. If not, it is added to S. Otherwise, a desired submatrix has been found.

The algorithm takes $O(mn + n^2)$ time because it visits each entry from $\mathbf{M}$ at most once and it adds at most $\binom{n}{2}$ pairs of column indexes into S. Because $m \geq n$, the total runtime is $O(mn)$. $\qquad\square$

Lemma 2. MIN $\left[\begin{smallmatrix} 1 & 0 \\ 1 & 1 \end{smallmatrix}\right]$ *is solvable in time* $O(mn)$ *for* m *by* n *Boolean matrices.*

Proof. Let $\mathbf{M}$ be an $m \times n$ Boolean matrix. The algorithm is based on the following claim: If $\mathbf{M}$ contains a block $B = \mathbf{M}[r, c; k, \ell]$ such that $\varkappa(B) = \left(\begin{smallmatrix} 1 & 0 \\ 1 & 1 \end{smallmatrix}\right)$, then it contains a block $B' = \mathbf{M}[r', c'; k', \ell']$ such that $\varkappa(B') = \left(\begin{smallmatrix} 1 & 0 \\ 1 & 1 \end{smallmatrix}\right)$, $B'_{i,1} = 0$ for all $i = 2, \ldots, k' - 1$ and $B'_{k',j} = 0$ for all $j = 2, \ldots, \ell' - 1$ (i.e., the left and bottom edge of B', excluding the corners, contain only 0 entries).

To see this, suppose without loss of generality that $B_{i,1} = 1$ for some $1 < i < k$. Let $B_1 = \mathbf{M}[r, c; i, \ell]$ and $B_2 = \mathbf{M}[r + i - 1, c; k - i + 1, \ell]$. Then, either $\varkappa(B_1) = \left(\begin{smallmatrix} 1 & 0 \\ 1 & 1 \end{smallmatrix}\right)$ (if $B_{i,\ell} = 1$) or $\varkappa(B_2) = \left(\begin{smallmatrix} 1 & 0 \\ 1 & 1 \end{smallmatrix}\right)$ (if $B_{i,\ell} = 0$). Since B_1 and B_2 are proper subsets of B, we have found a smaller block containing $\left(\begin{smallmatrix} 1 & 0 \\ 1 & 1 \end{smallmatrix}\right)$ as a submatrix.

Now, we present the algorithm. It creates a map σ where a key is a pair (i, j) such that $\mathbf{M}_{i,j} = 1$. The value associated with (i, j) is a pair (i', j') such that i' is the largest row index less than i such that $\mathbf{M}_{i',j} = 1$ (i.e., i' is the row index of the nearest entry 1 located upwards from the position (i, j)) and j' is the smallest column index greater than j such that $\mathbf{M}_{i,j'} = 1$ (i.e., the column index of the nearest entry 1 rightwards). Note that the value of i' or j' might be undefined if there is no such row index or column index, respectively.

It is possible to build σ in $O(mn)$ time by making two passes over $\mathbf{M}$. The first pass is to compute the i''s. The matrix $\mathbf{M}$ is scanned column by column. Each column index j is scanned from top to bottom. Whenever entry 1 is detected at a position (i', j), then i' is the first component of $\sigma(i, j)$ for the next detected entry 1 from position (i, j). Analogously, the second pass, scanning $\mathbf{M}$ row by row, is to compute the j''s.

Now, for each key (i, j) in the map σ, the algorithm takes its value (i', j') and checks if rows i, i' and columns j, j' form a desired submatrix matching $\left(\begin{smallmatrix} 1 & 0 \\ 1 & 1 \end{smallmatrix}\right)$. By doing this, every existing block with 0 entries on the left and bottom edges is checked. Among these blocks, a minimum-area block B such that $\varkappa(B) = \left(\begin{smallmatrix} 1 & 0 \\ 1 & 1 \end{smallmatrix}\right)$ is returned as the result.

Assuming constant time map operations, the algorithm runs in $O(mn)$ time (note that the map σ can be implemented by using an $m \times n$ array). $\square$

Lemma 3. MIN $\left[\begin{smallmatrix} 1 & 0 \\ 0 & 1 \end{smallmatrix}\right]$ *is solvable in time* $\widetilde{O}(mn)$ *for* m *by* n *Boolean matrices.*

Proof. **Case I (square matrices):** Let an $n \times n$ Boolean matrix $\mathbf{M}$ be given. We present a divide and conquer strategy. If $t(n)$ denotes the runtime of searching for a desired minimum submatrix in an $n \times n$ matrix, then we show that

$$t(n) = 4 \cdot t \left(\frac{n}{2} \right) + O \left(n^2 \right). \tag{1}$$

To accomplish this, we split $\mathbf{M}$ horizontally into matrices $\mathbf{M}_{\text{top}}$ and $\mathbf{M}_{\text{bottom}}$, where $\mathbf{M}_{\text{top}}$ is $\lceil \frac{n}{2} \rceil$ by n and $\mathbf{M}_{\text{bottom}}$ is $\lfloor \frac{n}{2} \rfloor$ by n. Next, we split $\mathbf{M}_{\text{top}}$ vertically into $\mathbf{M}_{\text{top,left}}$, which is $\lceil \frac{n}{2} \rceil$ by $\lceil \frac{n}{2} \rceil$, and $\mathbf{M}_{\text{top,right}}$, which is $\lceil \frac{n}{2} \rceil$ by $\lfloor \frac{n}{2} \rfloor$. Quite analogously, we split $\mathbf{M}_{\text{bottom}}$ vertically into $\mathbf{M}_{\text{bottom,left}}$ and $\mathbf{M}_{\text{bottom,right}}$. A desired minimum submatrix is either in one of the four $\frac{n}{2}$ by $\frac{n}{2}$ matrices or it spans the border between either $\mathbf{M}_{\text{top}}$ and $\mathbf{M}_{\text{bottom}}$, $\mathbf{M}_{\text{top,left}}$ and $\mathbf{M}_{\text{top,right}}$, or, $\mathbf{M}_{\text{bottom,left}}$ and $\mathbf{M}_{\text{bottom,right}}$. We propose a procedure running in $O(n^2)$ time that finds a desired minimum submatrix by the assumption that there is such a submatrix crossing the specified borders. Therefore, in $O(n^2)$ time, we reduce finding a desired submatrix in an n by n matrix $\mathbf{M}$ to finding a desired submatrix in one of four $\frac{n}{2}$ by $\frac{n}{2}$ matrices. If we solve recurrence (1), then we get $t(n) = O \left(n^2 \log(n) \right)$ [3].

Without loss of generality, let us deal only with the border between $\mathbf{M}_{\text{top}}$ and $\mathbf{M}_{\text{bottom}}$. We claim: if $B = \mathbf{M}[r, c; k, \ell]$ is a minimum-area block of $\mathbf{M}$ such that $\varkappa(B) = \left(\begin{smallmatrix} 1 & 0 \\ 0 & 1 \end{smallmatrix}\right)$, then $B_{i,1} = B_{i,\ell}$ for all $i = 2, \ldots, k-1$. Indeed, $B_{i,1} \neq B_{i,\ell}$ would clearly contradict the minimality of B.

Based on the claim, we create maps σ_{top} and σ_{bottom} such that $\sigma_{\text{top}}(\{i,j\})$ is the largest row index such that columns i and j differ in $\mathbf{M}_{\text{top}}$, and, analogously, $\sigma_{\text{bottom}}(\{i,j\})$ is the smallest row index such that columns i and j differ in $\mathbf{M}_{\text{bottom}}$. Once we have constructed σ_{top} and σ_{bottom} we go through each pair of column indexes $\{i,j\}$ and check if rows $\sigma_{\text{top}}(\{i,j\})$, $\sigma_{\text{bottom}}(\{i,j\})$ and columns i, j together create a desired submatrix of $\mathbf{M}$. A minimum submatrix among the detected submatrices is the candidate for the resulting submatrix returned by the procedure.

It remains to explain how we obtain the maps. Let us first give a construction for σ_{top}. We create a set X of pairwise disjoint sets of column indexes. Initially, X contains one set containing all column indexes. We repeat the following process for each row of $\mathbf{M}_{\text{top}}$, starting at the bottommost one and proceeding upwards: Create two disjoint sets A_0 and A_1 where A_0 contains all column indexes that are 0's and A_1 contains all column indexes that are 1's in the current row. For each set S in X, split S into two disjoint subsets $S_0 = S \cap A_0$ and $S_1 = S \cap A_1$. For every $\{i,j\}$ such that $i \in S_0$ and $j \in S_1$, set $\sigma_{\text{top}}(\{i,j\})$ to the current row index. Then, update X by replacing S with S_0 and S_1. Throw out any sets from

X that have less than two elements. Finish when X is empty or every row of $\mathbf{M}_{\text{top}}$ has been processed.

We similarly build σ_{bottom}, but we start at the top row of $\mathbf{M}_{\text{bottom}}$ going one row down at a time. It only takes $O(n^2)$ time to construct σ_{top} and σ_{bottom} because we do $O(n)$ work per row plus an additional constant amount of work for each pair of columns.

Case II (rectangular matrices): Let an $m \times n$ Boolean matrix $\mathbf{M}$ be given. Assume without loss of generality that $m > n$.

We perform horizontal splits to divide $\mathbf{M}$ into $d = \lceil \frac{m}{n} \rceil$ smaller matrices $\{\mathbf{M}_k\}_{k \in [d]}$ such that for each $k \in [d-1]$, $\mathbf{M}_k$ is n by n, and $\mathbf{M}_d$ is c by n for some $c \leq n$. A desired minimum submatrix is either in $\mathbf{M}_k$ for some $k \in [d]$ or it crosses the border between $\mathbf{M}_k$ and $\mathbf{M}_{k+1}$ for some $k \in [d-1]$. Then, the former cases in total take $O\left(\frac{m}{n} \cdot t(n)\right)$ time. We claim that the latter cases take $O(mn)$ time. For each $k \in [d-1]$, we construct maps $\sigma_{k,\text{top}}$ and $\sigma_{k,\text{bottom}}$ such that $\sigma_{k,\text{top}}(\{i,j\})$ is the smallest row index such that columns i and j differ in $\mathbf{M}_k$, and $\sigma_{k,\text{bottom}}(\{i,j\})$ is the largest row index such that columns i and j differ in $\mathbf{M}_k$. Following the same approach as for the square matrix case, we can construct all maps in total time $O(mn)$. Then, for each pair of column indexes i, j we have up to d cases to check. This results in total time $O(mn)$. Note that if a map is not defined at $\{i,j\}$, then we try the next map and combine the cases together since this means a submatrix might span across multiple horizontal splits. In total, our algorithm takes $O\left(\frac{m}{n} \cdot t(n) + mn\right) = O(mn\log(n))$ time. $\square$

Lemma 4. ANY $\left[\begin{smallmatrix}1 & 0 \\ 1 & 0\end{smallmatrix}\right]$ is solvable in time $\tilde{O}(mn)$ for m by n Boolean matrices.

Proof. Let an $m \times n$ Boolean matrix $\mathbf{M}$ be given.

Case I (tall matrices): We consider the case when $m \geq n$. We proceed in a similar manner as in the proof of Lemma 1. We create a set S of pairs of column indexes. Initially, the set is empty. The matrix is traversed row by row. For each row, we do the following. We create a set R. Initially, R is empty, but we will add column indexes to R. We scan entries from left to right in the row. When we encounter a 1 entry at column index i, we add i to R. When we encounter a 0 entry at column index j, we go through each column index i from R. If (i,j) is in S, then we found a desired submatrix. Otherwise, we add (i,j) to S. Since $m \geq n$, this takes $O(mn + n^2) = O(mn)$ time.

Case II (short matrices): We consider the case when $m < n$. We perform vertical splits to divide $\mathbf{M}$ into $d = \lceil \frac{n}{m} \rceil$ smaller matrices $\{\mathbf{M}_k\}_{k \in [d]}$ such that for each $k \in [d-1]$, $\mathbf{M}_k$ is m by m, and $\mathbf{M}_d$ is m by c for some $c \leq m$. The matrix $\mathbf{M}$ contains a desired submatrix if and only if some $\mathbf{M}_k$ contains a minimal submatrix for some $k \in [d]$ or there is a minimal submatrix that crosses the border between $\mathbf{M}_k$ and $\mathbf{M}_{k+1}$ for some $k \in [d-1]$.

Consider the former condition. Checking if a given $\mathbf{M}_k$ matrix contains a minimal desired submatrix takes $O(m^2)$ time by applying the approach from the first case. Checking all of the matrices in $\{\mathbf{M}_k\}_{k \in [d]}$ takes $O(d \cdot m^2) = O\left(\frac{n}{m} \cdot m^2\right) = O(mn)$ time.

Now, we focus on checking the latter condition. For each $k \in [d-1]$, we construct maps $\sigma_{k,\text{left}}$ and $\sigma_{k,\text{right}}$ such that $\sigma_{k,\text{left}}(\{i,j\})$ is the smallest column index such that rows i and j are equal in $\mathbf{M}_k$, and $\sigma_{k,\text{right}}(\{i,j\})$ is the largest column index such that rows i and j are equal in $\mathbf{M}_k$. Once we have constructed these maps, we consider each pair of rows i and j. We have up to $d-1$ cases to check where each case considers the border between $\mathbf{M}_k$ and $\mathbf{M}_{k+1}$ for some $k \in [d-1]$. We check each case by seeing if rows i and j along with columns $\sigma_{k,\text{right}}(\{i,j\})$ and $\sigma_{k+1,\text{left}}(\{i,j\})$ form a desired submatrix. It is sufficient to check these submatrices because we are only concerned with desired submatrices crossing the border that are minimal. Note if a map is not defined at $\{i,j\}$, then we try the next map and combine the cases together since this means a submatrix might span across multiple vertical splits. This takes $O(d \cdot m^2) = O\left(\frac{n}{m} \cdot m^2\right) = O(mn)$ time. It remains to describe how the maps are constructed. We claim that the maps can be constructed in $O(mn\log(m))$ time. Therefore, the total runtime is $O(mn\log(m))$.

Given $k \in [d-1]$, we describe how to construct $\sigma_{k,\text{left}}$ for the matrix $\mathbf{M}_k$. For each $\ell \in [\log(m)]$, we construct a matrix $\mathbf{M}_{k,\ell}$. The matrix $\mathbf{M}_{k,\ell}$ is obtained from $\mathbf{M}_k$ by negating all bits in each row i such that i's binary expansion has a 1 at position ℓ. Next, in a similar manner as described in the proof of Lemma 3, we construct a map σ_ℓ such that $\sigma_\ell(\{i,j\})$ is the smallest column index where rows i and j differ in $\mathbf{M}_{k,\ell}$. Now, we use these $\log(m)$ maps to construct $\sigma_{k,\text{left}}$. For each pair of rows i and j, there is some position ℓ in i and j's binary expansions where they differ. The smallest column index where rows i and j differ in $\mathbf{M}_{k,\ell}$ is exactly the same as the smallest column index where rows i and j are equal in $\mathbf{M}_k$. Hence, we make $\sigma_{k,\text{left}}(\{i,j\}) = \sigma_\ell(\{i,j\})$. It takes $O(m^2\log(m))$ time to construct $\sigma_{k,\text{left}}$. The map $\sigma_{k,\text{right}}$ can be constructed in a similar manner. In total, it takes $O(d \cdot m^2\log(m)) = O\left(\frac{n}{m} \cdot m^2\log(m)\right) = O(mn\log(m))$ time to construct all of the maps.   □

Theorem 1. *Problem* ANY $\begin{bmatrix} a & b \\ c & d \end{bmatrix}$ *is solvable in time* $\widetilde{O}(mn)$ *for* m *by* n *Boolean matrices and any* $a,b,c,d \in \{0,1\}$.

Proof. Consider the set of matrices $S = \{\left(\begin{smallmatrix} 1 & 1 \\ 1 & 1 \end{smallmatrix}\right), \left(\begin{smallmatrix} 1 & 0 \\ 1 & 1 \end{smallmatrix}\right), \left(\begin{smallmatrix} 1 & 0 \\ 0 & 1 \end{smallmatrix}\right), \left(\begin{smallmatrix} 1 & 0 \\ 1 & 0 \end{smallmatrix}\right)\}$. Every 2×2 Boolean matrix $\mathbf{A}$ is similar to a Boolean matrix $\mathbf{B} \in S$ in the sense that $\mathbf{B} = U(\mathbf{A})$ for an operation U that combines a rotation with an optional negation of all bits. Further, for every Boolean matrix $\mathbf{M}$, the matrix $\mathbf{M}$ contains $\mathbf{A}$ as a submatrix if and only if $U(\mathbf{M})$ contains $\mathbf{B}$ as a submatrix. Applying Lemmas 1, 2, 3, and 4, we can determine if $\mathbf{M}$ has $\mathbf{A}$ as a submatrix in $\widetilde{O}(mn)$ time.   □

4   Searching for a Minimum 2-by-2 Submatrix of 1's

In the previous section, we presented fast algorithms for minimization problems MIN $\begin{bmatrix} 1 & 0 \\ 1 & 1 \end{bmatrix}$ and MIN $\begin{bmatrix} 1 & 0 \\ 0 & 1 \end{bmatrix}$. Here, we use preceding results for ANY $\begin{bmatrix} 1 & 1 \\ 1 & 1 \end{bmatrix}$ and ANY $\begin{bmatrix} 1 & 0 \\ 1 & 0 \end{bmatrix}$ to also obtain fast algorithms for MIN $\begin{bmatrix} 1 & 1 \\ 1 & 1 \end{bmatrix}$ and MIN $\begin{bmatrix} 1 & 0 \\ 1 & 0 \end{bmatrix}$.

First, we introduce an algorithm for $\mathsf{MIN}\left[\begin{smallmatrix}1&1\\1&1\end{smallmatrix}\right]$. The technique we apply requires several preparatory steps: a characterization of Boolean matrices that do not have $\left(\begin{smallmatrix}1&1\\1&1\end{smallmatrix}\right)$ as a submatrix (Lemma 5), an algorithm solving $\mathsf{MIN}\left[\begin{smallmatrix}1&1\\1&1\end{smallmatrix}\right]$ whose complexity depends on the number of pairs of 1's within the same rows (Lemma 6), and a fast algorithm solving $\mathsf{MIN}\left[\begin{smallmatrix}1&1\\1&1\end{smallmatrix}\right]$ approximately (Lemma 8). Then, we can apply a similar approach to solve $\mathsf{MIN}\left[\begin{smallmatrix}1&0\\1&0\end{smallmatrix}\right]$.

Lemma 5. *Let* $\mathbf{A}$ *be an* m *by* n *Boolean matrix. Let* a_i *denote the number of 1's in the* i*-th row of* $\mathbf{A}$*. If* $\Sigma_{i=0}^{m}\binom{a_i}{2} > \binom{n}{2}$*, then* $\mathbf{A}$ *must contain a block whose corners are 1's.*

Proof. $\Sigma_{i=0}^{m}\binom{a_i}{2}$ is the size of the set $T = \{(i, \{j, k\}) \mid j \neq k \wedge \mathbf{A}_{ij} = \mathbf{A}_{ik} = 1\}$. If $|T| > \binom{n}{2}$, then there are $(i_1, \{j, k\}), (i_2, \{j, k\}) \in T$, where $i_1 \neq i_2$. This means that rows i_1, i_2 and columns j, k form a submatrix $\left(\begin{smallmatrix}1&1\\1&1\end{smallmatrix}\right)$. $\square$

Lemma 6. *Let* $\mathbf{M}$ *be an* m *by* n *Boolean matrix and* $T(\mathbf{M}) = \{(i, \{j, k\}) \mid j \neq k \wedge \mathbf{M}_{i,j} = \mathbf{M}_{i,k} = 1\}$*. There is an algorithm solving* $\mathsf{MIN}\left[\begin{smallmatrix}1&1\\1&1\end{smallmatrix}\right]$ *in* $O(|T(\mathbf{M})| + mn)$ *time.*

Proof. The algorithm uses a map σ with keys $\{j, k\}$, where $j \neq k$ are column indexes. The value of $\sigma(\{j, k\})$ is a row index. Initially, the map is empty.

The input Boolean matrix $\mathbf{M}$ is processed row by row. In the i-th row, the following actions are performed for each $(i, \{j, k\}) \in T(\mathbf{M})$. First, it is checked whether $\sigma(\{j, k\})$ is defined. If it is not, then $\sigma(\{j, k\})$ is set to i. Otherwise, the algorithm finds out whether the rectangle formed by rows i, $\sigma(\{j, k\})$ and columns j, k is the minimum one so far. Then, $\sigma(\{j, k\})$ is updated to be i. $\square$

For convenience, for each Boolean matrix $\mathbf{M}$ considered now until the end of this section, assume that the number of rows and the number of columns of $\mathbf{M}$ are powers of 2. Since any matrix of a general size $m \times n$ can be extended to a $2^{\lceil \log_2 m \rceil} \times 2^{\lceil \log_2 n \rceil}$ matrix (with the added entries set to "undefined" value), the assumption will not have any impact on the generality and asymptotic time complexity of the presented algorithms.

For $p \in \mathbb{N}^+$, let $\mathcal{S}(p) = \{2^i \mid i = 1, 2, \dots, \lfloor \log_2 p \rfloor\}$ be the set of powers of two greater than 1 and not greater than p. Let $\mathbf{M}$ be an $m \times n$ Boolean matrix. For $k \in \mathcal{S}(m)$ and $\ell \in \mathcal{S}(n)$, let $\mathcal{R}_{\mathbf{M}}(k, \ell)$ denote the set of all $k \times \ell$ blocks of $\mathbf{M}$ whose top left corner is located in $\mathbf{M}$ at a position $(1 + a \cdot \frac{k}{2}, 1 + b \cdot \frac{\ell}{2})$ for some $a, b \in \mathbb{N}$. Let $\mathcal{R}_{\mathbf{M}} = \bigcup_{k \in \mathcal{S}(m), \ell \in \mathcal{S}(n)} \mathcal{R}_{\mathbf{M}}(k, \ell)$.

Lemma 7. *Let* B *be a* p *by* q *block of* $\mathbf{M}$*. There are powers of 2, denoted by* k *and* ℓ*, such that* $k < 4p$*,* $\ell < 4q$ *and* B *is included in a block from* $\mathcal{R}_{\mathbf{M}}(k, \ell)$*.*

Proof. Assume $B = \mathbf{M}[r, c; p, q]$. Let $k = \min\{m, 2 \cdot 2^{\lceil \log_2 p \rceil}\}$, $\ell = \min\{n, 2 \cdot 2^{\lceil \log_2 q \rceil}\}$,

$$a = \max\{x \mid x \in \mathbb{N} \wedge x \cdot \frac{k}{2} + k \leq m \wedge 1 + x \cdot \frac{k}{2} \leq r\}, \text{ and} \tag{2}$$

$$b = \max\{y \mid y \in \mathbb{N} \wedge y \cdot \frac{\ell}{2} + \ell \leq n \wedge 1 + y \cdot \frac{\ell}{2} \leq c\}. \tag{3}$$

Then, B is included in the block $\mathbf{M}[1 + a \cdot \frac{k}{2}, 1 + b \cdot \frac{\ell}{2}; k, \ell] \in \mathcal{R}_{\mathbf{M}}(k, \ell)$. This is proved as follows. The definition of a ensures that $1 + a \cdot \frac{k}{2} \leq r$. It is also needed to verify that

$$r + p - 1 \leq a \cdot \frac{k}{2} + k. \tag{4}$$

Observe that this inequality is trivially fulfilled when $a \cdot \frac{k}{2} + k = m$. Hence, assume that

$$a \cdot \frac{k}{2} + k < m. \tag{5}$$

Since m is divisible by $\frac{k}{2}$, inequality (5) implies that

$$(a + 1) \cdot \frac{k}{2} + k \leq m. \tag{6}$$

It must thus hold that

$$1 + (a + 1) \cdot \frac{k}{2} > r \tag{7}$$

(otherwise the right-hand side of (2) is greater than a). Now, it suffices to combine (7) and $p \leq 2^{\lceil \log_2 p \rceil} = \frac{k}{2}$ to obtain inequality (4).

Quite analogously, the definition of b ensures that $1 + b \cdot \frac{\ell}{2} \leq c$ and it can be proved that $c + q - 1 \leq b \cdot \frac{\ell}{2} + \ell$.   □

Lemma 7 and the defined set of blocks $\mathcal{R}_{\mathbf{M}}$ provide a basis for designing a fast algorithm that solves MIN $\left[\begin{smallmatrix} 1 & 1 \\ 1 & 1 \end{smallmatrix}\right]$ approximately.

Lemma 8. *There is an algorithm that, for any m by n Boolean matrix $\mathbf{M}$, finds in $O(mn \log m \log n)$ time a block B of $\mathbf{M}$ such that $\varkappa(B) = \left(\begin{smallmatrix} 1 & 1 \\ 1 & 1 \end{smallmatrix}\right)$ and $\mathrm{a}(B) < 16 \cdot \mathrm{a}(B_{\min})$, where $B_{\min}$ is a minimum-area block of $\mathbf{M}$ fulfilling $\varkappa(B_{\min}) = \left(\begin{smallmatrix} 1 & 1 \\ 1 & 1 \end{smallmatrix}\right)$.*

Proof. The algorithm works as follows. For each block $B \in \mathcal{R}_{\mathbf{M}}$, it uses the algorithm of Lemma 1 to search inside B for a submatrix matching $\left(\begin{smallmatrix} 1 & 1 \\ 1 & 1 \end{smallmatrix}\right)$. Among all the detected submatrices, it outputs a minimal one.

By Lemma 7, $B_{\min}$ is a part of a block $B' \in \mathcal{R}_{\mathbf{M}}$ whose area is less than 16 times the area of $B_{\min}$, hence the algorithm of Lemma 1 running on B' finds a block of $\mathbf{M}$ fulfilling the lemma requirement.

For each $(k, \ell) \in \mathcal{S}(m) \times \mathcal{S}(n)$, the sum of the areas of the blocks in $\mathcal{R}_{\mathbf{M}}(k, \ell)$ is $O(mn)$, hence all these blocks are processed by the algorithm of Lemma 1 cumulatively in $O(mn)$ time. Finally, since $|\mathcal{S}(m) \times \mathcal{S}(n)| = O(\log m \log n)$, the proposed algorithm runs in $O(mn \log m \log n)$ time.   □

Theorem 2. MIN $\left[\begin{smallmatrix} 1 & 1 \\ 1 & 1 \end{smallmatrix}\right]$ *is solvable in time $\tilde{O}(mn)$ for m by n Boolean matrices.*

Proof. Let $\mathbf{M}$ be an input $m \times n$ Boolean matrix. Assume that the algorithm of Lemma 8 finds in $\mathbf{M}$ a block of an area S. The minimum area of a block of $\mathbf{M}$ containing $\left(\begin{smallmatrix} 1 & 1 \\ 1 & 1 \end{smallmatrix}\right)$ as a submatrix is in the range $(\frac{S}{16}, S]$.

Let us identify a suitable subset of blocks in $\mathcal{R}_{\mathbf{M}}$ such that each minimum-area block containing $\left(\begin{smallmatrix} 1 & 1 \\ 1 & 1 \end{smallmatrix}\right)$ as a submatrix is a part of a block from the subset. Define

$$P = \left\{ (k,\ell) \mid k \in \mathcal{S}(m) \wedge \ell = 2^{\lceil \log_2 \frac{16 \cdot S}{k} \rceil - 1} \wedge \ell \leq n \right\}, \text{ and}$$
$$\mathcal{R}'_{\mathbf{M}} = \bigcup_{(k,\ell) \in P} \mathcal{R}_{\mathbf{M}}(k,\ell).$$

The subset of blocks $\mathcal{R}'_{\mathbf{M}}$ satisfies the following properties.

Claim I: Every $k \times \ell$ block B in $\mathcal{R}'_M$ is of area less than $16 \cdot S$. Indeed, we can derive

$$\mathrm{a}(B) = k\ell = k \cdot 2^{\lceil \log_2 \frac{16 \cdot S}{k} \rceil - 1} < k \cdot 2^{\log_2 \frac{16 \cdot S}{k}} = 16 \cdot S.$$

Claim II: Every $p \times q$ block B of $\mathbf{M}$ such that $\mathrm{a}(B) \leq S$ is a subset of a block in $\mathcal{R}'_{\mathbf{M}}$. This is proved as follows. By Lemma 7, there is a $k \times \ell$ block B_1 such that $4p > k \in \mathcal{S}(m)$, $4q > \ell \in \mathcal{S}(n)$, and B is included in B_1. It holds that $k\ell < 16 \cdot pq \leq 16 \cdot S$, and hence $\ell < 2^{\lceil \log_2 \frac{16 \cdot S}{k} \rceil}$. Since l is a power of 2, we can write $\ell \leq 2^{\lceil \log_2 \frac{16 \cdot S}{k} \rceil - 1} = \ell'$, which implies that there is a block $B_2 \in \mathcal{R}_{\mathbf{M}}(k,\ell') \subseteq \mathcal{R}'_{\mathbf{M}}$ that includes B_1 as well as B.

Claim III: Every $k \times \ell$ block B in $\mathcal{R}'_M$ fulfills $\min\{|T(B)|, |T(B^{\mathsf{T}})|\} = O((\min\{k,\ell\})^2)$ (see Lemma 6 for the definition of $T(B)$). To show this, assume without loss of generality that $k \geq \ell$ and $k \geq 256$. Consider B to be split horizontally into 256 subblocks B_i of size $\frac{k}{256} \times \ell$. Hence, $\mathrm{a}(B_i) = \frac{k\ell}{256}$. By Claim I, it holds that $k\ell < 16 \cdot S$, and hence $\mathrm{a}(B_i) < \frac{16 \cdot S}{256} = \frac{S}{16}$. This means that B_i does not contain $\left(\begin{smallmatrix} 1 & 1 \\ 1 & 1 \end{smallmatrix}\right)$ as a submatrix, and hence Lemma 5 implies that $|T(B_i)| = O(\ell^2)$. Finally, we derive $|T(B)| = \sum_{i=1}^{256} |T(B_i)| = O(\ell^2)$.

Algorithm: We now have all prerequisites for describing the intended algorithm and deriving its time complexity. It works as follows. Call the algorithm of Lemma 8 to obtain S. For each $B \in \mathcal{R}'_{\mathbf{M}}$ of a size $k \times \ell$, call the algorithm of Lemma 6 either for B (if $k \geq \ell$) or B^{T} (if $k < \ell$) to find a minimum-area block within B containing $\left(\begin{smallmatrix} 1 & 1 \\ 1 & 1 \end{smallmatrix}\right)$ as a submatrix in time $O((\min\{k,\ell\})^2)$. A minimum-area block among all found blocks is returned as the final output.

For an ordered pair $(k,\ell) \in P$, the blocks of $\mathcal{R}_{\mathbf{M}}(k,\ell)$ are processed by the algorithm of Lemma 6 in cumulative time

$$O\left(\frac{mn}{k\ell} \cdot \left((\min\{k,\ell\})^2 + k\ell \right) \right) = O(mn).$$

Hence, assuming without loss of generality that $m \leq n$, all the blocks of $\mathcal{R}'_{\mathbf{M}}$ are processed in total time

$$O\left(\sum_{(k,\ell) \in P} mn \right) = O(mn \log m).$$

Since the other stages of the algorithm run in $\widetilde{O}(mn)$ time, the stated time complexity has been proven. $\qquad \square$

Theorem 3. MIN $\left[\begin{smallmatrix} 1 & 0 \\ 1 & 0 \end{smallmatrix}\right]$ *is solvable in time* $\widetilde{O}(mn)$ *for* m *by* n *Boolean matrices.*

Theorem 4. MIN $\left[\begin{smallmatrix} a & b \\ c & d \end{smallmatrix}\right]$ *is solvable in time* $\widetilde{O}(mn)$ *for* m *by* n *Boolean matrices and any* $a, b, c, d \in \{0, 1\}$.

5   Searching for a Maximum Matching Submatrix

We first prove that the problem MAX $\left[\begin{smallmatrix} a & b \\ c & d \end{smallmatrix}\right]$ is Triangle Finding-hard for any $a, b, c, d \in \{0, 1\}$ (Theorem 5). Then, we show how MAX $\left[\begin{smallmatrix} a & b \\ c & d \end{smallmatrix}\right]$ can be solved using rectangular matrix multiplication (Theorem 6).

Theorem 5. MAX $\left[\begin{smallmatrix} a & b \\ c & d \end{smallmatrix}\right]$ *is Triangle Finding-hard for any* $a, b, c, d \in \{0, 1\}$.

Proof. By the same reasoning given in the proof of Theorem 1, it suffices to prove Triangle Finding-hardness for problems MAX $\left[\begin{smallmatrix} 1 & 1 \\ 1 & 1 \end{smallmatrix}\right]$, MAX $\left[\begin{smallmatrix} 1 & 0 \\ 1 & 1 \end{smallmatrix}\right]$, MAX $\left[\begin{smallmatrix} 1 & 0 \\ 0 & 1 \end{smallmatrix}\right]$, and MAX $\left[\begin{smallmatrix} 1 & 0 \\ 1 & 0 \end{smallmatrix}\right]$. We first present a fine-grained reduction from Triangle Finding to MAX $\left[\begin{smallmatrix} 1 & 1 \\ 1 & 1 \end{smallmatrix}\right]$. We then adapt the reduction to the other three problems.

Let a graph $G = (V, E)$ be given. Let the set of vertices be $V = \{v_i \mid i \in \{1, \ldots, n\}\}$. Let $\mathbf{A}$ be an $n \times n$ lower triangular Boolean matrix derived from the adjacency matrix of G as follows: $\mathbf{A}_{i,j} = 1$ if and only if $i > j$ and $\{v_i, v_j\} \in E$. Observe that $\{v_i, v_j, v_k\}$, where $i < j < k$, is a triangle in G if and only if $\mathbf{A}_{j,i} = \mathbf{A}_{k,i} = \mathbf{A}_{k,j} = 1$.

Define the matrix

$$\mathbf{M}_1 = \begin{pmatrix} \mathbf{A}_1 & \mathbf{O} & \mathbf{I} \\ \mathbf{O} & \mathbf{O} & \mathbf{O} \\ \mathbf{A}_2 & \mathbf{O} & \mathbf{A}_3 \end{pmatrix}$$

where $\mathbf{O}$ is the $n \times n$ zero matrix, $\mathbf{I}$ is the $n \times n$ identity matrix, and $\mathbf{A}_1 = \mathbf{A}_2 = \mathbf{A}_3 = \mathbf{A}$. An example of a graph G and the induced matrices $\mathbf{A}$ and $\mathbf{M}_1$ is given in Fig. 1.

Triangle Finding-hardness of MAX $\left[\begin{smallmatrix} 1 & 1 \\ 1 & 1 \end{smallmatrix}\right]$ is implied by the following property.

Claim: G has a triangle if and only if there is a block B of $\mathbf{M}$ such that $\varkappa(B) = \left(\begin{smallmatrix} 1 & 1 \\ 1 & 1 \end{smallmatrix}\right)$ and $\mathrm{a}(B) \geq 3n^2$.

To prove this, let us investigate which blocks can exist in $\mathbf{M}_1$ where all four corners are 1's. It is easy to see that there are three types of such blocks:

– A block B included in one of the matrices $\mathbf{A}_i$, $i \in \{1, 2, 3\}$. Its area is not greater than n^2.
– A block B with two corners in $\mathbf{A}_2$ and the other two corners in either $\mathbf{A}_1$ or $\mathbf{A}_3$. Assume e.g. that the leftmost column of such a block is the k-th column of $\mathbf{M}_1$ and the rightmost column is in the ℓ-th column of $\mathbf{M}_1$, where $\ell > 2n$. The height of B is at most $n - k$ and it holds that $\ell < 3n$. Hence, $\mathrm{a}(B)$ is upper bounded by $(\ell - k + 1)(n - k) \leq (3n - k)(n - k) < 3n^2$.
– A block B that has one corner in each of the matrices $\mathbf{A}_1$, $\mathbf{A}_2$, $\mathbf{A}_3$, $\mathbf{I}$. Let the top left corner of B be in $\mathbf{M}_1$ at a position (k, ℓ), and the bottom right corner of B be at a position (s, t). Properties of the $\mathbf{A}_i$'s and $\mathbf{I}$ ensure that $|\ell - k| < n$, $t = 2n + k$, and $s > 2n + \ell$. Hence, $\mathrm{a}(B) = (s - k + 1)(t - \ell + 1) > (2n + \ell - k)(2n + k - \ell) = 4n^2 - (\ell - k)^2 \geq 3n^2$.

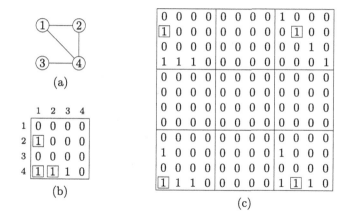

Fig. 1. (a) A sample graph G on the set of vertices $\{1, 2, 3, 4\}$ where the subset $T = \{1, 2, 4\}$ forms a triangle. (b) The lower triangular adjacency matrix of G. The framed entries indicate the existence of the triangle T. (c) Triangle Finding in G reduced to $\mathsf{MAX}\left[\begin{smallmatrix} 1 & 1 \\ 1 & 1 \end{smallmatrix}\right]$.

It is not difficult to verify that there is a one-to-one correspondence between blocks of the third type and triples $i < j < k$ such that $\mathbf{A}_{j,i} = \mathbf{A}_{k,i} = \mathbf{A}_{k,j} = 1$, hence representing triangles $\{v_i, v_j, v_k\}$ of G.

To reduce Triangle Finding to $\mathsf{MAX}\left[\begin{smallmatrix} 1 & 0 \\ 1 & 1 \end{smallmatrix}\right]$, $\mathsf{MAX}\left[\begin{smallmatrix} 1 & 0 \\ 0 & 1 \end{smallmatrix}\right]$ and $\mathsf{MAX}\left[\begin{smallmatrix} 1 & 0 \\ 1 & 0 \end{smallmatrix}\right]$, define the matrices $\mathbf{M}_2$, $\mathbf{M}_3$ and $\mathbf{M}_4$, respectively, as

$$\mathbf{M_2} = \begin{pmatrix} \mathbf{A_1} & \mathbf{O} & \bar{\mathbf{I}} \\ \mathbf{O} & \mathbf{O} & \mathbf{O} \\ \mathbf{A_2} & \mathbf{O} & \mathbf{A_3} \end{pmatrix}, \quad \mathbf{M_3} = \begin{pmatrix} \mathbf{A_1} & \mathbf{O} & \bar{\mathbf{I}} \\ \mathbf{O} & \mathbf{O} & \mathbf{O} \\ \overline{\mathbf{A_2}} & \mathbf{O} & \mathbf{A_3} \end{pmatrix}, \quad \mathbf{M_4} = \begin{pmatrix} \mathbf{A_1} & \overline{\mathbf{O}} & \bar{\mathbf{I}} \\ \mathbf{O} & \mathbf{O} & \mathbf{O} \\ \mathbf{A_2} & \mathbf{O} & \overline{\mathbf{A_3}} \end{pmatrix}.$$

Recall that $\overline{\mathbf{C}}$ denotes the matrix created from a Boolean matrix $\mathbf{C}$ by negating its entries. Figure 2 shows the matrices $\mathbf{M}_3$ and $\mathbf{M}_4$ constructed for the graph G from Fig. 1.

One can again verify that G has a triangle if and only if each of the constructed matrices contains a block B such that $\mathrm{a}(B) \geq 3n^2$ and $\varkappa(B)$ matches the desired 2×2 matrix. □

Now, let us focus on approaches for solving the maximization problems. Given an $m \times n$ Boolean matrix $\mathbf{M}$ and $p, q \in \{0, 1\}$, let $\sigma_{p,q}^{\mathbf{M}}$ denote the map whose keys are pairs (i, j), where i, j are row indexes of $\mathbf{M}$ such that $i < j$. For a key (i, j), the map value is defined as the smallest column index c such that $\mathbf{M}_{i,c} = p$ and $\mathbf{M}_{j,c} = q$.

Let $\mathbf{M}'$ denote the matrix $\mathbf{M}$ flipped left to right. It is easy to see that every problem $\mathsf{MAX}\left[\begin{smallmatrix} a & b \\ c & d \end{smallmatrix}\right]$, where $a, b, c, d \in \{0, 1\}$, can be solved based on the maps $\sigma_{a,c}^{\mathbf{M}}$ and $\sigma_{b,d}^{\mathbf{M}'}$. Conversely, this also shows that these maps are Triangle Finding-hard to build.

```
0 0 0 0 | 0 0 0 0 | 0 1 1 1
[1]0 0 0| 0 0 0 0 | 1 [0]1 1
0 0 0 0 | 0 0 0 0 | 1 1 0 1
1 1 1 0 | 0 0 0 0 | 1 1 1 0
0 0 0 0 | 0 0 0 0 | 0 0 0 0
0 0 0 0 | 0 0 0 0 | 0 0 0 0
0 0 0 0 | 0 0 0 0 | 0 0 0 0
0 0 0 0 | 0 0 0 0 | 0 0 0 0
1 1 1 1 | 0 0 0 0 | 0 0 0 0
0 1 1 1 | 0 0 0 0 | 1 0 0 0
1 1 1 1 | 0 0 0 0 | 0 0 0 0
[0]0 0 1| 0 0 0 0 | 1 [1]1 0
```
(a)

```
0 0 0 0 | 1 1 1 1 | 0 1 1 1
[1]0 0 0| 1 1 1 1 | 1 [0]1 1
0 0 0 0 | 1 1 1 1 | 1 1 0 1
1 1 1 0 | 1 1 1 1 | 1 1 1 0
0 0 0 0 | 0 0 0 0 | 0 0 0 0
0 0 0 0 | 0 0 0 0 | 0 0 0 0
0 0 0 0 | 0 0 0 0 | 0 0 0 0
0 0 0 0 | 0 0 0 0 | 1 1 1 1
1 0 0 0 | 0 0 0 0 | 0 1 1 1
0 0 0 0 | 0 0 0 0 | 1 1 1 1
[1]1 1 0| 0 0 0 0 | 0 [0]0 1
```
(b)

Fig. 2. Triangle Finding reduced to (a) MAX $\left[\begin{smallmatrix}1&0\\0&1\end{smallmatrix}\right]$ and (b) MAX $\left[\begin{smallmatrix}1&0\\1&0\end{smallmatrix}\right]$, respectively.

The maps can be computed based on a minimum witness for Boolean matrix multiplication [4], and hence the time complexity of solving MAX $\left[\begin{smallmatrix}a&b\\c&d\end{smallmatrix}\right]$ in this way coincides with the time complexity in [4] for the minimum witness problem.

Lemma 9. *There is an algorithm that, for any m by n Boolean matrix $\mathbf{M}$ where $m \le n$, and any $p, q \in \{0,1\}$, builds $\sigma_{p,q}^{\mathbf{M}}$ in time $O\left(mn \cdot m^{0.5302}\right)$.*

Theorem 6. *For any $a, b, c, d \in \{0,1\}$, there is an algorithm solving MAX $\left[\begin{smallmatrix}a&b\\c&d\end{smallmatrix}\right]$ in time $O\left(mn \cdot (\min\{m,n\})^{0.5302}\right)$ for m by n Boolean matrices.*

6   Conclusion

We investigated the complexity of Four Corner Problems over Boolean matrices. A Four Corner Problem is concerned with searching for a given 2×2 submatrix in a given Boolean matrix. We demonstrated that minimum-area Four Corner Problems over Boolean matrices are solvable in nearly linear time in the input matrix area (Theorem 4) and maximum-area Four Corner Problems over Boolean matrices are Triangle Finding-hard (Theorem 5). The algorithms that we presented for the former problems might lead to efficient implementations, while the results achieved for the latter problems give rise to an interesting unresolved theoretical question: Are the maximum-area Four Corner Problems harder than the Triangle Finding problem? Going further, we suggest that a possible future direction is to investigate the complexity of Four Corner Problems over matrices with entries from larger alphabets.

Acknowledgment. We greatly appreciate all of the help and suggestions that we received. We would especially like to thank Joseph Swernofsky for helping us obtain some preliminary results, for contributing to a preliminary version of this work, and for providing valuable feedback. We also thank the Czech Science Foundation for supporting the first author (grant no. 19-09967S).

References

1. Abboud, A., Backurs, A., Williams, V.V.: If the current clique algorithms are optimal, so is Valiant's parser. In: Guruswami, V. (ed.) IEEE 56th Annual Symposium on Foundations of Computer Science (FOCS 2015), pp. 98–117. IEEE Computer Society (2015)
2. Aho, A.V.: Algorithms for finding patterns in strings. In: van Leeuwen, J. (ed.) Algorithms and Complexity, Handbook of Theoretical Computer Science, vol. A, pp. 255–300. The MIT Press, Cambridge (1990)
3. Bentley, J.L., Haken, D., Saxe, J.B.: A general method for solving divide-and-conquer recurrences. SIGACT News **12**(3), 36–44 (1980)
4. Cohen, K., Yuster, R.: On minimum witnesses for boolean matrix multiplication. Algorithmica **69**(2), 431–442 (2014)
5. Giammarresi, D., Restivo, A.: Two-dimensional languages. In: Rozenberg, G., Salomaa, A. (eds.) Handbook of Formal Languages, pp. 215–267. Springer, Heidelberg (1997). https://doi.org/10.1007/978-3-642-59126-6_4
6. Itai, A., Rodeh, M.: Finding a minimum circuit in a graph. In: 9th Annual ACM Symposium on Theory of Computing (STOC 1977), pp. 1–10. ACM, New York (1977)
7. Lee, L.: Fast context-free grammar parsing requires fast boolean matrix multiplication. J. ACM **49**(1), 1–15 (2002)
8. Mráz, F., Průša, D., Wehar, M.: Two-dimensional pattern matching against basic picture languages. In: Hospodár, M., Jirásková, G. (eds.) CIAA 2019. LNCS, vol. 11601, pp. 209–221. Springer, Cham (2019). https://doi.org/10.1007/978-3-030-23679-3_17
9. de Oliveira Oliveira, M., Wehar, M.: Intersection non-emptiness and hardness within polynomial time. In: Hoshi, M., Seki, S. (eds.) DLT 2018. LNCS, vol. 11088, pp. 282–290. Springer, Cham (2018). https://doi.org/10.1007/978-3-319-98654-8_23
10. Peeters, R.: The maximum edge biclique problem is NP-complete. Discret. Appl. Math. **131**(3), 651–654 (2003)
11. Potechin, A., Shallit, J.: Lengths of words accepted by nondeterministic finite automata. CoRR abs/1802.04708 (2018)
12. Salomaa, A.: Jewels of Formal Language Theory. Computer Science Press, Rockville (1981)
13. Sun, X., Nobel, A.B.: On the size and recovery of submatrices of ones in a random binary matrix. J. Mach. Learn. Res. **9**(Nov), 2431–2453 (2008)
14. Williams, V.V.: Multiplying matrices faster than Coppersmith-Winograd. In: 44th Annual ACM Symposium on Theory of Computing (STOC 2012), pp. 887–898. ACM, New York (2012)
15. Williams, V.V.: Hardness of easy problems: basing hardness on popular conjectures such as the strong exponential time hypothesis (invited talk). In: Husfeldt, T., Kanj, I.A. (eds.) 10th International Symposium on Parameterized and Exact Computation (IPEC 2015). LIPIcs, vol. 43, pp. 17–29. Schloss Dagstuhl - Leibniz-Zentrum für Informatik (2015)

Avoiding 5/4-Powers on the Alphabet of Nonnegative Integers (Extended Abstract)

Eric Rowland[ID] and Manon Stipulanti[✉][ID]

Department of Mathematics, Hofstra University, Hempstead, NY 11549, USA
eric.rowland@hofstra.edu, m.stipulanti@uliege.be

Abstract. We identify the structure of the lexicographically least word avoiding 5/4-powers on the alphabet of nonnegative integers.

Keywords: Combinatorics on words · Power-freeness · Lexicographic-leastness

1 Introduction

For any (finite or infinite) alphabet Σ, we let Σ^* denote the set of finite words on Σ. We start indexing (finite and infinite) words at position 0.

A *morphism* on an alphabet Σ is a map $\varphi \colon \Sigma \to \Sigma^*$. It extends naturally to finite and infinite words by concatenation. We say that a morphism φ on Σ is *k-uniform* if $|\varphi(c)| = k$ for all $c \in \Sigma$. A 1-uniform morphism is also called a *coding*. If there exists a letter $c \in \Sigma$ such that $\varphi(c)$ starts with c, then iterating φ on c gives a word $\varphi^\omega(c)$, which is a fixed point of φ beginning with c.

A fractional power is a partial repetition, defined as follows. Let a and b be relatively prime positive integers. If $v = v_0 v_1 \cdots v_{\ell-1}$ is a nonempty word whose length ℓ is divisible by b, the *a/b-power* of v is the word

$$v^{a/b} := v^{\lfloor a/b \rfloor} v_0 v_1 \cdots v_{\ell \cdot \{a/b\}-1},$$

where $\{a/b\} = a/b - \lfloor a/b \rfloor$ is the fractional part of a/b. Note that $|v^{a/b}| = \frac{a}{b}|v|$. If $a/b > 1$, then a word w is an a/b-power if and only if w can be written $v^e u$ where e is a positive integer, u is a prefix of v, and $\frac{|w|}{|v|} = \frac{a}{b}$.

Example 1. The 3/2-power of the word 0111 is $(0111)^{3/2} = 011101$.

In general, a 5/4-power is a word of the form $(xy)^{5/4} = xyx$, where $|xy| = 4\ell$ and $|xyx| = 5\ell$ for some $\ell \geq 1$. It follows that $|x| = \ell$ and $|y| = 3\ell$.

Elsewhere in the literature, researchers have been interested in words with no α-power factors for all $\alpha \geq a/b$. In this paper, we consider a slightly different notion, and we say that a word is *a/b-power-free* if none of its factors is an (exact) a/b-power.

Supported in part by a Francqui Foundation Fellowship of the Belgian American Educational Foundation.

N. Jonoska and D. Savchuk (Eds.): DLT 2020, LNCS 12086, pp. 280–293, 2020.
https://doi.org/10.1007/978-3-030-48516-0_21

Notation 1. Let a and b be relatively prime positive integers such that $a/b > 1$. Define $\mathbf{w}_{a/b}$ to be the lexicographically least infinite word on $\mathbb{Z}_{\geq 0}$ avoiding a/b-powers.

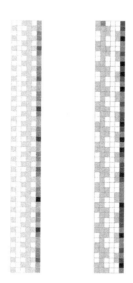

Fig. 1. Portions of $\mathbf{w}_{3/2}$ (left) and $\mathbf{w}_{5/4}$ (right), partitioned into rows of width 6. The letter 0 is represented by white cells, 1 by slightly darker cells, and so on. The word $\mathbf{w}_{3/2}$ is shown from the beginning. The word $\mathbf{w}_{5/4} = w(0)w(1) \cdots$ is shown beginning from $w(i)_{i \geq 6758}$; the term $w(6759)$ (top row, second column) is the last entry in $w(6i+3)_{i \geq 0}$ that is not 1.

Guay-Paquet and Shallit [1] identified the structure of $\mathbf{w}_a$ for each integer $a \geq 2$. In particular,

$$\mathbf{w}_2 = 01020103010201040102010301020105 \cdots$$

is the fixed point of the 2-uniform morphism μ on the alphabet of nonnegative integers defined by $\mu(n) = 0(n+1)$ for all $n \geq 0$. The first-named author and Shallit [3] studied the structure of

$$\mathbf{w}_{3/2} = 0011021001120011031001130011021001140011031 \cdots ,$$

which is the image under a coding of a fixed point of a 6-uniform morphism. A prefix of this word appears in Fig. 1. Pudwell and the first-named author [2] undertook a large study of $\mathbf{w}_{a/b}$ for rational numbers in the range $1 < \frac{a}{b} < 2$, and identified many of these words as images under codings of fixed points of morphisms.

The simplest number $\frac{a}{b}$ in this range for which the structure of $\mathbf{w}_{a/b}$ was not known is $\frac{5}{4}$. In this paper we give a morphic description for $\mathbf{w}_{5/4}$. Let $w(i)$ be the ith letter of $\mathbf{w}_{5/4}$. For the morphic description of $\mathbf{w}_{5/4}$, we need 8 types of letters, $n_0, n_1, \ldots, n_7$ for each integer $n \in \mathbb{Z}$. For example, $0_0, 0_1, \ldots, 0_7$ are the 8 different letters of the form 0_j. The subscript j of the letter n_j will determine the first 5 letters of $\varphi(n_j)$, which correspond to the first 5 columns in Fig. 1. The definition of φ in Notation 2 implies that these columns are eventually periodic with period length 1 or 4.

Notation 2. Let Σ_8 be the alphabet $\{n_j \mid n \in \mathbb{Z}, 0 \le j \le 7\}$. Let φ be the 6-uniform morphism defined on Σ_8 by

$$\varphi(n_0) = 0_0 1_1 0_2 0_3 1_4 (n+3)_5$$
$$\varphi(n_1) = 1_6 1_7 0_0 0_1 0_2 (n+2)_3$$
$$\varphi(n_2) = 1_4 1_5 1_6 0_7 0_0 (n+3)_1$$
$$\varphi(n_3) = 0_2 1_3 1_4 0_5 1_6 (n+2)_7$$
$$\varphi(n_4) = 0_0 1_1 0_2 0_3 1_4 (n+1)_5$$
$$\varphi(n_5) = 1_6 1_7 0_0 0_1 0_2 (n+2)_3$$
$$\varphi(n_6) = 1_4 1_5 1_6 0_7 0_0 (n+1)_1$$
$$\varphi(n_7) = 0_2 1_3 1_4 0_5 1_6 (n+2)_7.$$

We also define the coding $\tau(n_j) = n$ for all $n_j \in \Sigma_8$. In the rest of the paper, we think about the definitions of φ and $\tau \circ \varphi$ as 8×6 arrays of their letters. In particular, we will refer to letters in images of φ and $\tau \circ \varphi$ by their columns (first through sixth).

The following gives the structure of $\mathbf{w}_{5/4}$.

Theorem 1. *There exist words $\mathbf{p}, \mathbf{z}$ of lengths $|\mathbf{p}| = 6764$ and $|\mathbf{z}| = 20226$ such that $\mathbf{w}_{5/4} = \mathbf{p}\,\tau(\varphi(\mathbf{z})\varphi^2(\mathbf{z})\cdots)$.*

This word is more complicated than previously studied words $\mathbf{w}_{a/b}$ in two major ways. Unlike all words $\mathbf{w}_{a/b}$ whose structures were previously known, the natural description of $\mathbf{w}_{5/4}$ is not as an image under a coding of a fixed point of a morphism, since $\tau(\mathbf{z})$ is not a factor of $\mathbf{w}_{5/4}$. Additionally, the value of d in the image $\varphi(n_j) = u(n+d)_i$ varies with j. The sequence $3, 2, 3, 2, 1, 2, 1, 2, \ldots$ of d values is periodic with period length 8, hence the 8 types of letters. These properties make the proofs significantly more intricate.

We will define $\mathbf{s} = \mathbf{z}\varphi(\mathbf{z})\varphi^2(\mathbf{z})\cdots$. To prove Theorem 1, we must show that

1. $\mathbf{p}\,\tau(\varphi(\mathbf{s}))$ is 5/4-power-free, and
2. $\mathbf{p}\,\tau(\varphi(\mathbf{s}))$ is lexicographically least (by showing that decreasing any letter introduces a 5/4-power ending in that position).

We use *Mathematica* to carry out several computations required in the proofs.

Theorem 1 implies the following recurrence for letters sufficiently far out in $\mathbf{w}_{5/4}$ with positions $\equiv 1 \mod 6$.

Corollary 1. *Let $w(i)$ be the ith letter of $\mathbf{w}_{5/4}$. Then, for all $i \geq 0$,*

$$w(6i + 123061) = w(i + 5920) + \begin{cases} 3 & if\ i \equiv 0, 2 \mod 8 \\ 1 & if\ i \equiv 4, 6 \mod 8 \\ 2 & if\ i \equiv 1 \mod 2. \end{cases}$$

This paper is organized as follows. Section 2 gives preliminary properties about φ, $\mathbf{p}$, and $\mathbf{z}$. In Sect. 3, we introduce the concept of pre-5/4-power-freeness. We show that $\mathbf{w}_{5/4} = \mathbf{p}\,\tau(\varphi(\mathbf{s}))$ in two steps. First, in Sect. 4, we show that $\mathbf{p}\,\tau(\varphi(\mathbf{s}))$ is 5/4-power-free using the fact that $\mathbf{s}$ is pre-5/4-power-free. Second, in Sect. 5, we show that $\mathbf{p}\,\tau(\varphi(\mathbf{s}))$ is lexicographically least. Proofs of some supporting lemmas and propositions are omitted in this extended abstract due to space constraints and appear in the full version of the paper [4].

2   Preliminary Properties

Note that in Notation 2 the subscripts in each image under φ increase by 1 modulo 8 from one letter to the next and also from the end of each image to the beginning of the next.

Definition 1. *A (finite or infinite) word w on Σ_8 is subscript-increasing if the subscripts of the letters of w increase by 1 modulo 8 from one letter to the next.*

If w is a subscript-increasing word on Σ_8, then so is $\varphi(w)$. For every subscript-increasing word w on Σ_8, it follows from Notation 2 that the subsequence of letters with even subscripts in $\varphi(w)$ is a factor of $(0_0 0_2 1_4 1_6)^\omega$.

Iterating φ on any word on Σ_8 will eventually give a word containing letters n_j with arbitrarily large n. Indeed, after one iteration, we see a letter with subscript 3 or 7, so after two iterations we see a letter with subscript 7. Since $\varphi(n_7)$ contains $(n + 2)_7$, the alphabet is growing.

Before position 6764, we cannot expect the prefix of $\mathbf{w}_{5/4}$ to be the image of another word under the morphism φ because the five columns have not become periodic yet (recall Fig. 1 where $w(6759)$ is the last term of $w(6i + 3)$ before a periodic pattern appears).

One checks programmatically that there are two subscript-increasing preimages of $w(6764)w(6765) \cdots$ under $\tau \circ \varphi$. We choose the following for $\mathbf{s}$.

Definition 2. *Let $\mathbf{p}$ denote the length-6764 prefix of $\mathbf{w}_{5/4}$. We define*

$$\mathbf{z} = 0_2 0_3 3_4 0_5 1_6 1_7 (-1_0) 2_1 0_2 2_3 2_4 0_5 \cdots 0_0 1_1 0_2 0_3 1_4 2_5 1_6 2_7 0_0 0_1 0_2 3_3$$

to be the length-20226 subscript-increasing word on Σ_8 starting with 0_2 and satisfying

$$\tau(\varphi(\mathbf{z})) = w(6764)w(6765) \cdots w(6764 + 6|\mathbf{z}| - 1).$$

We define $\mathbf{s} = \mathbf{z}\varphi(\mathbf{z})\varphi^2(\mathbf{z}) \cdots$ as before.

Lemma 1. *Let $\Gamma \subset \Sigma_8$ be the finite alphabet*

$$\{-3_0, -3_2, -2_0, -2_1, -2_2, -2_3, -2_5, -2_7, -1_1, -1_3, -1_4, -1_5, -1_6, -1_7, 0_4, 0_6\}.$$

We have the following properties.

1. *The length-844 suffixes of* **p** *and* $\tau(\mathbf{z})$ *are equal.*
2. *The word* **z** *is a subscript-increasing finite word whose alphabet is the 32-letter set*

$$\mathrm{Alph}(\mathbf{z}) = \{-1_0, -1_2, 0_0, 0_1, 0_2, 0_3, 0_5, 0_7, 1_0, 1_1, 1_2, 1_3, 1_4, 1_5, 1_6, 1_7,$$
$$2_1, 2_3, 2_4, 2_5, 2_6, 2_7, 3_1, 3_3, 3_4, 3_5, 3_6, 3_7, 4_1, 4_3, 4_5, 4_7\}.$$

   In particular, **z** *contains no letters in* Γ. *The last letter of the form* -1_j *appears at position 80, and the subsequence of letters with even subscripts starting at position 86 is a finite prefix of* $(0_0 0_2 1_4 1_6)^\omega$.
3. *The word* **s** *is a subscript-increasing infinite word whose alphabet is* $\mathrm{Alph}(\mathbf{z}) \cup \{5_3, 6_3\} \cup \{n_7 \mid n \geq 5\}$. *In particular,* **s** *contains no letters in* Γ.
4. *For all words w on Σ_8, the only letters of $\varphi(w)$ of the form n_j with even j are 0_j and 1_j. More precisely, the only letter with subscript 0 (resp., 2; resp., 4; resp., 6) in $\varphi(w)$ is 0_0 (resp., 0_2; resp., 1_4; resp., 1_6).*
5. *As long as $n_j \in (\Sigma_8 \setminus \Gamma)$ with $0 \leq j \leq 7$, the last letter of $\varphi(n_j)$ is not of the form 0_i or 1_i. In particular, for all letters n_j of* **s**, *the last letter of $\varphi(n_j)$ is of the form n_i where $n \geq 2$.*

3   Pre-5/4-Power-Freeness

A morphism μ on Σ is *a/b-power-free* if μ preserves a/b-power-freeness, that is, for all a/b-power-free words w on Σ, $\mu(w)$ is also a/b-power-free. Previously studied a/b-power-free words [1–3] have all been described by a/b-power-free morphisms. However, the morphism φ defined in Notation 2 is not 5/4-power-free. Indeed for any integers $n, \bar{n} \in \mathbb{Z}$, the word $0_4 n_5 \bar{n}_6$ is 5/4-power-free, but $\varphi(0_4 n_5 \bar{n}_6)$ contains the length-10 factor $1_4 1_5 1_6 1_7 0_0 0_1 0_2 (n+2)_3 1_4 1_5$, which is a 5/4-power. Therefore, to prove that $\mathbf{w}_{5/4}$ is 5/4-power-free, we use a different approach. We still need to guarantee that there are no 5/4-powers in images of certain words under φ. Specifically, we would like all factors xyx' of a word with $|x| = \frac{1}{3}|y| = |x'|$ to satisfy $\varphi(x) \neq \varphi(x')$.

Definition 3. A *pre-5/4-power-free* word is a (finite or infinite) subscript-increasing word w on Σ_8 such that, for all factors xyx' of w with $|x| = \frac{1}{3}|y| = |x'|$, there exists $0 \leq m \leq |x| - 1$ such that

1. if the subscripts of $x(m)$ and $x'(m)$ are equal or odd, then $\tau(x(m)) \neq \tau(x'(m))$, and
2. if the subscripts of $x(m)$ and $x'(m)$ are even and differ by 4, then $\tau(x(m)) - \tau(x'(m)) \notin \{-2, 2\}$.

Note that if the subscripts of $x(m)$ and $x'(m)$ are not equal, then they differ by 4 since $|xy| = 4|x|$. Definition 3 involves the set $\{-2, 2\}$ because if the subscripts of $x(m)$ and $x'(m)$ are even and differ by 4 and $\tau(x(m)) - \tau(x'(m)) \in \{-2, 2\}$ then $\varphi(x(m)) = \varphi(x'(m))$.

For example, the word $0_0 n_1 \bar{n}_2 \bar{\bar{n}}_3 2_4$ is not pre-5/4-power-free because

$$\varphi(0_0 n_1 \bar{n}_2 \bar{\bar{n}}_3 2_4) = 0_0 1_1 0_2 0_3 1_4 3_5 \varphi(n_1 \bar{n}_2 \bar{\bar{n}}_3) 0_0 1_1 0_2 0_3 1_4 3_5$$

is a 5/4-power of length 30. On the other hand, the word $0_0 n_1 \bar{n}_2 \bar{\bar{n}}_3 0_4$ is pre-5/4-power-free because

$$\varphi(0_0 n_1 \bar{n}_2 \bar{\bar{n}}_3 0_4) = 0_0 1_1 0_2 0_3 1_4 3_5 \varphi(n_1 \bar{n}_2 \bar{\bar{n}}_3) 0_0 1_1 0_2 0_3 1_4 1_5$$

is not a 5/4-power. The word $0_0 n_1 \bar{n}_2 \bar{\bar{n}}_3 0_4$ is also 5/4-power-free, since 0_0 and 0_4 are different letters. The next proposition states that pre-5/4-power-freeness implies 5/4-power-freeness in general.

Proposition 1. *If a word is pre-5/4-power-free, then it is also 5/4-power-free.*

One checks programmatically that the finite word $\mathbf{z}$ is pre-5/4-power-free. As a consequence, we will show that $\mathbf{p}\tau(\varphi(\mathbf{s}))$ is 5/4-power-free in Sect. 4.

Lemma 2. *Let $\ell \geq 6$ be an integer. If w is a pre-5/4-power-free word on Σ_8, then $\varphi(w)$ contains no 5/4-power of length 5ℓ.*

Lemma 2 takes care of large 5/4-powers. Toward avoiding all 5/4-powers, the following proposition shows that φ preserves the property of pre-5/4-power-freeness.

Proposition 2. *Let Γ be the alphabet*

$$\{-3_0, -3_2, -2_0, -2_1, -2_2, -2_3, -2_5, -2_7, -1_1, -1_3, -1_4, -1_5, -1_6, -1_7, 0_4, 0_6\}.$$

For all pre-5/4-power-free words w on $\Sigma_8 \setminus \Gamma$, $\varphi(w)$ is pre-5/4-power-free.

Next we show that the word $\mathbf{s} = \mathbf{z}\varphi(\mathbf{z})\varphi^2(\mathbf{z}) \cdots$ is pre-5/4-power-free. The common length-4 suffix 0003 of $\mathbf{p}$ and $\tau(\mathbf{z})$ is a possible factor of φ, which requires extra consideration in Theorems 2 and 3, and in Proposition 3.

Theorem 2. *The word $\mathbf{s}$ is pre-5/4-power-free.*

Proof. We show that, for all $e \geq 1$, $\mathbf{z}\varphi(\mathbf{z}) \cdots \varphi^e(\mathbf{z})$ is pre-5/4-power-free. Note that $\mathbf{z}\varphi(\mathbf{z}) \cdots \varphi^e(\mathbf{z}) \in (\Sigma_8 \setminus \Gamma)^*$ for all $e \geq 1$. We proceed by induction on $e \geq 1$.

Base Case. Suppose that $e = 1$. We wrote code to check that $\mathbf{z}\varphi(\mathbf{z})$ is pre-5/4-power-free. The computation took about 13 hours.

Induction Step. We suppose that $\mathbf{z}\varphi(\mathbf{z}) \cdots \varphi^e(\mathbf{z})$ is pre-5/4-power-free for $e \geq 1$. We show that $\mathbf{z}\varphi(\mathbf{z}) \cdots \varphi^{e+1}(\mathbf{z})$ is also pre-5/4-power-free. First, the word $\mathbf{z}\varphi(\mathbf{z}) \cdots \varphi^{e+1}(\mathbf{z})$ is subscript-increasing. Let xyx' be a factor of $\mathbf{z}\varphi(\mathbf{z}) \cdots \varphi^{e+1}(\mathbf{z})$ with $|x| = \frac{1}{3}|y| = |x'|$.

Suppose that xyx' is a factor of either $\mathbf{z}$ or $\varphi(\mathbf{z}) \cdots \varphi^{e+1}(\mathbf{z})$. In the first case, Definition 3 holds since $\mathbf{z}$ is pre-5/4-power-free. In the second case, Definition 3 holds too. Indeed, it suffices to use Proposition 2 with $\mathbf{z}\varphi(\mathbf{z}) \cdots \varphi^e(\mathbf{z}) \in (\Sigma_8 \setminus \Gamma)^*$, which is pre-5/4-power-free by the induction hypothesis.

It remains to check the case where xyx' overlaps $\mathbf{z}$. If yx' overlaps $\mathbf{z}$, then x is a factor of $\mathbf{z}$ and $|xyx'| = 5|x| \leq 5|\mathbf{z}| < 6|\mathbf{z}| = |\varphi(\mathbf{z})|$. In particular, xyx' is a factor of $\mathbf{z}\varphi(\mathbf{z})$. The base case of the induction implies that Definition 3 holds in this case. Thus we may suppose that x overlaps $\mathbf{z}$.

If the overlap length is at least 5, then x contains the suffix $2_7 0_0 0_1 0_2 3_3$ of $\mathbf{z}$. If the subscripts of x and x' differ by 4, then x' being a factor of $\varphi(\mathbf{z}) \cdots \varphi^{e+1}(\mathbf{z})$ implies that the factor $2_7 0_0 0_1 0_2 3_3$ of x corresponds to a factor $n_3 1_4 \bar{n}_5 1_6 \bar{\bar{n}}_7$ of x'. So Definition 3 is fulfilled. If the subscripts of x and x' line up, then x' being a factor of $\varphi(\mathbf{z}) \cdots \varphi^{e+1}(\mathbf{z})$ implies that the factor $2_7 0_0 0_1 0_2 3_3$ of x corresponds to a factor $n_7 0_0 \bar{n}_1 0_2 \bar{\bar{n}}_3$ of x'. If $n_7 \neq 2_7$ or $\bar{n}_1 \neq 0_1$ or $\bar{\bar{n}}_3 \neq 3_3$, then Definition 3 holds. Otherwise, then x' contains $2_7 0_0 0_1 0_2 3_3$, which is impossible since this factor does not appear in $\varphi(\mathbf{z}) \cdots \varphi^{e+1}(\mathbf{z})$.

Suppose the overlap length is less than or equal to 4. If $|x| = |x'|$ is odd, then the subscripts of x and x' differ by 4. Since x contains the factor $3_3 1_4$, then x' being a factor of $\varphi(\mathbf{z}) \cdots \varphi^{e+1}(\mathbf{z})$ implies that the factor $3_3 1_4$ of x corresponds to a factor $n_7 0_0$ of x', and Definition 3 holds. If $|x| = |x'|$ is even, then the subscripts of x and x' line up. The words x and x' agree on even subscripts (because the length-4 suffix of $\mathbf{z}$ is $0_0 0_1 0_2 3_3$). For odd subscripts, if the corresponding letters of x and x' belong to different columns (that is, their positions are not congruent modulo 6), then Definition 3 holds by Part 5 of Lemma 1. If the corresponding letters belong to the same column, then x and x' agree everywhere except maybe in the sixth column. Toward a contradiction, suppose the words are equal in the sixth column, so $x = x'$. Since x' is a factor of $\varphi(\mathbf{z}) \cdots \varphi^{e+1}(\mathbf{z})$, there exists a word $w \in \Sigma_8^*$ such that x' is a factor of $\varphi(w)$ and w itself is a factor of $\mathbf{z}\varphi(\mathbf{z}) \cdots \varphi^e(\mathbf{z})$. We take w to be of minimal length.

If $|w| \leq 3$, then $|x| = |x'| \leq |\varphi(w)| \leq 18$, so $|xyx'| = 5|x| \leq 90 < |\mathbf{z}\varphi(\mathbf{z})|$, which means that xyx' is a factor of $\mathbf{z}\varphi(\mathbf{z})$. Due to the base case, we already know that xyx' satisfies the conditions of Definition 3.

If $|w| \geq 4$, then w can take two different forms:

$$w = \begin{cases} 1_1 0_2 0_3 3_4 \cdots \\ 1_5 2_6 0_7 1_0 \cdots \end{cases}$$

since $0_0 0_1 0_2 3_3 \varphi(\mathbf{z}) \varphi^2(\mathbf{z}) \cdots = 0_0 0_1 0_2 3_3 1_4 1_5 1_6 0_1 0_0 3_1 \cdots$. Due to the definition of the morphism φ, the letters 1_0 and 3_4 do not occur in $\varphi(\mathbf{z}) \cdots \varphi^e(\mathbf{z})$. Since w is a factor of $\mathbf{z}\varphi(\mathbf{z}) \cdots \varphi^e(\mathbf{z})$, they must belong to $\mathbf{z}$. But none of the words $1_1 0_2 0_3 3_4$ and $1_5 2_6 0_7 1_0$ belongs to $\mathbf{z}$. Indeed, the letter 3_4 occurs in $\mathbf{z}$ only at positions 2 and 66, and we have $z(63)z(64)z(65)z(66) = 0_1 0_2 0_3 3_4$. Similarly, the letter 1_0 only occurs in $\mathbf{z}$ only at positions 22 and 54 and 78, and we find

$$z(19)z(20)z(21)z(22) = 2_5 2_6 0_7 1_0,$$
$$z(51)z(52)z(53)z(54) = 1_5 2_6 2_7 1_0,$$
$$z(75)z(76)z(77)z(78) = 1_5 2_6 1_7 1_0.$$

This case is thus impossible.

4   5/4-power-freeness

In this section we show that $\mathbf{p}\tau(\varphi(\mathbf{s}))$ is 5/4-power-free. As a corollary of Theorem 2, we obtain the following.

Corollary 2. *The infinite word $\varphi(\mathbf{s})$ is 5/4-power-free.*

Proof. By Theorem 2, $\mathbf{s}$ is pre-5/4-power-free. Since $\mathbf{s} \in (\Sigma_8 \setminus \Gamma)^\omega$, Proposition 2 implies that $\varphi(\mathbf{s})$ is pre-5/4-power-free. By Proposition 1, $\varphi(\mathbf{s})$ is 5/4-power-free.

The following lemma asserts that applying the coding τ on $\varphi(\mathbf{s})$ preserves the 5/4-power-freeness.

Lemma 3. *The infinite word $\tau(\varphi(\mathbf{s}))$ is 5/4-power-free.*

The last step in showing 5/4-power-freeness is to prove that prepending $\mathbf{p}$ to $\tau(\varphi(\mathbf{s}))$ also yields a 5/4-power-free word. To that aim, we introduce the following notion.

Definition 4. Let N be a set of integers, and let $\alpha, \beta \in \mathbb{Z} \cup \{n+1, n+2, n+3\}$, where n is a symbol. Then α and β are *possibly equal with respect to N* if there exist $m, m' \in N$ such that $\alpha|_{n=m} = \beta|_{n=m'}$.

Two letters α, β are possibly equal with respect to N if we can make them equal by substituting integers from N for the symbol n. In particular, for every nonempty set N, two integers α, β are possibly equal if and only if $\alpha = \beta$. The definition of possibly equal letters extends to words in the natural way. The next two lemmas will be used to prove Theorem 3.

Lemma 4. *Let n be a symbol, and let $N \supseteq \{-3, -2, \ldots, 4\}$. Let α, β be elements of $\{0, 1, \ldots, 5\} \cup \{n+1, n+2, n+3\}$. If α and β are possibly equal with respect to N, then they are possibly equal with respect to $\{-3, -2, \ldots, 4\}$.*

Lemma 5. *Let n be a symbol, let $D = \{1, 2, 3\}$, and let $N = \{-3, -2, \ldots, 4\}$. Let v be the prefix of $\mathbf{p}\tau(\varphi(\mathbf{s}))$ of length $|\mathbf{p}| + 952 = 7716$. For each position $i \geq 0$ in the eventually periodic word*

$$v\tau\Big(\varphi(n_1)\varphi(n_2)\varphi(n_3)\varphi(n_4)\varphi(n_5)\varphi(n_6)\varphi(n_7)\varphi(n_0)\cdots\Big)$$

on $\{0, 1, \ldots, 5\} \cup \{n+1, n+2, n+3\}$, define the set

$$X_i = \begin{cases} \{c\} & \text{if the letter in position } i \text{ is } c \in \mathbb{Z} \\ N + d & \text{if the letter in position } i \text{ is } n+d \text{ where } d \in D. \end{cases}$$

For all $i, j \geq 0$, if the letters in positions i, j are possibly equal with respect to N, then $X_i \cap X_j$ is nonempty.

Proof. The set X_i is the set of integer letters c such that the letter in position i in $v\tau(\varphi(n_1)\varphi(n_2)\varphi(n_3)\cdots)$ is possibly equal to c with respect to N. Since v is a word on $\{0, 1, \ldots, 5\}$ and the integer letters of $\tau \circ \varphi$ are elements of $\{0, 1\} \subseteq \{0, 1, \ldots, 5\}$, the integer letters of $v\tau(\varphi(n_1)\varphi(n_2)\varphi(n_3)\cdots)$ are in $\{0, 1, \ldots, 5\}$. There are three cases to consider depending on the nature of the letters in positions i and j.

If both letters are integers, let c be the letter in position i and c' be the letter in position j. Since they are possibly equal with respect to N, then $c = c'$. So $c \in X_i \cap X_j$.

If one letter is an integer and the other is symbolic, without loss of generality, let c be the letter in position i and $n + d$ be the letter in position j with $d \in D$. By assumption, $c = n + d$ for some $n \in N$. So $X_i \cap X_j = \{c\} \cap (N + d) = \{c\}$.

If both letters are symbolic, let $n + d$ be the letter in position i and $n + d'$ be the letter in position j with $d, d' \in D$. By assumption, $X_i = N + d$ and $X_j = N + d'$, so $0 \in X_i \cap X_j$.

Theorem 3. *The infinite word* $\mathbf{p}\tau(\varphi(\mathbf{s}))$ *is 5/4-power-free.*

Proof. Since $\mathbf{p}$ is the prefix of $\mathbf{w}_{5/4}$ of length 6764, $\mathbf{p}$ is 5/4-power-free. By Lemma 3, $\tau(\varphi(\mathbf{s}))$ is also 5/4-power-free. So if $\mathbf{p}\tau(\varphi(\mathbf{s}))$ contains a 5/4-power, then it must overlap $\mathbf{p}$ and $\tau(\varphi(\mathbf{s}))$. We will show that there are no 5/4-powers xyx starting in $\mathbf{p}$.

For factors xyx with $|x| < 952$ starting in $\mathbf{p}$, note that $|x| < 952$ implies $|xyx| < 5 \cdot 952$, so it is enough to look for 5/4-powers in $\mathbf{p}\tau(\varphi(\mathbf{z}))$—as opposed to $\mathbf{p}\tau(\varphi(\mathbf{s}))$—starting in $\mathbf{p}$. We check programmatically that there is no such 5/4-power xyx. The computation took less than a minute.

For long factors, we show that each length-952 factor x starting in $\mathbf{p}$ only occurs once in $\mathbf{p}\tau(\varphi(\mathbf{s}))$. This will imply that there is no 5/4-power xyx in $\mathbf{p}\tau(\varphi(\mathbf{s}))$ starting in $\mathbf{p}$ such that $|x| \geq 952$. Since $\mathbf{s} = 0_2 0_3 3_4 \cdots$, the word $\mathbf{p}\tau(\varphi(\mathbf{s}))$ is of the form

$$\mathbf{p}\tau\Big(\varphi(n_2)\varphi(n_3)\varphi(n_4)\varphi(n_5)\varphi(n_6)\varphi(n_7)\varphi(n_0)\varphi(n_1)\cdots\Big).$$

Here we abuse notation; namely, the n's are not necessarily equal. Observe that $|\varphi(n_2)\varphi(n_3)\cdots\varphi(n_0)\varphi(n_1)| = 48$. We use a method based on [2, Section 6]. However, this is another situation in which the structure of the word $\mathbf{w}_{5/4}$ is more complex. The sequence $3, 2, 1, 2, 1, 2, 3, 2, \ldots$ of increments d in Notation 2 is periodic with period length 8, but each of the first five columns is periodic with period length at most 4. In other words, the factors starting at positions i and $i + 24$ are possibly equal when i is sufficiently large. Therefore we need to run the following procedure for two different sets of positions instead of one. These two sets are

$$S_1 = \{0, 1, \ldots, |\mathbf{p}| - 1\} \cup \{|\mathbf{p}|, |\mathbf{p}| + 1, \ldots, |\mathbf{p}| + 23\}$$

and

$$S_2 = \{0, 1, \ldots, |\mathbf{p}| - 1\} \cup \{|\mathbf{p}| + 24, |\mathbf{p}| + 25, \ldots, |\mathbf{p}| + 47\}.$$

The positions in $\{0, 1, \ldots, |\mathbf{p}| - 1\}$ represent factors starting in the prefix $\mathbf{p}$ of $\mathbf{p}\tau(\varphi(\mathbf{s}))$, while the other positions are representatives of general positions modulo 48 in the suffix $\tau(\varphi(\mathbf{s}))$. We also need to specify a set N of integers that, roughly speaking, represent the possible values that each symbolic n can take.

Let S be a set of positions, and let N be a set of integers. As in Lemma 5, the set N represents values of n such that the last letter of $\tau(\varphi(n_j))$, namely $n + d$ for some $d \in \{1, 2, 3\}$, is a letter in $\mathbf{p}\tau(\varphi(\mathbf{s}))$. We maintain classes of positions corresponding to possibly equal factors starting at those positions. Start with $\ell = 0$, for which all length-0 factors are equal. Then all positions belong to the same class S. At each step, we increase ℓ by 1, and for each position i we consider the factor of length ℓ starting at position i, extended from the previous step by one letter to the right. We break each class into new classes according to the last letter of each extended factor, as described in the following paragraph. We stop once each class contains exactly one position, because then each factor occurs at most once. Note that this procedure does not necessarily terminate, depending on the inputs. If it terminates, then the output is ℓ.

We define the *value* of each letter in Σ_8 to be its image under τ. For each class $\mathcal{I}$, we build subclasses $\mathcal{I}_c$ indexed by integers c. For each position $i \in \mathcal{I}$, we consider the extended factor of length ℓ starting at position i. If $0 \leq i \leq |\mathbf{p}| - 1$, then the new letter is in $\mathbb{Z}$ because $i + \ell - 1$ represents a particular position in $\mathbf{p}\tau(\varphi(\mathbf{s}))$. If $i \geq |\mathbf{p}|$, then the new letter is either in $\mathbb{Z}$ or symbolic in n because $i + \ell - 1$ represents all sufficiently large positions congruent to $i + \ell - 1$ modulo 48. Now there are two cases. If the new letter is an integer c, we add the position i to the class $\mathcal{I}_c$. If the new letter is $n + d$ where $d \in \{1, 2, 3\}$, then we add the position i to the class $\mathcal{I}_{n'+d}$ for each $n' \in N$. We do this for all classes $\mathcal{I}$ and we use the union

$$\bigcup_{\mathcal{I}} \{\mathcal{I}_c : c \in \mathbb{Z}\}$$

as our new set of classes in the next step.

For the sets S_1 and S_2, we use $N = \{0, 1, 2, 3, 4\}$. The prefix of $\mathbf{p}\tau(\varphi(\mathbf{s}))$ of length $|\mathbf{p}| + 952$ is a word on the alphabet $\{0, 1, \ldots, 5\}$. Therefore, since $d \in \{1, 2, 3\}$, at most the eight classes $\mathcal{I}_0, \mathcal{I}_1, \ldots, \mathcal{I}_7$ arise in each step of the procedure, since $\{0, 1, \ldots, 5\} \cup (N + \{1, 2, 3\}) = \{0, 1, \ldots, 7\}$. We wrote code that implements this procedure. For both sets, it terminates and gives $\ell = 952$. Our implementation took about 20 s for each set.

It remains to show that using the set $N = \{0, 1, 2, 3, 4\}$ is sufficient to guarantee that if the procedure terminates then each length-952 factor x starting in $\mathbf{p}$ only occurs once in $\mathbf{p}\tau(\varphi(\mathbf{s}))$. Since $\mathbf{p}\tau(\varphi(\mathbf{s}))$ is a word on the alphabet $\mathbb{N}$, it suffices to choose a subset N of $\mathbb{N} - \{1, 2, 3\} = \{-3, -2, \ldots\}$.

There exist letters in $\mathbf{p}\tau(\varphi(\mathbf{s}))$ that arise as the last letter of $\tau(\varphi(n_j))$ for arbitrarily large n, but the procedure cannot use an infinite set N. We use Lemmas 4 and 5 to show that N need not contain any integer greater than 4. Let v be the prefix of $\mathbf{p}\tau(\varphi(\mathbf{s}))$ of length $|\mathbf{p}|+952$. The procedure examines factors of $v\tau(\varphi(n_1)\varphi(n_2)\varphi(n_3)\cdots)$, which is a word on the alphabet $\{0, 1, \ldots, 5\} \cup \{n + 1, n + 2, n + 3\}$. Let $N \supseteq \{0, 1, \ldots, 5\} - \{1, 2, 3\} = \{-3, -2, \ldots, 4\}$. On the step corresponding to length ℓ in the procedure, suppose the length-ℓ factors of $v\tau(\varphi(n_1)\varphi(n_2)\varphi(n_3)\cdots)$ starting at positions i and j are possibly equal with respect to N. By Lemma 4, the two factors are possibly equal with respect to $\{-3, -2, \ldots, 4\}$. In particular, the last two letters, which have positions $i + \ell - 1$ and $j + \ell - 1$, are possibly equal with respect to $\{-3, -2, \ldots, 4\}$. By Lemma 5, there exists $c \in X_{i+\ell-1} \cap X_{j+\ell-1}$. Therefore the letters in positions $i + \ell - 1$ and $j + \ell - 1$ are both possibly equal to c with respect to $\{-3, -2, \ldots, 4\}$, so i and j are both added to $\mathcal{I}_c$.

We have shown that $N \subseteq \{-3, -2, \ldots, 4\}$ suffices. Next we remove -3 and -2. We continue to consider letters in $\mathbf{p}\tau(\varphi(\mathbf{s}))$ that arise as the last letter of $\tau(\varphi(n_j))$ for $n \in N$. Since $\tau(\mathbf{s})$ does not contain the letter -3, this implies $n + 3$ and 0 are not possibly equal with respect to the alphabet of $\tau(\mathbf{s})$, so N need not contain -3. Similarly, $\tau(\mathbf{s})$ does not contain the letter -2; therefore $n + 2$ and 0 are not possibly equal, and $n + 3$ and 1 are not possibly equal, so N need not contain -2. Therefore $N \subseteq \{-1, 0, 1, 2, 3, 4\}$ suffices.

To remove -1, we run the procedure on both sets S_1 and S_2 again but with the set $\{-1, 0, 1, 2, 3, 4\}$. However, we artificially stop the procedure at $\ell = 952$.

For the set S_1, there are 4 nonempty classes of positions remaining, namely $\{6760, 6784\}$, $\{6761, 6785\}$, $\{6762, 6786\}$, $\{6763, 6787\}$. The smallest position in each class is one of the last 4 positions in $\mathbf{p}$. As length-952 factors of $\mathbf{p}\tau(\varphi(n_2)\varphi(n_3)\cdots)$, each pair of factors starting at those positions are possibly equal with respect to N. For instance, consider the two factors

$000 \quad\quad 3\,11100 \quad\quad 3\,01101 \quad\quad 2\,01001 \quad\quad 4\,11000 \quad\quad 2\,11100 \quad\quad 2\cdots,$

$000(n+2)11100(n+1)01101(n+2)01001(n+3)11000(n+2)11100(n+3)\cdots$

starting at positions $6760, 6784$. The first factor is $0003\tau(\varphi(\mathbf{s}))$ and the second is $000(n+2)\tau(\varphi(n_6)\varphi(n_7)\cdots)$, which occurs every 48 positions in the repetition period $\tau(\varphi(n_2)\varphi(n_3)\cdots)$. For these two factors to be equal, the pair of letters 2 and $n+3$ have to be equal, and solving $2 = n+3$ gives $n = -1$. Similarly, the other three pairs of factors are only equal if the same pair of letters are equal, which again gives $n = -1$. But letters -1 only appear in $\mathbf{s}$ in its prefix $\mathbf{z}$ and only with subscripts 0 and 2 and only at the nine positions $6, 14, 16, 32, 40, 48, 56, 70, 80$. A finite check shows that the factors in each pair are different.

For S_2, there are also 4 nonempty classes of positions remaining. To show that the factors in each pair are different, we use slightly longer prefixes of $0003\tau(\varphi(\mathbf{s}))$ and $000(n+2)\tau(\varphi(n_2)\varphi(n_3)\cdots)$ than we used for S_1, and we again find a pair of letters 2 and $n+3$. This again implies $n = -1$ for each class.

Therefore we can remove -1 from N. We are left with $N \subseteq \{0, 1, 2, 3, 4\}$, so $N = \{0, 1, 2, 3, 4\}$ suffices.

5   Lexicographic-Leastness

In this section we show that $\mathbf{p}\tau(\varphi(\mathbf{s}))$ is lexicographically least by showing that decreasing any nonzero letter introduces a 5/4-power ending in that position.

Recall that, for all $e \geq 1$, the subsequence of letters with even subscripts in $\varphi^e(\mathbf{z})$ is a factor of $(0_0 0_2 1_4 1_6)^\omega$ by Lemma 1.

Lemma 6. *If 0_0 (resp., 0_2; resp., 1_4; resp., 1_6) appears in $\mathbf{s}$ at a position $i \geq 90$, then the letter at position $i - 4$ in $\mathbf{s}$ is 1_4 (resp., 1_6; resp., 0_0; resp., 0_2).*

Proposition 3. *Decreasing any nonzero letter of $\mathbf{p}\tau(\varphi(\mathbf{s}))$ introduces a 5/4-power ending in that position.*

Proof. We proceed by induction on the positions i of letters in $\mathbf{p}\tau(\varphi(\mathbf{s}))$. As a base case, since we have computed a long enough common prefix of $\mathbf{w}_{5/4}$ and $\mathbf{p}\tau(\varphi(\mathbf{s}))$, decreasing any nonzero letter of $\mathbf{p}\tau(\varphi(\mathbf{s}))$ at position $i \in \{0, 1, \ldots, 331039\}$ introduces a 5/4-power in $\mathbf{p}\tau(\varphi(\mathbf{s}))$ ending in that position.

Now suppose that $i \geq 331040$ and assume that decreasing any nonzero letter in any position $< i$ in $\mathbf{p}\tau(\varphi(\mathbf{s}))$ introduces a 5/4-power in $\mathbf{p}\tau(\varphi(\mathbf{s}))$ ending in that position. We will show that decreasing the letter in position i in $\mathbf{p}\tau(\varphi(\mathbf{s}))$ introduces a 5/4-power in $\mathbf{p}\tau(\varphi(\mathbf{s}))$ ending in that position. Since $|\mathbf{p}| = 6764$ and $|\mathbf{z}| = 20226$, observe that the letter in position $i \geq 331040$ actually belongs to the suffix $\tau(\varphi^2(\mathbf{s}))$, and $i - |\mathbf{p}\tau(\varphi(\mathbf{z}))| = i - |\mathbf{p}| - 6|\mathbf{z}|$ gives its position in $\tau(\varphi^2(\mathbf{s}))$. Recall that every such letter is a factor of $\tau(\varphi(n_j))$ for some $n \in \mathbb{N}$ and $j \in \{0, 1, \ldots, 7\}$. We make use of the array in Notation 2 of letters of φ.

If $i - |\mathbf{p}| - 6|\mathbf{z}|$ is not congruent to 5 modulo 6, then the letter in position i belongs to one of the first five columns. Any 0 letters cannot be decreased. Observe that the fourth column is made of letters 0. Since the second column contains only letters 1, decreasing any letter 1 to 0 in the second column produces a new 5/4-power of length 5 between the fourth and second columns. Since the even-subscript letters in $\varphi(\mathbf{s})$ form the word $(1_4 1_6 0_0 0_2)^\omega$, then decreasing any letter 1 to 0 in the first, third or fifth column introduces a 5/4-power of length 5 of the form $0y0$.

It remains to consider positions i such that $i - |\mathbf{p}| - 6|\mathbf{z}| \equiv 5 \mod 6$, that is, letters in the sixth column. By Part 5 of Lemma 1, note that the letters in the sixth column belong to $\{n_j \mid n \geq 2, 0 \leq j \leq 7\}$, that is, their values are greater than or equal to 2. If we decrease the letter in position i to 0, then we create one of the following 5/4-powers of length 10:

$$
\begin{array}{ll}
10 \cdot 1(n+2)0100 \cdot 10, & 10 \cdot 1(n+2)0100 \cdot 10, \\
00 \cdot 1(n+3)1100 \cdot 00, & 00 \cdot 1(n+1)1100 \cdot 00, \\
00 \cdot 0(n+2)1110 \cdot 00, & 00 \cdot 0(n+2)1110 \cdot 00, \\
10 \cdot 0(n+3)0110 \cdot 10, & 10 \cdot 0(n+1)0110 \cdot 10.
\end{array}
$$

If we decrease the letter in position i to 1, then we create a new 5/4-power of length 5 because each letter in the second column is 1.

It remains to show that decreasing the letter in position i to some letter c with $c \geq 2$ introduces a 5/4-power ending in that position. Intuitively, this operation corresponds, under $\tau \circ \varphi$, to decreasing a letter n_j of $\varphi(\mathbf{s})$ to $(c - d)_j$ with

$0 \leq j \leq 7$ and $d \in \{1, 2, 3\}$. To be more precise, the last letter of $\tau(\varphi((c-d)_j))$ is c. We examine three cases according to the value of d.

Case 1. If $d = 1$, then the corresponding letter in position i in $\varphi^2(\mathbf{s})$ is of the form $(n+1)_1$ or $(n+1)_5$. From Notation 2, we see that $(n+1)_1$ and $(n+1)_5$ appear in the images of n_4 and n_6 under φ. By Part 4 of Lemma 1, the only letters with subscripts 4 and 6 in $\varphi(\mathbf{s})$ are 1_4 and 1_6. The images of these letters under $\tau \circ \varphi$ contain $\tau((n+1)_1) = 2$ and $\tau((n+1)_5) = 2$. So there is nothing to check since the value of the letters is too small.

Case 3. If $d = 3$, then the corresponding letter in position i in $\varphi^2(\mathbf{s})$ is of the form $(n+3)_1$ or $(n+3)_5$, which appear in the images of n_0 and n_2 under φ. From Part 4 of Lemma 1, the only letters with subscripts 0 or 2 in $\varphi(\mathbf{s})$ are 0_0 and 0_2, so the only relevant value of c is 2. Lemma 6 tells us that decreasing the value of a letter $(n+3)_1$ or $(n+3)_5$ to $c = 2$ introduces one of the following $5/4$-powers of length 30:

$$\tau(\varphi(1_4 n_5 \bar{n}_6 \bar{\bar{n}}_7))01001(3-1) = 010012\tau(\varphi(n_5 \bar{n}_6 \bar{\bar{n}}_7))010012,$$
$$\tau(\varphi(1_6 n_7 \bar{n}_0 \bar{\bar{n}}_1))11100(3-1) = 111002\tau(\varphi(n_7 \bar{n}_0 \bar{\bar{n}}_1))111002.$$

Case 2. If $d = 2$, then the corresponding letter in position i in $\varphi^2(\mathbf{s})$ is of the form $(n+2)_3$ or $(n+2)_7$, which appear in the images of n_1, n_3, n_5, and n_7 under φ. Let w be the length-$(i - |\mathbf{p}| - 6|\mathbf{z}| + 1)$ prefix of $\tau(\varphi^2(\mathbf{s}))$ with last letter $n+2$. Since $i - |\mathbf{p}| - 6|\mathbf{z}| \equiv 5 \mod 6$, let u be the prefix of $\varphi(\mathbf{s})$ of length $\frac{i - |\mathbf{p}| - 6|\mathbf{z}| + 1}{6}$. Then $\tau(\varphi(u)) = w$ and $\tau(u)$ ends with n as pictured.

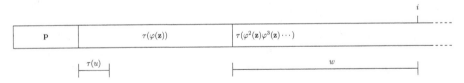

Let w' be the word obtained by decreasing the last letter $n+2$ to c in w, and let u' be the word obtained by decreasing the last letter with value n of u to $c - 2$. Then $\tau(\varphi(u')) = w'$. By the induction hypothesis, $\mathbf{p}\tau(u')$ contains a $5/4$-power suffix xyx. Now we consider two subcases, depending on where xyx starts in $\mathbf{p}\tau(u')$ as depicted below. (We show that the middle case does not actually occur).

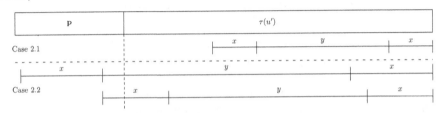

Case 2.1. Suppose that xyx starts after $\mathbf{p}$ in $\mathbf{p}\tau(u')$, that is, xyx is a suffix of $\tau(u')$. Write $x = \tau(x') = \tau(x'')$ and $y = \tau(y')$ where $x'y'x''$ is the corresponding subscript-increasing suffix of u'. Since $|xy|$ is divisible by 4, the subscripts of x'

and x'' are either equal or differ by 4. Since $d = 2$, the subscripts of the last letters of x' and x'' are odd. If $|x| = 1$, then $x = c - 2$ and $\varphi(x') = \varphi(x'')$ by definition of φ. Now if $|x| \geq 2$, the subscripts of the penultimate letters of x' and x'' are even and either equal each other or differ by 4. They cannot differ by 4; otherwise $\tau(x') \neq \tau(x'')$ since the subsequence of letters with even subscripts in $\varphi(\mathbf{s})$ is a factor of $(0_0 0_2 1_4 1_6)^\omega$. So they must be equal and then $\varphi(x') = \varphi(x'')$. Then $\mathbf{p}\tau(\varphi(\mathbf{z}))w' = \mathbf{p}\tau(\varphi(\mathbf{z}))\tau(\varphi(u'))$ contains the 5/4-power $\tau(\varphi(x'x''))$ as a suffix. Therefore, decreasing that letter $n + 2$ to c introduces a 5/4-power in $\mathbf{p}\tau(\varphi(\mathbf{s}))$ ending in position i.

Case 2.2. Suppose that xyx starts before $\tau(u')$ in $\mathbf{p}\tau(u')$. In particular, $\tau(u')$ is a suffix of xyx. Since $i \geq 331040$, the first x is not a factor of $\mathbf{p}$. Otherwise, we have

$$\frac{i - |\mathbf{p}| - 6|\mathbf{z}| + 1}{6} = |u| = |\tau(u')| \leq |xyx| = 5|x| \leq 5|\mathbf{p}|,$$

which implies $i \leq 31|\mathbf{p}| + 6|\mathbf{z}| - 1 = 331039$. Therefore, the first x overlaps $\mathbf{p}$ but is not a factor of $\mathbf{p}$. Suppose the overlap length is at least 5. Then the first x contains 20003 as a factor. But since 20003 is never a factor of $\tau \circ \varphi$, this case does not happen. Now suppose that the overlap length is less than or equal to 4. Then the first factor $x = sv$ is made of a nonempty suffix s of 0003 followed by a nonempty prefix v of $\tau(u')$ such that $vyx = \tau(u')$. Write $v = \tau(v')$, $x = \tau(x'')$ and $y = \tau(y')$ where $v'y'x'' = u'$. To get around the fact that $\mathbf{p}$ does not have subscripts, we use $\mathbf{z}$ instead. Recall that 0003 is a common suffix of $\mathbf{p}$ and $\tau(\mathbf{z})$, and the corresponding suffix in $\mathbf{z}$ is $0_0 0_1 0_2 3_3$. So let s' be the suffix of $0_0 0_1 0_2 3_3$ such that $s = \tau(s')$. Now observe that $x' = s'v'$ is a subscript-increasing factor of $\mathbf{z}u'$, overlapping $\mathbf{z}$. Thus $\varphi(x'y'x'')$ is a subscript-increasing suffix of $\varphi(\mathbf{z}u')$, overlapping $\varphi(\mathbf{z})$. Similarly to Case 2.1, since $\tau(x') = \tau(s'v') = sv = x = \tau(x'')$ and $|x| \geq 2$, the subscripts of x' and x'' are equal. Consequently, $x' = x''$ and $\varphi(x') = \varphi(x'')$. Then $\mathbf{p}\tau(\varphi(\mathbf{z}))w' = \mathbf{p}\tau(\varphi(\mathbf{z}u'))$ contains the 5/4-power $\tau(\varphi(x'y'x''))$ as a suffix. Therefore, decreasing that letter $n + 2$ to c introduces a 5/4-power in $\mathbf{p}\tau(\varphi(\mathbf{s}))$ ending in that position i.

References

1. Guay-Paquet, M., Shallit, J.: Avoiding squares and overlaps over the natural numbers. Discrete Math. **309**(21), 6245–6254 (2009)
2. Pudwell, L., Rowland, E.: Avoiding fractional powers over the natural numbers. Electron. J. Comb. **25**(2), Paper 2.27 (2018). 46 pages
3. Rowland, E., Shallit, J.: Avoiding 3/2-powers over the natural numbers. Discrete Math. **312**(6), 1282–1288 (2012)
4. Rowland, E., Stipulanti, M.: Avoiding 5/4-powers on the alphabet of non-negative integers. https://arxiv.org/abs/2005.03158

Transition Property for α-Power Free Languages with $\alpha \geq 2$ and $k \geq 3$ Letters

Josef Rukavicka[✉]

Department of Mathematics, Faculty of Nuclear Sciences and Physical Engineering,
Czech Technical University in Prague, Prague, Czech Republic
josef.rukavicka@seznam.cz

Abstract. In 1985, Restivo and Salemi presented a list of five problems concerning power free languages. Problem 4 states: Given α-power-free words u and v, decide whether there is a transition from u to v. Problem 5 states: Given α-power-free words u and v, find a transition word w, if it exists.

Let Σ_k denote an alphabet with k letters. Let $L_{k,\alpha}$ denote the α-power free language over the alphabet Σ_k, where α is a rational number or a rational "number with $+$". If α is a "number with $+$" then suppose $k \geq 3$ and $\alpha \geq 2$. If α is "only" a number then suppose $k = 3$ and $\alpha > 2$ or $k > 3$ and $\alpha \geq 2$. We show that: If $u \in L_{k,\alpha}$ is a right extendable word in $L_{k,\alpha}$ and $v \in L_{k,\alpha}$ is a left extendable word in $L_{k,\alpha}$ then there is a (transition) word w such that $uwv \in L_{k,\alpha}$. We also show a construction of the word w.

Keywords: Power free languages · Transition property · Dejean's conjecture

1 Introduction

The power free words are one of the major themes in the area of combinatorics on words. An α-*power* of a word r is the word $r^\alpha = rr \ldots rt$ such that $\frac{|r^\alpha|}{|r|} = \alpha$ and t is a prefix of r, where $\alpha \geq 1$ is a rational number. For example $(1234)^3 = 123412341234$ and $(1234)^{\frac{7}{4}} = 1234123$. We say that a finite or infinite word w is α-*power free* if w has no factors that are β-powers for $\beta \geq \alpha$ and we say that a finite or infinite word w is α^+-power free if w has no factors that are β-powers for $\beta > \alpha$, where $\alpha, \beta \geq 1$ are rational numbers. In the following, when we write "α-power free" then α denotes a number or a "number with $+$". The power free words, also called repetitions free words, include well known square free (2-power free), overlap free (2^+-power free), and cube free words (3-power free). Two surveys on the topic of power free words can be found in [8] and [13].

One of the questions being researched is the construction of infinite power free words. We define the *repetition threshold* $\mathrm{RT}(k)$ to be the infimum of all rational numbers α such that there exists an infinite α-power-free word over an alphabet with k letters. Dejean's conjecture states that $\mathrm{RT}(2) = 2$, $\mathrm{RT}(3) = \frac{7}{4}$,

© Springer Nature Switzerland AG 2020
N. Jonoska and D. Savchuk (Eds.): DLT 2020, LNCS 12086, pp. 294–303, 2020.
https://doi.org/10.1007/978-3-030-48516-0_22

$RT(4) = \frac{7}{5}$, and $RT(k) = \frac{k}{k-1}$ for each $k > 4$ [3]. Dejean's conjecture has been proved with the aid of several articles [1–3,5,6,9].

It is easy to see that α-power free words form a factorial language [13]; it means that all factors of a α-power free word are also α-power free words. Then Dejean's conjecture implies that there are infinitely many finite α-power free words over Σ_k, where $\alpha > RT(k)$.

In [10], Restivo and Salemi presented a list of five problems that deal with the question of extendability of power free words. In the current paper we investigate Problem 4 and Problem 5:

- Problem 4: Given α-power-free words u and v, decide whether there is a transition word w, such that uwu is α-power free.
- Problem 5: Given α-power-free words u and v, find a transition word w, if it exists.

A recent survey on the progress of solving all the five problems can be found in [7]; in particular, the problems 4 and 5 are solved for some overlap free (2^+-power free) binary words. In addition, in [7] the authors prove that: For every pair (u, v) of cube free words (3-power free) over an alphabet with k letters, if u can be infinitely extended to the right and v can be infinitely extended to the left respecting the cube-freeness property, then there exists a "transition" word w over the same alphabet such that uwv is cube free.

In 2009, a conjecture related to Problems 4 and Problem 5 of Restivo and Salemi appeared in [12]:

Conjecture 1. [12, Conjecture 1] Let L be a power-free language and let $e(L) \subseteq L$ be the set of words of L that can be extended to a bi-infinite word respecting the given power-freeness. If $u, v \in e(L)$ then $uwv \in e(L)$ for some word w.

In 2018, Conjecture 1 was presented also in [11] in a slightly different form.

Let $\mathbb{N}$ denote the set of natural numbers and let $\mathbb{Q}$ denote the set of rational numbers.

Definition 1. *Let*

$$\Upsilon = \{(k, \alpha) \mid k \in \mathbb{N} \text{ and } \alpha \in \mathbb{Q} \text{ and } k = 3 \text{ and } \alpha > 2\}$$
$$\cup \{(k, \alpha) \mid k \in \mathbb{N} \text{ and } \alpha \in \mathbb{Q} \text{ and } k > 3 \text{ and } \alpha \geq 2\}$$
$$\cup \{(k, \alpha^+) \mid k \in \mathbb{N} \text{ and } \alpha \in \mathbb{Q} \text{ and } k \geq 3 \text{ and } \alpha \geq 2\}.$$

Remark 1. The definition of Υ says that: If $(k, \alpha) \in \Upsilon$ and α is a "number with +" then $k \geq 3$ and $\alpha \geq 2$. If $(k, \alpha) \in \Upsilon$ and α is "just" a number then $k = 3$ and $\alpha > 2$ or $k > 3$ and $\alpha \geq 2$.

Let L be a language. A finite word $w \in L$ is called *left extendable* (resp., *right extendable*) in L if for every $n \in \mathbb{N}$ there is a word $u \in L$ with $|u| = n$ such that $uw \in L$ (resp., $wu \in L$).

In the current article we improve the results addressing Problems 4 and Problem 5 of Restivo and Salemi from [7] as follows. Let Σ_k denote an alphabet with k letters. Let $L_{k,\alpha}$ denote the α-power free language over the alphabet Σ_k. We show that if $(k, \alpha) \in \Upsilon$, $u \in L_{k,\alpha}$ is a right extendable word in $L_{k,\alpha}$, and $v \in L_{k,\alpha}$ is a left extendable word in $L_{k,\alpha}$ then there is a word w such that $uwv \in L_{k,\alpha}$. We also show a construction of the word w.

We sketch briefly our construction of a "transition" word. Let u be a right extendable α-power free word and let v be a left extendable α-power free word over Σ_k with $k > 2$ letters. Let $\bar{u}$ be a right infinite α-power free word having u as a prefix and let $\bar{v}$ be a left infinite α-power free word having v as a suffix. Let x be a letter that is recurrent in both $\bar{u}$ and $\bar{v}$. We show that we may suppose that $\bar{u}$ and $\bar{v}$ have a common recurrent letter. Let t be a right infinite α-power free word over $\Sigma_k \setminus \{x\}$. Let $\bar{t}$ be a left infinite α-power free word such that the set of factors of $\bar{t}$ is a subset of the set of recurrent factors of t. We show that such $\bar{t}$ exists. We identify a prefix $\tilde{u}xg$ of $\bar{u}$ such that g is a prefix of t and $\tilde{u}xt$ is a right infinite α-power free word. Analogously we identify a suffix $\bar{g}x\tilde{v}$ of $\bar{v}$ such that $\bar{g}$ is a suffix of $\bar{t}$ and $\bar{t}x\tilde{v}$ is a left infinite α-power free word. Moreover our construction guarantees that u is a prefix of $\tilde{u}xt$ and v is a suffix of $\bar{t}x\tilde{v}$. Then we find a prefix hp of t such that $px\tilde{v}$ is a suffix of $\bar{t}x\tilde{v}$ and such that both h and p are "sufficiently long". Then we show that $\tilde{u}xhpx\tilde{v}$ is an α-power free word having u as a prefix and v as a suffix.

The very basic idea of our proof is that if u, v are α-power free words and x is a letter such that x is not a factor of both u and v, then clearly uxv is α-power free on condition that $\alpha \geq 2$. Just note that there cannot be a factor in uxv which is an α-power and contains x, because x has only one occurrence in uxv. Our constructed words $\tilde{u}xt$, $\bar{t}x\tilde{v}$, and $\tilde{u}xhpx\tilde{v}$ have "long" factors which does not contain a letter x. This will allow us to apply a similar approach to show that the constructed words do not contain square factor rr such that r contains the letter x.

Another key observation is that if $k \geq 3$ and $\alpha > \mathrm{RT}(k-1)$ then there is an infinite α-power free word $\bar{w}$ over $\Sigma_k \setminus \{x\}$, where $x \in \Sigma_k$. This is an implication of Dejean's conjecture. Less formally said, if u, v are α-power free words over an alphabet with k letters, then we construct a "transition" word w over an alphabet with $k - 1$ letters such that uwv is α-power free.

Dejean's conjecture imposes also the limit to possible improvement of our construction. The construction cannot be used for $\mathrm{RT}(k) \leq \alpha < \mathrm{RT}(k-1)$, where $k \geq 3$, because every infinite (or "sufficiently long") word w over an alphabet with $k - 1$ letters contains a factor which is an α-power. Also for $k = 2$ and $\alpha \geq 1$ our technique fails. On the other hand, based on our research, it seems that our technique, with some adjustments, could be applied also for $\mathrm{RT}(k-1) \leq \alpha \leq 2$ and $k \geq 3$. Moreover it seems to be possible to generalize our technique to bi-infinite words and consequently to prove Conjecture 1 for $k \geq 3$ and $\alpha \geq \mathrm{RT}(k-1)$.

2   Preliminaries

Recall that Σ_k denotes an alphabet with k letters. Let ϵ denote the empty word. Let Σ_k^* denote the set of all finite words over Σ_k including the empty word ϵ, let

$\Sigma_k^{\mathrm{N},R}$ denote the set of all right infinite words over Σ_k, and let $\Sigma_k^{\mathrm{N},L}$ denote the set of all left infinite words over Σ_k. Let $\Sigma_k^{\mathrm{N}} = \Sigma_k^{\mathrm{N},L} \cup \Sigma_k^{\mathrm{N},R}$. We call $w \in \Sigma_k^{\mathrm{N}}$ an infinite word.

Let $\mathrm{occur}(w,t)$ denote the number of occurrences of the nonempty factor $t \in \Sigma_k^* \setminus \{\epsilon\}$ in the word $w \in \Sigma_k^* \cup \Sigma_k^{\mathrm{N}}$. If $w \in \Sigma_k^{\mathrm{N}}$ and $\mathrm{occur}(w,t) = \infty$, then we call t a *recurrent* factor in w.

Let $\mathrm{F}(w)$ denote the set of all finite factors of a finite or infinite word $w \in \Sigma_k^* \cup \Sigma_k^{\mathrm{N}}$. The set $\mathrm{F}(w)$ contains the empty word and if w is finite then also $w \in \mathrm{F}(w)$. Let $\mathrm{F}_r(w) \subseteq \mathrm{F}(w)$ denote the set of all recurrent nonempty factors of $w \in \Sigma_k^{\mathrm{N}}$.

Let $\mathrm{Prf}(w) \subseteq \mathrm{F}(w)$ denote the set of all prefixes of $w \in \Sigma_k^* \cup \Sigma_k^{\mathrm{N},R}$ and let $\mathrm{Suf}(w) \subseteq \mathrm{F}(w)$ denote the set of all suffixes of $w \in \Sigma_k^* \cup \Sigma_k^{\mathrm{N},L}$. We define that $\epsilon \in \mathrm{Prf}(w) \cap \mathrm{Suf}(w)$ and if w is finite then also $w \in \mathrm{Prf}(w) \cap \mathrm{Suf}(w)$.

We have that $\mathrm{L}_{k,\alpha} \subseteq \Sigma_k^*$. Let $\mathrm{L}_{k,\alpha}^{\mathrm{N}} \subseteq \Sigma_k^{\mathrm{N}}$ denote the set of all infinite α-power free words over Σ_k. Obviously $\mathrm{L}_{k,\alpha}^{\mathrm{N}} = \{w \in \Sigma_k^{\mathrm{N}} \mid \mathrm{F}(w) \subseteq \mathrm{L}_{k,\alpha}\}$. In addition we define $\mathrm{L}_{k,\alpha}^{\mathrm{N},R} = \mathrm{L}_{k,\alpha}^{\mathrm{N}} \cap \Sigma_k^{\mathrm{N},R}$ and $\mathrm{L}_{k,\alpha}^{\mathrm{N},L} = \mathrm{L}_{k,\alpha}^{\mathrm{N}} \cap \Sigma_k^{\mathrm{N},L}$; it means the sets of right infinite and left infinite α-power free words.

3   Power Free Languages

Let $(k,\alpha) \in \Upsilon$ and let u, v be α-power free words. The first lemma says that uv is α-power free if there are no word r and no nonempty prefix $\bar{v}$ of v such that rr is a suffix of $u\bar{v}$ and rr is longer than $\bar{v}$.

Lemma 1. *Suppose* $(k,\alpha) \in \Upsilon$, $u \in \mathrm{L}_{k,\alpha}$, *and* $v \in \mathrm{L}_{k,\alpha} \cup \mathrm{L}_{k,\alpha}^{\mathrm{N},R}$. *Let*

$$\Pi = \{(r,\bar{v}) \mid r \in \Sigma_k^* \setminus \{\epsilon\} \text{ and } \bar{v} \in \mathrm{Prf}(v) \setminus \{\epsilon\} \text{ and}$$
$$rr \in \mathrm{Suf}(u\bar{v}) \text{ and } |rr| > |\bar{v}|\}.$$

If $\Pi = \emptyset$ *then* $uv \in \mathrm{L}_{k,\alpha} \cup \mathrm{L}_{k,\alpha}^{\mathrm{N},R}$.

Proof. Suppose that uv is not α-power free. Since u is α-power free, then there are $t \in \Sigma_k^*$ and $x \in \Sigma_k$ such that $tx \in \mathrm{Prf}(v)$, $ut \in \mathrm{L}_{k,\alpha}$ and $utx \notin \mathrm{L}_{k,\alpha}$. It means that there is $r \in \mathrm{Suf}(utx)$ such that $r^\beta \in \mathrm{Suf}(utx)$ for some $\beta \geq \alpha$ or $\beta > \alpha$ if α is a "number with $+$"; recall Definition 1 of Υ. Because $\alpha \geq 2$, this implies that $rr \in \mathrm{Suf}(r^\beta)$. If follows that $(tx,r) \in \Pi$. We proved that $uv \notin \mathrm{L}_{k,\alpha} \cup \mathrm{L}_{k,\alpha}^{\mathrm{N},R}$ implies that $\Pi \neq \emptyset$. The lemma follows. $\square$

The following technical set $\Gamma(k,\alpha)$ of 5-tuples (w_1, w_2, x, g, t) will simplify our propositions.

Definition 2. *Given* $(k,\alpha) \in \Upsilon$, *we define that* $(w_1, w_2, x, g, t) \in \Gamma(k,\alpha)$ *if*

1. $w_1, w_2, g \in \Sigma_k^*$,
2. $x \in \Sigma_k$,
3. $w_1 w_2 x g \in \mathrm{L}_{k,\alpha}$,

4. $t \in L_{k,\alpha}^{N,R}$,
5. $\text{occur}(t, x) = 0$,
6. $g \in \text{Prf}(t)$,
7. $\text{occur}(w_2 x g y, x g y) = 1$, *where $y \in \Sigma_k$ is such that $g y \in \text{Prf}(t)$, and*
8. $\text{occur}(w_2, x) \geq \text{occur}(w_1, x)$.

Remark 2. Less formally said, the 5-tuple (w_1, w_2, x, g, t) is in $\Gamma(k, \alpha)$ if $w_1 w_2 x g$ is α-power free word over Σ_k, t is a right infinite α-power free word over Σ_k, t has no occurrence of x (thus t is a word over $\Sigma_k \setminus \{x\}$), g is a prefix of t, $x g y$ has only one occurrence in $w_2 x g y$, where y is a letter such that $g y$ is a prefix of t, and the number of occurrences of x in w_2 is bigger than the number of occurrences of x in w_1, where w_1, w_2, g are finite words and x is a letter.

The next proposition shows that if (w_1, w_2, x, g, t) is from the set $\Gamma(k, \alpha)$ then $w_1 w_2 x t$ is a right infinite α-power free word, where (k, α) is from the set Υ.

Proposition 1. *If $(k, \alpha) \in \Upsilon$ and $(w_1, w_2, x, g, t) \in \Gamma(k, \alpha)$ then $w_1 w_2 x t \in L_{k,\alpha}^{N,R}$.*

Proof. Lemma 1 implies that it suffices to show that there are no $u \in \text{Prf}(t)$ with $|u| > |g|$ and no $r \in \Sigma_k^* \setminus \{\epsilon\}$ such that $rr \in \text{Suf}(w_1 w_2 x u)$ and $|rr| > |u|$. Recall that $w_1 w_2 x g$ is an α-power free word, hence we consider $|u| > |g|$. To get a contradiction, suppose that such r, u exist. We distinguish the following distinct cases.

- If $|r| \leq |u|$ then: Since $u \in \text{Prf}(t) \subseteq L_{k,\alpha}$ it follows that $xu \in \text{Suf}(r^2)$ and hence $x \in F(r^2)$. It is clear that $\text{occur}(r^2, x) \geq 1$ if and only if $\text{occur}(r, x) \geq 1$. Since $x \notin F(u)$ and thus $x \notin F(r)$, this is a contradiction.
- If $|r| > |u|$ and $rr \in \text{Suf}(w_2 x u)$ then: Let $y \in \Sigma_k$ be such that $gy \in \text{Prf}(t)$. Since $|u| > |g|$ we have that $gy \in \text{Prf}(u)$ and $xgy \in \text{Prf}(xu)$. Since $|r| > |u|$ we have that $xgy \in F(r)$. In consequence $\text{occur}(rr, xgy) \geq 2$. But Property 7 of Definition 2 states that $\text{occur}(w_2 x g y, x g y) = 1$. Since $rr \in \text{Suf}(w_2 x u)$, this is a contradiction.
- If $|r| > |u|$ and $rr \notin \text{Suf}(w_2 x u)$ and $r \in \text{Suf}(w_2 x u)$ then: Let $w_{11}, w_{12}, w_{13}, w_{21}, w_{22} \in \Sigma_k^*$ be such that $w_1 = w_{11} w_{12} w_{13}$, $w_2 = w_{21} w_{22}$, $w_{12} w_{13} w_{21} = r$, $w_{12} w_{13} w_2 x u = rr$, and $w_{13} w_{21} = xu$; see Figure below.

		xu				
w_{11}	w_{12}	w_{13}	w_{21}	w_{22}	x	u
		r		r		

It follows that $w_{22} x u = r$ and $w_{22} = w_{12}$. It is easy to see that $w_{13} w_{21} = xu$. From $\text{occur}(u, x) = 0$ we have that $\text{occur}(w_2, x) = \text{occur}(w_{22}, x)$ and $\text{occur}(w_{13}, x) = 1$. From $w_{22} = w_{12}$ it follows that $\text{occur}(w_1, x) > \text{occur}(w_2, x)$. This is a contradiction to Property 8 of Definition 2.

– If $|r| > |u|$ and $rr \notin \mathrm{Suf}(w_2xu)$ and $r \notin \mathrm{Suf}(w_2xu)$ then: Let $w_{11}, w_{12}, w_{13} \in \Sigma_k^*$ be such that $w_1 = w_{11}w_{12}w_{13}$, $w_{12} = r$ and $w_{13}w_2xu = r$; see Figure below.

w_{11}	w_{12}	w_{13}	w_2	x	u
	r		r		

It follows that

$$\mathrm{occur}(w_{12}, x) = \mathrm{occur}(w_{13}, x) + \mathrm{occur}(w_2, x) + \mathrm{occur}(xu, x).$$

This is a contradiction to Property 8 of Definition 2.

We proved that the assumption of existence of r, u leads to a contradiction. Thus we proved that for each prefix $u \in \mathrm{Prf}(t)$ we have that $w_1w_2xu \in \mathrm{L}_{k,\alpha}$. The proposition follows. $\qquad\square$

We prove that if $(k, \alpha) \in \Upsilon$ then there is a right infinite α-power free word over Σ_{k-1}. In the introduction we showed that this observation could be deduced from Dejean's conjecture. Here additionally, to be able to address Problem 5 from the list of Restivo and Salemi, we present in the proof also examples of such words.

Lemma 2. *If $(k, \alpha) \in \Upsilon$ then the set $\mathrm{L}_{k-1,\alpha}^{N,R}$ is not empty.*

Proof. If $k = 3$ then $|\Sigma_{k-1}| = 2$. It is well known that the Thue Morse word is a right infinite 2^+-power free word over an alphabet with 2 letters [11]. It follows that the Thue Morse word is α-power free for each $\alpha > 2$.

If $k > 3$ then $|\Sigma_{k-1}| \geq 3$. It is well known that there are infinite 2-power free words over an alphabet with 3 letters [11]. Suppose $0, 1, 2 \in \Sigma_k$. An example is the fixed point of the morphism θ defined by $\theta(0) = 012$, $\theta(1) = 02$, and $\theta(2) = 1$ [11]. If an infinite word t is 2-power free then obviously t is α-power free and α^+-power free for each $\alpha \geq 2$.

This completes the proof. $\qquad\square$

We define the sets of extendable words.

Definition 3. *Let $\mathrm{L} \subseteq \Sigma_k^*$. We define*

$$\mathrm{lext}(\mathrm{L}) = \{w \in \mathrm{L} \mid w \text{ is left extendable in } \mathrm{L}\}$$

and

$$\mathrm{rext}(\mathrm{L}) = \{w \in \mathrm{L} \mid w \text{ is right extendable in } \mathrm{L}\}.$$

If $u \in \mathrm{lext}(\mathrm{L})$ then let $\mathrm{lext}(u, \mathrm{L})$ be the set of all left infinite words $\bar{u}$ such that $\mathrm{Suf}(\bar{u}) \subseteq \mathrm{L}$ and $u \in \mathrm{Suf}(\bar{u})$. Analogously if $u \in \mathrm{rext}(\mathrm{L})$ then let $\mathrm{rext}(u, \mathrm{L})$ be the set of all right infinite words $\bar{u}$ such that $\mathrm{Prf}(\bar{u}) \subseteq \mathrm{L}$ and $u \in \mathrm{Prf}(\bar{u})$.

We show the sets $\text{lext}(u, \mathrm{L})$ and $\text{rext}(v, \mathrm{L})$ are nonempty for left extendable and right extendable words.

Lemma 3. *If* $\mathrm{L} \subseteq \Sigma_k^*$ *and* $u \in \text{lext}(\mathrm{L})$ *(resp.,* $v \in \text{rext}(\mathrm{L})$*) then* $\text{lext}(u, \mathrm{L}) \neq \emptyset$ *(resp.,* $\text{rext}(v, \mathrm{L}) \neq \emptyset$*).*

Proof. Realize that $u \in \text{lext}(\mathrm{L})$ (resp., $v \in \text{rext}(\mathrm{L})$) implies that there are infinitely many finite words in L having u as a suffix (resp., v as a prefix). Then the lemma follows from König's Infinity Lemma [4,8].   □

The next proposition proves that if $(k, \alpha) \in \Upsilon$, w is a right extendable α-power free word, $\bar{w}$ is a right infinite α-power free word having the letter x as a recurrent factor and having w as a prefix, and t is a right infinite α-power free word over $\Sigma_k \backslash \{x\}$, then there are finite words w_1, w_2, g such that the 5-tuple (w_1, w_2, x, g, t) is in the set $\Gamma(k, \alpha)$ and w is a prefix of $w_1 w_2 x g$.

Proposition 2. *If* $(k, \alpha) \in \Upsilon$, $w \in \text{rext}(\mathrm{L}_{k,\alpha})$, $\bar{w} \in \text{rext}(w, \mathrm{L}_{k,\alpha})$, $x \in \mathrm{F}_r(\bar{w}) \cap \Sigma_k$, $t \in \mathrm{L}_{k,\alpha}^{\mathbb{N},R}$, *and* $\text{occur}(t, x) = 0$ *then there are finite words* w_1, w_2, g *such that* $(w_1, w_2, x, g, t) \in \Gamma(k, \alpha)$ *and* $w \in \text{Prf}(w_1 w_2 x g)$.

Proof. Let $\omega = \mathrm{F}(\bar{w}) \cap \text{Prf}(xt)$ be the set of factors of $\bar{w}$ that are also prefixes of the word xt. Based on the size of the set ω we construct the words w_1, w_2, g and we show that $(w_1, w_2, x, g, t) \in \Gamma(k, \alpha)$ and $w_1 w_2 x g \in \text{Prf}(\bar{w}) \subseteq \mathrm{L}_{k,\alpha}$. The Properties 1, 2, 3, 4, 5, and 6 of Definition 2 are easy to verify. Hence we explicitly prove only properties 7 and 8 and that $w \in \text{Prf}(w_1 w_2 x g)$.

- If ω is an infinite set. It follows that $\text{Prf}(xt) = \omega$. Let $g \in \text{Prf}(t)$ be such that $|g| = |w|$; recall that t is infinite and hence such g exists. Let $w_2 \in \text{Prf}(\bar{w})$ be such that $w_2 x g \in \text{Prf}(\bar{w})$ and $\text{occur}(w_2 x g, x g) = 1$. Let $w_1 = \epsilon$.
  Property 7 of Definition 2 follows from $\text{occur}(w_2 x g, x g) = 1$. Property 8 of Definition 2 is obvious, because w_1 is the empty word. Since $|g| = |w|$ and $w \in \text{Prf}(\bar{w})$ we have that $w \in \text{Prf}(w_1 w_2 x g)$.
- If ω is a finite set. Let $\bar{\omega} = \omega \cap \mathrm{F}_r(\bar{w})$ be the set of prefixes of xt that are recurrent in $\bar{w}$. Since x is recurrent in $\bar{w}$ we have that $x \in \bar{\omega}$ and thus $\bar{\omega}$ is not empty. Let $g \in \text{Prf}(t)$ be such that xg is the longest element in $\bar{\omega}$. Let $w_1 \in \text{Prf}(w)$ be the shortest prefix of $\bar{w}$ such that if $u \in \omega \backslash \bar{\omega}$ is a non-recurrent prefix of xt in $\bar{w}$ then $\text{occur}(w_1, u) = \text{occur}(\bar{w}, u)$. Such w_1 obviously exists, because ω is a finite set and non-recurrent factors have only a finite number of occurrences. Let w_2 be the shortest factor of $\bar{w}$ such that $w_1 w_2 x g \in \text{Prf}(\bar{w})$, $\text{occur}(w_1, x) < \text{occur}(w_2, x)$, and $w \in \text{Prf}(w_1 w_2 x g)$. Since xg is recurrent in $\bar{w}$ and $w \in \text{Prf}(\bar{w})$ it is clear such w_2 exists.
  We show that Property 7 of Definition 2 holds. Let $y \in \Sigma_k$ be such that $gy \in \text{Prf}(t)$. Suppose that $\text{occur}(w_2 x g, x g y) > 0$. It would imply that xgy is recurrent in $\bar{w}$, since all occurrences of non-recurrent words from ω are in w_1. But we defined xg to be the longest recurrent word ω. Hence it is contradiction to our assumption that $\text{occur}(w_2 x g, x g y) > 0$.
  Property 8 of Definition 2 and $w \in \text{Prf}(w_1 w_2 x g)$ are obvious from the construction of w_2.

This completes the proof. □

We define the *reversal* w^R of a finite or infinite word $w = \Sigma_k^* \cup \Sigma_k^N$ as follows: If $w \in \Sigma_k^*$ and $w = w_1 w_2 \ldots w_m$, where $w_i \in \Sigma_k$ and $1 \leq i \leq m$, then $w^R = w_m w_{m-1} \ldots w_2 w_1$. If $w \in \Sigma_k^{N,L}$ and $w = \ldots w_2 w_1$, where $w_i \in \Sigma_k$ and $i \in \mathbb{N}$, then $w^R = w_1 w_2 \cdots \in \Sigma_k^{N,R}$. Analogously if $w \in \Sigma_k^{N,R}$ and $w = w_1 w_2 \ldots$, where $w_i \in \Sigma_k$ and $i \in \mathbb{N}$, then $w^R = \ldots w_2 w_1 \in \Sigma_k^{N,L}$.

Proposition 1 allows one to construct a right infinite α-power free word with a given prefix. The next simple corollary shows that in the same way we can construct a left infinite α-power free word with a given suffix.

Corollary 1. *If* $(k, \alpha) \in \Upsilon$, $w \in \text{lext}(L_{k,\alpha})$, $\bar{w} \in \text{lext}(w, L_{k,\alpha})$, $x \in F_r(\bar{w}) \cap \Sigma_k$, $t \in L_{k,\alpha}^{N,L}$, *and* $\text{occur}(t,x) = 0$ *then there are finite words* w_1, w_2, g *such that* $(w_1^R, w_2^R, x, g^R, t^R) \in \Gamma(k, \alpha)$, $w \in \text{Suf}(gxw_2w_1)$, *and* $txw_2w_1 \in L_{k,\alpha}^{N,L}$.

Proof. Let $u \in \Sigma_k^* \cup \Sigma_k^N$. Realize that $u \in L_{k,\alpha} \cup L_{k,\alpha}^N$ if and only if $u^R \in L_{k,\alpha} \cup L_{k,\alpha}^N$. Then the corollary follows from Proposition 1 and Proposition 2. □

Given $k \in \mathbb{N}$ and a right infinite word $t \in \Sigma_k^{N,R}$, let $\Phi(t)$ be the set of all left infinite words $\bar{t} \in \Sigma_k^{N,L}$ such that $F(\bar{t}) \subseteq F_r(t)$. It means that all factors of $\bar{t} \in \Phi(t)$ are recurrent factors of t. We show that the set $\Phi(t)$ is not empty.

Lemma 4. *If* $k \in \mathbb{N}$ *and* $t \in \Sigma_k^{N,R}$ *then* $\Phi(t) \neq \emptyset$.

Proof. Since t is an infinite word, the set of recurrent factors of t is not empty. Let g be a recurrent nonempty factor of t; g may be a letter. Obviously there is $x \in \Sigma_k$ such that xg is also recurrent in t. This implies that the set $\{h \mid hg \in F_r(t)\}$ is infinite. The lemma follows from König's Infinity Lemma [4,8]. □

The next lemma shows that if u is a right extendable α-power free word then for each letter x there is a right infinite α-power free word $\bar{u}$ such that x is recurrent in $\bar{u}$ and u is a prefix of $\bar{u}$.

Lemma 5. *If* $(k, \alpha) \in \Upsilon$, $u \in \text{rext}(L_{k,\alpha})$, *and* $x \in \Sigma_k$ *then there is* $\bar{u} \in \text{rext}(u, L_{k,\alpha})$ *such that* $x \in F_r(\bar{u})$.

Proof. Let $w \in \text{rext}(u, L_{k,\alpha})$; Lemma 3 implies that $\text{rext}(u, L_{k,\alpha})$ is not empty. If $x \in F_r(w)$ then we are done. Suppose that $x \notin F_r(w)$. Let $y \in F_r(w) \cap \Sigma_k$. Clearly $x \neq y$. Proposition 2 implies that there is $(w_1, w_2, y, g, t) \in \Gamma(k, \alpha)$ such that $u \in \text{Prf}(w_1 w_2 yg)$. The proof of Lemma 2 implies that we can choose t in such a way that x is recurrent in t. Then $w_1 w_2 yt \in \text{rext}(u, L_{k,\alpha})$ and $x \in F_r(w_1 w_2 yt)$. This completes the proof. □

The next proposition shows that if u is left extendable and v is right extendable then there are finite words $\tilde{u}, \tilde{v}$, a letter x, a right infinite word t, and a left infinite word $\bar{t}$ such that $\tilde{u}xt, \bar{t}x\tilde{v}$ are infinite α-power free words, t has no occurrence of x, every factor of $\bar{t}$ is a recurrent factor in t, u is a prefix of $\tilde{u}xt$, and v is a suffix of $\bar{t}x\tilde{v}$.

Proposition 3. *If $(k,\alpha) \in \Upsilon$, $u \in \mathrm{rext}(L_{k,\alpha})$, and $v \in \mathrm{lext}(L_{k,\alpha})$ then there are $\tilde{u}, \tilde{v} \in \Sigma_k^*$, $x \in \Sigma_k$, $t \in \Sigma_k^{\mathbb{N},R}$, and $\bar{t} \in \Sigma_k^{\mathbb{N},L}$ such that $\tilde{u}xt \in L_{k,\alpha}^{\mathbb{N},R}$, $\bar{t}x\tilde{v} \in L_{k,\alpha}^{\mathbb{N},L}$, $\mathrm{occur}(t,x) = 0$, $\mathrm{F}(\bar{t}) \subseteq \mathrm{F}_r(t)$, $u \in \mathrm{Prf}(\tilde{u}xt)$, and $v \in \mathrm{Suf}(\bar{t}x\tilde{v})$.*

Proof. Let $\bar{u} \in \mathrm{rext}(u, L_{k,\alpha})$ and $\bar{v} \in \mathrm{lext}(v, L_{k,\alpha})$ be such that $\mathrm{F}_r(\bar{u}) \cap \mathrm{F}_r(\bar{v}) \cap \Sigma_k \neq \emptyset$. Lemma 5 implies that such $\bar{u}, \bar{v}$ exist. Let $x \in \mathrm{F}_r(\bar{u}) \cap \mathrm{F}_r(\bar{v}) \cap \Sigma_k$. It means that the letter x is recurrent in both $\bar{u}$ and $\bar{v}$.

Let t be a right infinite α-power free word over $\Sigma_k \setminus \{x\}$. Lemma 2 asserts that such t exists. Let $\bar{t} \in \Phi(t)$; Lemma 4 shows that $\Phi(t) \neq \emptyset$. It is easy to see that $\bar{t} \in L_{k,\alpha}^{\mathbb{N},L}$, because $\mathrm{F}(\bar{t}) \subseteq \mathrm{F}_r(t)$ and $t \in L_{k,\alpha}^{\mathbb{N},R}$.

Proposition 2 and Corollary 1 imply that there are $u_1, u_2, g, v_1, v_2, \bar{g} \in L_{k,\alpha}$ such that

- $(u_1, u_2, x, g, t) \in \Gamma(k, \alpha)$,
- $(v_1^R, v_2^R, x, \bar{g}^R, \bar{t}^R) \in \Gamma(k, \alpha)$,
- $u \in \mathrm{Prf}(u_1 u_2 xg)$, and
- $v^R \in \mathrm{Prf}(v_1^R v_2^R x \bar{g}^R)$; it follows that $v \in \mathrm{Suf}(\bar{g}xv_2v_1)$.

Proposition 1 implies that $u_1 u_2 xt, v_1^R v_2^R x \bar{t}^R \in L_{k,\alpha}^{\mathbb{N},R}$. It follows that $\bar{t}xv_2v_1 \in L_{k,\alpha}^{\mathbb{N},L}$. Let $\tilde{u} = u_1 u_2$ and $\tilde{v} = v_2 v_1$. This completes the proof. $\square$

The main theorem of the article shows that if u is a right extendable α-power free word and v is a left extendable α-power free word then there is a word w such that uwv is α-power free. The proof of the theorem shows also a construction of the word w.

Theorem 1. *If $(k, \alpha) \in \Upsilon$, $u \in \mathrm{rext}(L_{k,\alpha})$, and $v \in \mathrm{lext}(L_{k,\alpha})$ then there is $w \in L_{k,\alpha}$ such that $uwv \in L_{k,\alpha}$.*

Proof. Let $\tilde{u}, \tilde{v}, x, t, \bar{t}$ be as in Proposition 3. Let $p \in \mathrm{Suf}(\bar{t})$ be the shortest suffix such that $|p| > \max\{|\tilde{u}x|, |x\tilde{v}|, |u|, |v|\}$. Let $h \in \mathrm{Prf}(t)$ be the shortest prefix such that $hp \in \mathrm{Prf}(t)$ and $|h| > |p|$; such h exists, because p is a recurrent factor of t; see Proposition 3. We show that $\tilde{u}xhpx\tilde{v} \in L_{k,\alpha}$.

We have that $\tilde{u}xhp \in L_{k,\alpha}$, since $hp \in \mathrm{Prf}(t)$ and Proposition 3 states that $\tilde{u}xt \in L_{k,\alpha}^{\mathbb{N},R}$. Lemma 1 implies that it suffices to show that there are no $g \in \mathrm{Prf}(\tilde{v})$ and no $r \in \Sigma_k^* \setminus \{\epsilon\}$ such that $rr \in \mathrm{Suf}(\tilde{u}xhpxg)$ and $|rr| > |xg|$. To get a contradiction, suppose there are such r, g. We distinguish the following cases.

- If $|r| \leq |xg|$ then $rr \in \mathrm{Suf}(pxg)$, because $|p| > |x\tilde{v}|$ and $xg \in \mathrm{Prf}(x\tilde{v})$. This is a contradiction, since $px\tilde{v} \in \mathrm{Suf}(\bar{t}x\tilde{v})$ and $\bar{t}x\tilde{v} \in L_{k,\alpha}^{\mathbb{N},L}$; see Proposition 3.
- If $|r| > |xg|$ then $|r| \leq \frac{1}{2}|\tilde{u}xhpxg|$, otherwise rr cannot be a suffix of $\tilde{u}xhpxg$. Because $|h| > |p| > \max\{|\tilde{u}x|, |x\tilde{v}|\}$ we have that $r \in \mathrm{Suf}(hpxg)$. Since $\mathrm{occur}(hp, x) = 0$, $|h| > |p| > |x\tilde{v}|$, and $xg \in \mathrm{Suf}(r)$ it follows that there are words h_1, h_2 such that $\tilde{u}xhpxg = \tilde{u}xh_1h_2pxg$, $r = h_2pxg$ and $r \in \mathrm{Suf}(\tilde{u}xh_1)$. It follows that $xg \in \mathrm{Suf}(\tilde{u}xh_1)$ and because $\mathrm{occur}(h_1, x) = 0$ we have that $|h_1| \leq |g|$. Since $|p| > |\tilde{u}x|$ we get that $|h_2pxg| > |\tilde{u}xg| \geq |\tilde{u}xh_1|$; hence $|r| > |\tilde{u}xh_1|$. This is a contradiction.

We conclude that there is no word r and no prefix $g \in \mathrm{Prf}(\tilde{v})$ such that $rr \in \mathrm{Suf}(\tilde{u}xhpxg)$. Hence $\tilde{u}xhpx\tilde{v} \in L_{k,\alpha}$. Due to the construction of p and h we have that $u \in \mathrm{Prf}(\tilde{u}xhpx\tilde{v})$ and $v \in \mathrm{Suf}(\tilde{u}xhpx\tilde{v})$. This completes the proof.   □

Acknowledgments. The author acknowledges support by the Czech Science Foundation grant GAČR 13-03538S and by the Grant Agency of the Czech Technical University in Prague, grant No. SGS14/205/OHK4/3T/14.

References

1. Carpi, A.: On Dejean's conjecture over largealphabets. Theor. Comput. Sci. **385**, 137–151 (2007)
2. Currie, J., Rampersad, N.: A proof of Dejean's conjecture. Math. Comp. **80**, 1063–1070 (2011)
3. Dejean, F.: Sur un théorème de Thue. J. Comb. Theor. Series A **13**, 90–99 (1972)
4. König, D.: Sur les correspondances multivoques des ensembles. Fundamenta Math. **8**, 114–134 (1926)
5. Ollagnier, J.M.: Proof of Dejean's conjecture for alphabets with 5, 6, 7, 8, 9, 10 and 11 letters. Theor. Comput. Sci. **95**, 187–205 (1992)
6. Pansiot, J.-J.: A propos d'une conjecture de F. Dejean sur les répétitions dans les mots. Discrete Appl. Math. **7**, 297–311 (1984)
7. Petrova, E.A., Shur, A.M.: Transition property for cube-free words. In: van Bevern, R., Kucherov, G. (eds.) CSR 2019. LNCS, vol. 11532, pp. 311–324. Springer, Cham (2019). https://doi.org/10.1007/978-3-030-19955-5_27
8. Rampersad, N.: Overlap-free words and generalizations. A thesis, University of Waterloo (2007)
9. Rao, M.: Last cases of Dejean's conjecture. Theor. Comput. Sci. **412**, 3010–3018 (2011). Combinatorics on Words (WORDS 2009)
10. Restivo, A., Salemi, S.: Some decision results on nonrepetitive words. In: Apostolico, A., Galil, Z. (eds.) Combinatorial Algorithms on Words, pp. 289–295. Springer, Heidelberg (1985). https://doi.org/10.1007/978-3-642-82456-2_20
11. Shallit, J., Shur, A.: Subword complexity and power avoidance. Theor. Comput. Sci. **792**, 96–116 (2019). Special issue in honor of the 70th birthday of Prof. Wojciech Rytter
12. Shur, A.M.: Two-sided bounds for the growth rates of power-free languages. In: Diekert, V., Nowotka, D. (eds.) DLT 2009. LNCS, vol. 5583, pp. 466–477. Springer, Heidelberg (2009). https://doi.org/10.1007/978-3-642-02737-6_38
13. Shur, A.M.: Growth properties of power-free languages. In: Mauri, G., Leporati, A. (eds.) DLT 2011. LNCS, vol. 6795, pp. 28–43. Springer, Heidelberg (2011). https://doi.org/10.1007/978-3-642-22321-1_3

Context-Freeness of Word-MIX Languages

Ryoma Sin'Ya[(✉)]

Akita University, Akita, Japan
ryoma@math.akita-u.ac.jp

Abstract. In this paper we provide a decidable characterisation of the context-freeness of a Word-MIX language $L_A(w_1, \ldots, w_k)$, where $L_A(w_1, \ldots, w_k)$ is the set of all words over A that contain the same number of subword occurrences of parameter words $w_1, \ldots, w_k$.

1 Introduction

Counting occurrences of letters in words is a major topic in formal language theory. In particular, much ink has been spent on investigating the counting ability of some language classes. For example, Joshi et al. [1] suggested that the language $\text{MIX} = \{w \in \{a, b, c\}^* \mid |w|_a = |w|_b = |w|_c\}$ should not be in the class of so-called mildly context-sensitive languages since it allows too much freedom in word order, so that relations between MIX and several language classes have been investigated (*e.g.*, indexed languages [2], range concatenation languages [3], tree-adjoining languages [4], multiple context-free languages [5], *etc.*). The Parikh map is another rich example on this topic (counting occurrences of letters) [6].

In the recent work [7] by Colbourn et al., the counting feature of MIX is generalised from counting *letter* occurrences to counting *word* occurrences. They considered several problems for languages of the form $L_A(w_1, \ldots, w_k) = \{w \in A^* \mid |w|_{w_1} = \cdots = |w|_{w_k}\}$ (where $|u|_v$ is the number of occurrences of v in w) which we call *Word-MIX languages* (WMIX for short) in this paper. While $L_A(w_1, w_2)$ is always deterministic context-free, it can also be regular ($L_A(ab, ba)$) is regular if $A = \{a, b\}$, while it is not regular if $A = \{a, b, c\}$, for example) [7]. This kind of generalisation – from letter occurrences to word occurrences – is also considered in the context of the Parikh map through so-called Parikh matrices [8] and subword histories [9,10] (in this setting they have considered *scattered* subword occurrences instead of subword occurrences).

Colbourn et al. [7] provided a necessary and sufficient condition for w_1 and w_2 for the WMIX language $L_A(w_1, w_2)$ to be regular, and gave a polynomial time algorithm for testing that condition. For the fully general case, the decidability of the regularity problem for WMIX languages can be derived from some known results on *unambiguous constrained automata* (UnCA for short), since

---

The author is also with RIKEN AIP.

© Springer Nature Switzerland AG 2020
N. Jonoska and D. Savchuk (Eds.): DLT 2020, LNCS 12086, pp. 304–318, 2020.
https://doi.org/10.1007/978-3-030-48516-0_23

$L_A(w_1, \ldots, w_k)$ is always recognised by an UnCA, and the regularity for UnCA languages is decidable due to [11].

In this paper, we show that *context-freeness is decidable* for WMIX languages. We also give an alternative decidability proof for the regularity of WMIX languages. As we mentioned above, the regularity for WMIX languages is already known to be decidable thanks to the decidability results on UnCA languages (which include all WMIX languages) given by Cadilhac et al. [11]. But the alternative proof of the regularity for WMIX languages given in this paper gives more *structural information* of WMIX languages, and the proof can be naturally extended into the context-freeness. We introduce a new notion called *dimension*, which represents certain structural information of WMIX languages, and prove that a WMIX language is (1) regular if and only if its dimension is at most one, and (2) context-free if and only if its dimension is at most two. To the best of our knowledge, there has been no research on the context-freeness for WMIX languages or UnCA languages. As far as we know, a language class with such a decidable context-freeness property is very rare. We are only aware of such examples in some subclasses of bounded languages [12–14] and languages associated with vector addition systems [15].

For the space restriction, we omit some definitions and proofs; see the full version [16] for details.

2   Preliminaries

For a set X, we denote by $\#(X)$ the cardinality of X. We denote by $\mathbb{N}$ the set of natural numbers including 0. We call a mapping $M : X \to \mathbb{N}$ multiset over X. For a set X, we write 2^X for the power set of X.

We assume that the reader has a basic understanding of automata and linear algebra.

2.1   Words and Word-MIX Languages

For an alphabet A, we denote the set of all words (resp. all non-empty words) over A by A^* (resp. A^+). We write A^n (resp. $A^{<n}$) for the set of all words of length n (resp. less than n), and write $\mathbb{N}^{\leq c}$ for the set of all natural numbers less than or equal c for $c \in \mathbb{N}$. For a pair of words $v, w \in A^*$, $|w|_v$ denotes the number of subword occurrences of v in w

$$|w|_v \stackrel{\text{def}}{=} \#(\{(w_1, w_2) \in A^* \times A^* \mid w_1 v w_2 = w\}).$$

We write $u \sqsubseteq v$ if u is a subword of v, and write $u \sqsubseteq_{\text{sc}} v$ if u is a scattered subword of v. For words $w_1, \ldots, w_k \in A^*$, we define

$$L_A(w_1, \ldots, w_k) \stackrel{\text{def}}{=} \{w \in A^* \mid |w|_{w_1} = \cdots = |w|_{w_k}\}$$

and call it the *Word-MIX* (WMIX for short) *language of k-parameter words* $w_1, \ldots, w_k$ *over* A. For a word $w \in A^*$, we denote the set of prefixes and suffixes of w by $\text{pref}(w)$ and $\text{suff}(w)$, and denote the length-n ($n \leq |w|$) prefix and suffix of w by $\text{pref}_n(w)$ and $\text{suff}_n(w)$, respectively.

2.2 Graphs and Walks

Let $\mathcal{G} = (V, E)$ be a (directed) graph. We call a sequence of vertices $\omega = (v_1, \ldots, v_n) \in V^n$ $(n \geq 1)$ *walk* (from v_1 into v_n in $\mathcal{G}$) if $(v_i, v_{i+1}) \in E$ for each $i \in \{1, \ldots, n-1\}$, and define the length of ω as $n-1$ and denote it by $|\omega|$. We denote by $\mathtt{from}(\omega)$ and $\mathtt{into}(\omega)$ the source $\mathtt{from}(\omega) \stackrel{\text{def}}{=} v_1$ and the target $\mathtt{into}(\omega) \stackrel{\text{def}}{=} v_n$ of ω. ω is called an *empty walk* if $|\omega| = 0$. If two walks $\omega_1 = (v_1, \ldots, v_m), \omega_2 = (v_1', \ldots, v_n')$ are connectable (*i.e.*, $\mathtt{into}(\omega_1) = \mathtt{from}(\omega_2)$), we write $\omega_1 \odot \omega_2$ for the connecting walk $\omega_1 \odot \omega_2 \stackrel{\text{def}}{=} (v_1, \ldots, v_m, v_2', \ldots, v_n')$. A non-empty walk ω is called *loop* (on $\mathtt{from}(\omega)$) if $\mathtt{from}(\omega) = \mathtt{into}(\omega)$. A walk $(v_1, \ldots, v_n)$ is called *path* if $v_i \neq v_j$ for every $i, j \in \{1, \ldots, n\}$ with $i \neq j$. A loop $(v, v_1, \ldots, v_n, v)$ is called *cycle* if $(v, v_1, \ldots, v_n)$ is a path. We use the metavariable π for a path, and the metavariable γ for a cycle. For a cycle γ and $n \geq 1$, we write γ^n for the loop which is an n-times repetition of γ. We denote by $\mathcal{W}(\mathcal{G}), \mathcal{P}(\mathcal{G})$, and by $\mathcal{C}(\mathcal{G})$ the set of all walks, paths and cycles in $\mathcal{G}$. Note that $\mathcal{W}(\mathcal{G})$ is infinite in general, but $\mathcal{P}(\mathcal{G})$ and $\mathcal{C}(\mathcal{G})$ are both finite if $\mathcal{G}$ is finite.

The N-*dimensional de Bruijn graph* $\mathcal{G}_A^N = (A^N, E)$ over A is a graph whose vertex set A^N is the set of words of length N and the edge set E is defined by

$$E \stackrel{\text{def}}{=} \{(av, vb) \mid a, b \in A, v \in A^{N-1}\}.$$

The case $N = 2$ is depicted in Fig. 1.

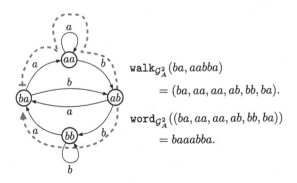

$$\mathtt{walk}_{\mathcal{G}_A^2}(ba, aabba)$$
$$= (ba, aa, aa, ab, bb, ba).$$

$$\mathtt{word}_{\mathcal{G}_A^2}((ba, aa, aa, ab, bb, ba))$$
$$= baaabba.$$

Fig. 1. The 2-dimensional de Bruijn graph $\mathcal{G}_A^2$ over $A = \{a, b\}$, a walk (ba, aa, aa, ab, bb, ba) (dotted red arrow) on $\mathcal{G}_A^2$ and its corresponding word $baaabba$. (Color figure online)

Let v be a vertex of $\mathcal{G}_A^N$. A word $w = a_1 \cdots a_m \in A^+$ induces the walk $(v, v_1, \ldots, v_m)$ (where $v_i = \mathtt{suff}_n(v\,\mathtt{pref}_i(w))$) in $\mathcal{G}_A^N$, and we denote it by $\mathtt{walk}_{\mathcal{G}_A^N}(v, w)$. Conversely, a walk $\omega = (v_1, \ldots, v_n)$ in $\mathcal{G}_A^N$ induces the word $v_1\mathtt{suff}_1(v_2) \cdots \mathtt{suff}_1(v_n) \in A^*$, and we denote it by $\mathtt{word}_{\mathcal{G}_A^N}(\omega)$ (see Fig. 1). For words $w, w_1, \ldots, w_k \in A^*$ and a walk $\omega = (v_0, v_1, \ldots, v_n) \in \mathcal{W}(\mathcal{G}_A^N)$, we define the following vectors in $\mathbb{N}^k$:

$$|w|_{(w_1,\ldots,w_k)} \overset{\text{def}}{=} (|w|_{w_1},\ldots,|w|_{w_k})$$

$$|\omega|_{(w_1,\ldots,w_k)} \overset{\text{def}}{=} \sum_{i=1}^{n}(c_{i,1},\ldots,c_{i,k}) \text{ where } c_{i,j} = 1 \text{ if } w_j \in \text{suff}(v_i), c_{i,j} = 0 \text{ otherwise.}$$

We call $|w|_{(w_1,\ldots,w_k)}$ (resp. $|\omega|_{(w_1,\ldots,w_k)}$) the *occurrence vector* of w (resp. ω). We notice that the range of the summation in the above definition of $|\omega|_{(w_1,\ldots,w_k)}$ *does not contain* 0, hence $|\omega|_{(w_1,\ldots,w_k)} = (0,\ldots,0)$ if ω is an empty walk $\omega = (v_0)$. The next proposition states a basic property of $\mathcal{G}_A^N$, which can be shown by a straightforward induction on the length of w.

Proposition 1. *Let* $w_1,\ldots,w_k \in A^*$ *and* $N = \max(|w_1|,\ldots,|w_k|)$. *For any pair of words* $v,w \in A^*$ *such that* $|v| = N$ *and* $\omega = \texttt{walk}_{\mathcal{G}_A^N}(v,w)$, *we have*

$$|vw|_{(w_1,\ldots,w_k)} = |v|_{(w_1,\ldots,w_k)} + |\omega|_{(w_1,\ldots,w_k)}.$$

2.3   Well-Quasi-Orders

A quasi order $\leq$ on a set X is called *well-quasi-order* (*wqo* for short) if any infinite sequence $(x_i)_{i\in\mathbb{N}}$ ($x_i \in X$) contains an increasing pair $x_i \leq x_j$ with $i < j$. Let $\leq_1$ be a quasi order on a set X_1 and $\leq_2$ be a quasi order on a set X_2. The *product order* $\leq_{1,2}$ is a quasi order on $X_1 \times X_2$ defined by

$$(x_1, y_1) \leq_{1,2} (x_2, y_2) \overset{\text{def}}{\Longleftrightarrow} x_1 \leq_1 x_2 \text{ and } y_1 \leq_2 y_2.$$

Proposition 2 (*cf.* **Proposition 6.1.1 in** [17]). *Let* $\leq_1$ *be a wqo on a set* X_1 *and* $\leq_2$ *be a wqo on a set* X_2. *The product order* $\leq_{1,2}$ *is again a wqo on* $X_1 \times X_2$.

We list some examples of wqos below:

(1) The identity relation $=$ on any finite set X is a wqo (*the pigeonhole principle*).
(2) The usual order $\leq$ on $\mathbb{N}$ is a wqo.
(3) The product order $\leq_m$ on $\mathbb{N}^m$ is a wqo for any $m \geq 1$ (*Dickson's lemma*), which is a direct corollary of Proposition 2.
(4) The point-wise order $\leq_{\text{pt}}$ on the multisets $\mathbb{N}^X$ ($M \leq_{\text{pt}} M' \overset{\text{def}}{\Longleftrightarrow} M(x) \leq M'(x)$ for all $x \in X$) over a finite set X is a wqo (just a paraphrase of Dickson's lemma).

3   Path-Cycle Decomposition of Walks

In this section, we provide a simple method which decomposes, in left-to-right manner, a walk ω into a (possibly empty) path π and a sequence of cycles Γ (Fig. 2). This decomposition, and its inverse operation (composition), are probably folklore, and the contents in this section appeared already in the author's unpublished note [18]. A similar method is also used in [11].

Let $\mathcal{G} = (V, E)$ be a graph. For a pair of sequences of cycles $\Gamma_1 = (\gamma_1, \ldots, \gamma_n)$, $\Gamma_2 = (\gamma_1', \ldots, \gamma_m')$, we write $\Gamma_1.\Gamma_2$ for the concatenation $(\gamma_1, \ldots, \gamma_n, \gamma_1', \ldots, \gamma_m')$. When $\Gamma_1 = (\gamma)$ we simply write $\gamma.\Gamma_2$ for $\Gamma_1.\Gamma_2$ We write $\emptyset$ for the empty sequence of cycles. For $\Gamma = (\gamma_1, \ldots, \gamma_n)$, we denote by $\Gamma(i)$ for the i-th component γ_i of Γ, and denote by $|\Gamma|_\gamma$ the number $\#(\{i \mid \Gamma(i) = \gamma\})$ of occurrences of γ in Γ. For a walk $\omega = (v_1, \ldots, v_n)$, we denote by $V(\omega)$ the set of all vertices appearing in ω: $V(\omega) \overset{\text{def}}{=} \{v_1, \ldots, v_n\}$.

$$\Phi_{\mathcal{K}_4} \begin{cases} \omega = (1, \underline{2, 3, 2}, 3, 4, 3, 4, 2, 4) \\ (1, \underline{2, 3, 4, 3}, 4, 2, 4) \& (\overline{(2, 3, 2)}) \\ (1, \underline{2, 3, 4, 2}, 4) \& ((2, 3, 2), \overline{(3, 4, 3)}) \\ (1, 2, 4) \& ((2, 3, 2), (3, 4, 3), \overline{(2, 3, 4, 2)}) = \Phi_{\mathcal{K}_4}(\omega) \\ (1, \overline{2, 3, 4, 2}, 4) \& ((2, 3, 2), \underline{(3, 4, 3)}) \\ (1, 2, \overline{3, 4, 3}, 4, 2, 4) \& (\underline{(2, 3, 2)}) \\ (1, \overline{2, 3, 2}, 3, 4, 3, 4, 2, 4) = \Psi_{\mathcal{K}_4}(\Phi_{\mathcal{K}_4}(\omega)) = \omega \end{cases} \Bigg\} \Psi_{\mathcal{K}_4}$$

Fig. 2. Computation of $\Phi_{\mathcal{K}_4}$ and $\Psi_{\mathcal{K}_4}$

We then define a decomposition function $\Phi_\mathcal{G}$ inductively as follows: $\Phi_\mathcal{G}((v)) \overset{\text{def}}{=} ((v), \emptyset)$ and

$$\Phi_\mathcal{G}(\omega \odot (v, v')) \overset{\text{def}}{=} \begin{cases} (\pi \odot (v, v'),\ \Gamma) & \text{if } v' \notin V(\pi), \\ (\pi_1,\ \Gamma.(\pi_2 \odot (v, v'))) & \text{if } \pi = \pi_1 \odot (v') \odot \pi_2 \end{cases}$$

$$\text{where } (\pi, \Gamma) = \Phi_\mathcal{G}(\omega).$$

It is clear by definition that, for any ω and $(\omega', \Gamma) = \Phi_\mathcal{G}(\omega)$, ω' is a path and Γ is a sequence of cycles, i.e., $\Phi_\mathcal{G} : \mathcal{W}(\mathcal{G}) \to \mathcal{P}(\mathcal{G}) \times \mathcal{C}(\mathcal{G})^*$. Conversely, we can define a composition (partial) function $\Psi_\mathcal{G}$ as an inverse of $\Phi_\mathcal{G}$, i.e., $\omega = \Psi_\mathcal{G}(\Phi_\mathcal{G}(\omega))$. The formal definition of $\Psi_\mathcal{G}$ can be found in the full version [16].

Example 1. Consider the complete graph $\mathcal{K}_4 = (V_4 = \{1, 2, 3, 4\}, E_4 = V_4 \times V_4)$ of order 4 and a walk $\omega = (1, 2, 3, 2, 3, 4, 3, 4, 2, 4)$. The result of decomposition is $\Phi_{\mathcal{K}_4}(\omega) = (\pi = (1, 2, 4), \Gamma = ((2, 3, 2), (3, 4, 3), (2, 3, 4, 2)))$. All intermediate computation steps of $\Phi_{\mathcal{K}_4}(\omega)$ and $\Psi_{\mathcal{K}_4}(\Phi_{\mathcal{K}_4}(\omega))$ are drawn in Fig. 2 (in the figure we denote by $\pi \& \Gamma$ a pair (π, Γ) for visibility).

3.1 Multi-traces and Traces

For a walk ω in a graph $\mathcal{G}$, we define the *multi-trace* $\mathrm{NTr}(\omega) : \mathcal{P}(\mathcal{G}) \cup \mathcal{C}(\mathcal{G}) \to \mathbb{N}$ of a walk ω as the following multiset over paths and cycles:

$$(\mathrm{NTr}(\omega))(\pi) \overset{\mathrm{def}}{=} \begin{cases} 1 & \text{if } \pi = \pi_\omega \\ 0 & \text{otherwise} \end{cases} \qquad (\mathrm{NTr}(\omega))(\gamma) \overset{\mathrm{def}}{=} |\Gamma|_\gamma$$

$$\text{where } (\pi_\omega, \Gamma) = \Phi_{\mathcal{G}}(\omega).$$

We define the *trace* $\mathrm{Tr}(\omega)$ of a walk ω in $\mathcal{G}$ as the following set of paths and cycles:

$$\mathrm{Tr}(\omega) \overset{\mathrm{def}}{=} \{\pi \in \mathcal{P}(\mathcal{G}) \mid (\mathrm{NTr}(\omega))(\pi) \neq 0\} \cup \{\gamma \in \mathcal{C}(\mathcal{G}) \mid (\mathrm{NTr}(\omega))(\gamma) \neq 0\}.$$

Intuitively, the multi-trace of ω in $\mathcal{G}$ is obtained by forgetting the ordering of the decomposition result $(\omega, \Gamma) = \Phi_{\mathcal{G}}(\omega)$ of ω, and the trace of ω is obtained by forgetting the multiplicity from the original multi-trace (see Fig. 3 for the relation).

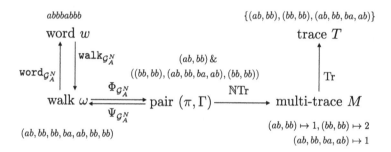

Fig. 3. Relations between words, walks and (multi-)traces ($N = 2$ for the examples).

The following proposition states that the occurrences of parameter words of a walk are completely determined from its multi-trace (see the full version [16] for details).

Proposition 3. *Let $w_1, \ldots, w_k \in A^*$ and $N = \max(|w_1|, \ldots, |w_k|)$. For any ω in $\mathcal{W}(\mathcal{G}_A^N)$, we have*

$$|\omega|_{(w_1, \ldots, w_k)} = \sum_{\pi \in \mathcal{P}(\mathcal{G}_A^N)} (\mathrm{NTr}(\omega))(\pi) \cdot |\pi|_{(w_1, \ldots, w_k)} + \sum_{\gamma \in \mathcal{C}(\mathcal{G}_A^N)} (\mathrm{NTr}(\omega))(\gamma) \cdot |\gamma|_{(w_1, \ldots, w_k)}.$$

4 Main Results

In this section we first introduce a new notion for WMIX languages called *dimension*. Afterwards, we state our main results that characterise both regularity and context-freeness of WMIX languages.

Definition 1. Let $w_1, \ldots, w_k \in A^*$ and $N = \max(|w_1|, \ldots, |w_k|)$. Let $T = \{\pi\} \cup \{\gamma_1, \ldots, \gamma_m\}$ be a trace of a walk ω in $\mathcal{G}_A^N$. A subset S of $\{\gamma_1, \ldots, \gamma_m\}$ is called *pumpable in T of* $L_A(w_1, \ldots, w_k)$ if, for any number $n \geq 1$, there exists a word $uv \in L_A(w_1, \ldots, w_k)$ with $\omega = \mathtt{walk}_{\mathcal{G}_A^N}(u, v)$ such that (1) $\mathrm{Tr}(\omega) = T$ and (2) $(\mathrm{NTr}(\omega))(\gamma) \geq n$ for each $\gamma \in S$. We further say S is *maximal* if no proper superset of S included in $\{\gamma_1, \ldots, \gamma_m\}$ is pumpable.

Remark 1. The emptyset $\emptyset$ is always pumpable in a trace T of $L_A(w_1, \ldots, w_k)$ such that $\mathrm{Tr}(\mathtt{walk}_{\mathcal{G}_A^N}(u, v)) = T$ for some $uv \in L_A(w_1, \ldots, w_k)$. Moreover, it is decidable whether S is pumpable or not in T of $L_A(w_1, \ldots, w_k)$ (see the full version [16]).

Recall that a *vector space* is a set $V \subseteq \mathbb{R}^k$ such that $\mathbf{0} \in V, V + V \subseteq V$ and $\mathbb{R}V = \{\alpha \cdot v \mid v \in V, \alpha \in \mathbb{R}\} \subseteq V$ where $\mathbf{0}$ is the vector with all zeros.

Definition 2. Let $w_1, \ldots, w_k \in A^*$, $N = \max(|w_1|, \ldots, |w_k|)$. The *dimension* of $L = L_A(w_1, \ldots, w_k)$ is the natural number defined as

$$\max\{\dim(V) \mid V = \mathrm{span}(\{|\gamma|_{(w_1, \ldots, w_k)} \mid \gamma \in S\}), S \text{ is pumpable in some } T \text{ of } L\}$$

where $\dim(V)$ is the dimension of the vector space V and $\mathrm{span}(B)$ is the vector space spanned by B (where $\mathrm{span}(\emptyset) \overset{\text{def}}{=} \{\mathbf{0}\}$).

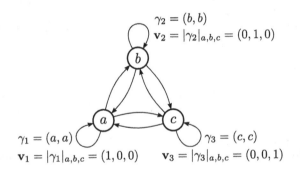

Fig. 4. The 1-dimensional de Bruijn graph $\mathcal{G}_A^1$ over $A = \{a, b, c\}$.

The dimension of a WMIX language L is, roughly speaking, the minimum number of cycles (in the de Bruijn graph) that should be counted *independently*. We describe this intuition more rigorously by using MIX $= L_A(a, b, c)$ for $A = \{a, b, c\}$ as a simple example.

Example 2. Since $\max(|a|, |b|, |c|) = 1$, it is enough to consider the 1-dimensional de Bruijn graph $\mathcal{G}_A^1$ over $A = \{a, b, c\}$ (see Fig. 4). One can easily observe that the set of cycles $S = \{\gamma_1 = (a, a), \gamma_2 = (b, b), \gamma_3 = (c, c)\}$, each γ_i is depicted in Fig. 4, is pumpable in the trace $T = \{(a, b, c)\} \cup S$: for any $n > 0$, the word $aw_n =$

$a^{n+1}b^{n+1}c^{n+1}$ is in MIX and it satisfies the two conditions in the Definition 1 as (1) $\mathrm{Tr}(\mathtt{walk}_{\mathcal{G}_A^N}(a, w_n)) = T$ and (2) $(\mathbb{N}\mathrm{Tr}(\mathtt{walk}_{\mathcal{G}_A^N}(a, w_n)))(\gamma_i) = n$ for each $\gamma_i \in S$. The occurrence vectors corresponding to $\gamma_1, \gamma_2, \gamma_3$ are $\boldsymbol{v_1} = (1, 0, 0)$, $\boldsymbol{v_2} = (0, 1, 0)$, $\boldsymbol{v_3} = (0, 0, 1)$, respectively. Since those occurrence vectors are linearly independent, the vector space spanned by them is $\mathbb{R}^3$ and thus the dimension of MIX is three.

By considering dimensions of WMIX languages, we can nicely characterise both regularity and context-freeness as follows.

Theorem 1 (regularity). $L_A(w_1, \ldots, w_k)$ is regular if and only if its dimension is at most one.

Theorem 2 (context-freeness). $L_A(w_1, \ldots, w_k)$ is context-free if and only if its dimension is at most two.

Some pushdown automaton $\mathcal{A}$ can recognise $L_A(a, b)$ since, by using its stack, $\mathcal{A}$ can track the number $|w|_a - |w|_b$. However, no pushdown automaton $\mathcal{A}$ can recognise MIX $= L_A(a, b, c)$ since, for that purpose, one should track the numbers $|w|_a - |w|_b$ and $|w|_b - |w|_c$ simultaneously. This is a rough intuition why a language with dimension greater than or equal three is never to be context-free (the formal proof is in the next section).

The set $\mathcal{P}(\mathcal{G}_A^N) \cup \mathcal{C}(\mathcal{G}_A^N)$ of paths and cycles in the N-dimensional de Bruijn graph is finite, hence we can effectively enumerate all traces of all walks in $\mathcal{G}_A^N$ (see the full version [16] for details). Moreover, as we mentioned in Remark 1, we can also effectively enumerate all pumpable sets in a trace. For a pumpable set S, computing the dimension of the vector space spanned by the occurrence vectors S is just counting the maximum number of linearly independent ones from the occurrence vectors of S. Combining these facts and Theorem 1–2, we can effectively compute the dimension of $L_A(w_1, \ldots, w_k)$ and hence we have the following decidability result.

Corollary 1. *Regularity and context-freeness are decidable for WMIX languages.*

5   Proof of the Main Results

The proof structure of Theorem 1 is similar with one of Theorem 2, albeit that the latter is more complicated. In this section, we firstly investigate some structural properties of pumpable sets, which play crucial role in the main proof. We secondly give a proof of Theorem 1 which would give a good intuition for the latter proof. Finally, we give a proof of Theorem 2.

5.1   Properties of Pumpable Sets

For a vector $\boldsymbol{v} = (c_1, \ldots, c_k) \in \mathbb{R}^k$, we define $\mathtt{diff}(\boldsymbol{v}) \overset{\text{def}}{=} \sum_{i=1}^k (\max\{c_1, \ldots, c_k\} - c_i)$. Observe that $w \in L_A(w_1, \ldots, w_k)$ if and only if $\mathtt{diff}(|w|_{(w_1, \ldots, w_k)}) = 0$.

Lemma 1. *Let* $w_1, \ldots, w_k \in A^*$ *and* $N = \max(|w_1|, \ldots, |w_k|)$. *For any maximum pumpable set* S *in* $T = \{\pi\} \cup S'$ *of* $L_A(w_1, \ldots, w_k)$, *if* $V = \mathrm{span}(\{|\gamma|_{(w_1, \ldots, w_k)} \mid \gamma \in S\})$ *has a non-zero dimension, then* V *contains the vector* $\mathbf{1}$.

Proof. Let $u = \mathtt{from}(\pi)$. Since S is pumpable, there exists an infinite sequence $(uv_i)_{i \in \mathbb{N}}$, where $u \in A^N$, of words that satisfies:

(1) $\mathrm{Tr}(\mathtt{walk}_{\mathcal{G}_A^N}(u, v_i)) = T$ for all $i \in \mathbb{N}$.
(2) $uv_i \in L_A(w_1, \ldots, w_k)$ for all $i \in \mathbb{N}$.
(3) $\mathrm{NTr}(\mathtt{walk}_{\mathcal{G}_A^N}(u, v_i))(\gamma) < \mathrm{NTr}(\mathtt{walk}_{\mathcal{G}_A^N}(u, v_j))(\gamma)$ for all $i, j \in \mathbb{N}$ with $i < j$ and for all $\gamma \in S$.

Now consider an infinite sequence of multi-traces of the above sequence

$$(M_i)_{i \in \mathbb{N}} \stackrel{\mathtt{def}}{=} (\mathrm{NTr}(\mathtt{walk}_{\mathcal{G}_A^N}(u, v_i)))_{i \in \mathbb{N}}.$$

Since the point-wise order on the multisets over any finite set is a wqo (thanks to Dickson's lemma) and $\mathcal{P}(\mathcal{G}_A^N) \cup \mathcal{C}(\mathcal{G}_A^N)$ is finite, $(M_i)_{i \in \mathbb{N}}$ contains an infinite increasing subsequence $(M_j)_{j \in J}$ $(J \subseteq \mathbb{N})$. Let $\overline{S} = (S' \setminus S)$. Because S is maximum, the number of maximum occurrence of any non-pumpable cycle $\gamma \in \overline{S}$ is bounded, *i.e.*, there is some constant $c \in \mathbb{N}$ such that $(\mathrm{NTr}(\mathtt{walk}_{\mathcal{G}_A^N}(u, v_i)))(\gamma) < c$ for any $\gamma \in \overline{S}$ and $i \in \mathbb{N}$. By using pigeonhole principle, we can deduce that, in the infinite sequence $(M_j)_{j \in J}$, there exists a pair $(i_1, i_2) \in J^2$ with $i_1 < i_2$ such that $(M_{i_1})(\gamma) = (M_{i_2})(\gamma)$ for all $\gamma \in \overline{S}$. Let $C = \sum_{\gamma \in \overline{S}} M_{i_1}(\gamma) \cdot |\gamma|_{(w_1, \ldots, w_k)}$. Combining the above observation and the condition (3) of $(uv_i)_{i \in \mathbb{N}}$, we have

$$M_{i_1}(\gamma) = M_{i_2}(\gamma) \text{ for all } \gamma \in \overline{S} \quad M_{i_1}(\gamma) < M_{i_2}(\gamma) \text{ for all } \gamma \in S. \quad (\bigstar)$$

Because $uv_{i_1}, uv_{i_2} \in L_A(w_1, \ldots, w_k)$, by Proposition 1 and Proposition 3, we have

$$\mathtt{diff}(|uv_{i_1}|_{(w_1, \ldots, w_k)}) = \mathtt{diff}(|uv_{i_2}|_{(w_1, \ldots, w_k)}) = 0$$

$$= \mathtt{diff}\left(|u|_{(w_1, \ldots, w_k)} + |\pi|_{(w_1, \ldots, w_k)} + C + \sum_{\gamma \in S} M_{i_1}(\gamma) \cdot |\gamma|_{(w_1, \ldots, w_k)}\right)$$

$$= \mathtt{diff}\left(|u|_{(w_1, \ldots, w_k)} + |\pi|_{(w_1, \ldots, w_k)} + C + \sum_{\gamma \in S} M_{i_2}(\gamma) \cdot |\gamma|_{(w_1, \ldots, w_k)}\right).$$

Moreover, from the above equation we obtain

$$\mathtt{diff}\left(\sum_{\gamma \in S} (M_{i_2}(\gamma) - M_{i_1}(\gamma)) \cdot |\gamma|_{(w_1, \ldots, w_k)}\right) = 0 \qquad (1)$$

because for any v such that $\texttt{diff}(v) = 0$, $\texttt{diff}(v + v') = 0$ if and only if $\texttt{diff}(v') = 0$. By Condition ($\bigstar$), the vector

$$v = \sum_{\gamma \in S}(M_{i_2}(\gamma) - M_{i_1}(\gamma)) \cdot |\gamma|_{(w_1, \ldots, w_k)}$$

is not the zero vector $\mathbf{0}$. Thus v is of the form $n \cdot \mathbf{1}$ $(n \neq 0)$, i.e., $\mathbf{1} \in \text{span}(V)$. $\square$

Lemma 2. *Let* $w_1, \ldots, w_k \in A^*$ *and* $N = \max(|w_1|, \ldots, |w_k|)$. *For any trace* T *of some walk in* $\mathcal{G}_A^N$, *if* $\text{Tr}(\texttt{walk}_{\mathcal{G}_A^N}(u, v)) = T$ *for some* $uv \in L_A(w_1, \ldots, w_k)$, *then there exists a unique maximal (i.e., the maximum) pumpable set* S *in* T *of* $L_A(w_1, \ldots, w_k)$.

Proof. Let S_1, S_2 be two maximal pumpable sets in T of $L_A(w_1, \ldots, w_k)$ and $S_1 = \{\gamma_1, \ldots, \gamma_m\}$. We now prove that $S_1 \cup S_2$ is also pumpable in T, which implies $S_1 = S_2$ by the maximality of S_1 and S_2. By Condition ($\bigstar$) and Equation (1) in the proof of Lemma 1, we can deduce that there exist $n_1, \ldots, n_m \in \mathbb{N}$ such that $n_i > 0$ for all $i \in \{1, \ldots, m\}$ and $\texttt{diff}(\sum_{i=1}^m n_i \cdot |\gamma_i|_{(w_1, \ldots, w_k)}) = 0$. Let $(uv_i)_{i \in \mathbb{N}}$ be an infinite sequence that ensures the pumpability of S_2, namely,

(1) $\text{Tr}(\texttt{walk}_{\mathcal{G}_A^N}(u, v_i)) = T$ for all $i \in \mathbb{N}$.
(2) $uv_i \in L_A(w_1, \ldots, w_k)$ for all $i \in \mathbb{N}$.
(3) $\text{NTr}(\texttt{walk}_{\mathcal{G}_A^N}(u, v_i))(\gamma) \geq i$ for all $i \in \mathbb{N}$ and for all $\gamma \in S_2$.

Let uv'_i be a word that satisfying $\text{Tr}(\texttt{walk}_{\mathcal{G}_A^N}(u, v'_i)) = T$, $\text{NTr}(\texttt{walk}_{\mathcal{G}_A^N}(u, v'_i))(\gamma_j)$ $= \text{NTr}(\texttt{walk}_{\mathcal{G}_A^N}(u, v_i))(\gamma_j) + i \times n_j$ for all $i \in \mathbb{N}$ and $\gamma_j \in S_1$. Such word uv'_i always exists because we can just pump an occurrence of $\gamma_j \in S_1$ in $\texttt{walk}_{\mathcal{G}_A^N}(u, v_i)$ $(i \times n_j)$-times repeatedly. Then the infinite sequence $(uv'_i)_{i \in \mathbb{N}}$ satisfies $uv'_i \in L_A(w_1, \ldots, w_k)$ and $\text{NTr}(\texttt{walk}_{\mathcal{G}_A^N}(u, v'_i))(\gamma) \geq i$ for all $i \in \mathbb{N}$ and for all $\gamma \in S_1 \cup S_2$, because $\texttt{diff}(\sum_{j=1}^m n_j \cdot |\gamma_j|_{(w_1, \ldots, w_k)}) = 0$. Which means that $(uv'_i)_{i \in \mathbb{N}}$ ensures the pumpability of $S_1 \cup S_2$, this ends the proof. $\square$

Lemma 3. *Let* $w_1, \ldots, w_k \in A^*$ *and* $N = \max(|w_1|, \ldots, |w_k|)$. *For any maximum pumpable set* S *of* $L_A(w_1, \ldots, w_k)$,

(1) if the vector space V *spanned by the occurrence vectors of* S *is of dimension one, then* $V = \text{span}(\{\mathbf{1}\})$ *where* $\mathbf{1}$ *is the* k*-dimensional vector with entries all 1, i.e., any occurrence vector* v *of* S *satisfies* $\texttt{diff}(v) = 0$.
(2) if the vector space V *spanned by the occurrence vectors of* S *is of dimension greater than or equal two, then we can choose a basis* $B \subseteq \{|\gamma|_{(w_1, \ldots, w_k)} \mid \gamma \in S\}$ *of* V *such that any element* v *of* B *satisfies* $\texttt{diff}(v) \neq 0$.

Proof. Condition (1) is a direct consequence of Lemma 1. Condition (2) is also from Lemma 1. Let $\gamma \in S$ be a pumpable cycle such that $\texttt{diff}(\gamma) \neq 0$. Such γ always exists since S contains at least two cycles whose occurrence vectors are linearly independent. Moreover, by Condition ($\bigstar$) in the proof of Lemma 1, we can deduce that there exists $B' \subseteq S$ such that the occurrence vectors of $B' \cup \{\gamma\}$ are linearly independent and $\mathbf{1} \in \text{span}(B' \cup \{\gamma\})$. Thus any vector of the form $n \cdot \mathbf{1}$ $(n \neq 0)$ is not in the occurrence vectors of $B' \cup \{\gamma\}$, we can take a desired basis B as an extension of $B' \cup \{\gamma\}$ $(B' \cup \{\gamma\} \subseteq B)$. $\square$

5.2 Proof of Theorem 1

To prove "only if" part, we modify standard Pumping Lemma as follows and call it Shrinking Lemma. Shrinking Lemma (see the full version [16] for the proof).

Lemma 4 (Shrinking Lemma for regular languages). *Let $L \subseteq A^*$ be a regular language. Then there exists a constant $c \in \mathbb{N}$ such that, for any number $n \geq c$ and for any word $w \in L$ with $|w| \geq n$, for any factorisation $w = xyz$ such that $|y| = n \geq c$, there exists a word y' such that (1) $y' \sqsubseteq_{sc} y$, (2) $|y'| \leq c$ and (3) $xy'z \in L$.*

Now we prove Theorem 1. Let $N = \max(|w_1|, \ldots, |w_k|)$. The "only if" part is shown by contraposition. Assume that the dimension of $L = L_A(w_1, \ldots, w_k)$ is two (higher-dimensional case can be shown similarly). Because L is of dimension two, there exists a maximum pumpable set $S = \{\gamma_{i_1}, \ldots, \gamma_{i_j}\}$ in some trace $T = \{\pi\} \cup \{\gamma_1, \ldots, \gamma_m\}$ in $\mathcal{G}_A^N$ such that two occurrence vectors $|\gamma_\alpha|_{(w_1,\ldots,w_k)}$ and $|\gamma_\beta|_{(w_1,\ldots,w_k)}$ of two cycles γ_α and γ_β in S are linearly independent and any occurrence vector of an element of S can be represented as a linear combination of $|\gamma_\alpha|_{(w_1,\ldots,w_k)}$ and $|\gamma_\beta|_{(w_1,\ldots,w_k)}$. By Condition (2) of Lemma 3, we can assume that $\mathrm{diff}(|\gamma_\alpha|_{(w_1,\ldots,w_k)}) \neq 0$ and $\mathrm{diff}(|\gamma_\beta|_{(w_1,\ldots,w_k)}) \neq 0$. Since S is a maximum pumpable set and the dimension of L is two, there exists a constant $c_T \in \mathbb{N}$ such that for any $n \in \mathbb{N}$ there exists a word $uv_n \in L$ with $\mathrm{Tr}(\mathtt{walk}_{\mathcal{G}_A^N}(u, v_n)) = T$, $(\mathbb{N}\mathrm{Tr}(u, v_n))(\gamma_\alpha) = n_\alpha$, $(\mathbb{N}\mathrm{Tr}(u, v))(\gamma_\beta) = n_\beta \geq n$ and $(\mathbb{N}\mathrm{Tr}(u, v_n))(\gamma_i) \leq c_T$ for each $i \in (\{1, \ldots, m\} \setminus \{\alpha, \beta\})$. By Proposition 3, we can assume that the walk $\mathtt{walk}_{\mathcal{G}_A^N}(u, v_n)$ is of the form

$$\mathtt{walk}_{\mathcal{G}_A^N}(u, v_n) = \omega_1 \odot \gamma_\alpha^{n_\alpha} \odot \omega_2 \odot \gamma_\beta^{n_\beta} \odot \omega_3.$$

Intuitively, $\mathtt{walk}_{\mathcal{G}_A^N}(u, v_n)$ firstly moves to $\mathtt{from}(\gamma_\alpha)$ (part of ω_1), and secondly passes γ_α repeatedly n_α-times and moves to $\mathtt{from}(\gamma_\beta)$ (part of $\gamma_\alpha^{n_\alpha} \odot \omega_2$), and lastly passes γ_β repeatedly n_β-times and moves to the end (part of $\gamma_\beta^{n_\beta} \odot \omega_3$). If L is regular, then by Lemma 4, there exists a constant c such that for any $n \geq c$ and the factorisation $uv_n = xy_nz_n$, where x, y_n and z_n are words corresponding to the first, second and last part of walks described above, there exists a word y_n' satisfying conditions (1)–(3) in Lemma 4. Because $\mathrm{diff}(|\gamma_\beta|_{(w_1,\ldots,w_k)}) \neq 0$, we have $|\gamma_\beta|_{w_j} < |\gamma_\beta|_{w_{j'}}$ for some $1 \leq j, j' \leq k$. However, since the length of x and y_n' are fixed by constant but z_n can be arbitrarily large, the gap of the occurrences $|z_n|_{w_{j'}} - |z_n|_{w_j}$ can be arbitrarily large (thus $|xy_n'z_n|_{w_{j'}} - |xy_n'z_n|_{w_j}$ can be arbitrarily large, too). It means that $xy_n'z_n \notin L$ for sufficiently large n, a contradiction.

The "if" part is achieved by showing that the language $L_T = \{uv \in L \mid |u| = N, \mathrm{Tr}(\mathtt{walk}_{\mathcal{G}_A^N}(u, v)) = T\}$ is regular for each trace $T = \{\pi\} \cup \{\gamma_1, \ldots, \gamma_m\}$ in $\mathcal{G}_A^N$. It implies that L is regular because $L = L^{<N} \cup \bigcup_{T:\mathrm{trace}} L_T$ (notice that $L^{<N} = \{w \in L \mid |w| < N\}$ is finite and thus regular). One can observe that $L = \{w \in L \mid |w| < N\} \cup \bigcup_{T:\ \mathrm{trace\ in}\ \mathcal{G}_A^N} L_T$, hence if every L_T is regular then L is also regular. To achieve it, we construct a deterministic automaton $\mathcal{A}_{T,S}$, where

S is the maximum pumpable set in T, so that $L_T = L(\mathcal{A}_{T,S})$. Let $\overline{S} = (T \backslash S \backslash \{\pi\})$ and define

$$c_T \stackrel{\text{def}}{=} \max\{(\text{NTr}(\text{walk}_{\mathcal{G}_A^N}(u,v)))(\gamma) \in \mathbb{N} \mid uv \in L, \text{Tr}(\text{walk}_{\mathcal{G}_A^N}(u,v)) = T, \gamma \in \overline{S}\}$$

(notice that $\max \emptyset \stackrel{\text{def}}{=} 0$ as usual). c_T is well-defined natural number, because, by the definition of pumpable set and S being maximum, for any cycle γ in T but not in S, the maximum number of occurrences of γ in a walk of some word in L is bounded. We denote by $\mathcal{F}$ the set of all functions from $\overline{S}$ to $\mathbb{N}^{\leq c_T}$. Notice that both $\overline{S}$ and $\mathbb{N}^{\leq c_T}$ are finite, $\mathcal{F}$ is also finite. Let $f_0 \in \mathcal{F}$ be the constant map to 0. Then the construction is as follows: $\mathcal{A}_{T,S} = (Q, \delta, \varepsilon, F)$ where each component is defined in Fig. 5.

$Q \stackrel{\text{def}}{=} A^{<N} \cup \{q_{\text{rej}}\} \cup Q'$ where $Q' = (\mathcal{P}(\mathcal{G}_A^N) \times 2^S \times \mathcal{F})$

$\delta \stackrel{\text{def}}{=} \{(u, a, ua) \mid u \in A^{<(N-1)}, a \in A\} \cup \{(u, a, ((ua), \emptyset, f_0)) \mid u \in A^{(N-1)}, a \in A\}$

$\cup \{(q_{\text{rej}}, a, q_{\text{rej}}) \mid a \in A\}$

① $\cup \{((\pi_1 \odot (aw), P, f), b, (\pi_1 \odot (aw, wb), P, f)) \mid a, b \in A, wb \notin V(\pi_1)\}$

② $\cup \{((\pi_1 \odot (wb) \odot \pi_2 \odot (aw), P, f), b, (\pi_1, P', f)) \mid a, b \in A, \pi_2 \odot (aw, wb) \in S,$
$P' = P \cup \{\pi_2 \odot (aw, wb)\}\}$

③ $\cup \{((\pi_1 \odot (wb) \odot \pi_2 \odot (aw), P, f), b, q_{\text{rej}}) \mid a, b \in A, \pi_2 \odot (aw, wb) \notin T\}$

④ $\cup \{((\pi_1 \odot (wb) \odot \pi_2 \odot (aw), P, f), b, (\pi_1, P, f')) \mid a, b \in A, \gamma = \pi_2 \odot (aw, wb) \in \overline{S},$
$f(\gamma) < c_T, f'(\gamma) = f(\gamma) + 1, f'(\gamma') = f(\gamma') \text{ for all } \gamma' \in (\overline{S} \setminus \{\gamma\})\}$

⑤ $\cup \{((\pi_1 \odot (wb) \odot \pi_2 \odot (aw), P, f), b, q_{\text{rej}}) \mid a, b \in A, \pi_2 \odot (aw, wb) \in \overline{S},$
$f(\pi_2 \odot (aw, wb)) \geq c_T\}$

$F \stackrel{\text{def}}{=} \{(\pi, S, f) \in Q' \mid f(\gamma) \geq 1 \text{ for all } \gamma \in \overline{S}, \text{diff}(\boldsymbol{v}(\pi, f)) = 0\}$

where $\boldsymbol{v}(\pi, f) = |\text{from}(\pi)|_{(w_1, \ldots, w_k)} + |\pi|_{(w_1, \ldots, w_k)} + \sum_{\gamma \in \overline{S}} f(\gamma) \cdot |\gamma|_{(w_1, \ldots, w_k)}$

Fig. 5. The construction of $\mathcal{A}_{T,S} = (Q, \delta, \varepsilon, F)$.

Although the formal definition in Fig. 5 could look complex, the behavior of $\mathcal{A}_{T,S}$ is simple: it computes path-cycle decomposition and counts the number of occurrences of each non-pumpable cycle $\gamma \in \overline{S}$. The main part of states is Q' which consists of the path part $\mathcal{P}(\mathcal{G}_A^N)$, pumpable-cycles part 2^S and non-pumpable-cycles part $\mathcal{F}$. While reading an input word w, $\mathcal{A}_{T,S}$ extends the path part (Case ①) if the next vertex wb is not in the current path. If the next vertex wb is already in the current path, there are four possibilities (Case ②–⑤). If the induced cycle γ on wb is in S (Case ②), $\mathcal{A}_{T,S}$ updates the pumpable cycle part. The number of occurrences of such cycle $\gamma \in S$ is not necessary to be memorised, since by Condition (1) of Lemma 3 $\text{diff}(|\gamma|_{(w_1, \ldots, w_k)}) = 0$. If γ is not in T (Case ③), $\mathcal{A}_{T,S}$ goes to the rejecting state q_{rej}, since the trace of w is never to be T. If γ is in $\overline{S}$, there are two possibilities further: if the current number of occurrences of γ is less than c_T (Case ④), $\mathcal{A}_{T,S}$ increments it, otherwise (Case ⑤), $\mathcal{A}_{T,S}$ goes to $q_{\texttt{rej}}$ because w is never to be in L by the definition of c_T.   □

5.3   Proof of Theorem 2

The proof structure is similar with the regular case (Theorem 1). The following lemma is a context-free variant of Lemma 4. Lemma 4 (see the full version [16] for the proof).

Lemma 5 (Shrinking Lemma for context-free languages). *Let $L \subseteq A^*$ be a context-free language. Then there exists a constant $c \in \mathbb{N}$ such that, for any number $n \geq c$ and for any word $w \in L$ with $|w| \geq n$, there exists a factorisation $w = xyz$ and a word y' such that (0) $2n > |y| \geq n \geq c$, (1) $y' \sqsubseteq_{\text{sc}} y$, (2) $|y'| \leq c$ and (3) $xy'z \in L$.*

Now we prove Theorem 2. Let $N = \max(|w_1|, \ldots, |w_k|)$. The "only if" part is shown by contraposition. Assume that the dimension of $L = L_A(w_1, \ldots, w_k)$ is three (higher-dimensional case can be shown similarly). Because L is of dimension three, there exists a maximum pumpable set $S = \{\gamma_{i_1}, \ldots, \gamma_{i_j}\}$ in some trace $T = \{\pi\} \cup \{\gamma_1, \ldots, \gamma_m\}$ in $\mathcal{G}_A^N$ such that three occurrence vectors $B = \{|\gamma_\alpha|_{(w_1, \ldots, w_k)}, |\gamma_\beta|_{(w_1, \ldots, w_k)}, |\gamma_\delta|_{(w_1, \ldots, w_k)}\}$ of three cycles γ_α, γ_β and γ_δ in S are linearly independent and any occurrence vector of an element of S can be represented as a linear combination of B. By Condition (2) of Lemma 3, we can assume that any vector $\boldsymbol{v}$ in B satisfies $\texttt{diff}(\boldsymbol{v}) \neq 0$. Since S is a maximum pumpable set and the dimension of L is three, there exists a constant $c_T \in \mathbb{N}$ such that for any $n \in \mathbb{N}$ there exists a word $uv_n \in L$ with $\text{Tr}(\texttt{walk}_{\mathcal{G}_A^N}(u, v_n)) = T$, $(\text{NTr}(\texttt{walk}_{\mathcal{G}_A^N}(u, v_n)))(\gamma_i) = n_i \geq n$ for each $i \in \{\alpha, \beta, \delta\}$ and $(\text{NTr}(\texttt{walk}_{\mathcal{G}_A^N}(u, v_n)))(\gamma_i) \leq c$ for each $i \in (\{1, \ldots, m\} \setminus \{\alpha, \beta, \delta\})$. By Proposition 3, we can assume that the walk $\texttt{walk}_{\mathcal{G}_A^N}(u, v_n)$ is of the form

$$\texttt{walk}_{\mathcal{G}_A^N}(u, v_n) = \omega_1 \odot \gamma_\alpha^{n_\alpha} \odot \omega_2 \odot \gamma_\beta^{n_\beta} \odot \omega_3 \odot \gamma_\delta^{n_\delta} \odot \omega_4.$$

Let $u_{n,1}, u_{n,2}$ and $u_{n,3}$ be words corresponding to $\omega_1 \odot \gamma_\alpha^{n_\alpha}, \omega_2 \odot \gamma_\beta^{n_\beta}$ and $\omega_3 \odot \gamma_\delta^{n_\delta} \odot \omega_4$, respectively (thus $uv_n = u_{n,1}u_{n,2}u_{n,3}$). Let $M_n = \min\{n_\alpha \cdot |\gamma_\alpha|, n_\beta \cdot |\gamma_\beta|, n_\delta \cdot |\gamma_\delta|\}$. If L is context-free, then by Lemma 5, there exists a constant c such that for any $n \geq c$, there is a factorisation $uv_n = x_n y_n z_n$ and a word y_n' satisfying conditions (0)–(3) in Lemma 5. Take $n \in \mathbb{N}$ that satisfies $M_n \geq c$. Then, the word y in the factorisation $uv_n = x_n y_n z_n$ above can cross at most two words from $u_{n,1}, u_{n,2}, u_{n,3}$. It means that $x_n y_n' z_n \notin L$ for sufficiently large n, a contradiction.

The "if" part is achieved in a similar way as the regular case: we can construct a pushdown automaton $\mathcal{A}_{T,S}$, where S is the maximum pumpable set in T, so that $L_T = L(\mathcal{A}_{T,S})$. The only difference is that $\mathcal{A}_{T,S}$ uses its stack for checking the consistency the occurrences of two linearly independent occurrence vectors. $\mathcal{A}_{T,S}$ achieves it as some pushdown automaton recognises $L_A(a, b)$.   □

6   Conclusion and Future Work

In this paper, we provided decidable, necessary and sufficient conditions of the regularity and context-freeness for WMIX languages by using the notion of dimensions. Complexity issues on these problems (tight lower/upper bounds, more efficient algorithm, *etc.*) are untouched and could be future work.

The author's main interest is how to generalise the main result into more richer language classes, *e.g.*, UnCA languages [11]. From WMIX languages (represented by de Bruijn graphs and diagonals $\{n \cdot \mathbf{1} \mid n \in \mathbb{N}\}$) into UnCA languages (represented by unambiguous automata and semilinear sets), although we should modify the notion of dimensions and some part of the proof strategy, the author conjectures that the context-freeness is still decidable for UnCA languages.

Acknowledgement. The author would like to thank Thomas Finn Lidbetter for telling me this topic in DLT 2018. Special thanks also go to my colleague Fazekas Szilard whose helpful discussion were an enormous help to me. The author also thank to anonymous reviewers for many valuable comments. This work was supported by JSPS KAKENHI Grant Number JP19K14582.

References

1. Joshi, A., Vijay-Shanker, K., Weir, D.: The convergence of mildly context-sensitive grammar formalisms. Foundational Issues in Natural Language Processing, pp. 31–82 (1991)
2. Marsh, W.: Some conjectures on indexed languages. Abstract Appears J. Symb. Log. **51**(3), 849 (1985)
3. Boullier, P.: Chinese numbers, mix, scrambling, and concatenation grammars range. In: EACL 1999, 9th Conference of the European Chapter of the Association for Computational Linguistics, 8–12 June 1999, University of Bergen, Bergen, Norway, pp. 53–60. The Association for Computer Linguistics (1999)
4. Kanazawa, M., Salvati, S.: MIX is not a tree-adjoining language. In: The 50th Annual Meeting of the Association for Computational Linguistics, Proceedings of the Conference, 8–14 July 2012, Jeju Island, Korea - Volume 1: Long Papers, pp. 666–674. The Association for Computer Linguistics (2012)
5. Salvati, S.: MIX is a 2-MCFL and the word problem in $\mathbb{Z}^2$ is captured by the IO and the OI hierarchies. J. Comput. Syst. Sci. **81**(7), 1252–1277 (2015)
6. Parikh, R.: On context-free languages. J. ACM **13**(4), 570–581 (1966)
7. Colbourn, C.J., Dougherty, R.E., Lidbetter, T.F., Shallit, J.: Counting subwords and regular languages. In: Hoshi, M., Seki, S. (eds.) DLT 2018. LNCS, vol. 11088, pp. 231–242. Springer, Cham (2018). https://doi.org/10.1007/978-3-319-98654-8_19
8. Mateescu, A., Salomaa, A., Salomaa, K., Yu, S.: A sharpening of the Parikh mapping. Theor. Inf. Appl. **35**(6), 551–564 (2001)
9. Mateescu, A., Salomaa, A., Yu, S.: Subword histories and Parikh matrices. J. Comput. Syst. Sci. **68**(1), 1–21 (2004)
10. Seki, S.: Absoluteness of subword inequality is undecidable. Theoret. Comput. Sci. **418**, 116–120 (2012)

11. Cadilhac, M., Finkel, A., McKenzie, P.: Unambiguous constrained automata. In: Yen, H.-C., Ibarra, O.H. (eds.) DLT 2012. LNCS, vol. 7410, pp. 239–250. Springer, Heidelberg (2012). https://doi.org/10.1007/978-3-642-31653-1_22

12. Ginsburg, S.: The Mathematical Theory of Context-Free Languages. McGraw-Hill, Inc., New York (1966)

13. Kászonyi, L.: A pumping lemma for DLI-languages. Discrete Math. **258**(1), 105–122 (2002)

14. Leroux, J., Penelle, V., Sutre, G.: The context-freeness problem is coNP-complete for flat counter systems. In: Cassez, F., Raskin, J.-F. (eds.) ATVA 2014. LNCS, vol. 8837, pp. 248–263. Springer, Cham (2014). https://doi.org/10.1007/978-3-319-11936-6_19

15. Schwer, S.R.: The context-freeness of the languages associated with vector addition systems is decidable. Theoret. Comput. Sci. **98**(2), 199–247 (1992)

16. Sin'ya, R.: Context-freeness of word-mix languages (full version) (2020). http://www.math.akita-u.ac.jp/~ryoma/misc/dlt2020full.pdf

17. de Luca, A., Varricchio, S.: Finiteness and Regularity in Semigroups and Formal Languages. Monographs in Theoretical Computer Science. An EATCS Series. Springer, Heidelberg (1999). https://doi.org/10.1007/978-3-642-59849-4

18. Sin'ya, R.: Note on the infiniteness of $L(w_1, \ldots, w_k)$. CoRR abs/1812.02600 (2018)

The Characterization of Rational Numbers Belonging to a Minimal Path in the Stern-Brocot Tree According to a Second Order Balancedness

Andrea Frosini[1] and Lama Tarsissi[2,3](✉)

[1] Dipartimento di Matematica e Informatica, Università di Firenze,
Viale Morgagni 65, 50134 Firenze, Italy
`andrea.frosini@unifi.it`
[2] LAMA, Univ Gustave Eiffel, UPEC, CNRS, 77454 Marne-la-Vallée, France
[3] LIGM, Univ Gustave Eiffel, CNRS, ESIEE Paris, 77454 Marne-la-Vallée, France
`lama.tarsissi@esiee.fr`

Abstract. In 1842, Dirichlet observed that any real number α can be obtained as the limit of a sequence $(\frac{p_n}{q_n})$ of irreducible rational numbers. Few years later, M. Stern (1858) and A. Brocot (1861) defined a tree-like arrangement of all the (irreducible) rational numbers whose infinite paths are the Dirichlet sequences of the real numbers and are characterized by their continued fraction representations. The Stern-Brocot tree is equivalent to the Christoffel tree obtained by ordering the Christoffel words according to their standard factorization. We remark that the Fibonacci word's prefixes belong to a minimal path in the Christoffel tree with respect to the second order balancedness parameter defined on Christoffel words. This alows us to switch back to the Stern-Brocot tree, in order to give a characterization of the continued fraction representation for all the rational numbers belonging to minimal paths with respect to the growth of the second order balancedness.

Keywords: Continued fractions · Stern-Brocot tree · Christoffel words · Balance property · Minimal path

1 Introduction

It appears that balancedness [2] is a crucial notion for the recently developed research area of combinatorics on words. A widely studied class of binary balanced words are the Christoffel words, whose interest arises from their strong connections with geometry, algebra, and number theory (see [2] for the definitions, first properties and extensive bibliography). Each Christoffel word is the

This work was partly funded by the French Programme d'Investissements d'Avenir (LabEx Bézout, ANR-10-LABX-58) and ANR-15-CE40-0006.

N. Jonoska and D. Savchuk (Eds.): DLT 2020, LNCS 12086, pp. 319–331, 2020.
https://doi.org/10.1007/978-3-030-48516-0_24

discretization of a line segment of rational slope, and it can be uniquely factorized into two different Christoffel words. Such standard factorization allows us to arrange all the Christoffel words in a binary tree structure, called Christoffel tree, whose root is the word $w = 01 = (0,1)$. The notion of balancedness allows us to define, for each binary word w, a balance matrix B_w whose maximal element provides its order of balancedness, noted by $\delta(w)$. The present study focuses on Christoffel words that are balanced words, i.e. whose order of balancedness is 1. In this case, each row of the balance matrix is again a binary word, whose order of balancedness can be computed again, obtaining a second order balance matrix U_w whose maximal element is defined as the second order balance value $\delta^2(w)$. This value can also be obtained using the link between the abelian complexity and balanced words represented by Zamboni ([6], Section 4), using the Parikh vectors.

The matrix U_w has several properties that allow a recursive construction by referring to three $U_{w'}$ matrices, of some specific ancestors w' of w in the Christoffel tree [12]. The authors used this notion to study the distribution of 1's in Christoffel words and relate it to the longstanding problem of their synchronization [7,8].

The correspondence between Christoffel word and continued fraction representation of its rational slope allows us to transfer the notion of balancedness to the Stern-Brocot tree. This tree, indicated hereafter as SB-tree, is a binary tree where the irreducible rational numbers are arranged according to a specific operator called Farey sum, defined independently in [4] and in [11], and that perfectly mimes the notion of Standard decomposition of a Christoffel word. In our study, we first establish a connection between the second order balance value of a Christoffel word and the form of the continued fraction representation of the slope of the related discrete segment. Then, we present the sequence of the ratios of consecutive Fibonacci numbers constitutes, in the SB-tree, as being an example of a path which minimizes the growth of the second order balancedness parameter of the related Christoffel words. Finally, relying on this result, we characterize the minimal paths (w.r.t the second order balance value) by mixing algebraic and arithmetic techniques on the continued fraction representations. In fact, in the final theorem, we determine the rational numbers belonging to a minimal path with respect to the form of their continued fractions.

The paper is structured as follows: in Sect. 2, we recall the definitions of Christoffel words, Christoffel tree and the Stern-Brocot tree. In Sect. 2.4, we introduce the notion of balancedness and we define the balance matrix of a binary word. Then we generalize the notion to the second order balance matrix. Finally, in Sect. 3, we show how the second order balancedness parameter is spread on the SB-tree, providing the example of the Fibonacci sequence. Here we provide our main result: the characterization of the minimal paths in the SB-tree according to the δ^2 parameter. We prove that a rational number on the SB-tree belongs to a minimal path, according to the growth of a second order balancedness, if the elements of its continued fractions start by $0,1$ and end by 2, while the middle terms are only made of blocks of $(1,1,1); (2,1); (1,2)$ or (3).

2   Definitions and Previous Results

We refer to the book [5] for the standard terminology in combinatorics of words: *alphabet, word, length of a word, occurrence of a letter, factor, prefix, suffix, period, conjugate, primitive, reversal, palindrome etc.* The related notation will be recalled when used.

Christoffel Paths and Christoffel Words. In discrete geometry, the theory of Christoffel words has been considered during this last few decades and has acquired a prominent role in the study of the discretization of segments and shapes. Concerning the former, let a, b be two co-prime numbers; the *Christoffel path* of slope $\frac{a}{b}$ is defined as the connected path joining the origin $O(0,0)$ to the point (b, a) in the integer lattice $\mathbb{Z} \times \mathbb{Z}$; such that it is the nearest path below the Euclidean line segment joining these two points, as shown in Fig. 1. So, there are no points of the discrete plane between the path and the line segment.

Still in Fig. 1, we can see the coding of a Christoffel path by a binary word, say Christoffel word, whose letters 0 and 1 represent a horizontal and a vertical step in the path, respectively.

The Christoffel word related to the path reaching the point (b, a) is indicated by $C(\frac{a}{b})$, and its slope is $\frac{a}{b} = \frac{|w|_1}{|w|_0}$, where the notation $|w|_x$ stands for the number of occurrences of the letter x in w.

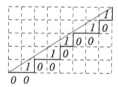

Fig. 1. The Christoffel path of the line segment of slope $\frac{5}{8}$, and the corresponding Christoffel word $C(\frac{5}{8}) = 0010010100101$.

A binary word w is *k-balanced*, with $k > 0$, if, for any two factors u and v of its conjugates such that $|u| = |v|$, it follows that $||u|_1 - |v|_1| \leq k$. One of the most important properties of Christoffel words is that they are 1-balanced binary words, or simply balanced words, as shown in [3].

2.1   Christoffel Tree

Let us recall the definition of the Christoffel tree, i.e. a well-known arrangement of the Christoffel words as a binary tree (see [1]). The root of the Christoffel tree is labeled by the pair $(0, 1)$, representing the Christoffel word 01 of slope $\frac{1}{1}$. Each node (u, v) generates two children according to two functions ϕ_0 and ϕ_1 such that: $\phi_0(u, v) = (u, uv)$ produces the left-child, and $\phi_1(u, v) = (uv, v)$

produces the right-child. Each node (u, v) corresponds to the Christoffel word uv and represents its *standard factorization form* [5]. Figure 2 shows the first levels of the Christoffel tree.

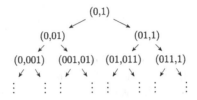

Fig. 2. The first levels of the Christoffel tree.

Let $w = C(\frac{a}{b})$ be a Christoffel word lying at level k of the tree. The *directive sequence* of w is the word $\Delta(\frac{a}{b}) = i_1 i_2 \ldots i_k$ such that $w = (\phi_{i_k} \circ \ldots \circ \phi_{i_2} \circ \phi_{i_1})(0, 1)$. According to the definition of ϕ_0 and ϕ_1, the elements of $\Delta(\frac{a}{b})$ also show, step by step, the directions to reach the word w in the tree starting from its root at level 1. If $i_n = 0$, we must move to the left, otherwise, to the right.

As an example, the directive sequence of the Christoffel word $C(\frac{4}{3}) = 0101011$ at level 4 of the Christoffel tree is $\Delta(\frac{4}{3}) = 100$: according to the definition it holds $(\phi_0 \circ \phi_0 \circ \phi_1)(0, 1) = (\phi_0 \circ \phi_0)(01, 1) = \phi_0(01, 011) = (01, 01011)$.

2.2   Stern-Brocot Tree

The Christoffel tree is known to be isomorphic to the SB-tree, that was introduced by M. Stern [11] and A. Brocot [4] as a binary-tree arrangement of the irreducible fractions. Such arrangement relies on the Farey sum operator, indicated by $\oplus$, and defined on two generic rational numbers as $\frac{a}{b} \oplus \frac{c}{d} = \frac{a+c}{b+d}$. The root of the SB-tree is labeled with the fraction $\frac{1}{1}$ and each node at level $n > 1$ is labeled with the Farey sum of its nearest left and right ancestors, i.e., the nodes lying on the greatest level of the tree and having k in its left and right subtree, respectively. The left and right ancestors of the root are considered to be the fractions $\frac{0}{1}$ and $\frac{1}{0}$, even if not present in the tree. The first levels of the SB-tree are depicted in Fig. 3.

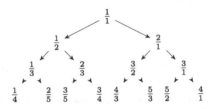

Fig. 3. The first levels of the Stern-Brocot tree.

As an example, we compute the left-child and right-child of the node $\frac{3}{4}$ by Farey summing it with its left-ancestor $\frac{2}{3}$ and with its right-ancestor $\frac{1}{1}$, and obtaining $\frac{5}{7}$ and $\frac{4}{5}$, respectively. The SB-tree contains once each irreducible fraction and over the years has attracted researchers for its many interesting properties.

Remark 1. By the construction of the SB-tree, the following hold

i) The tree is divided into two symmetric parts. The left part contains all the reduced fractions less than 1, while the right part contains the reduced fractions bigger than 1.

ii) For each level k of the SB tree, the rational numbers appear in increasing order from left to right, and their values are bounded by $\frac{1}{k}$ and $\frac{k}{1}$.

From our perspective, the SB-tree's appeal reveals in its close relationship with the Christoffel tree: in fact, we can note that if we replace the label of each node (u, v) of this last with the rational number $\frac{|uv|_1}{|uv|_0}$ representing the slope of the corresponding Christoffel word, we obtain the SB-tree. This correspondence is interesting but not surprising once we observe that the slope of the concatenation of two Christoffel words is the Farey sum of their slopes. For instance, take the rational number $\frac{3}{5}$ on the SB-tree, it is at the same position of the Christoffel word 00100101 of slope $\frac{3}{5}$ on the Christoffel tree.

2.3   Continued Fraction Representation

Our study requires to recall a final notion to identify the nodes of the SB-tree. Let $\frac{a}{b}$ be a positive fraction, we define its *continued fraction representation* as the sequence of integers $[a_0, \ldots, a_z]$, with $a_0 \geq 0$, represented below, and such that for each $1 \leq i \leq z$, $a_i \geq 1$:

$$\frac{a}{b} = a_0 + \cfrac{1}{a_1 + \cfrac{1}{\ddots\, a_{z-1} + \frac{1}{a_z}}}$$

In order to obtain a unique continued fraction representation of each rational number, it is also commonly required that if $z \geq 2$, then $a_z \geq 2$. Otherwise, with a simple calculation, we can remark that if $a_z = 1$, it is sufficient to reduce the sequence to a_{z-1} and add $+1$, for example: $\frac{2}{3} = [0, 1, 1, 1] = [0, 1, 2]$. The continued fraction representation of a rational number is always finite while the representation is infinite in the case of an irrational number (among the vast literature on continued fraction representation, we refer the reader to [10] for the main properties).

Equivalently to the Christoffel tree, the directive sequence of the Christoffel word of slope $\frac{a}{b}$ can be read in the SB-tree by using the terms of the continued fraction representation of $\frac{a}{b}$. The following lemma presents this connection (a proof can be found in [9] and [12]).

Lemma 1. [9] *Let w be the Christoffel word of slope $\frac{a}{b}$ whose continued fraction representation is $[a_0, a_1, \ldots, a_z]$. The directive sequence of w has the following form: $\Delta(\frac{a}{b}) = 1^{a_0} 0^{a_1} 1^{a_2} \ldots p^{a_z - 1}$, where $p \in \{0, 1\}$ according to the parity of z.*

Referring again to the continued fraction representation $[0, 1, 2, 2]$ of $\frac{5}{7}$, the directive sequence $\Delta(\frac{5}{7})$ is exactly $1^0 0^1 1^2 0^1$.

2.4   The Second Order Balance Matrix

The second order of balancedness of a Christoffel word w, denoted $\delta^2(w)$, provides an idea of how uniformly the factors of each abelian class, as defined in [6], are distributed in w. It is the maximal value of the second order balance matrix, denoted U_w, introduced in [12]. In order to construct this matrix, the author had to define the *balance matrix* B_w, of a binary word w. This matrix calculates the order of balancedness, that is usually obtained by computing $||u|_1 - |v|_1| \leq k$ for any factors u and v of same length, of any binary word $w = w_1 w_2 \ldots w_n$ in an explicit way. To do that, we first compute the matrix S_w of dimension $(n-1) \times n$ whose generic element $S_w[i, j]$ counts the number of elements 1 in the i-length prefix of the conjugate of w starting in position j. We let $M[i]$ represents the ith row of any matrix M.

The balance matrix B_w is computed from S_w by subtracting the minimum of a given row to each of its element, i.e.,

$$B_w[i, j] = S_w[i, j] - \min\{S_w[i]\}.$$

By construction, we can note the following lemma:

Lemma 2. *Let $w = w_1 w_2 \ldots w_n$, the maximum element of B_w is equal to the balance orer of w.*

We denote the balance order by $\delta(w)$. A simple example will clarify the construction.

Example 1. Let us consider the Christoffel word $w = 00100101$ of slope $\frac{3}{5}$. We provide the matrices S_w and B_w:

$$S_w = \begin{pmatrix} 0 & 0 & 1 & 0 & 0 & 1 & 0 & 1 \\ 0 & 1 & 1 & 0 & 1 & 1 & 1 & 1 \\ 1 & 1 & 1 & 1 & 1 & 2 & 1 & 1 \\ 1 & 1 & 2 & 1 & 2 & 2 & 1 & 2 \\ 1 & 2 & 2 & 2 & 2 & 2 & 2 & 2 \\ 2 & 2 & 3 & 2 & 2 & 3 & 2 & 2 \\ 2 & 3 & 3 & 2 & 3 & 3 & 2 & 3 \end{pmatrix}, B_w = \begin{pmatrix} 0 & 0 & 1 & 0 & 0 & 1 & 0 & 1 \\ 0 & 1 & 1 & 0 & 1 & 1 & 1 & 1 \\ 0 & 0 & 0 & 0 & 0 & 1 & 0 & 0 \\ 0 & 0 & 1 & 0 & 1 & 1 & 0 & 1 \\ 0 & 1 & 1 & 1 & 1 & 1 & 1 & 1 \\ 0 & 0 & 1 & 0 & 0 & 1 & 0 & 0 \\ 0 & 1 & 1 & 0 & 1 & 1 & 0 & 1 \end{pmatrix}.$$

Since each Christoffel word w is 1-balanced, then the rows of its balance matrix B_w are binary words too. So, we can push the notion of balancedness to a second level by studying the balancedness of the rows of B_w. Let us define the *second order balance matrix* U_w of dimension $(n-1) \times (n-1)$, with $n = |w|$, as $U_w[i, j] = \max(B_{B_w[i]}[j])$. In words, the element $U_w[i, j]$ shows the balancedness of the i-th row of B_w with respect to its proper factors of length j. More details and properties about this matrix can be found in [12].

Example 2. Let us consider the 1-balanced word w of Example 1. The fourth row of the matrix U_w is obtained by computing the balance value of the fourth row of B_w with respect to all the possible factors' lengths from 1 to 8. We explicitly determine the element $U_w[4, 3]$: the eight factors of the conjugates of $B_w[4]$ of length 3 are $(001, 010, 101, 011, 110, 101, 101, 100)$, and the maximum difference of the number of elements 1 between any two of them is 1, so $U_w[4, 3] = 1$. Doing an analogous computation for all the lengths from 1 to 7, we obtain $U_w[4] = (1, 2, 1, 2, 1, 2, 1)$.

So, the second order of balancedness of a Christoffel word w provides an idea of how uniformly the factors of each abelian class, as defined in [6], are distributed in w.

3   Path Minimality in SB-Tree

In this section, we prove some new properties of the second order balance parameter for Christoffel words and we show how it is distributed on the Christoffel tree. Finally, we characterize the paths of the tree where the parameter's growth is minimal. So, let us consider the following two ancestors of $\frac{a}{b} = [a_0, \ldots, a_z]$ in the SB-tree:

$\frac{u}{v}$ whose continued fraction expansion is $= [a_0, \ldots, a_{z-1} + 1]$
$\frac{\rho}{\theta}$ whose continued fraction expansion is either $[a_0, \ldots, a_z - 2]$ if $a_z > 2$, or $[a_0, \ldots, a_{z-2}]$ if $a_z = 2$.

In [12], Section 6.1, the authors proved that the matrix U_w, with $w = C(\frac{a}{b})$, can be decomposed into blocks belonging either to U_{w_1} or to $U_{w_2} + 1$, with $w_1 = C(\frac{u}{v})$ and $w_2 = C(\frac{\rho}{\theta})$. From this result it immediately follows:

Theorem 1. *The second order balance value of $C(\frac{a}{b})$ is:*

$$\delta^2 \left(C \left(\frac{a}{b} \right) \right) = \max \left(\delta^2 \left(C \left(\frac{u}{v} \right) \right), \delta^2 \left(C \left(\frac{\rho}{\theta} \right) \right) + 1 \right).$$

Using the previous result, we set a lower bound to the growth of the second order balanced value inside the Christoffel tree. The next lemma states that $\delta^2(C(\frac{a}{b}))$ increases according to the distance between $\frac{a}{b}$ and $\frac{u}{v}$ in the SB-tree.

Lemma 3. *Let $\frac{a}{b} = [a_0, \ldots, a_z]$ and belongs to the level k on the SB-tree, where $\delta^2(C(\frac{u}{v})) = t$ and $\delta^2(C(\frac{\rho}{\theta})) = n$.*

i) If $a_z = 2$, then $\delta^2(C(\frac{a}{b})) \in \{t; t + 1\}$.
ii) If $a_z > 2$, then $\delta^2(C(\frac{a}{b})) = n + 1$, i.e it is increased by one value each 2 levels after $\frac{u}{v}$.

Proof. Let us consider these two rational numbers $\frac{a}{b} = [a_0, \ldots, a_z]$ and $u/v = [a_0, \ldots, a_{z-1} + 1]$, where $\delta^2(C(\frac{u}{v})) = t$.
If $a_z = 2$, then $\frac{\rho}{\theta} = [a_0, \ldots, a_{z-2}]$, in this case, its second order value is equal to $n \leq t$. In all the cases, from Theorem 1, $\delta^2(C(\frac{a}{b})) = \max(\delta^2(C(\frac{u}{v})), \delta^2(C(\frac{\rho}{\theta})) + 1)$ that can be either t or $t + 1$.
If $a_z \geq 3$, then we must consider two cases for the rational number $\frac{\rho}{\theta}$.

- If $a_z = 3$, in this case $\frac{\rho}{\theta} = \frac{u}{v}$ and $n = t$, then:
  $\delta^2(C(\frac{a}{b})) = \max(\delta^2(C(\frac{u}{v})), \delta^2(C(\frac{\rho}{\theta}) + 1) = n + 1$.
- If $a_z \geq 4$, in this case $\frac{\rho}{\theta} = [a_0, a_1, \cdots, a_z - 2]$, and $n \geq t$ then:
  $\delta^2(C(\frac{a}{b})) = \delta^2(C(\frac{\rho}{\theta})) + 1 = n + 1$.

Finally, we provide a lower bound, denoted by δ_k^2, for the second order balance value of any element at level k in the Christoffel tree.

Theorem 2. *For each level k in the SB tree, we have:* $\delta_k^2 \geq \lceil \frac{k}{3} \rceil$.

Proof. By induction on the levels of the Stern-Brocot tree, we have $\delta_1^2 = \delta_2^2 = 1$; and $\delta_3^2 \in \{1, 2\}$ while $\delta_4^2 = 2$. Hence the minimal value of δ_3^2 (denoted $\min \delta_3^2$) is equal to 1 and $\min \delta_4^2 = 2$. Which means that $\delta_i^2 \geq \lceil \frac{i}{3} \rceil$ with $1 \leq i \leq 4$.

Suppose that it is true for all the levels till $k = 3k'$; i.e $\min \delta_k^2 = \min \delta_{k-1}^2 = \min \delta_{k-2}^2 = k'$. We prove that $\delta_{k+1}^2 \geq \lceil \frac{k+1}{3} \rceil = k' + 1$.

All the children of the fractions on level k, with $\delta^2 > k'$, have $\delta^2 \geq k' + 1$. It remains to study the case where $\delta_k^2 = k'$. For that, we let $\frac{u}{v}$ be a fraction at level k with $\delta^2(C(\frac{u}{v})) = k'$. By contradiction, we let $\frac{c}{d}$ be the fraction at level $k + 1$ with $\delta^2(C(\frac{c}{d})) = k'$. This fraction is either in the same, or opposite direction of $\frac{u}{v}$. If $\frac{c}{d}$ is in the same direction of $\frac{u}{v}$, then by Lemma 3, we get $\delta^2(C(\frac{a}{b})) = k' - 1$, where $\frac{a}{b}$ is the fraction at level $k - 1$. If $\frac{c}{d}$ is in the opposite direction, then the fraction $\frac{u}{v}$ can be either in the same direction or opposite to $\frac{a}{b}$. In this case, we let $\frac{m}{n}$ be the fraction at level $(k - 2)$ and we have:

- If $u/v = (a/b)$ then by Lemma 3, $\delta^2(C(\frac{m}{n})) = k' - 1$.
- If $u/v = (a/b)$, then if $\delta^2(C(\frac{a}{b})) = k'$, we get:
  $\delta^2(C(\frac{c}{d})) = \max(\delta^2(C(\frac{u}{v})), \delta^2(C(\frac{m}{n})) + 1) = k'$. If $\delta^2(C(\frac{m}{n})) = k'$ then $\delta^2(C(\frac{c}{d})) = k' + 1$; and if $\delta^2(C(\frac{m}{n})) = k' - 1$ we get a contradiction since $\delta_{(k-2)}^2 \geq \lceil \frac{k-2}{3} \rceil = k'$.

Therefore $\delta^2 \geq k' + 1$ for all the fractions on the level $k + 1$ thus

$$\delta_{k+1}^2 \geq \lceil \frac{k+1}{3} \rceil = \lceil \frac{3k'+1}{3} \rceil = k' + 1.$$

$\square$

The remaining part of the section, is devoted to the study of the paths of the SB-tree whose related Christoffel words realize the lower bound. The symmetry of the tree allows us to consider, without loss of generality, its left part only, i.e., those fractions less than or equal to 1 and whose continued fractions have $a_0 = 0$. We denote the subtree SB_L.

3.1 Minimal Paths in the SB-Tree

Let us assign to each fraction of SB_L the δ^2 value of its related Christoffel word starting from the root $\frac{1}{1}$: Fig. 4 shows the computation till the fifth level of the tree.

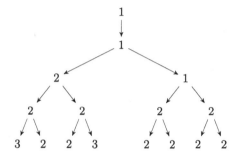

Fig. 4. The second order balancedness of the rational numbers belonging to the first five levels of SB_L.

Let us indicate by $P(\frac{a}{b})$ the sequence of all the fractions in the path from the root of SB_L to the element $\frac{a}{b}$, and by $P_{\delta^2}(\frac{a}{b})$ the sequence of the related δ^2 values. From Lemma 3, it follows that $P_{\delta^2}(\frac{a}{b})$ is a weakly increasing integer sequence.

As an example, let us consider the fraction $\frac{7}{12}$, lying in the sixth level of SB_L. We have $P(\frac{7}{12}) = (\frac{1}{1}, \frac{1}{2}, \frac{2}{3}, \frac{3}{5}, \frac{4}{7}, \frac{7}{12})$, and the related δ^2 sequence is $P_{\delta^2}(\frac{7}{12}) = (1, 1, 1, 2, 2, 2)$.

We give now the definition of a *minimal path* at a certain level k of SB_L.

Definition 1. *For each level k of the SB_L, the minimal path MP_k is given by:*
$(1, 1, 1, 2, 2, 2, 3, \ldots, \lceil \frac{k}{3} \rceil)$.

In order to give an example, we consider the Fibonacci sequence $\{f_k\}_{k=0}^{\infty} = 1, 1, 2, 3, 5, 8, 13, \ldots$, and define the *zig-zag* path to be the sequence of rational numbers obtained by the ratio of two consecutive elements. It is well known that the golden number is the limit of the sequence of the rational numbers $\frac{f_k}{f_{k+1}}$, with $k \geq 0$. The continued fraction representations of these rational numbers are of the form $[0, 1, 1 \ldots, 1, 2]$, and they constitute a (infinite) path in SB_L. This path realizes the lower bound stated in Theorem 2 assuring that, for each k, $P_{\delta^2}(\frac{f_k}{f_{k+1}})$ is minimal, and it equals MP_k.

Now we focus our study on the questions: For each level k, is the *zig-zag* path the only minimal one? If no, can we characterize all the minimal paths in SB_L?

3.2 A Connection Between δ^2 Value and the Continued Fraction Representation

Let us consider the following continued fraction representations:
$\frac{a}{b} = [a_0, a_1, \cdots, a_z]; \frac{u}{v} = [a_0, a_1, \cdots, a_{z-1} + 1]; \frac{p}{q} = [a_0, \cdots, a_{z-2} + 1];$
$\frac{s}{t} = [a_0, \cdots, a_{z-2}];$ and $\frac{\rho}{\theta} = \begin{cases} [a_0, a_1, \cdots, a_z - 2] & \text{if } a_z \geq 4 \\ [a_0, a_1, \cdots, a_{z-1} + 1] & \text{if } a_z = 3 \\ [a_0, a_1, \cdots, a_{z-2}] & \text{if } a_z = 2. \end{cases}$

From Lemma 3, we know that $\delta^2 \left(C \left(\frac{a}{b} \right) \right)$ is related to $\delta^2 \left(C \left(\frac{u}{v} \right) \right)$ and $\delta^2 \left(C \left(\frac{\rho}{\theta} \right) \right)$. In the final part of the section, we will show how this relation reflects on their continued fraction representations.

Theorem 3. *Let* $\frac{a}{b} = [a_0, a_1, \cdots, a_z]$, *we have:*

$$\delta^2 \left(C \left(\frac{a}{b} \right) \right) = \begin{cases} \delta^2 \left(C \left(\frac{u}{v} \right) \right) & \text{if } a_{z-1} \geq 2 \text{ and } a_z = 2 \\ \delta^2 \left(C \left(\frac{\rho}{\theta} \right) \right) + 1 & \text{elsewhere.} \end{cases}$$

Proof. Without loss of generality and in order to enlighten the notation, we denote $\delta^2 \left(C \left(\frac{a}{b} \right) \right)$ by D_a, $\delta^2 \left(C \left(\frac{u}{v} \right) \right) = D_u$, $\delta^2 \left(C \left(\frac{\rho}{\theta} \right) \right) = D_r$, $\delta^2 \left(C \left(\frac{\rho}{q} \right) \right) = D_p$ and $\delta^2 \left(C \left(\frac{s}{t} \right) \right) = D_s$. Equivalently, by symmetry, we consider z to be an even number which means that the last element of $\Delta(\frac{a}{b})$ is equal to 1. We start by considering separately the cases $a_z \geq 4$ and $a_z = 3$, then we discuss the case $a_z = 2$.

1. If $a_z \geq 4$, we let $D_u = d$, then $D_r \geq d$ since $\frac{\rho}{\theta} > \frac{u}{v}$. Hence $D_a = \max(D_u, D_r + 1) = D_r + 1$ (see Fig. 5, on the left).
2. If $a_z = 3$, in this case the fractions $\frac{u}{v}$ and $\frac{\rho}{\theta}$ are the same and we have: $D_a = \max(D_u, D_r + 1) = \max(D_r, D_r + 1) = D_r + 1$ (see Fig. 5, on the right).
3. If $a_z = 2$, in this case we get several sub-cases depending on the values of a_{z-1} and a_{z-2}.
   (a) If $a_{z-1} = 1$, we get 4 sub-cases that are summarized in the table of Fig. 6, where fraction $\frac{u}{v} = [a_0, \cdots, 2]$ and $D_r = d$. If $D_p = d$, then D_u can be equal to d or $d+1$. In the former case, it holds $D_a = \max(D_u, D_r + 1) = d + 1 = D_r + 1$, while, in the latter, it holds then $D_a = \max(d+1, d+1) = d+1 = D_r + 1$. Acting similarly in the case $D_p = d+1$, it holds $D_a = d + 1 = D_r + 1$.
   (b) If $a_{z-1} = 2$, the cases are represented in Fig. 7. If $D_p = d$, then by Lemma 3, $D_u = d + 1$ and $D_a = d + 1 = D_u$. On the other hand, if $D_p = d+1$, then $D_u = d + 2$ and $D_a = d + 2 = D_u$.
   (c) If $a_{z-1} = 3$, the rational numbers are represented in Fig. 8 where the fraction $\frac{k}{l} = [a_0, \cdots, a_{z-2}, 2]$ with $\delta^2 \left(C \left(\frac{k}{l} \right) \right) = D_k$ must be considered. If $D_k = d$, by Lemma 3, it holds $D_u = d + 1$ and $D_a = d + 1 = D_u$, while if $D_k = d+1$, it holds $D_u = d + 2$ and $D_a = d + 2 = D_u$.
   (d) Finally, if $a_{z-1} > 3$, it holds $D_u > D_r + 1$ therefore, $D_a = D_u$. $\square$

$$\frac{u}{v} \qquad \frac{\rho}{\theta} \qquad \frac{a}{b} = [a_0, \ldots, 4] \qquad\qquad \frac{u}{v} = \frac{\rho}{\theta} \qquad \frac{a}{b} = [a_0, \ldots, 3]$$

Fig. 5. Positions of the fractions $\frac{a}{b}, \frac{u}{v}, \frac{\rho}{\theta}$ on SB_L in case where the last element of the continued fraction of $\frac{a}{b} = a_z = 4$ or $a_z = 3$ respectively.

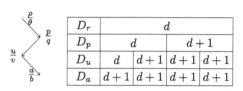

D_r	d			
D_p	d	$d+1$		
D_u	d	$d+1$	$d+1$	$d+1$
D_a	$d+1$	$d+1$	$d+1$	$d+1$

Fig. 6. Position of the fractions $\frac{p}{\theta}, \frac{p}{q}, \frac{u}{v}$ and $\frac{a}{b}$, when $a_z = 2$, and $a_{z-1} = 1$, and the table showing the 4 possible subcases

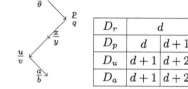

D_r	d	
D_p	d	$d+1$
D_u	$d+1$	$d+2$
D_a	$d+1$	$d+2$

Fig. 7. Position of the fractions $\frac{p}{\theta}, \frac{p}{q}, \frac{u}{v}$ and $\frac{a}{b}$, when $a_z = 2$, and $a_{z-1} = 2$ $a_z = 2$ and the table representing the 2 sub-cases.

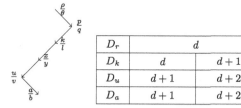

D_r	d	
D_k	d	$d+1$
D_u	$d+1$	$d+2$
D_a	$d+1$	$d+2$

Fig. 8. Position of the fractions $\frac{p}{\theta}, \frac{p}{q}, \frac{k}{l}, \frac{u}{v}$ and $\frac{a}{b}$, when $a_z = 2$, *and* $a_{z-1} = 3$ and the table representing the 2 sub-cases.

3.3 General Form of a Minimal Path

In this section we give the final result of the paper, by providing a characterization on the elements of the continued fraction of $\frac{a}{b}$ in order to consider $P_{\delta^2}(\frac{a}{b})$ as a minimal path in SB_L.

At level 4 of the SB_L, only two rational numbers out of 4 belong to a minimal path as we can see in Table 1. We have $P_{\delta^2}(\frac{3}{5}) = P_{\delta^2}(\frac{3}{4}) = (1,1,1,2)$, where $\frac{3}{5} = [0,1,1,2]$ and $\frac{3}{4} = [0,1,3]$. This insures that the continued fraction of any rational number in SB_L, who belongs to a minimal path must starts with $[0,1]$. At level 5, we note that these two rational numbers generate 4 rational numbers that belong to a minimal path. While at level 6, we get again 4 rational numbers that are also represented in Table 1. By Definition 1, we can notice that three consecutive elements p_i, p_{i+1}, and p_{i+2}, of a minimal path are such that $p_{i+1} - p_i \leq 1$. Hence, all the children of the rational numbers of SB_L belonging to MP_6 maintain the minimality at levels 7 and 8. While at level 9, there exist only 16 $(= 4^2)$ fractions that belong to MP_9 and are represented in Table 2.

Considering these first levels, we note that the continued fraction representations of the elements of MP_9 must start with $0, 1$ and end with 2. On the other hand, all the remaining δ^2 values of each element of MP_9 can be grouped into the blocks $(1,1,1)$; $(1,2)$; $(2,1)$ or 3 as the red and blue colored sequences in Table 2 witness. Aware of this, we can characterize the elements of SB_L that belongs to a minimal path.

Table 1. The rational numbers $\frac{a}{b}$ at level 4 of SB_L with their continued fraction representation, $P(\frac{a}{b})$ and $P_{\delta 2}(\frac{a}{b})$. The four rational numbers at level 6 that belong to a minimal path with their continued fraction representation, and sequences P and $P_{\delta 2}$.

Level	Fraction $\frac{a}{b}$	Continued fraction	$P(\frac{a}{b})$	$P_{\delta 2}(\frac{a}{b})$
4	$\frac{1}{4}$	$[0,4]$	$(\frac{1}{1},\frac{1}{2},\frac{1}{3},\frac{1}{4})$	$(1,1,2,2)$
4	$\frac{2}{5}$	$[0,2,2]$	$(\frac{1}{1},\frac{1}{2},\frac{1}{3},\frac{2}{5})$	$(1,1,2,2)$
4	$\frac{3}{5}$	$[0,1,1,2]$	$(\frac{1}{1},\frac{1}{2},\frac{2}{3},\frac{3}{5})$	$(1,1,1,2)$
4	$\frac{3}{4}$	$[0,1,3]$	$(\frac{1}{1},\frac{1}{2},\frac{2}{3},\frac{3}{4})$	$(1,1,1,2)$
6	$\frac{8}{13}$	$[0,1,1,1,1,2]$	$(\frac{1}{1},\frac{1}{2},\frac{2}{3},\frac{3}{5},\frac{5}{8},\frac{8}{13})$	$(1,1,1,2,2,2)$
6	$\frac{8}{11}$	$[0,1,2,1,2]$	$(\frac{1}{1},\frac{1}{2},\frac{2}{3},\frac{3}{4},\frac{5}{7},\frac{8}{11})$	$(1,1,1,2,2,2)$
6	$\frac{7}{12}$	$[0,1,1,2,2]$	$(\frac{1}{1},\frac{1}{2},\frac{2}{3},\frac{3}{5},\frac{4}{7},\frac{7}{12})$	$(1,1,1,2,2,2)$
6	$\frac{7}{9}$	$[0,1,3,2]$	$(\frac{1}{1},\frac{1}{2},\frac{2}{3},\frac{3}{4},\frac{4}{5},\frac{7}{9})$	$(1,1,1,2,2,2)$

Table 2. The 16 rational numbers that belong to MP_9 on SB_L.

$\frac{29}{50}$	$\frac{27}{46}$	$\frac{26}{45}$	$\frac{23}{39}$
$[0,1,\mathbf{1,2},1,1,1,2]$	$[0,1,\mathbf{1,2},2,1,2]$	$[0,1,\mathbf{1,2},1,2,2]$	$[0,1,\mathbf{1,2},3,2]$
$\frac{34}{55}$	$\frac{30}{49}$	$\frac{31}{50}$	$\frac{25}{41}$
$[0,1,\mathbf{1,1,1},1,1,1,2]$	$[0,1,\mathbf{1,1,1},2,1,2]$	$[0,1,\mathbf{1,1,1},1,2,2]$	$[0,1,\mathbf{1,1,1},3,2]$
$\frac{34}{47}$	$\frac{30}{41}$	$\frac{31}{43}$	$\frac{25}{34}$
$[0,1,\mathbf{2,1},1,1,1,2]$	$[0,1,\mathbf{2,1},2,1,2]$	$[0,1,\mathbf{2,1},1,2,2]$	$[0,1,\mathbf{2,1},3,2]$
$\frac{29}{37}$	$\frac{27}{35}$	$\frac{26}{33}$	$\frac{23}{30}$
$[0,1,\mathbf{3},1,1,1,2]$	$[0,1,\mathbf{3},2,1,2]$	$[0,1,\mathbf{3},1,2,2]$	$[0,1,\mathbf{3},3,2]$

Theorem 4. *Let $\frac{a}{b}$ be an element of SB_L at a certain level $k \geq 6$ and whose continued fraction representation is $[a_0, a_1, \ldots, a_z]$. If the a_i's respect the following conditions: $a_0 = 0$, $a_1 = 1$, $a_z = 2$ and the elements $a_2, \ldots, a_{z-1}$ are obtained by the concatenation of the blocks $(1,1,1)$, $(2,1)$ $(1,2)$ and (3), then $\frac{a}{b} \in MP_k$.*

Proof. By the computation of the first minimal paths of SB_L we have that $a_0 = 0$ and $a_1 = 1$. From Lemma 3, we know that a_z must be equal to 2 in order to realize the minimal growth. Now, from Definition 1, we know that the rational number $\frac{a}{b}$ belongs to a minimal path if the values of $P_{\delta 2}(\frac{a}{b})$ are increased by 1, each three steps. Relying on that, we consider all the possible ways to pass from a level t to a level $t+3$ in SB_L keeping the minimal growth of one, and we easily realize that these elements can only have one the following forms: $(1,1,1)$, $(2,1)$ $(1,2)$ or (3), i.e. all the possible integer decompositions of the number three. Note that the first case, is equivalent to the *zig-zag* path, while the other cases are obtained by Theorem 3 and Lemma 3. □

Finally, some simple computations lead to the following:

Corollary 1. *Let k be a certain level in SB_L, the number of MP_k is equal to:*

$$\begin{cases} 4^{\frac{k}{3}-1} & if \ k \equiv_3 0 \\ 2 \cdot 4^{\lfloor \frac{k}{3} \rfloor - 1} & if \ k \equiv_3 1 \\ 4 \cdot 4^{\lfloor \frac{k}{3} \rfloor - 1} & if \ k \equiv_3 2 \end{cases}$$

This last theorem determines the form of the minimal paths inside the SB-tree according to the δ^2-paths, and from a geometrical perspective, it defines a new family of Christoffel words that deserves to be investigated. A further generalization of the order of balancedness may also identify special paths of the SB-tree showing a fixed point property.

References

1. Berstel, J., De Luca, A.: Sturmian words, Lyndon words and trees. Theor. Comput. Sci. **178**(1), 171–203 (1997)
2. Berstel, J., Lauve, A., Reutenauer, C., Saliola, F.: Combinatorics on words: Christoffel words and repetition in words, Université de Montréal et American Mathematical Society (2008)
3. Borel, J.-P., Laubie, F.: Quelques mots sur la droite projective réelle. Journal de théorie des nombres de Bordeaux **5**(1), 23–51 (1993)
4. Brocot, A.: Calcul des rouages par approximation, Revue chronométrique. Journal des horlogers, scientifique et pratique **3**, 186–194 (1861)
5. Lothaire, M.: Algebraic Combinatorics on Words, Encyclopedia of Mathematics and its Applications, vol. 90. Cambridge University Press, Cambridge (2002)
6. Richomme, G., Saari, K., Zamboni, L.-Q.: Abelian properties of words, arXiv preprint arXiv:0904.2925 (2009)
7. Fraenkel, A.: The bracket function and complementary sets of integers. Can. J. Math. **21**, 6–27 (1969)
8. Fraenkel, A., Levitt, J., Shimshoni, M.: Characterization of the set of values $f(n) = [na], n = 1, 2, \ldots$. Discrete Math. **2**(4), 335–345 (1972)
9. Graham, R.E., Knuth, D.E., Patashnik, O.: Concrete Mathematics. A Foundation for Computer Science, 2nd edn. Addison-Wesley, Reading (1994)
10. Pettofrezzo, A.J., Byrkit, D.R.: Elements of Number Theory. Prentice Hall, Englewood Cliffs (1970)
11. Stern, M.A.: Ueber eine zahlentheoretische Funktion. Journal für die reine und angewandte Mathematik **55**, 193–220 (1858)
12. Tarsissi, L., Vuillon, L.: Second order balance property on Christoffel words. In: Slamanig, D., Tsigaridas, E., Zafeirakopoulos, Z. (eds.) MACIS 2019. LNCS, vol. 11989, pp. 295–312. Springer, Cham (2020). https://doi.org/10.1007/978-3-030-43120-4_23

Author Index

Printed in the United States
By Bookmasters